Disciplines, Disasters and Emergency Management

DISCIPLINES, DISASTERS AND EMERGENCY MANAGEMENT

The Convergence and Divergence of Concepts, Issues and Trends from the Research Literature

Edited by

DAVID A. McENTIRE

CHARLES C THOMAS • PUBLISHER, LTD.
Springfield • Illinois • U.S.A.

Published and Distributed Throughout the World by

CHARLES C THOMAS • PUBLISHER, LTD.
2600 South First Street
Springfield, Illinois 62704

ISBN 978-0-398-07743-3 (hard)
ISBN 978-0-398-07744-0 (paper)

Library of Congress Catalog Card Number: 2007005672

With THOMAS BOOKS *careful attention is given to all details of manufacturing and design. It is the Publisher's desire to present books that are satisfactory as to their physical qualities and artistic possibilities and appropriate for their particular use.* THOMAS BOOKS *will be true to those laws of quality that assure a good name and good will.*

Printed in the United States of America
UB-R-3

Library of Congress Cataloging-in-Publication Data

Disciplines, disasters, and emergency management : the convergence and divergence of conepts, issues and trends from the research literature / editor, David A. McEntire.

p. cm.

Includes bibliographical references and index.
ISBN-13: 978-0-398-07743-3 (hardbound)
ISBN-10: 0-398-07743-6 (hardbound)
ISBN-13: 978-0-398-07744-0 (paperbound)
ISBN-10: 0-398-07744-4 (paperbound)
1. Emergency management--Research. 2. Disaster relief--Research. 3. Crisis management. I. McEntire, David A. II. Title.

HV551.2.D583 2007
364.34'8--dc22

2007005672

For Kimberly and our children.

Contributors

Richard Bissell, Ph.D.
Associate Professor and Graduate Program Director
Department of Emergency Health Services
University of Maryland Baltimore County
ACIV Rm 316
Baltimore, MD 21250
(410) 455-3776

Dr. Rick Bissell is an Associate Professor of Emergency Health Services, and Graduate Program Director at the University of Maryland Baltimore County. He has worked as an emergency health services researcher and educator since the mid-1970s. A former paramedic with training in preventive medicine/epidemiology, health services administration, emergency management and international relations, his work has encompassed the intersection of emergencies and public health in more than a dozen countries, and numerous states and localities. His research in disaster epidemiology and disaster health services, encompassing more than 20 peer-reviewed articles and book chapters, has been well recognized. For more than five years, Dr. Bissell headed the team that developed and implemented training for medical and logistical response teams within the U.S. Public Health Service's (later DHS) National Disaster Medical System. He has worked with and consulted to the Centers for Disease Control and Prevention, the U.S. Office of Foreign Disaster Assistance, the German National Committee on Global Change Research, the FEMA Emergency Management Institute, the American Red Cross, and numerous local, state, and national governments. Dr. Bissell is currently conducting research on major challenges to human survival, using a combined public health and emergency management paradigm to both understand threatening dynamics and seek potential mitigation and response strategies.

Lori A. Byers, Ph.D.
Assistant Professor, University of North Texas
225 Ave. B, GAB 309
Denton, TX 76203-5268
(940) 565-2588

Lori A. Byers earned her Ph.D. in Communication Studies at Ohio University. Her research examines narrative sensemaking during times of crisis, including times of serious illness. Dr. Byers has coauthored several book chapters, and has articles in *Alzheimer's Care Quarterly* and *Texas Speech Communication Journal*. Dr. Byers recently earned a "Top Three" paper award presented by the Communication and Aging division of the National Communication Association.

Terry L. Clower, Ph.D.
Associate Director – Center for Economic Development and Research
University of North Texas
P.O. Box 310469
Denton, Texas 76203
(940) 565-4049

Terry L. Clower is associate director of the University of North Texas Center for Economic Development and Research. The Center provides economic and public policy consulting services to clients in the private, non-profit and public sectors. Prior to joining UNT in January 1992, Dr. Clower was employed in private industry in logistics and transportation management positions. Dr. Clower's research interests include economic and community development, economic and fiscal impact analysis, land use planning, transportation, and economic forecasting. He serves as the Center's resident expert on telecommunications focusing on policy issues regarding infrastructure development. Drawing upon nearly a decade of experience in logistics management, Dr. Clower also leads the Center's transportation research efforts.

Ana Maria Cruz, Ph.D.
Senior Scientific Officer
European Commission, Joint Research Centre
Major Accident Hazards Bureau and NEDIES
TP 670, Via Fermi 1
I-21020 Ispra (VA), Italy
(39) 0332-785381 E-mail: ana-maria.cruz@jrc.it

Dr. Cruz has completed research concerning assessment of risk management and emergency response practices for natural disaster triggered technological (Natech) disasters in the United States, Europe and Japan. Dr. Cruz's research interests include developing methodologies for measuring preparedness capacity and establishing indicators for preparedness for multiple and simultaneous technological accidents; analysis of cascading failures at industrial establishments during natural disasters; and analysis of risk management of soil contamination problems in areas subject to high natural hazard risk. Dr. Cruz is currently developing a rapid assessment tool for diagnosis of Natech risk in urban areas in Japan, and is working with the Joint Research Centre of the European Commission on development of a typology of Natech disasters, as well as development of Natech hazard maps for selected European countries. Her research findings have been presented at national and international conferences, including the recent United Nations' International Conference on Disaster Reduction in Davos, Switzerland in August 2006 and the World Conference on Disaster Management held in Toronto, Canada in June 2006. Dr. Cruz has received several fellowships and grants from the National Science Foundation, Tulane University, the Japan Society for Promotion of Research and the United Nations' International Strategy for Disaster Reduction to conduct research on Natech disasters in Louisiana and California, Turkey, Japan, and Europe. Dr. Cruz is the author of several articles in *Natural Hazards Review*, *Earthquake Spectra*, *Journal of Risk Research* and *Emergency Management Canada*. In addition, she recently contributed with a book chapter to an instructor guide for the Federal Emergency Management Agency. Prior to joining the European Commission, Dr. Cruz was a faculty member at University of North Texas and a research fellow at the Disaster Prevention Research Institute at Kyoto University.

Thomas E. Drabek, Ph.D.
John Evans Professor, Emeritus
Department of Sociology and Criminology
University of Denver
Denver, Colorado 80208-2948
(303) 771-6889

Thomas E. Drabek is an Emeritus Professor of Sociology and Criminology at the University of Denver. He was a faculty member there since 1965 when he completed his graduate training at the Ohio State University, Disaster Research Center. He was department chair (1974-1979 and 1985-1987). Upon his retirement in 2004, he was awarded Emeritus status and continues his research on a part-time basis. His research has examined group and organizational responses to large-scale disasters. He has authored numerous books and articles including *Strategies for Coordinating Disaster Responses* (2003); *Disaster-Induced Employee Evacuation* (1999); *Disaster Evacuation Behavior: Tourists and Other Transients* (1996); *Disaster Evacuation in the Tourist Industry* (1994); *Microcomputers in Emergency Management: Implementation of Computer Technology* (1991); *Emergency Management: Principles and Practice for Local Government* (1991, co-edited with Gerard J. Hoetmer); *Emergency Management: Strategies for Maintaining Organizational Integrity* (1990); *The Professional Emergency Manager: Structures and Strategies for Success* (1987); and *Human System Responses to Disaster: An Inventory of Sociological Findings* (1986). He served as the co-editor of the *International Journal of Mass Emergencies and Disasters* (1986-1990) and was elected President of the International Sociological Association's Research Committee on Disasters (1990-1994). He prepared four Instructor Guides for the Emergency Management Institute, Federal Emergency Management Agency: *Sociology of Disaster* (1996); *The Social Dimensions of Disaster* (1996); *Emergency Management Principles and Application for Tourism, Hospitality, and Travel Management Industries* (2000, co-authored with Chuck Y. Gee); and *Social Dimensions of Disaster, 2nd ed.* (2004). Currently he is writing a general text entitled *The Human Side of Disaster.*

Kathy Dreyer, M.S.
Program Coordinator, Texas Institute for Research and Education on Aging
University of North Texas
P.O. Box 310919 – TIREA
Denton, Texas, 76203-0919
(940) 565.3450

Kathy Dreyer, M.S., is a program coordinator at the Texas Institute for Research and Education on Aging at the University of North Texas. Her academic interests include caregiver burden and coping mechanisms, Alzheimer's disease, aging and ageism, HIV/AIDS awareness in older adults, and health care expenditures. She has received grants to work on improving rates of childhood immunization by engaging senior volunteers to educate families in Texas and other states. Ms. Dreyer has secured funding to educate and support caregivers whose family members have Alzheimer's disease. She has also worked on projects collaborating with independent school districts in north Texas to create and sustain school health advisory councils and provide information about healthy living and careers in the health field, in addition to projects related to incorporating gerontology at the classroom level. Ms. Dreyer has worked on research projects related to state funding levels for Medicaid nursing facilities and the role of self-managed work teams in nursing facilities. She is currently earning a Ph.D. in Applied Gerontology at the University of North Texas.

Margaret Gibbs, Ph.D.
School of Psychology
Fairleigh Dickinson University
Teaneck, NJ 07666
(201) 692 2302

Margaret Gibbs is a Professor of Psychology at Fairleigh Dickinson University. She received her Ph.D. in Clinical Psychology from the Department of Social Relations at Harvard University, and her interests have usually combined the clinical and the social. She has explored the effects of disasters, especially technological disasters, on psychological functioning, and has also investigated the impact of domestic violence and sexual harassment on psychological functioning. She has co-edited three books, published more than 30 chapters and articles, and made more than 100 professional presentations. Articles relevant to this volume have appeared in *Environment and Behavior*, *The Journal of Traumatic Stress*, *The Journal of Social Behavior and Personality*, *Clinical Psychology Review*, *Sex Roles*, and *International Journal of Mass Emergencies and Disasters*. She also has a private practice and consults at Alternatives to Domestic Violence in Hackensack, NJ.

Doug Henry, Ph.D.
Assistant Professor
Department of Anthropology
University of North Texas
P.O. Box 310409
Denton, TX 76203
Tel: 940-565-3836

Doug Henry is a Medical Anthropologist and Assistant Professor at the University of North Texas in Denton, Texas. He earned a Ph.D. in Anthropology from Southern Methodist University. His academic interests include West African refugees, violence, population displacement, and the unique patterns of morbidity that occur around violent conflict. He has received several research grants allowing him to conduct research on complex emergencies and their aftermath in Sierra Leone and the Republic of Guinea, and has worked as consultant to the United Nations High Commissioner for Refugees (UNHCR). He has received National Science Foundation funding to assess the roles of poverty and disability among displaced Katrina evacuees.

James M. Kendra, Ph.D.
Assistant Professor and Program Coordinator
Emergency Administration and Planning Program
Department of Public Administration
P.O. Box 310617
University of North Texas
Denton, Texas 76203-0167 USA
(940) 565-2213

Dr. Kendra is the coordinator of the Emergency Administration and Planning Program. Prior to joining the faculty at UNT, Dr. Kendra was the Research Coordinator at the University of Delaware Disaster Research Center, managing a number of projects focusing on disasters and emergency planning. His research interests include individual and organizational responses to risk and hazard, emergency and crisis management, and organizational improvisation and resilience.

John R. Labadie, Ph.D.
Senior Environmental Analyst
Seattle Public Utilities
Suite 4900
700 Fifth Avenue
P. O. Box 34018
Seattle, WA 98104
(206) 684-8311

John R. Labadie earned a B.S. from the Massachusetts Institute of Technology and an M.A. and Ph.D. from the University of Washington. He has thirty years' experience in the multidisciplined technical and business management of environmental services, hazardous waste management, and emergency preparedness planning for Federal agencies, state and local governments, and private industry. He has developed Comprehensive Emergency Management plans and Hazard Mitigation Plans for cities in Washington State. Dr. Labadie prepared a Comprehensive Emergency Management Plan for the Washington State Department of Health and Emergency Response Plans for WDOH facilities, and he conducted Disaster Planning training courses for public utilities, sponsored by Green River Community College. He served as the Director of Disaster and Hazardous Materials Management for the City of Salem, OR. His responsibilities included: emergency management and disaster preparedness planning; management of the City's purchase, use, and disposal of hazardous materials; and management of a 27-member Hazardous Materials Response Team. As Vice-President and Manager of the Environmental Services Division for JAYCOR, Dr. Labadie directed and managed a number of projects for government clients, including Remedial Investigation and Feasibility Studies (RI/FS), environmental compliance assessments, and Environmental Impact Statements. In his current position at Seattle Public Utilities, Dr. Labadie has concentrated on environmental management, hazardous materials management, environmental compliance auditing, risk assessment, quality assurance, and internal performance auditing. Dr. Labadie is also a Certified Hazardous Materials Manager and a Certified Environmental Auditor.

Robert J. Louden, Ph.D.
Georgian Court University
900 Lakewood Avenue
Lakewood, NJ 08701
(732) 987-2711

Robert Louden is Professor and Program Director, Criminal Justice at Georgian Court University, Lakewood, NJ. Prior to his current appointment he was a Professor in the Department of Public Management at John Jay College of Criminal Justice/CUNY where he taught graduate courses in Protection Management and in Criminal Justice. His research interests include police organization and administration, protection management, terrorism, hostage negotiation, disaster response and decision-making. He had accepted his position at John Jay College in 1987 upon his retirement as Lieutenant, Commander of Detective Squad and Chief Hostage Negotiator for the NYPD.

In 1993 he was appointed by the U.S. Department of Justice and the U.S. Department of the Treasury to present prospective recommendations on how federal law enforcement agencies should manage complex hostage/barricade situations such as the standoff that occurred near Waco, Texas. Dr. Louden has testified about police practices and procedures before the NYC Council Committee on Public Safety and the US Commission on Civil Rights. He served as a Co-convener of *Urban Hazards Forum Two (2003); Homeland Security After 9/11*, and a presenter at *Urban Hazards Forum One (2002); Terrorism-Catastrophic Events-Mitigation*, a series of conferences sponsored by FEMA-Region II and John Jay College of Criminal Justice. Robert J. Louden earned a Bachelor of Business Administration degree from Baruch College and an MA from John Jay College of Criminal Justice. He later received the Master of Philosophy and Doctor of Philosophy degrees from the Graduate School of CUNY.

Sarah Mathis, B.S.
Sumners Graduate Research Assistant
Department of Public Administration
P.O. Box 310617
University of North Texas
Denton, Texas 76203-0167 USA
(940) 565-3292

Sarah Mathis earned her B.S. in Emergency Management in 2005 from the University of North Texas. She has spent time working on development projects in Africa, Asia, Europe, Australia, New Zealand, and the Caribbean. Mathis joined the response to Hurricane Katrina in 2005, serving on the Mississippi coast. Her research interests center on the relationship of disasters to development and the influence of disasters on diplomacy. She will complete her Master of Public Administration in 2007.

David A. McEntire, Ph.D.
Associate Professor
Emergency Administration and Planning
Department of Public Administration
University of North Texas
P.O. Box 310617
Denton, TX 76203-0617
(940) 565-2996

David A. McEntire's academic interests include emergency management theory, international disasters, community preparedness, response coordination, and vulnerability reduction. He has received several grants which allowed him to conduct research on disasters in Peru, the Dominican Republic, Texas, New York and California. Dr. McEntire is the author of *Disaster Response and Recovery* (Wiley) and numerous articles in *Public Administration Review*, the *Australian Journal of Emergency Management*, *Disasters*, the *International Journal of Mass Emergencies and Disasters*, *Journal of Emergency Management*, *Journal of the Environment and Sustainable Development*, *Sustainable Communities Review*, *International Journal of Emergency Management*, *Towson Journal of International Affairs*, *Journal of the American Society of Professional Emergency Planners*, and the *Journal of International and Public Affairs*. Dr. McEntire recently completed an instructor guide for the Federal Emergency Management Agency. He has received grants to conduct terrorism response training for FEMA in Arkansas and Oklahoma. Dr. McEntire is a former Coordinator of the EADP program. Prior to coming to the University of North Texas, he worked for the International and Emergency Services Departments at the American Red Cross in Denver, Colorado.

Kent M. McGregor, Ph.D.
Associate Professor
Department of Geography
University of North Texas
P.O. Box 305279
Denton, TX 76203-5279
(940) 565-2380

Kent M. McGregor holds a Ph.D. from the Department of Geography and Meteorology at the University of Kansas. His principal academic interests are in climate variability and water resources, especial extreme events such as floods and droughts. His current research focus is El Niño phenomena and the impact on water resources in Texas and the Southwest. He has published articles in *Physical Geography, Water Resources Bulletin, Journal of Hydrology, Agricultural History, History of Geography,* and *Geographical Perspectives.* He authored book chapters in the *Changing Climate of Texas, Water for Texas: 2000 and Beyond,* and *Current Trends in Remote Sensing Education.* He has served as a consultant to FEMA Region VI and the Texas Air Control Board.

Kimberly Montagnino, Ph.D.
Counseling Services
Seton Hall University
400 South Orange Ave.
South Orange, NJ 07079

Kimberly Montagnino earned a Ph.D. in Clinical Psychology from Fairleigh Dickinson University, New Jersey. Her interests include the treatment of post-traumatic and other anxiety disorders, and the influence of gender and culture on psychological functioning and psychopathology. Dr. Montagnino has presented research on feminist identity development in Latina women, covert racism and the Racist Argument Scale, the appeal of feminist therapy among psychologists, and emotional abuse in intimate relationships. She currently provides psychological services and psychoeducational programming as a Staff Psychologist at Seton Hall University Counseling Services.

William C. Nicholson, Esquire
Assistant Professor
Department of Criminal Justice
301 Whiting Criminal Justice Building
North Carolina Central University
Durham, NC 27707
(919) 530-7501

Bill Nicholson is an internationally known terrorism and emergency law expert. He is an Assistant Professor in North Carolina Central University's Criminal Justice Department, where he teaches graduate and undergraduate courses including Introduction to

Homeland Security; Homeland Security Law and Policy; Emergency Management and Recovery; as well as Criminal Justice Management and Organizational Theory, (SEMA), Indiana Department of Fire and Building Services (DFBS) and Public Safety Training Institute (PSTI). He publishes numerous articles and speaks nationwide on terrorism and emergency law issues. Nicholson is also a Member of the Editorial Board, *Best Practices in Emergency Services: Today's Tips for Tomorrow's Success* as well as the Editorial Board, *Journal of Emergency Management.* Recent notable publications include: two books, *Emergency Response and Emergency Management Law* (2003) and *Homeland Security Law and Policy* (William C. Nicholson ed., 2005), and a major law review article *Legal Issues in Emergency Response to Terrorism Incidents Involving Hazardous Materials: The Hazardous Waste Operations and Emergency Response* ("HAZWOPER") *Standard, Standard Operating Procedures, Mutual Aid and the Incident Command System*, 9 Widener L. Symp. J. 295 (2003). Nicholson earned a B.A. from Reed College in Portland, Oregon and a Juris Doctor from Washington and Lee University's School of Law in Lexington, Virginia (he may be reached at: wnicholson@nccu.edu).

John C. Pine, Ed.D.
Research Professor and Director
Disaster Science and Management
Department of Environmental Studies
School of the Coast and Environment; Department of Geography and Anthropology College of Arts and Sciences
Louisiana State University
227 Howe Russell Hall
Baton Rouge, LA 70803
(225) 578-1075

John C. Pine is Director, Disaster Science and Management (see http://www.dsm.lsu.edu) and Professor-Research with the Department of Environmental Studies and the Department of Geography and Anthropology at LSU. He is also currently serving as co-director of the LSU Hurricane Katrina and Rita Clearinghouse which is providing permanent storage of data associated with these hurricanes (http://www.katrina.lsu.edu). He received his Doctorate from the University of Georgia in 1979 and came to LSU in 1980. His research involves using hazard modeling and mapping to understand the social, environmental and economic impacts of disasters and indicators of sustainable communities with funding provided by the National Science Foundation, NOAA, FEMA and the Department of Homeland Security, Department of Interior, the Environmental Protection Agency, and Sea Grant. His book *Technology and Emergency Management* was published by John Wiley & Sons in 2006. His publications focus on hazards analysis and emergency management. His book *Tort Liability Today: Claims against State and Local Governments* (published by the Public Risk Management Association and the National League of Cities) is in its tenth edition. He also has numerous book chapters and articles concerning risk assessment published by the *Social Science Quarterly, Hazardous Materials Journal, Oceanography, Journal of Homeland Security and Emergency Management, Natural Disaster Review, American Society of Professional Emergency Planners Journal, Journal of Environmental Health.* He was also the lead developer of FEMA's Higher Education Project courses on Technology and Emergency Management and Hazard Modeling and Mapping (see: http://www.training.fema.gov/EMIWeb/edu/coursesunderdev.asp).

Brian K. Richardson, PH.D.
Department of Communication Studies
University of North Texas
Denton, TX 76203-5268
(940) 565-4748

Brian K. Richardson earned a PH.D. in Communication Studies from the University of North Texas in 2001. His area of specialty is Organizational Communication and his academic interests include whistle-blowing, sexual harassment, peer reporting of unethical behavior, organizational crises and disasters. Dr. Richardson has published articles in the *International Journal of Mass Emergencies and Disasters, Human Communication Research, Management Communication Quarterly*, the *International and Intercultural Communication Annual*, and *Case Studies in Organizational Communication.* Prior to coming to the University of North Texas, Dr. Richardson conducted media crisis and crisis communication training workshops in the private sector.

Joseph Scanlon
Emergency Communications Research Unit
3rd Floor
St. Patrick's
College
Carleton University
1125 Colonel By Drive
Ottawa, Ont., K1S 5B6
Canada.
1-613-730-9239

Joseph Scanlon is Professor Emeritus and Director of the Emergency Communcations Research Unit (ECRU) at Carleton University in Ottawa, Canada. After graduating in Journalism he joined the staff of Canada's largest newspaper, the *Toronto Daily Star*, serving it as both Washington and Parliamentary correspondent. He also took time out to complete a graduate degree in Politics at Queen's University. He returned to Carleton as Director of the School in Journalism. His interest in disaster was sparked by a study of rumours for Canada's Defence Research Board. Then over the next 20 years he supervised dozens of case studies of Canadian emergency incidents. He has shared the findings of his research in lectures, monographs, chapters in 26 books and hundreds of articles in academic and professional journals including the *Australian Journal of Emergency Management, Disaster Prevention and Management*, the *International Journal of Mass Emergencies and Disasters, Journal of Contingencies and Crisis Management, Journal of Hazardous Materials, Northern Mariner, Applied Behavioral Science Review, English Studies in Canada, Journalism & Mass Communication Editor, British Columbia Medical Journal, Canadian Journal of Communication, Disaster Management, Public Administration Review, Newspaper Research Journal, Canadian Journal of Criminology, Journalism Educator, Journalism Quarterly, Ekistics, Gazette and Canadian Public Administation.* Most recently, he was part of a team funded by NSF to study the handling of the dead from the 2004 Indian Ocean tsunami and has with colleagues received funding from the Social Science and Humanities Research Council of Canada to study the Canadian mass death network.

Gregory L. Shaw, D.Sc.
Senior Research Scientist, Associate Professor, and Graduate Coordinator
Institute for Crisis, Disaster and Risk Management
VCU Homeland Security and Emergency Preparedness Program
GWU/EMSE
1776 G Street, NW, Suite 101
Washington, DC 20052
202-994-6736

Dr. Gregory Shaw is a Senior Research Scientist at The George Washington University Institute for Crisis, Disaster and Risk Management and an Associate Professor in the Virginia Commonwealth University's Homeland Security and Emergency Preparedness program. Over the past five years he has conducted research in support of Operation Safe Commerce, the organizational response to the 9-11 attacks, mass casualty preparedness and response, the impact of the 9-11 attacks on corporate America, the identification of competencies for Business crisis and Continuity Management executives, a risk assessment of the San Francisco Ferry System, and the development of a system for managing volunteers during disaster response and recovery operations. Gregory Shaw is an adjunct faculty member for Florida Atlantic University and the University of Maryland University College teaching graduate level courses in Homeland Security, Crisis and Emergency Management and Risk Management. Previously, Gregory Shaw served as an officer in the United States Coast Guard, retiring as a Captain in October 1996. While on active duty, he commanded four cutters and the Coast Guard's largest shore unit, served as Liaison Officer to the United States Navy Mine Warfare Command, and as a senior analyst for the Congressionally chartered Commission on Roles and Missions of the Armed Forces. Gregory Shaw has earned his Doctor of Science in Engineering Management at GWU and Masters Degrees in Physics at Wesleyan University, Education and Human Development at GWU, and Business Administration at Webster University. He is a Certified Business Continuity Professional (CBCP) through Disaster Recovery Institute International.

Tisha Slagle Pipes, M.S.
University of North Texas
School of Library and Information Sciences
P.O. Box 311068
Denton, Texas, 76203-1068
(940) 565-3568

Tisha Slagle Pipes earned a Master of Science in Information Science and a Master of Science in Computer Education and Cognitive Sciences from the University of North Texas (UNT). She is currently a doctoral candidate in the UNT School of Library and Information Sciences (SLIS) Interdisciplinary Information Science Ph.D. Program. Her academic interests include information behavior, information poverty, and information-communication technologies, especially in relationship to post-disaster information. She received a Quick Response Grant from the Natural Hazard Center (NHC) and traveled to Slidell, Louisiana to study the information behavior following Hurricane Katrina. She has written a chapter, "Katrina Bankrupts the Information-Rich: Information Poverty in Slidell" that has been accepted for publication in the NHC's 2006 Katrina anthology, *Learning From Catastrophe.* Mrs. Pipes has worked at

UNT as a Research Assistant, Teaching Assistant and is currently working as a Teaching Fellow for SLIS. Before coming to UNT, she taught computer applications for Region XI Education Service Center, as well as for several independent school districts and colleges in the North Texas area.

Sarah B. Smith, MPA
Senior Consultant, Booz Allen Hamilton
8283 Greensboro Drive
McLean, VA 22102
(703) 902-5744-w

Sarah B. Smith earned her Masters in Public Administration from the School of Public Affairs and Community Service at the University of North Texas. Her academic interests include international disasters, national preparedness, response coordination, and vulnerability reduction. Mrs. Smith is currently working at Booz Allen Hamilton on their CBRNE (Chemical, Biological, Radiological, Nuclear, and Explosive) Team where she designs, develops, and executes disaster exercises all over the world for clients such as the Department of Homeland Security and the Department of State. Prior to working for Booz Allen Hamilton, Mrs. Smith worked as a firefighter/paramedic for the City of Eau Claire, Wisconsin.

Richard T. Sylves, Ph.D.
Professor of Political Science and International Relations
Senior Policy Fellow, Center for Energy and Environmental Policy
Dept. of Political Science and International Relations
University of Delaware
Newark, DE 19716
(302) 831-1943 (work voice)
(302) 831-4452 (fax work)

Richard Sylves has researched presidential disaster declarations for more than fifteen years, the last two supported by the PERI Foundation. He has co-edited with W. Waugh, *Disaster Management in the United States and Canada: Politics, Policy, Administration, Study* and *Instruction of Emergency Management* (Springfield, IL: Charles C Thomas Publishers, 1996) and *Cities and Disaster: North American Studies in Emergency Management* (Springfield, IL.: Charles C Thomas Publishers, 1990). He also authored the book, *The Nuclear Oracles* with Iowa State University Press in 1986. He is completing a textbook with Congressional Quarterly Press and a trade book with Elsevier Butterworth/Heinemann. From 1995-1999, Sylves completed two research grant projects for the U.S. FEMA. He has served on the National Academy of Science (NAS) panel, "Estimating the Costs of Natural Disasters," and he completed a three-year term as an appointed member of the NAS Disasters Roundtable in 2005.

William L. Waugh, Jr., Ph.D.
Professor
Department of Public Administration and Urban Planning
Georgia State University
Altanta, GA 30303-0000
(404) 651-4592

William L. Waugh, Jr., is Professor of Public Administration, Urban Studies, and Political Science in the Andrew Young School of Policy Studies at Georgia State University. He taught at Mississippi State University and Kansas State University before moving to GSU in 1985. He is the author of *Living with Hazards, Dealing with Disasters* (2000), *Terrorism and Emergency Management* (1990), and *International Terrorism* (1982); co-author of *State and Local Tax Policies* (1995); coeditor of *Disaster Management in the US and Canada* (1996), *Cities and Disaster* (1990), and *Handbook of Emergency Management* (1990); and editor of *The Future of Emergency Management* (2006), as well as the author of over a hundred journal articles, chapters, and reports published in the U.S., Canada, Europe, and Asia. He is editor-in-chief of the *Journal of Emergency Management* and is on the editorial boards of *Public Administration Review* and other journals. Currently, he serves on the Certified Emergency Manager (CEM) Commission and on the Emergency Management Accreditation Program (EMAP) Commission. Dr. Waugh has been a consultant and provided training for emergency management, law enforcement, military, non-profit, private, and international organizations on leadership, strategic management, disaster management, and anti- and counter-terrorism policies. His consulting work ranges from developing a strategic management program for Solidarity trade union in Poland to providing training on the design of anti- and counter-terrorism programs for emergency management, law enforcement, and military organizations.

Michael J. Zakour, Ph.D.
Tulane University
School of Social Work
New Orleans, LA 70118
(504) 862-3495

Michael J. Zakour earned an M.S.W. and Ph.D. from the George Warren Brown School of Social Work at Washington University in St. Louis. His academic interests include disaster volunteerism, disaster preparedness and response, interorganizational networks and coordination, political ecology of disaster, and vulnerability theory and practice. He has received several research grants to examine networks of volunteer organizations in disaster social service delivery in the U.S., as well as to evaluate volunteer social service programs after Hurricane Katrina. He has published in the *Journal of Social Service Research*, *Social Work Research*, the *Nonprofit and Voluntary Sector Quarterly*, the *Journal of Volunteer Administration*, and the *Journal of Social Work Education*. He edited Volumes 21-22 of *Tulane Studies in Social Welfare*, which is a special issue on Disaster and Traumatic Stress Research and Intervention. This double issue was published as a monograph in 2000. In 1994, he founded and was the original chairperson of the Disaster & Traumatic Stress Symposium, a symposium of the Council on Social Work Education's annual program meeting. He is the current chairperson of the Disaster & Traumatic Stress Symposium, and is Director of the Disaster & Volunteerism Research Center at Tulane University's School of Social Work. He is also a founder and coordinator of the dual M.S.W./M.P.H. in Emergency Management and Disaster Response, co-sponsored by Tulane University Schools of Social Work and Public Health.

Preface

"How does one conduct research on disasters at the Graduate School of International Studies?" This was the question posed several years ago after I introduced my academic interests to other students on the first day of a seminar on development.

I admit – rather reluctantly – that the professor's inquiry took me by surprise. Not only did I fumble unsuccessfully through my attempt to satiate his curiosity, but the resulting incredulity to my response made me contemplate switching schools as well as fields. The experience proved to be valuable, however, in that subsequent reflection and further research has convinced me that disasters can and should be studied by those interested in international relations, comparative politics and policy analysis.

The first of these fields provides the context for the creation of emergency management in the United States (e.g., the impact of the Cold War on civil defense) and it enables scholars to understand the actors involved in international humanitarian activity as well as the unacceptable barriers that inhibit disaster mitigation and preparedness across national borders. The second field helps students comprehend the plethora of problems that must be overcome if disasters are to be reduced in developing nations. And the latter academic area is beneficial as it provides the tools necessary to assess the strengths and weaknesses of disaster policies at the domestic and international levels. Thus, international studies *may* certainly offer unique contributions to the rapidly growing disaster studies field.

Beyond this, it can be argued that international studies *must* add to the vital knowledge base about natural and human-induced catastrophes. The 9/11 terrorists attacks dramatically altered the nature and direction of emergency management in North America, and the current emphasis on homeland security stresses the importance of addressing international grievances and doing more to prevent or prepare for the possible use of weapons of mass destruction. In addition, little is known about disasters in other countries (comparatively speaking), which hinders the transfer of lessons learned and suggests a bleak future for the vast majority of the planets inhabitants. Furthermore, calamitous events have a variety of direct and indirect consequences on all countries, and growing interdependence will ensure that catastrophes in distant locations will be felt in one way or another around the world. Scholars in these branches of the social sciences therefore have a responsibility to generate knowledge about disasters in all nations, and alert the citizens and leaders of the United States to the fact that immunity from the consequences of calamity in developed or developing nations is a fallacy.

If it is true that international studies can and should participate in the ongoing discussion about how to reduce disasters, it is only a reflection of the state of disaster research as a whole. This important area of investigation has always been examined from various disciplines. Besides natural/physical scientists and engineers, other key participants include sociologists, political scientists, psychologists, anthropologists, urban planners, development scholars, students of emergency management and many others from diverse academic backgrounds. Because of this disparate set of contributors, there has never before been as great a need to integrate research findings for practitioners.

Accordingly, this edited volume attempts to do that: synthesize what is known about calamities in order to assist those policymakers and emergency managers who seek to reverse the disturbing trends of disasters in the United States and elsewhere around the world. Nonetheless, it is hoped that this work will also foster further discussion among the academic community. Considerable effort has been given to the assessment of past and current research findings as well as anticipated needs within and across the most salient fields of study related to disasters. In this sense, the book may help solidify multidisciplinary research in the disaster studies field and serve as a springboard for truly interdisciplinary scholarship for the future. The following work should therefore be read with the above issues and goals in mind.

Acknowledgments

The authors would like to express gratitude to the Federal Emergency Management Agency for grants that funded the development of chapters in this book. The contributors to this text also wish to thank Wayne Blanchard, FEMA Higher Education Program Manager, for his review of the manuscripts and insightful comments that improved our presentation of research on the many disciplines related to disasters and emergency management. While we alone are responsible for this book's content, the support of FEMA and guidance from Dr. Blanchard have been invaluable and are duly recognized.

Contents

Disciplines, Disasters and Emergency Management

Chapter 1

The Importance of Multi- and Interdisciplinary Research on Disasters and for Emergency Management

David A. McEntire

ABSTRACT

This introductory chapter discusses the emerging consensus among scholars and practitioners for multi- and interdisciplinary approaches to disasters and emergency management. It explains why such this strategy is deemed necessary and highlights the benefits of moving beyond explanations emanating from single or separate fields of study. The chapter then outlines what the reader can expect from the book and concludes with a discussion about barriers inhibiting disciplinary convergence and how they might be overcome.

INTRODUCTION

In any given emergency or disaster, numerous actors from the public, private and non-profit sectors arrive at the affected area to protect life, minimize human suffering, overcome social disruption, deal with the destruction of property and clean up a degraded environment. This convergence, as it is widely known, is not limited to post-disaster activities or the profession of emergency management. Disaster scholarship is increasingly multi- and interdisciplinary.[1] Researchers from various disciplines study natural, technological and civil/conflict hazards, and explore their interaction with the causes and consequences of vulnerability. The following edited volume discusses research findings and issues important to each discipline in the hope of finding points of intersection as well as gaps in the literature. In so doing, the contributing authors also generate recommendations to more effectively reduce the impact of disasters.

This introductory chapter discusses a growing consensus among scholars and practitioners for multi- and interdisciplinary approaches to disasters and emergency management. It explains why this strategy is deemed necessary and highlights the benefits of moving beyond explanations emanating from single or separate fields of study. The chapter then outlines what the reader can expect from the book and concludes with a discussion about barriers inhibiting disciplinary convergence and how they might be overcome.

A GROWING CONSENSUS

There appears to be much agreement that multi- and interdisciplinary approaches are needed to understand and effectively deal with the complex problems

1. Multidisciplinary research includes studies from various disciplines that are not always synthesized in a holistic and unified fashion. Interdisciplinary research, on the other hand, includes findings from diverse fields of study that are integrated in a complex but more coherent manner. The first is easier, but limited in theoretical and practical rewards; the latter is much more difficult, but is most likely to generate new knowledge for the solution of problems facing emergency management.

of our day. This is the case in academia in general but practitioners also appear to value inclusiveness of divergent viewpoints. Edward O. Wilson's book, *Conscilience: The Unity of Knowledge* (1999), is a great example of this trend in scholarship. Writing from the perspective of a Scientific Materialist who is interested in environmental conservation, Wilson asserts that we will be unable to resolve the problems we are faced with if we do not integrate knowledge from the natural and social sciences. We accordingly must rely on "conscilience," or the jumping together or blending of facts and theory from several disciplines. He states, "as we cross [the boundaries of several disciplines] . . . we find ourselves in an increasingly unstable and disorienting region. The ring closest to the intersection [of various disciplines], where most real-world problems exist, is the one in which fundamental analysis is most needed" (Wilson 1999, p. 10). Wilson therefore believes multidisciplinary perspectives take into account reality and are most apt to generate solutions for complicated challenges. His research is typical of many efforts among scholars to span conceptual issues and diverse fields of study (e.g., information sciences, environmental studies, bio-engineering and chemistry, etc.).

Practitioners in a variety of professions also share an affinity in synthesizing knowledge and bridging gaps across functional areas. For instance, those working in Public administration must have an understanding of politics, economics, and management as well as the issues pertaining to transportation, public health, human resources and urban development, among other things. The current concern about terrorism also involves several areas of expertise. According to Richard A. Falkenrath:

> Men and women from dozens of different disciplines – regional experts, terrorism analysts, law enforcement officials, intelligence officers, privacy specialists, diplomats, military officers, immigration specialists, customs inspectors, specific industry experts, regulatory lawyers, doctors and epidemiologists, research scientists, chemists, nuclear physicists, information technologists, emergency managers, firefighters, communications specialists, and politicians, to name a few – are currently involved in homeland security (in Damien 2006, xxvi).

Many careers now require employees to be ever-learning, willing to seek out valuable information about subjects and topics previously believed to be foreign or irrelevant. And more individuals are finding it in their benefit to do so. It is reported that Wayne Hale, an engineer and Deputy Space Shuttle Program Manager at NASA's Space Center in Houston, said, "you laugh, but when you talk about culture and how people subconsciously deal with hierarchy and where they fit in within an organization and whether they feel comfortable in bringing things up. . . . I'm wishing I'd taken more sociology courses in college." Knowledge bases that were once held sacrosanct and sufficient are now believed to be isolated and incomplete.

Such views about the importance of integrated research activities are especially prevalent in disaster studies and emergency management. Several decades ago Gilbert White and Eugene Haas recognized that "little attempt had been made to tap the social sciences to better understand the economic, social, and political ramifications of extreme natural events" (cited by Mileti 1999, 1). However, today, Ehren Ngo asserts "ideally, disaster research is multidisciplinary, and understanding the impact of disasters . . . requires a synthesis of various disciplines" (2001, 81). For instance, Mileti observes that "hazards research now encompasses disciplines such as climatology, economics, engineering, geography, geology, law, meteorology, planning, seismology, and sociology" (1999, 2), and his book, *Disasters by Design*, is a notable example of combining diverse knowledge sets from an eclectic group of well-known scholars. Britton also states "disaster research and its close companions (hazard research and risk research) and their application in the emergency management context is becoming more multidisciplinary" (1999, 229). Cutter and her colleagues agree that the study of disaster "is an interdisciplinary endeavor and spans the divide between the social, natural, engineering and health sciences" (2003, 7).

Conference panels, including one comprised of Earnest Paylor, Dennis Wenger and David Applegate, have been devoted to "A Holistic Assessment of Hazards" (see the 2004 *Natural Hazards Workshop*). In that session, Havidán Rodriguez examined the "role, contributions and complexities of interdisciplinary research" (2004). Others have likewise tried to take an interdisciplinary approach in their research, albeit with a slightly different focus. McEntire gives priority to the concept of vulnerability along with its attendant components, and he has illustrated their unique relation to several hazards, phases, actors, functions, and variables that influence the impact of disasters[2] (2004; 2003; 2002; 2002) (see Table 1.1). His work also illustrates a close relation to several disciplines (see Table 1.2). Acknowledging the presence of interdisciplinary research in the field, Brenda Phillips (2003) asks an interesting question to which there may be no clear or definitive answer: "is emergency management a discipline or a multidisciplinary endeavor?" Gruntfest and Weber seem to agree with the latter view – that "emergency managers are of no one particular discipline; likewise, the information they need is not limited to the purview of any one scientific discipline" (1998, 59).

Those working in the disaster field share sentiments similar to scholars. In response to the tragic Tsunami in Southeast Asia, the Public Entity Risk Institute held a conference for risk managers in 2005. It was entitled "Early Warning Systems: Interdisciplinary Observations and Policies from a Local Government Perspective." Business continuity planners also appear to value the varied activities of their disaster partners. The theme for the 2005 *Contingency Planning and Management Conference* in Las Vegas was "The Future is Convergence: Discover the Synergy among Business Continuity, Emergency Management and Security." Emergency managers, too, share interest in expanding the number of agencies participating in disaster reduction and response.

The need for multi- and interdisciplinary research is not limited to scholars and practitioners in the United States. An edited book by Mario Garza Salinas and Daniel Rodríguez (1998) bears the title *The Disasters of Mexico: A Multidisciplinary Perspective.* At the 2003 *FEMA Higher Education Conference*, Neil Britton, a scholar and practitioner respected around the Pacific Rim, declared "theory has to transcend disciplines." Empirical studies from around the world also suggest a growing interest in collective research methodologies. Ronan et al. (2000) assert that "dialogue needs to involve members of the volcanological community and its multidisciplinary team colleagues." Moving beyond a single disciplinary approach is undoubtedly gaining global acceptance.

As a result of this agreement, there is a concomitant realization that we must utilize multi-and interdisciplinary approaches in emergency management education. Bob Reed (one of the first faculty members in the Emergency Administration and Planning Program at the University of North Texas) is reported to have said virtually every discipline is related to disasters, perhaps with the exception of modern dance[3] (Neal, 2000, 429). Mileti believes "education in hazard mitigation and preparedness should therefore expand to include interdisciplinary and holistic

Table 1.1
ENVIRONMENTS

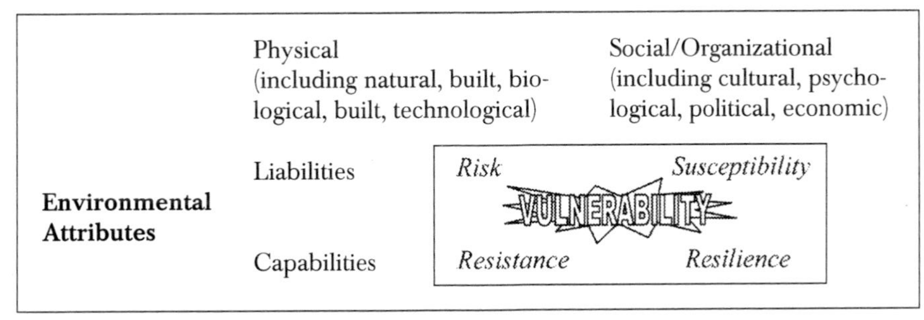

		Physical (including natural, built, biological, built, technological)	Social/Organizational (including cultural, psychological, political, economic)
Environmental Attributes	Liabilities	*Risk*	*Susceptibility*
		VULNERABILITY	
	Capabilities	*Resistance*	*Resilience*

Adapted from McEntire, David A. 2001. "Triggering Agents, Vulnerabilities and Disaster Reduction: Towards a Holistic Paradigm." *Disaster Prevention and Management 10*(3): 189–196.

2. McEntire (2005) believes we are able to influence and determine our vulnerability to hazards, and not necessarily control the hazards themselves.
3. It could even be argued that modern dance is related to disasters, because emergency medical care might be needed if one is not coordinated!

Table 1.2

Discipline	*View(s) of Vulnerability*	*Recommendation(s)*
Geography	Vulnerability is determined by the use of hazard-prone areas	Land-use planning that takes into account hazards to reduce risk
Meteorology	Vulnerability is due to a lack of advanced warning of severe weather	Acquisition, creation and effective use of warning systems
Engineering	Vulnerability occurs when structures and infrastructure cannot withstand the forces of hazards	Design and construction of buildings and infrastructure that promotes disaster resistance
Anthropology	Vulnerability emanates from constraining values, attitudes and practices	Alter attitudes to discourage risk-taking practices and susceptibility
Economics	Vulnerability is related to poverty and results in an inability to prevent, prepare for or recover from a disaster	Improve the distribution of wealth and purchase insurance to minimize losses and promote resilience
Sociology	Vulnerability is a product of inaccurate assumptions about disaster behavior and is related to race, gender, age, disability, etc.	Understand behavioral patterns in disasters and pay attention to needs of special populations
Psychology	Vulnerability is a function of overlooking or minimizing risk and not being able to cope emotionally with stress and/or loss	Help people to recognize risk and provide crisis counseling to enable resilience
Epidemiology	Vulnerability is susceptibility to disease or injury and is related to malnutrition and other health factors	Improve provision of public health/emergency medical care before, during and after disasters
Environmental Science	Vulnerability is proneness to environmental degradation, which may change weather patterns and produce long-term disasters	Conserve natural resources, protect green space areas, and ensure that debris management is performed in an environmentally conscious manner
Political Science	Vulnerability is produced by the political structure and incorrect decision making	Alter structure of political system and educate politicians and legislators about disasters
Public Administration	Vulnerability results from misguided laws, the failure to implement policies effectively, and an inability to enforce regulations	Strengthen response and recovery capabilities through preparedness measures, improved policy implementation and increased code enforcement
Law	Vulnerability results from negligence, which is a failure to act as reason or legal statutes dictate	Understand the law, alter statutes, and ensure compliance to widely accepted ethical practices in emergency management
Journalism	Vulnerability is a result of insufficient public awareness about hazards and how to respond to disasters	Dispel myths about disasters, foster increased media capabilities, and educate the public about hazards
Emergency Management	Vulnerability is the lack of capacity to perform important functions before and after disaster strikes (e.g., evacuation, search and rescue, public information, etc.)	Foster public awareness about disasters and build capacities through hazard and vulnerability analyses, resource acquisition, planning, training and exercises
Homeland Security	Vulnerability is due to cultural misunderstandings, permeable borders and fragile infrastructure, and weak disaster management institutions	Correct domestic and foreign policy mistakes, enhance counter-terrorism measures, protect borders and infrastructure, and improve WMD capabilities

Adapted from McEntire, David A. 2003. "Searching for a Holistic Paradigm and Policy Guide." *International Journal of Emergency Management* 1(3): 298–308.

degree programs" (1999, 13). He continues, "interdisciplinary problem-focused degree programs would provide professionals with the tools needed to access new knowledge from those educated in more traditional ways and would facilitate the application of interdisciplinary solutions to tomorrow's problems" (1999, 259). Many academic programs in emergency management have heeded or preceded this counsel (see one example in Table 1.3).

THE NEED FOR MULTI- AND INTERDISCIPLINARY RESEARCH

Studying disasters from the perspective of different disciplines and assimilating their findings should not be viewed as an end unto itself. Instead, multi- and interdisciplinary research should be regarded as the means to better understand disasters and more effectively formulate and implement disaster policies. There are a number of reasons why this is the case.

First, scholars and practitioners are increasingly aware that we are experiencing more hazards today in terms of number and diversity. There are a number of natural hazards that may affect us including, but certainly not limited to, earthquakes, tsunamis, tornadoes, hurricanes, floods, droughts, wildfires, landslides, avalanches and other events triggered above, on, or below the earth. There is also the possibility of more anthropogenic-related incidents such as computer disasters, infrastructure failures (blackout), hazmat releases, industrial explosions, railroad derailments, and intentional disasters such as plane hijackings, anthrax attacks or suicide bombings. Beyond

Table 1.3
CURRICULUM OF THE EADP PROGRAM

<table>
<tr><td>University Core, including:
• Technical Writing
• Earth Science, Introduction to Physical Geology or Environmental Science
• Interpersonal Communications or Public Speaking</td><td rowspan="3">Electives Outside the Major:
• Public Administration
• American Intergovernmental Relations
• Public Policy Analysis
• Biological Resource Conservation and Management
• Community and Public Service
• Introduction to Philanthropy and Fundraising
• Social Evolution of Contemporary Fundraising
• Proposal Writing and Grants Administration
• Volunteer Management Concepts and Applications
• Community Resource Mapping and Collaboration
• Volunteer Program Planning and Evaluation
• Cartography and Graphics
• Medical Geography
• Map-Air Photo Analysis
• Meteorology
• Introduction to Geographic Information Systems
• Environmental Geology
• Risk Management
• Workplace Health and Safety
• Collective Behavior
• Sociology of Disaster</td></tr>
<tr><td>Major Core:
• Introduction to Emergency Management
• Hazard Mitigation and Preparedness
• Disaster Response and Recovery
• Leadership and Organizational Behavior (or Public Management)
• Capstone Course in Emergency Management
• Financial Aspects of Government
• Internship Preparation
• Internship Practicum</td></tr>
<tr><td>Electives Within the Major:
• Images of Disasters in Film and Media
• Hazardous Materials Planning and Management
• Private Sector Issues
• Special Populations and Disasters
• Technology and Emergency Management
• Terrorism and Emergency Management
• The Federal Government and Disasters
• Flood Plain Management
• Public Health and Disasters</td></tr>
</table>

these hazards, humans may be faced with biological threats such as SARS, Avian Flu, West Nile, AIDS, Hoof and Mouth disease, etc. With this in mind, Thomas and Mileti assert that the "hazards managers of the future will require an understanding of a wider variety of hazards. Few will have the option of only considering a single hazard, but instead must be more broadly trained to consider the full range of hazards that exist in a given area, including natural, technological and terrorist hazards" (2003, 18). Of course, we must take into account the fact that hazards are not isolated and they often interact one with another. For instance, an earthquake may cause a landslide or the breach of a dam. A wildfire may threaten a nuclear power plant or an industrial facility. A terrorist attack may include the sabotage of infrastructure, or the use of chemical or biological weapons. Future emergency managers must have an appreciation for complex, compound or cascading disasters.

Second, emergency management includes various functions across many phases. Activities in this profession include: hazard and vulnerability assessments, land-use planning, structural mitigation, the passing of laws and ordinances, code enforcement, education of politicians and citizens, planning, training, exercises, warning, evacuation, sheltering, debris management, and donations management. Other measures to be taken are continuity of government, volunteer management, traffic control, fire suppression, damage assessment, disaster declaration, mass fatality management, emergency medical care, public information, individual assistance, public assistance, decontamination, WMD detection, environmental restoration, etc. Such steps are integral to emergency management, although finding the proper balance among mitigation, preparedness, response and recovery activities is difficult to obtain (Thomas and Mileti, 2003, 17).

A third reason why it is important to take a multi- or interdisciplinary approach is because there are so many actors involved in emergency management. At the *Designing Educational Opportunities for Emergency Managers Workshop* in Denver in 2003, Ellis Stanley, the Emergency Manager of Los Angeles, stated that in his city departments from Aging to the Zoo have an important role in preventing or responding to disasters. Emergency managers are undoubtedly not the only participants in emergency management, although they do play a central role. Additional actors include politicians, flood plain managers, fire and police officials, building code inspectors, meteorologists, representatives of the American Red Cross, business continuity planners, and volunteers of religious organizations. There are numerous others in state and federal government. In many ways, the lines and boundaries among the levels of government and all departments, agencies and organizations in the public, private and non-profit sectors are blurring (e.g., homeland security needs the support and involvement of local jurisdictions and businesses to be effective). Expertise and experience in any given sector is not enough due to the current disaster setting.

These points suggest that the traditional disciplines involved in emergency management may not be able to address – in spite of their long history of excellent contributions to the field – every issue or answer question relating to disasters. Furthermore, the traditional approach to the study of emergency management is incomplete or does not work. It is incorrect to assume that we can study hazards and problems of vulnerability, develop alternative policy options concerning what functions and phases to address, choose one route to pursue, and move on to the next concern (Mileti 1999, 27). In most cases, challenges are interrelated and complex, and solutions are multifaceted with both advantages and disadvantages. "Buy in" and collaboration among all participants are also vital for success.

THE MERIT OF DIVERSE AND INTEGRATED FINDINGS

The advantages of multi- and interdisciplinary studies are noteworthy. Collective research projects have the distinct benefit of recognizing the unique and evolving world of disasters. Thomas and Mileti declare:

> Emergency management is different than it was a decade ago, and not yet what it will be in the future. It is more complex and includes many more topics than it did just a few years ago. Emergency/hazards management includes mitigation, preparedness, response and recovery. It demands knowledge and skills in the natural and physical sciences, the social and behavioral sciences, aspects of engineering, and technology. Emergency/hazards

> management is, without question, interdisciplinary in nature, since it requires drawing on knowledge now housed in various disciplines. And still, some continue to even debate the question of whether or not emergency/hazards management is a discipline at all. Some consider it to be an activity that lives in the cracks between several 'real' disciplines, similar to programs such as environmental or women's studies. Others analogize emergency/hazards management today to the experience of urban and regional planning over the last several decades. Regardless of one's perspective, those who work in emergency/hazards management today must break down traditional academic and professional boundaries (2003, 17).

Multi- and interdisciplinary research also helps to fill in the gaps in academia. Although there remains much to be learned in any given area of focus, the major holes in disaster studies today exist across disciplines and not necessarily within them. Larry Brandt, a Program Manager at the National Science Foundation, commented during a meeting to review disaster-related grants that Congress is becoming more interested in funding cross-disciplinary studies. Such projects are able to move beyond simplistic descriptions of phenomena, provide explanations that are rarely self-evident, and span and show applicability to multiple fields of knowledge as well as different groups of practitioners.

Utilizing the unique methods and analysis from different disciplines also allows flexibility in approach. In *Geographical Dimensions of Terrorism*, Cutter, Richardson and Wilbanks declare ". . . the prime requisite is versatility and the ability to think without allowing oneself to be constrained by disciplinary boundaries" (2003, 14). In other words, it is more appropriate to find suitable methods to answer inquiries instead of the other way around. This may be another argument why interdisciplinary research is often regarded as cutting-edge (Mileti, 1999, 241).

Integrating the research of scholars from several disciplines also permits a holistic understanding of the unique and multifaceted disaster problems we are facing today. Mileti notes that "researchers have called for a broader view of the disaster problem" (1999, 35). Havidán Rodriguez likewise believes "an interdisciplinary approach is extremely important and necessary in order to generate a comprehensive . . . understanding of disasters. . . . Consequently, we will be able to generate scientific knowledge with 'better value and use' to our society" (2004).

This brings up a final benefit of multi- and interdisciplinary research to be discussed here: findings from many fields of study may help to generate and implement better policies for practitioners. A common view in the past was that disaster problems could be resolved in isolation from one another or the broader impact of societal activity (Mileti, 1999). For example, new laws pertaining to land use were regarded to be the solution for rising flood losses, preparedness was regarded as a function of creating plans and holding exercises, advanced warning systems were seen as the means to keep people safe in time of disaster, modern technology was believed to overcome communications difficulties during response operations, and insurance was seen as the way to promote recovery. However, it is becoming common knowledge that:

- new laws require enforcement and a change in political culture;
- being ready to deal with disaster entails building capabilities and not just going through planning motions or drills with no follow-through on evaluations;
- warning systems are useless unless they are based on sound principles of human behavior;
- interorganizational rivalry is often a greater problem than communications equipment that fails;
- and some cannot afford insurance premiums so their resilience in the aftermath of disasters is limited.

In regards to the challenge of terrorism, Falkenrath reminds us that:

> the tendency to organize around disciplines, to adopt "stovepiped" approaches to problems, and to optimize solutions for part but not all of the problem is too strong among loose collections of unadulterated specialists. Only a team of individuals with genuine cross-cutting knowledge and experience will be able to understand the complexity of any particular homeland security challenge, devise an efficient and viable strategy for dealing with the problem, and implement this strategy effectively (in Damien 2006, xxvi).

Summarizing many of these points, Britton declares that multidisciplinary research is a major step forward and he points out that "there is now a greater likelihood that research and practice can better capture the reality of relevant issues and their particular social contexts. The field is gaining confidence that it can identify relevant universals pertaining to disaster as a phenomenon and, with it, developing more appropriate methods for managing them" (1999, 229).

Thus, we can no longer accept simplistic views of disasters because they do not correspond to reality. Emergency managers, if they are to truly be effective, must have a sound understanding of the "science" of several disciplines. But, because disasters are often and incorrectly viewed as uncommon events that are separated from daily human activities, emergency managers should also gain skills in the "art" of their craft as well (e.g., sales, marketing, interpersonal communication, persuasion, argument, public speaking, networking, political posturing, cajoling, societal mobilization, etc.). Therefore, multifaceted disaster problems require intricate assessments and interrelated solutions promoted by professionals who are respected for their wide-range of knowledge, skills and abilities.

A PREVIEW OF THIS VOLUME

If the necessity for and advantages of multi- and interdisciplinary disaster research are warranted as argued above, there can be much justification for this edited volume entitled *Disciplines, Disasters and Emergency Management: The Convergence and Divergence of Concepts, Issues and Trends from the Research Literature.* The book includes contributions from a number of scholars who bring their knowledge to bear on the study of disasters and profession of emergency management. Many of the authors have long-standing interest in these subjects and are well known for their work in this area. Others have become interested in disasters and emergency management recently, but are nonetheless experts in their respective fields. Disciplines represented in this book are numerous and include those from both the hard and soft sciences. Many of these disciplines have been fundamental in the development of knowledge about disasters, while others are only beginning to explore their relationship with emergency management.

Although the scope of this book is broad, the edited collection of chapters should not be construed as a complete discussion of the topics and issues subsumed under the heading of multi- or interdisciplinary approaches to the study of disasters and emergency management. Difficult choices had to be made about the scope of the project before it began and as it proceeded. Geography, for instance, is comprised of several sub-disciplines (including geology, volcanology, hydrology, climatology, etc.) that could have had their own dedicated chapter. There are also different types of engineering (e.g., civil, structural, chemical, etc.) that had to be integrated in a single chapter on the built environment and other engineering/technological applications. The discipline of history could have also been incorporated into the book, but there is already an excellent review of this discipline by Bankoff (2004). Constraints on space therefore made it difficult permit the inclusion of these and other areas of specialization. In other cases, the call for experts to write a chapter about computer science went unheeded. There were also discussions during the 2005 *FEMA Higher Education Conference* about what constitutes an academic "discipline," and so some subjects like social marketing or safety had to be omitted because they were not viewed as a widely recognized discipline or because regarded to be more vocational in nature. The decisions made could at times be conceived as haphazard in that criminal justice was included but not fire science (even though criminal justice is more common on college campuses and fire science plays an integral role in emergency management). In spite of these weaknesses, the book may complement important multi- and interdisciplinary research today and provide a road map of issues that need to be addressed in the future.

In an attempt to bring order to the discussion, the authors in this volume were asked to respond to several important questions. These included:

1. What is the history of your discipline as it relates to disasters?
2. How does your discipline view or define disasters, hazards, vulnerability and emergency management?
3. What disaster-related issues and concerns are prevalent in your field of study?

4. What contributions does your discipline make to the knowledge base of these disruptive and destructive events? (e.g., methods).
5. How do the findings from your discipline overlap with those of other disciplines?
6. What are the gaps in knowledge in your discipline?
7. What suggestions does your discipline offer to improve emergency management?
8. What recommendations do you have for your discipline and others in the future?

Each author took discretion in answering these questions, and some chapters cover them more fully and in a direct manner than others. But the information regarding each discipline and its relation to others and disasters is impressive and helps to generate new knowledge, identify gaps in the literature, and provide recommendations for emergency management.

The book starts off with several disciplines that have laid the foundational concepts, theory and principles in emergency management. Jim Kendra discusses the positive impact geography has had on the study of disasters but points out that scholars in this area need to ensure that their findings are integrated into policies that impact human activity. Kendra also keenly observes an interesting tension between the hazards and vulnerability concepts that must be discussed in further research. Writing about meteorology, Kent McGregor discusses various weather-related hazards, and provides the state of knowledge regarding hurricanes, tornadoes, floods, droughts, heat waves and ice storms. His summative research suggests that prediction models will become more important in the future and that practitioners must give more attention to global warming and the transfer of early warning technology to developing nations. Ana Maria Cruz has a chapter covering the contributions of engineering to emergency management. She states that engineering has helped set standards and build safer structures, and she astutely points out that we must look at the interface of the natural environment and infrastructure. Thomas Drabek traces the long history of sociology in emergency management and acknowledges that we still do not have a widely accepted definition of disaster. His excellent review of the literature summarizes many findings pertaining to social behavior, the applicability of different methodologies, and the need to apply lessons that have been uncovered in many disasters.

After the discussion of these founding disciplines, the book shifts attention to those disciplines that have become more important over time. Joe Scanlon explores the relation of journalism to emergency management, and he utilizes his unique experience in this area to encourage special attention to the vital and somewhat troubling role of the media in disasters. He argues that the media should avoid perpetuating myths and take its role in communications more seriously. Margaret Gibbs, coming from the discipline of psychology, looks at the emotional impact of disasters. She provides a great overview of the nature of psychopathology and vulnerability, and highlights different types of interventions and their relative merit. An anthropologist, Doug Henry, delves into the human causes of disasters, and examines those who are most likely to be adversely affected. Henry's research focuses on disasters outside of the United States, and his research appropriately questions top-down approaches to recovery assistance. Michael Zakour is the author of a chapter on social work. He notes that scholars in this area are highly concerned about coordination efforts among relief organizations and he exposes the structural relations that create social vulnerability. Zakour integrates several studies about his discipline in an in-depth manner, and his work has many notable parallels to that of Gibbs and Henry. Rick Sylves and Bill Waugh have chapters on political science and public administration respectively. Sylves relies on his extensive research background to uncover the political dimensions of disasters. He points out that homeland security policies are inherently political and that emergency managers cannot afford to ignore this fact. Bill Waugh admits that scholars in his discipline have been slow to address disasters, but he is no doubt correct to point out that emergency management is the quintessential government function. Waugh's chapter covers several topics, but he provides a great discussion about the tensions between homeland security and emergency management.

At this point, the book turns attention to various disciplines that have not received adequate recognition in the disaster literature. David McEntire illustrates that international relations has close ties to the birth of emergency management and that its findings have relevance to the security dilemma owing to global terrorism. His chapter reiterates that we must learn more about radical Islamic fundamentalists, and explains that his field of study can help promote better responses to disasters because of the lessons gleaned regarding decision making. McEntire also includes a chapter on comparative politics with a UNT graduate student, Sarah Mathis, arguing that scholars have failed to value the benefit of research that contrasts emergency management in different societies. He also notes the different types of disasters and impacts in developing and developed nations, and asserts that more must be done to help the poorer countries of the world. John Pine, a respected scholar in the field, writes about the field of management – a discipline that is ironically underrepresented in emergency management. His chapter helps practitioners understand the parts of the emergency management system, and he advocates more strategic planning in the field. Kathy Dreyer is an expert in gerontology. She explores the unrecognized links between her discipline and disasters. Her research stresses the vulnerability of older adults, and suggests that more attention needs to be given to disaster planning of nursing facilities.

The penultimate section of the book covers several disciplines that are becoming much more important than they were in the past. Rick Bissell and Robert Louden provide chapters that have direct relation to the terrorist attacks on 9/11. Bissell identifies some issues pertinent to emergency medical care, but spends most of his time discussing public health and diseases. His valuable study reiterates the complexity of health issues in disasters, and recommends that emergency managers spend more time interacting with public health officials. Louden, a scholar of criminal justice, notes the vital role of the police in disasters in terms of traffic control, and the investigation and prosecution of terrorists. He also concludes that more emphasis needs to be given to emergency management planning for correctional institutions. Terry Clower, an applied economist, identifies how economics can help practitioners with damage assessment, disaster declarations and insurance policies. Besides covering methods in detail, Clower notes that rising disaster losses may not always be as bad as advertised because of the continual growth of the national economy. Clower is aware disasters produce winners and losers though. Bill Nicholson, a national expert in disaster law, relates the history of emergency management as a function of congressional legislation. He points out that emergency managers must develop stronger ties to the legal counsel in their communities, and find ways to implement "litigation mitigation" – an effort to reduce liability in the community. Coming from environmental management, John Labadie compares the nexus between emergency management and the environment. His research, undoubtedly vital in light of ongoing degradation, concludes that more attention needs to be given to slow onset disasters and the concept of sustainable development. Brian Richardson and Lori Byers, two knowledgeable students of communication studies, trace the history of this field and identify its relation to emergency management. They reveal that research needs to focus more on "sensemaking" while practitioners must employ persuasion strategies and be aware of the impact of communication on disaster victims/survivors. Drawing from his insightful instructor guide on business continuity (published by the Federal Emergency Management Agency), Greg Shaw shows why being prepared should be important to those in the private sector. He defines what business continuity means, shares concerns about the concept of risk management, and advocates compliance with NFPA 1600. Tisha Slagle Pipes, a budding scholar finishing up her Ph.D. in information sciences, notes the complex relation among technology, data storage and communications. Her research indicates an increased potential for computer-related disasters, and encourages emergency managers to pay more attention to information as a way to promote disaster reduction and improve emergency responses.

The concluding chapter is written by David McEntire and Sarah Smith, another graduate student at the University of North Texas. It summarizes the content of the book, paying special attention to the points of intersection across different fields of study as well as gaps in interdisciplinary knowledge. Several lessons for practitioners are also highlighted, and recommendations for future research and practical application are identified.

BARRIERS TO IMPLEMENTATION

Before continuing on with this text, one final comment is in order. There are certainly significant barriers inhibiting multi- and interdisciplinary research and drawbacks are equally possible. Scholars from different disciplines do not speak the same language, which poses substantial communication challenges. For instance, sociologists study "emergent phenomena" while engineers explore "linear elastic analysis." In addition, values also complicate the sharing of knowledge and the application of information across different disciplines (e.g., some scholars want to foster land-use planning while others want to improve warning system effectiveness). There is likewise a difference of opinion regarding the domain boundaries of different fields of study. David Neal's review of emergency management education (2000) includes an important, perplexing, and perhaps controversial question about where to house such programs in academia. Should emergency management degrees be placed in the traditional departments (e.g., geography or sociology) only? The prestige of certain (and perhaps all) disciplines likewise limits the interaction of scholars from various backgrounds. Regarding this statement, Neil Britton notes that some engineers were reluctant to share findings with those from other disciplines in one multidisciplinary endeavor (2004). Even the reward structure for tenure complicates multi- and interdisciplinary research, although many universities are now seeking such studies of their faculty members.[4] Summarizing these and other problems, Mileti comments:

> The academic community recognizes individuals but is hard pressed to do the same for interdisciplinary groups; promotion committees have difficulty ascertaining the relative contributions in multi-authored publications; the overhead structures of many institutions discourage cross-institutional research teams; and graduate students are restricted to the department and university in which they are enrolled (1999, 260).

These challenges are undoubtedly formidable and may even be impossible to overcome.

Quarantelli and Dynes also remind us that:

> The past history of interdisciplinary research, including efforts in the disaster area, is not supportive of the ideas that better research results are obtained or that applications of findings are more easily accomplished by taking an interdisciplinary stance. In such an approach, contributions of different disciplines are often reduced to the lowest common denominator, which is sometimes only slightly, if at all, a common-sense level (as cited by Phillips, 2003, 18).

Thus, multi- and interdisciplinary research may not adequately capture all knowledge pertaining to disasters, and there is always the chance that findings will be regarded as irrelevant, incomplete, erroneous, or even offensive to some because of different epistemological assumptions.

At the same time, Drabek makes some interesting observations about the inescapable breadth of emergency management research:

> Today, emergency management studies are conducted by research with various specialties, including the physical and natural sciences as well as sociology, psychology, anthropology, geography, economics, political science, and public administration. The study topics reflect the particular disciplines of the researchers. A psychologist might ask how well victims are sleeping after a tornado; a geographer might map the rebuilt environment and ask whether the new spatial patterns will place the community at greater risk; a political scientist might explore the process by which a community makes mitigation decisions (1991, 21).

He also reiterates the need to integrate studies in that:

> Emergency management requires research of many types. Some research should reflect the strengths of the theories and methods of single disciplines; other investigations will require more interdisciplinary approaches. Although research based on single disciplines will continue to enrich the understanding of emergency managers, they must become increasingly skilled at making cross-disciplinary syntheses and applications (Drabek, 1991, 21).

In conclusion, the editor and contributing authors of this book are aware that they are exploring uncharted waters, and that there are undoubtedly stormy seas, misty views, and jagged rocks posing a threat to the multi- or interdisciplinary vessel. Havidán Rodriquez reports that "the path to interdisciplinary research is complex and often difficult to navigate" (2004). However, successfully searching the increasingly coveted harbor of increased knowledge and reduced

---

4. The University of North Texas has included multi- and interdisciplinary studies as part of the strategic plan it adopted in 2005.

disasters through collective research projects will be more likely if we will follow Britton's admonition to find consensus about goals and collaborate with others (2004). It is hoped that this book may, in some small way, encourage such joint approaches by requesting "all hands on deck." Reaching the promised land of understanding and reducing disaster through multi- and interdisciplinary research may not necessarily be guaranteed, but the embarkation may be worth the exploration nonetheless.

REFERENCES

Bankoff, G. (2004). "Time is of the Essence: Disasters, Vulnerability and History." *International Journal of Mass Emergencies and Disasters 22*(3): 23–42.

Britton, N. (2004). "Multidisciplinary, Multinational Research Projects: Challenges and Benefits." Paper presented at the *29th Annual Natural Hazards Workshop*, July 13, Boulder, Colorado.

Britton, N. (1999). "Whither the Emergency Manager?" *International Journal of Mass Emergencies and Disasters 17*(2): 223–235.

Cutter, S.L. Richardson, D.R., and Wilbanks, T.J. (2003). *The Geographical Dimensions of Terrorism.* New York: Routledge.

Damien, D.G. (2006). *The McGraw-Hill Homeland Security Handbook.* McGraw Hill, New York.

Drabek, T.E. (1991). "Evolution of Emergency Management." In Drabek, T.E. and G.G. Hoetmer (eds.) *Emergency Management: Principles and Practice for Local Government.* ICMA, Washington, D.C.

Gruntfest, E. and Weber, M. (1998). "Internet and Emergency Management: Prospects for the Future." *International Journal of Mass Emergencies and Disasters 16*(1): 55–72.

McEntire, D.A. (2005). "Revisiting the Meaning of Hazards and the Importance of Reducing Vulnerability." *Journal of Emergency Management 3*(4): 9–12.

McEntire, D.A. (2004). "Tenets of Vulnerability: An Assessment of a Fundamental Concept." *Journal of Emergency Management 2*(2): 23–29.

McEntire, D.A. (2003a). "Causation of Catastrophe." *Journal of Emergency Management 1*(2): 22–29.

McEntire, D.A. (2003b). "Searching for a Holistic Paradigm and Policy Guide." *International Journal of Emergency Management 1*(3): 298–308.

McEntire, D.A. and Fuller, C. (2002). "The Need for a Holistic Theorectical Approach: An Examination from the El Niño Disasters in Peru." *Disaster Prevention and Management 11*(2): 128–140.

McEntire, D.A., Fuller, C., Johnston, C.W., and Weber, R. (2002). "A Comparison of Disaster Paradigms: The Search for a Holistic Policy Guide." *Public Administration Review 62*(3): 267–281.

McEntire, D.A. (2001). "Triggering Agents, Vulnerabilities and Disaster Reduction: Towards a Holistic Paradigm." *Disaster Prevention and Management 10*(3): 189–196.

Mileti, D.S. (1999). *Disasters by Design: A Reassessment of Natural Hazards in the United States.* Joseph Henry Press, Washington, D.C.

Neal, D.M. (2000). "Developing Degree Programs in Disaster Management: Some Reflections and Observations." *International Journal of Mass Emergencies and Disasters 18*(3): 417–437.

Ngo, E. (2001). "When Disasters and Age Collide: Reviewing Vulnerability of the Elderly." *Natural Hazards Review 2*(2): 80–89.

Phillips, B. (2003). "Disasters by Discipline: Necessary Dialogue for Emergency Management Education." Paper presented at the Workshop, *Creating Educational Opportunities for the Hazards Manager of the 21st Century.* Denver, Colorado, October 22.

Rodriguez, H. (2004). "The Role, Contributions, and Complexities of Interdisciplinary Research: A Holistic Approach to Hazards and Disasters." Paper presented at the *29th Annual Natural Hazards Workshop*, July 14, Boulder, Colorado.

Ronan, K.R., Douglas Paton, David M. Johnson and Bruce F. Houghton. (2000). "Managing Societal Uncertainty in Volcanic Hazards: A Multidisciplinary Perspective." *Disaster Prevention and Management 9*(5): 339.

Salinas, M.G. and Rodríguez, D. (1998). *The Disasters of Mexico: A Multi-disciplinary Perspective. (Los Desastres en México: Una Perspective Multidisciplinaria).* Universidad IberoAmericana, Ciudad de Mexico.

Wilson, E.O. (1999). *Conscilience: The Unity of Knowledge.* First Vintage Books, New York.

Chapter 2

Geography's Contributions to Understanding Hazards and Disasters

James M. Kendra

ABSTRACT

Geography has a many-decades-long record of research and practical application in understanding and managing hazards and disasters. This chapter reviews some of the major points of geographic connection to these areas. It examines some of the earlier conceptions of geography's interest in human-environment interactions and discusses shifts in understanding the nature of hazard, including recent emphasis on vulnerability. The chapter notes some possible future directions and research needs from across the social sciences, and concludes by arguing that changing and elusive hazards create a need for continued robust research.[1]

INTRODUCTION

Geographers have had a longstanding role to play in understanding the full range of crises brought on through interactions of natural and social systems, and the discipline is generally recognized as one of the founding disciplines of hazard as a field of study. Topics for research have included, on the natural hazards side, the full range of geological and atmospheric agents, such as earthquakes, hurricanes, riverine and coastal flooding, drought, and, increasingly, global warming. As for technological hazards, research has included studies of response to nuclear accidents and the siting and distribution of hazardous waste storage facilities and their proximity to other land uses, such as residential areas. While it is often convenient for discussion to place hazards into discreet categories: natural and technological (or anthropogenic), researchers have argued that it is not possible to make a bold division between the two (Alexander, 1993; Cutter, 1994).

Mitchell (1990), for example, argues that hazards can be seen as *mismatches* of human and environmental or technological systems. An axiom in geography is that "Hazards are threats to humans and what they value" (Harriss, Hohenemser, and Kates, 1978:6); hence if people or their goods are not "in the way" of powerful geophysical or climatological agents, there is no hazard, because there is no one there to be threatened. Extending this theme, hazards don't exist as things in and of themselves; rather, they are created by people who place themselves, and that which they value, in places that are subject to climatological, geophysical, or technological extremes. They are the products of particular social, political, and economic decisions that are made either without sufficient knowledge of the environment,

1. Earlier portions of this chapter were presented at the Disaster Research Center 40th Anniversary Conference, Newark, Delaware, "Disaster Research and the Social Sciences: Lessons Learned and Future Trajectories," April 30–May 1, 2004, and at The National Academies Disasters Roundtable. "The Emergency Manager of the Future. The Next Generation." Washington, D.C. June 13, 2003. The author thanks Rutherford H. Platt, David A. McEntire, and James K. Mitchell for comments on earlier drafts of this chapter. The views expressed here are those of the author.

without the capacity to make different decisions, or are made with some calculation or hope that "it" (flood, earthquake, wildfire) won't happen in a timeframe that will negatively affect the decision-maker.

WHO IS A GEOGRAPHER?

Sometimes identifying who is a geographer is difficult; people whose academic degrees are not in geography may be doing work that is geographic, while people with training in geography may hold jobs that do not appear to be geographic (though the person might or might not use geographic skills). An old joke held that "geography is what geographers do," suggesting that people gathered around the name or the theme of geographic work even though there was disagreement about what constituted truly geographic methods. Certainly, geographers are concerned with the distribution of various kinds of social, biological, and geomorphological phenomena over space. *Where* something happens, and why it happens there, are the questions that distinguish geographic approaches to understanding the world from the approaches of other scientific disciplines, and concerning risk, hazard, and disaster. Geographers are interested in the interaction of social, physical, technological, and political/legal systems.

Geographers generally trace the foundation of the geographic approach to hazard to Harlan Barrows' (1923) conception of geography as "human ecology," the interaction of people and the natural environment (Alexander, 1993; Cutter et al., 2000). In trying to ground geography in a set of perspectives and approaches that would distinguish it from other sciences – some of which, such as sociology and psychology – were still comparatively young themselves, and seeking their own disciplinary identity, Barrows argued that the human ecological perspective offered an approach that resonated with geographers' interest in both the natural environment and human activity in relation to it. At the same time, Barrows set the stage for maintaining intellectual contact with the environment while shedding the perspective of *environmental determinism*, popular at the turn of the last century, that held the view that the moral qualities and cultural characteristics of groups are determined by their environment.

> . . . geography is the science of *human ecology*. The implications of the term "human ecology" make evident at once what I believe will be the future objective of geographic inquiry. Geography will aim to make clear the relationships existing between natural environments and the distribution and activities of man. Geographers will, I think, be wise to view this problem in general from the standpoint of man's adjustment to environment, rather than from that of environmental influence. The former approach is much more likely to result in the recognition and proper valuation of all the factors involved, and especially to minimize the danger of assigning to the environmental factors a determinative influence which they do not exert (Barrows, 1923: 3).

Over time, of course, geographers' interests shifted, so that Barrows' conception of geography as principally human ecology never became the defining principle, though in many of the discipline's subfields it remains at least an implicit orientation. Moreover some areas, such as climatology, that Barrows thought should stand alone or join some other discipline (and which he referred to as "peripheral specialisms") have become prominent branches of contemporary geography even though they share interests and methods with other fields of study.

Most geographers credit Barrows' student, Gilbert White, with inaugurating the essential research directions and fundamental methodological approaches that continue to inform contemporary thinking about hazards (Mitchell, 1990; Cutter, 2001a). White (1973: 198) summarized his work on the perception of floodplain occupants to the risks of living there and the differential adjustments *(modifying the cause, modifying the loss, or distributing the loss)* that they adopted in different places and at different times. This work conducted by White and his colleagues had two basic goals, one that was policy-oriented and one oriented toward understanding the cognitive aspect of human behavior with respect to the creation or mitigation of hazard.

The policy aspect centered on evaluating the efficacy of several billion dollars of flood control projects on the Western rivers. Congress approved the Flood Control Act of 1936 after multiple severe floods; White and other geographers sought to learn if these had in fact been successful in reducing flood losses (White, 1973).

> In brief, it was found that while flood-control expenditures had multiplied, the level of flood damages had risen, and that the national purpose of reducing the toll of flood losses by building flood-control projects had not been realized. . . . The findings also indicated that because of the Federal government's concentration upon flood-control works and upstream water-management activities to the exclusion of other obvious but relatively unpracticed types of adjustments, the situation was becoming progressively worse and showed no promise of being improved by a continuation of the prevailing policies (White, 1973: 198–199).

These findings exposed a key paradox in understanding human exposure to environmental extremes – that structural mitigation measures may conceal hazard and may even foster development in supposedly-protected areas, so that more property is exposed if the mitigation measures are overwhelmed. The findings thus pointed the way to the spectrum of non-structural mitigation measures that are now considered to be part of the standard package of mitigation tools, especially land-use planning accompanied by incentives and penalties to encourage adoption of mitigation measures. For example, Platt (2004: 389) notes several methods for reducing flood losses that were recommended by Gilbert White's task force beyond structural flood-control, including land-use management and flood insurance. Though the task force expressed several cautions regarding the implementation of a low-cost flood insurance system, his report provided the basis for the National Flood Insurance Program, which requires that communities adopt certain floodplain management standards in exchange for access to the flood insurance that is not available in the private insurance market. The National Flood Insurance Program, though controversial, remains one of the most important national-level mitigation initiatives.

From the cognitive or behavioral standpoint, White found, people "did not behave as it had been expected they would" when the large system of dams and levees had been initiated. Answering the question "why not" in turn pointed to a need to understand how people perceived the flood hazard and how they perceived options for adjustment (White, 1973). According to White's historical review, understanding people's poor locational choices required geographers to draw on the work of psychologists and economists who studied decision-making, but the intersection of questions of choice – depending as they did on knowledge and personality issues – with attempts to understand natural processes (including as they are modified by human action) led to the development of systems models of hazards, systems theory then being ascendant (see Kates, 1971). A basic model grew out of this work, and it dominated thinking about hazards in both programmatic and academic ways for some 15 years. Stated briefly, the model holds that people make decisions – that is, expose themselves to danger – based on imperfect knowledge of the nature, magnitude, and return period of extreme events or inaccurate assessments of their capacity to endure these events. Policy prescriptions follow, and these are fundamentally to get people to understand the true danger and keep them away from it. These goals would preferably be accomplished largely through institutional mechanisms such as land-use planning. Virtually the entire hazards-reduction enterprise, whether in academia or in the public or private sector, is oriented around these dimensions: (1) identify the threat; (2) communicate its nature; and (3) persuade people to avoid it. These are not settled areas, of course, and they are tightly interrelated, but these propositions guide most of the research and policymaking in the field. The last dimension has proved elusive to accomplish, as the inducements for living in dangerous places are powerful, not only for individuals, but also for larger social systems even to the national level. Coastal areas provide a principal example of such inducements, where the amenity value of beaches and the commercial significance of ports and harbors provide powerful economic reasons for development.

TECHNOLOGICAL HAZARDS

While researchers are increasingly recognizing a blending of the categories of natural and technological hazards, that was not always the case, nor are we yet able to definitively say that a distinction between these two kingdoms of peril is entirely valueless as a way of organizing kinds of danger and potential responses. By the late 1970s, and especially after the accident at the Three Mile Island nuclear power plant, geographers turned their attention to industrial hazards, initially simply extending the same approach they had taken to natural hazards but then looking at the origins of technical failures as well. Geographers have made most of their contributions to technological hazards research in the zones of overlap between geography with its spatial focus, and psychology and sociology with their behavioral and perceptual foci.

One of the particular strengths of the geographic method is its capacity to draw theories and empirical findings from other disciplines and meld them with a spatial perspective to explain, or to predict, distributions and relationships of phenomena. Platt (2004: 33–41) identifies four "organizing themes" which are emblematic of the geographic method: spatial organization, scale, function, and externalities. Each of these themes is also integral to research into technological hazards. Spatial organization is perhaps the most obvious dimension: where to locate a facility, for example, to minimize harms. In the event of an accident, where will contaminants go? Where will evacuated populations flee? Scale is a related dimension: Is one large facility better than smaller, associated facilities? What complications extend from location near a large population center? Function addresses the role in a social-economic system of the activity or enterprise creating the hazard. Does all of society receive benefits, or only a subset? Externalities are the impacts of particular decisions, for example certain land uses, that may impact those other than the decision makers or those who obtain the benefits of those decisions. Pollution is the emblematic externality. Working within these four themes obviously requires the diverse expertise of many other disciplines, but it is the art and the science of the geographer to assemble them to explain the creation and distribution of hazard and to suggest possible mitigation.

Much work in geography has centered upon, first, distinguishing and classifying natural and technological hazards, and then further subclassifying technological hazards. Kasperson and Pijawka (1985) compared and contrasted natural and technological hazards on several points. Natural hazards, for example, tend to be obvious: natural disasters have clear beginnings and endings. They are familiar, and there is a history of dealing with them. Furthermore, the disaster represents a "commonality" (p. 16) among both victims and non-victims, resulting in mutual cooperation which helps the recovery process. Technological hazards, however, are frequently novel: there is no history of dealing with them. These hazards are not readily observable, but the effects may take years to appear. And because the hazard is of long duration, abandonment of the area, rather than rebuilding, becomes the more attractive option. Furthermore, typical social support mechanisms do not appear: people may even shun the afflicted area, a phenomenon Kasperson et al. (1988) would label "stigmatization."

Geographers also studied not just the nature and social ramifications of technological hazards, but how such hazards evolve in the first place. Hohenemser, Kasperson, and Kates (1985) developed a "causal structure" of technological hazards which emphasized possibilities for intervention at each point of a sequence of events: from the recognition of a human need, through the development of a technology to meet that need, to the event precipitated through the use of that technology. In a companion article, Hohenemser, Kates, and Slovic (1985) propose a "causal taxonomy" for grouping hazards by common features. Their taxonomy is linked to the causal sequence and is intended to help in the comparison and selection of technologies suitable for a particular purpose, as well as in comparing new technological hazards to those which may be somewhat familiar as a way of suggesting management strategies. The structural and taxonomic nomenclautre can also provide a direction for research.

A question which geographers have touched, but not firmly grappled with, is that of technology itself as a hazard: i.e., as a *social hazard.* Not a hazard which is detrimental to society by causing physical harm, with attendant social disruption, but a hazard which is detrimental by disrupting some necessary or agreeable component of society. Lurking round the edges of the literature, however, is a suspicion that the *process* of technological advancement, and not just its by-products, is harmful.

Chapter 1 of Susan Cutter's (1993) *Living with Risk* shows the tension between technological hazards manifest as social disruption, and technological hazards manifest as physical harms, by opening with a quote from Jacques Ellul: "In the modern world, the most dangerous form of determinism is the technological phenomenon. It is not a question of getting rid of it, but, by an act of freedom, of transcending it" (Ellul 1964, p. xxxiii).

Ellul's book *The Technological Society* is the philosophical starting point for most recent critiques of technology, such as Langdon Winners's work and then Neil Postman's (1992). However, the next line in *Living with Risk* is a list of the most notorious failures of technological systems: "Bhopal, Chernobyl, Love Canal . . ." Thus we shift from social hazard back to nuts-and-bolts physical hazard, within three sentences.

Zeigler, Johnson, and Brunn (1983) addressed the issue more explicitly, but largely as an afterthought. They suggest an analogy between conventional hazards, which release materials or energy, and high-tech futuristic hazards, involving the release of information into the interconnected computer and communications network. Computer software viruses, of course, fit this analogy, and their ability to inflict serious financial losses is appreciated by computer-security experts. Recent concern about cyberterrorism is an extension of their ideas.

In an imaginative endeavor, Graham and Kasperson (1985) directly examined television as a social hazard, examining the propensity for televised violence to incite actual violence in children. While this study also focused on directly tangible, physically manifest phenomena – death and injury – by examining them as products of behavior influenced by TV, Graham and Kasperson drew closer to a critique of the technology, by arguing that though television content had a "weak toxicity," the sustained exposure actually created a large "dose." However, they noted that no study had drawn conclusive links between television and such undesirable effects as reduced intelligence, diminished creativity, or "breakdown of community." They noted, however, that concern about those effects lingered, still not definitively resolved.

Their critique was not entirely well-received by reviewers. Otway and Cannell (1987) doubted whether the rational model of hazard – a framework representing real-world phenomena and pointing to corrective actions – could be applied to television. And they ridiculed the "information release" concept which Graham and Kasperson employed in a manner similar to Zeigler et al. (1983). Nevertheless, it is generally accepted that technological advancement has outpaced society's ability to develop philosophical structures, with attendant public policy, to accommodate unwanted side-effects. Geographers have not yet reconciled, if that could ever be possible, the spatial problems and opportunities raised by technologies currently available, let alone potential future breakthroughs. For example, what are the implications for privacy and liberty of burgeoning data-gathering, mapping, and spatial analysis technologies? (See Pickles [ed.], 1995.) Postman, a communications theorist, observed (1992) that before the Industrial Revolution, centuries might elapse between developments in science or engineering or philosophy that could "revolutionize" society. Since the Industrial Revolution, that interval has become even more shortened, all the more so with the advent of computers and applications such as the Internet. Which space is relevant in analyzing these broad transformations? Social space? Economic space? Will there be populations that are particularly disadvantaged? Where will they be? There is definitely interest in geography in debate about technological development and the distribution of benefits and harms. Nevertheless, these concerns are dispersed throughout geography's many subfields and are not a principal focus of hazards geographers.

THEORETICAL SHIFTS

A major departure in thinking was Hewitt's (1983) *Interpretations of Calamity*, which criticized previous emphases on decision-making (Alexander, 1993: 12; Cutter, 2001a: 5), emphasizing instead the socioeconomic aspects of vulnerability to extreme events. Though mainly emphasizing natural hazards, geographers also found that vulnerability principles extended well to industrial hazards. Being exposed to hazard, then, could not be entirely ascribed to the consequences of bad decisions, but of choices that were constrained by the social and economic conditions of people in dangerous places. The poor, for example, lacking resources for better locations, often live in perilous and unsanitary conditions, placing them at greater risk and moreover impeding their recovery prospects. For instance, the poor tend to have no insurance or only inadequate insurance, so that when they experience disaster they have fewer reserve resources than the wealthy. Hewitt's work pointed the way to studies of what factors are associated with vulnerability, such as gender, race, ethnicity, and poverty. Equipped with this perspective, geographers were able to broaden their view of both the causes of hazards and the consequences and to more fully understand the often-limited options available to people exposed to dangerous conditions. This perspective allows us to look at events such as the Indian Ocean tsunami and to see how much of the disaster lies not in the initial death and destruction. Staggering as they are, they are not the end of the story. The psychological, economic, and social recovery of the survivors is closely tied to pre-existing social and economic conditions in place before the event. Many of the affected communities were small-scale fishing communities dependent on boats that were destroyed. With no insurance, boat owners must continue to repay loans on vessels that are gone as well as take out loans for new equipment (see Rodríguez, Wachtendorf, Trainor, and Kendra (2005) for a more detailed discussion). But though Hewitt and subsequent scholars took a largely international approach to the vulnerability question, the approach is valid in the United States, as well, since the less-expensive land and housing within the price range of poor people is often in dangerous locations, too.

Interest in vulnerability, though important in expanding the range of theoretical tools available both for studying hazard and for providing policy guidance, brought its own set of challenges. In particular, a single definition of vulnerability remains elusive. Fundamentally, vulnerability means the possibility of loss or harm, but many researchers working with the same concepts have used different language, or have used similar language for concepts that they understand and present differently. Cutter (1996: 531–532) identified nearly twenty definitions or conceptions of vulnerability, generally similar, but nevertheless offering variations, such as conceptualizing vulnerability as risk, as differences in capacity to reduce risk, or as an integrated quality of both the hazards in a place and the social characteristics of the people who live there. Moreover, there often are not single words or economical phrases to accommodate concepts that include a complex functional and iterative relationship, a difficulty that has emerged with definitions of *disaster* in the sociological field and, in fact, with *hazard* in geography. For example, though hazards are normally seen as a result of certain kinds of human decisionmaking, geographers also include their impact (Cutter, 2001a: 2) as part of an overall conceptualization. Burton, Kates, and White (1993) are careful to refer to *hazard events* when referring to a tornado or some other agent, but also refer to them merely as hazards, thus illustrating the difficulty of managing the concept which has several components requiring different emphases in different contexts. Similarly, because vulnerabilities are also interactive phenomena, they can't really be discussed without including some of the elements that are presumed to be constitutive, complicating the definitional task.

Nevertheless, vulnerability has become a central component in thinking about hazard. For example, in defining hazard, Mitchell considers the following function:

> Hazard = f(Risk x Exposure x Vulnerability x Response) where risk is defined as the probability of a damaging event or circumstance. Exposure is a measure of the population at risk. Vulnerability is the potential for experiencing loss. Response is the degree to which society acts to reduce, avoid, or prevent loss. Hazards result from various combinations of these factors (Mitchell, 1990: 132).

Based on this conception, Mitchell, Devine, and Jagger (1989) developed a *contextual model* of hazard, a nested model which encapsulates risk, exposure and vulnerability, response, and cost within modifying contexts of economic, social, and technological processes. Applying the model to the October 1987 windstorm in Britain, they identified several points at which particular physical, social, and political circumstances combined to alter the effects of the storm, with respect to both damage and post-storm evaluations of its consequences.

For example, at the outset, weather forecasters in Britain accepted French forecasts, not accounting for differences in modeling methodologies. The loss of electric power necessitated closure of the London Stock Exchange, which may have contributed to the worldwide stock market crash by preventing communication between markets. Subsequently, the economic concerns brought about by the crash and by the prospect of the Channel Tunnel (which could be expected to cost jobs in some of the same communities affected by the storm) overshadowed the storm's impacts and preempted a thorough examination of hazard response and management. Overall damage, especially to woodlands, was intensified by ground-soaking rains falling prior to the windstorm. Assistance from the Thatcher government was forthcoming only because some Conservative Members of Parliament represented constituencies hard-hit by the storm. Thus, many non-hazard factors, competing priorities, and conflicting interests created outcomes independent of the storm's magnitude and the characteristics of the built environment.

Dow (1999: 76), however, takes a somewhat different approach, defining vulnerability as "the differential susceptibility of ecosystems, households, or social groups to losses." Dow further defines vulnerability as being comprised of three elements. "*Exposure* is the degree of risk of an event experienced in everyday life, from the probability of a hazard to actual occurrences of events of all sizes." "*Resistance* is the ability to withstand the impacts and continue to function." "*Resilience* is the ability to recover, ranging in degree from simply achieving stability at any level of functioning to recovering the full range of resources and positive momentum that existed prior to the event" (italics in original).

A complete reconciliation of the various directions of thinking on vulnerability appears unlikely, owing to the widespread and growing use of the concept both inside and outside of geography, so that some final definition is really a moving target. Nevertheless, a long step in stabilizing the meaning was made by Cutter (1996), who integrated various expressions of vulnerability in a *hazards of place* model, expressing the net vulnerability of a place as an interaction of both social vulnerability (such as demographic characteristics and risk perception) and biophysical vulnerability (characteristics of the place).

Palm (1990) sought to place hazards research within an inclusive structural framework that accounts for interactions among three levels of society: individuals and households (micro-level), governments and economies (macro-scale) and the zone of intermediaries: corporate officers, local officials, and bureaucrats (meso-level). It is only by examining the interplay of these levels, Palm asserts, that researchers can examine the social component of vulnerability. Though ultimately limited by social, economic, and political circumstances, individual action is also a function of needs, desires, and opportunities within those limits. Palm (1990) argues that hazards researchers in geography, sociology, and other disciplines must integrate individual and socioeconomic responses to natural hazards in order to understand the full scope of vulnerability and to suggest appropriate remedial actions.

Recognizing the importance of vulnerability, in turn, pointed the way toward consideration of how people take steps to lessen their vulnerability, or how they enhance their resilience to extreme variations in their environment. The debate on this continues throughout the field, not only in geography but also in other fields as well. Are vulnerability and resilience merely the obverses of the same coin? Or are they different in some way, addressing dimensions of the problem of human survival that do not necessarily intersect?

TECHNOLOGICAL DEVELOPMENTS

Recent developments have moved geography – and geographers – much closer to actual emergency and disaster management tasks than previously. The widening use of Geographic Information Systems has brought significant mapping and analytical capability to the desktop and, therefore, these systems can facilitate the organization and management of response operations after an event, so that geographic methods are useful not merely for long-term planning but also for short-term decision-making. Emergency managers, of course, have always used maps, but GIS brought mapmaking capacity directly into the hands of emergency responders. Of course, the knowledge base established by geographers was relevant at all stages of disaster, but the principal thrust had been directed more toward hazard, and hence to the early or causal factors of disaster, and less so in the actual emergency response phase. There were exceptions, of course, particularly in the areas of warning and evacuation behavior – areas where emergency managers could directly apply geographic knowledge of risk perception and decision-making during environmental extremes. Nevertheless, geographers' emphasis was on phenomena that unfolded over years or tens of years: where people lived, why they lived there, and how their coping capacity was affected by their conditions of life. Certain technological developments changed that, especially GIS. For example, areas that need to be closed to vehicular traffic can be mapped and the maps distributed to law enforcement personnel who are establishing roadblocks, or the locations for temporary housing can be plotted with respect to relevant services such as water or electricity. Following the 2001 attack on the World Trade Center, thousands of maps were created – often by geographers who volunteered their services to the Office of Emergency Management. These maps depicted areas with utility outages, road closures, the locations of supply caches and staging areas, and other relevant information needed by emergency responders. Moreover, as the weeks passed and the "secured area" around the Trade Center site grew continually smaller, locations for critical facilities had to be continually updated. GIS allowed this to be done quickly, with new, updated maps distributed into the user community (see Kendra and Wachtendorf, 2003). After the 1994 Northridge earthquake, GIS applications included coding damaged structures with reference to severity of damage and combining that data with the location of field teams, while after the Willow Incident, a serious California wildfire, GIS was used to track the location and movement of fire with respect to houses, equipment, and fire crews (Amdahl, 2001).

In other applications, GIS has proved to have simultaneous practical value, providing information for decision-makers, and theoretical value, helping to validate models of human environment interaction. The hazard of place model proposed by Cutter (1996) was tested using GIS (Cutter, Mitchell, and Scott, 2000) in a community in South Carolina. In that example, social and demographic data was mapped, and then these map layers were overlaid with local weather and climate data (e.g., prevalence of hurricanes) and the proximity of industrial facilities. The result of this analysis was an estimate of the community's net "hazardousness" that could be depicted in different zones on maps of the county. Apart from producing maps that represent various vulnerable areas (useful to emergency planners in the event of, for example, evacuations or specialized warnings, such as day care facilities), Cutter et al. (2000) also found that the greatest vulnerability occurred in areas with *both* moderate-high social and moderate biophysical vulnerability, rather than in places registering at the extremes of these elements. Applications of models as in this example is important to ensure their reliability for planning and policy purposes.

CURRENT NEEDS FOR KNOWLEDGE

In spite of the progress that geographers have made in the last several decades in comprehending the creation and distribution of hazard, some significant gaps remain. Some of these are fundamental scientific questions that remain unanswered; others are more prosaic but no less important.

One significant scientific issue engaging geographers is that of global warming. Most climatologists agree that the earth's atmosphere is warming largely due to human action, though the extent of human contribution remains uncertain. The potential degree of warming and over what time period remains uncertain and, as a consequence, so to the geographical implications for human societies. For example, warming of the atmosphere will probably cause sea-level rise, which may in turn flood low-lying areas. But how much will the sea rise, and what extent of coastline and current or future development there will be affected? Policy decisions regarding such development depend on answers that are still burdened by persistent uncertainties. This question tends to attract the most attention in public policy debates, and is highly charged politically. When scientific issues take on policy significance, there is a tendency – by media, policymakers, observers, and sometimes scientists themselves – to squeeze issues into two opposed clusters that the various antagonists argue correspond with the political orientations of the participants in the disagreement. Lost in the noise of these arguments is a pressing question: if the earth is indeed warming, whatever the cause, what are the implications for human habitation? While the ramifications appear to be negative, the available data provide only broad ranges of, for example, the amount of shoreline to be inundated over the next 50 to 100 years. Such a timeframe is far beyond the planning horizon of commercial interests, and sustained public-sector attention has been lacking at all scales of political activity. Moreover the consequences vary dramatically across the globe. Small island states may suffer serious inundation and loss of habitable area, while other countries, especially with Arctic borders, may see new opportunities for settlement, commerce, and navigation (Middleton, 1999).

Broadly speaking, at the national level, we know that disaster losses are increasing but we really don't have a clear understanding of the extent of hazard or the degree to which hazardousness may be increasing or shifting between locations (Thomas, 2001; Cutter, 2001b). The principal reason for this is a lack of consistent data. For example, monetary losses can be assessed using actual damages, replacement value for buildings and infrastructure, or losses of economic activity in a certain area in the post-impact time. Even in situations wherein certain data, say economic loss, is available, it may not be comparable from place to place or even in the same place from time to time. Record-keeping has varied in different locales over many decades, so that precise longitudinal assessments are simply not possible (Thomas, 2001; Cutter, 2001b).

It should be noted that while more research is important, what is also needed is the sustained application of what is already known. Mitchell (1990), at the outset of the International Decade for Natural Disaster Reduction, a United Nations initiative, argued that while the goal of halving worldwide disaster losses was worthy, the need was not for more research but for sustained and thoughtful application of what was already known. To a large extent, this situation remains the case for many hazards. Areas prone to hurricanes, flooding, earthquakes, landslides, and so on are well-known by now, and the structural and non-structural mitigation measures are well established. Generally a lack of "political will" is cited as the principal obstruction to the kinds of mitigation measures that would be necessary. Certainly this is true, but this view understates the powerful economic and social forces that hold people in particular places, and the difficult set of decisions that people make in their locational decision-making. Palm's (1981) study of the California housing market showed that earthquake hazard information had little effect on potential homebuyers; rather, considering the house as an investment is the primary motivator, reinforced by a need to buy an affordable home "whenever and wherever it becomes available" (399). There are many risks that people face in buying a house apart from the geophysical or climatological risk of the area. Will the value appreciate or depreciate? Is the house in good repair? What is the likely economic trajectory of the community? There are many potential losses and vulnerabilities in the homebuying process. In areas with less obvious risks than California, the risk calculation is even more difficult.

POLICY GUIDANCE

Geographic work has been readily applicable to policymaking needs, expressed at two broad scales: guidance for the direction policies should take, given what is known about the nature and distribution of various kinds of hazards, and specific recommendations.

In terms of general policy guidance, geographers and others have become increasingly concerned about a growing transference of responsibility for disaster response to the Federal government, an incentive for communities to do less to reduce hazards and to prepare appropriately. Some scholars and practitioners (e.g., Platt, 1999) have questioned whether the possibility of federal assistance creates a "moral hazard:" inducing people to take risks they would not take in the absence of such assistance. This concern was always present, reflected in the cautions of the White task force in the late 1960s, but burgeoning disaster losses in the 1980s and 1990s, especially Hurricane Andrew in 1992 and the Midwest floods in 1993, intensified the concern that localities were abdicating their responsibility and counting on Federal assistance. In terms of suggestions for overall policy guidance, Cutter (2001b: 164) presents several suggestions, including

> a strategic plan for hazards reduction at all levels of government. This plan should be based on tangible goals and specific indicators of accountability. Do expenditures for hazards reduction programs make a difference, or do they facilitate the movement of people into increasingly hazardous areas? We need audits of our national disaster aid, recovery, and insurance programs to assess their effectiveness in reducing losses and overall vulnerability.

She also argues (2001b: 165):

> Reducing the nation's vulnerability to environmental hazards will take public support and political will and needs to be addressed within the confines of local community growth and economic development constraints and national priorities for hazard mitigation. The federal government should not bail out communities and individuals when they make foolish locational decisions. Instead, the burdens will be placed locally, and so will the solutions.

Platt makes a similar argument and sets out some ways this might be done, including restricting construction in the most hazardous areas. "Managed Retreat" is one such option: systematically and over time discouraging or prohibiting further development in threatened areas, especially areas threatened by erosion (Platt et al., 1992). Platt (1999: 297) also argues that "FEMA should work with the U.S. Department of Justice to seek a Supreme Court ruling that affirms the constitutionality of public land use regulations in cases of extreme hazard *regardless of the economic impact on the landowner and the community*" (italics in original). He further suggests (1999: 298) that if state and local governments continue to make poor land use decisions, "their eligibility for flood insurance and public assistance . . . under the Stafford Act should be suspended, or provided under less favorable financial terms . . ."

More recently, Platt has fostered something of a social movement, one grounded in a unity of various disciplines that takes a comprehensive look at how cities should, not merely "fit" in the environment, but how they can healthfully be a part of their physical setting. The Ecological Cities program takes a broadly multidisciplinary perspective on the ecological health of urban areas, where good ecological health means that land-uses take into account terrestrial features and dynamic patterns of a built/natural ecosystem (see Platt, 2004).

OVERLAPS WITH OTHER DISCIPLINES

In many ways geography is quintessentially interdisciplinary; its concerns with the intersection of social, physical, and technological and political/legal systems means that it shares areas of interest, knowledge, and methods with many other fields of study. Sociology, with its focus on collective behavior and organizational activity, may be the most closely allied of the various disciplines having, as Cutter (2001a) stated, taken *disasters* as its specialty. Sociologists and geographers share qualitative methods, such as interviews and focus groups, and also the quantitative methods of questionaires and analysis of demographic data.

The demarcation between hazards research and disaster research, though plain in definition, is not as clear in the pattern of the literature. Alexander (1991, 1993) identified six "schools of thought" into which different approaches to both hazards and disaster research may be grouped, while Mitchell (1990) divides hazards research into three subfields, one of which is disaster research. According to Alexander (1993), the *geographical approach* draws on the human-ecological perspective, or man-nature interaction, as its entry point, growing out of the research of Harlan Barrows and, later, Gilbert White. The *sociological approach* focuses on the impacts of disaster on social organization. The *anthropological approach* studies the impact of disaster on social development and especially the response of people in developing countries. *Development studies* focuses on management of relief efforts, and providing nutrition and medical services. The *medical* or *epidemiological* approach studies disaster response and treatment of casualties, and transmission of disease. And the *technical approach* studies, measures, and predicts geophysical phenomena. What is important about this classification is that geographers may participate in any of these approaches, so that these represent more areas of emphasis rather than disciplinary boundaries.

Meanwhile, Mitchell (1990) has cut the field of hazards research along three lines. *Disaster research* is primarily cast as a branch of sociology, concentrating on the response of people and organizations, and often using the disaster as the context for exploring stress reactions. *Natural hazards research* is grounded in the human-ecological perspective, as in Alexander (above). *Risk analysis* studies not only the components of a system to discover sources of failure, but also social processes of risk perception and risk communication. More recently, Cutter (2001a) reinforced this basic delineation.

Thus geography shares interests with many other fields. Probably the most disciplinary affinities, at least in the area of hazards and disaster, are seen with sociology which, as discussed elsewhere in this volume, was the founding field for disaster research. In a sense, in the early years of systematic hazard and disaster research, geographers and sociologists established a division of labor that temporally bracketed the disastrous event – geographers focusing on the decisions that led to the creation of hazard, with sociologists looking principally at the organizational aspects of responding to the impact of the hazard agent – the disaster. This temporal bracketing was not rigidly exclusive, of course, but over time became even less so. For example, sociologists looked at pre-disaster preparations, while geographers studied post-event evacuation. Moreover, there has been a ready use of methods common to all of the social sciences, such as interviews and questionnaires, and more recently sociologists have made increasing use of Geographic Information Systems (Dash, Peacock, and Morrow, 2000), producing work that is clearly geographic in tenor. Mileti (1999) in fact notes that though sociologists, too, were interested in the human ecological approach that Barrows claimed was the hallmark of geography, they did not pursue this field as intensively as the geographers. Barrows (1923: 6) himself noted that sociologists had an interest in "the relation of society to the natural environment," but he anticipated that its interest would remain principally in the "social environment." Peacock et al. (2000) have undertaken work with a definite geographic flair. In particular, their theory of a "socio-political ecology," though emphasizing social relationships and interactions, takes explicit account of the social, political, economic, *and* environmental conditions of places. It is virtually impossible to look at the body of research on emergencies and disasters without seeing the intellectual contributions of geographers, sociologists, and the other disciplines represented in this book.

And geographers, sociologists, and many other disciplines have considered cultural, political, or economic impediments to proper land-use decision making and the implementation of other mitigation measures. In other words, increasingly the distinctions in the approaches of geographers and sociologists have become blurred. Geographers and sociologists are interested in how people perceive risk, the decisions they make based on those perceptions, and how to more effectively communicate information about risk and suitable actions to take. For example, the Disaster Research Center's assessment of Project Impact (e.g., Wachtendorf et al., 2002), with choice of methods, analysis, and conclusions, might have been accomplished by geographers as well as sociologists. Geographers work directly with scholars from other disciplines. Platt (2004) for example, routinely brings together scholars and practitioners from diverse disciplines such as law and planning (see also Platt et al. [1987]).

A further example of such interdisciplinary collaboration is the "social amplification of risk" theory of Kasperson et al. (1988). This is a major and continuing effort to unite psychometric research within a framework of formal and informal social institutions. This construction is then itself referenced to metaphors drawn from communications theory – namely, "amplification" and "attenuation:" the strengthening or weakening, respectively, of a signal as it is transmitted and received. The authors suggest that an analogous process of amplification or attenuation occurs during the transmission of risk information from a source, possibly via intermediaries, to a receiver. The final strength of the risk "signal" varies according to a number of social and cultural factors, which can include the credibility of the institutions sending the risk information, the responses and attitudes of intermediaries (e.g., family or friends) and finally the personal "bias" of the receiver.

FUTURE DIRECTIONS FOR GEOGRAPHY, AND FOR GEOGRAPHY IN EMERGENCY MANAGEMENT

At the 2001 meeting of the Association of American Geographers, the science writer John Noble Wilford took geographers to task for not grappling, or at least not seeming to grapple, with questions of fundamental human interest. Astronomers are interested in the origin of the cosmos; archaeologists are concerned with the origin of civilization; psychologists study the mind and how we think and understand – grand subjects that stretch our imaginations and that inspire and are inspired by our curiosity about origins, futures, and directions. Set against those questions, Wilford argued, it was no wonder that geographers' recurrent concerns about methods and disciplinary legitimacy never made it to the *New York Times* science page. What are geographers interested in that would excite the wonder, interest, and awe of people outside of the field? What questions do geographers answer and how do those questions supply some kind of elemental knowledge?

Mitchell (2001) and Cutter et al. (2002) supplied at least the beginnings of some answers to Wilford's provocative question. For Mitchell, geographers are interested in human survival. This most elemental challenge includes the full range of responses to environmental shifts and the challenges of human occupancy of a dynamic planet. But survival to a thoughtful society should include some consideration of how that survival is to be won. It doesn't mean mere survival, but preserving, conserving, sustaining, and renewing the bases for life. Survival, by definition, can't stop, but must be a continuous process of renegotiating changing human and environmental capacities. Cutter and her colleagues identified ten "big questions in geography," at least four of which, such as "how and why do sustainability and vulnerability change from place to place and over time," directly relate to human-environment interactions. Geographic methods and approaches are uniquely positioned to provide the knowledge we need for understanding both the natural processes of the earth and the social processes of human beings, and hopefully to bring about some harmony of these that can reduce both environmental degradation and disaster losses. Geographers are not solely concerned with intersecting natural and social systems, however, when it comes to the most salient recent hazard: terrorism. For example, geographers such as Cutter have examined terror as a geopolitical phenomenon, rooted in the intersection of social systems and strategic choices of many different potential antagonists.

And it is with respect to such burgeoning, complex, and elusive threats that we might turn to postmodernism as a philosophical orientation that can provide some useful insight. The evolution of postmodernist/poststructuralist philosophies which question the foundation of science and the production of knowledge shook traditional research perspectives in the social sciences, including human geography. Postmodern thinking is characterized by a desire "for a philosophic culture freed from the search for ultimate foundations or the final justification" (Dear, 1988: 265). Such a search, wrote Dear, can stifle alternative discourse, that is, alternative explanations for observed phenomena. Postmodernists reject "metanarrative," and proclaim the validity of local knowledge (Mitchell, 1994: 25) that is rooted in the particular experiences of people in certain places, at certain times, and in different social systems. Though postmodern thinking is not widespread in the hazards field, Mitchell (1994) notes that some researchers have employed postmodern perspectives in critiques of hazards research. Postmodernism is popularly understood in negative terms, as a philosophy of mere relativism advocated by know-nothing nihilists who would deny morals and truth. This is, however, an extravagant oversimplification of this vast body of philosophic work, and there are various touchstones of this intellectual orientation that continue to have value for researchers *and practitioners* in the area of hazards, emergencies, and disasters. Mitchell (1999) for example questioned how hazards researchers could work with some of the broad tenets that could widen our imagination in looking at sets of problems while avoiding some of the seemingly relativist tendencies that undercut the viability of policy prescriptions. Bearing in mind that one of the main tenets of this intellectual movement was that ultimate knowledge or grand explanations were impossible, it is possible to see postmodern characteristics in certain of the "new" hazards, such as mass terror attacks with nuclear, biological, or cyber weapons. They defy understanding with respect to time, place, magnitude, frequency, duration, and origin. Along those lines, Mitchell turned his attention to such features of hazards as *ambiguity* and *surprise.* These postmodern qualities have even more relevance now than when he first wrote. By themselves these are not strictly geographic concepts – researchers in other fields such as the philosophy of science and mathematics (Funtowicz and Ravetz, 1990), have worked with these or allied terms. But it is certainly worthwhile to consider how the latest hazard (at least in the US), terror attacks, has qualities of surprise and ambiguity.

The September 11 attacks on the World Trade Center and the Pentagon confirmed the emergence of a new kind of complex threat. It is not that the threat wasn't always there, but it was apparently nebulous and ephemeral, defying the classical definition of a risk as a probability multiplied by magnitude, and instead more characterized by ambiguity.

> When hazards are framed in terms of risk, objective measurement and rational decision-making are conceivable. By shifting the frame to one of uncertainty, subjective choice comes into view and decision-making becomes only boundedly rational at best. Under conditions of ambiguity the rules of choice are themselves unstable. Ambiguity connotes circumstances of indecision – where customary guides to choice are missing, non-functioning, undependable or so deeply conflicted that decision making is effectively paralyzed (Mitchell, 1999: 11).

A number of researchers have grappled with some of the themes that are relevant for understanding 9-11 and the subsequent challenges that confront us now. Mitchell defined a number of different kinds of surprises, including events that were thought to be impossible or events thought to be unlikely, that nevertheless happened anyway. Karl Weick, a psychologist, refers (1993) to *cosmology episodes*, breakdowns in expectations in which the usual rules for understanding the world no longer make sense. Threats that are ambiguous set the stage for surprises, or cosmology episodes, because the real nature of the threat is concealed until it manifests as disaster. While there is an ambiguous threat, there is a chance that one has understood it and applied the appropriate action. But when disaster occurs, it is obvious that one has misjudged – 9-11 counts as such an event – a surprise, though perhaps it shouldn't have been. But the operative quality of surprise is not that no one ever thought that something might happen, but how the event registers and resonates in the affected social systems. On a vast scale, all the short cuts and mental models that people had employed for assessing their risks in a variety of activities no longer seemed viable but were instead destabilized. For a period of days, maybe weeks, anything seemed possible and no threat could be deemed unlikely. The later anthrax attacks contributed to a persistent sense of threat. Previously understood indices of what was risky, such as air travel or high-rise buildings, had to be revised. The emergence of a terrorism with training, communications, and technical skill, and with elements appearing everywhere, is now added to the range of familiar but no less threatening hazards that have always plagued us.

We now confront an amalgam of threats, some of which are familiar and understandable, others less so, in which, particularly at the outset, it may not even be clear precisely what is happening or what the cause is. Consider the massive electrical power failure across the Northeast in August, 2003, in which officials hastened to assure the public that it was not an act of terror. It is interesting to realize that a massive failure of the electrical distribution system, on its own, does not seem to be as alarming as if it were initiated by terrorists (see, for example, a report by Elizabeth Vargas on *ABC News Primetime*, August 14, 2003). In terms of understanding the public response to this, one might well apply the amplification of risk model – a terror attack on the electric grid would have a much greater *signal value* (Kasperson et al., 1988), indicating sustained vulnerability to such assaults.

There are ambiguities beyond the nature of an event as well. According to information available, as in *The 9/11 Commission Report* (2004), the attacks on 9/11 began ambiguously. It is possible that subsequent attacks, especially if biological warfare or cyberwarfare agents are involved, may begin ambiguously as well. There are ambiguities in defining the post-event stage. When were the 9-11 attacks over? When was the emergency over? And these basic considerations encompass only the tactical or short-term scale. As more information becomes available, it appears that the prodromal stage (Fink, 1986) of the September 11 attacks extended over many years, perhaps beginning with the February 1993 bombing, or before, in the planning stage for *that* event. Events of this magnitude disrupt the spatial, temporal, and organizational scales in which we are used to working. The arguments that played out before Congress and in public dialogue, over which presidential administration is most to blame, are merely the political reflection of an inability to cope with a temporal and spatial scale of threats that is longer than we have seen previously with most threats. There is experience with the Cold War, but in that sustained conflict the goal was for the protagonists not to shoot directly at each other, and the centers of gravity of political interests could be plotted more reliably. Certainly there are other matters that confront officials that require multigenerational, multijurisdictional approaches – pollution, greenhouse gases, space-exploration – but the erratic handling of these challenges is not encouraging.

Thus we can see substantial shifts in the risk milieu, wherein ambiguous threats may in some instances resemble familiar threats and in other instances lie dormant for many years, extending across very difficult scales of time, space, and organization. Mitchell's call for attention to such features as surprise and ambiguity should thus have more than merely scholarly interest. If the hazards, especially those of terrorism, now confronting us do indeed have these features, what policy guidance is available to us? What steps can be taken at all phases of the disaster cycle, but especially when the nature of potential threats is vague and amorphous, that is, during mitigation and preparedness? Shortly after the 9/11 attacks, geographers considered how their disciplinary expertise might be applied to these suddenly salient, if not exactly new, threats (Cutter, Richardson, and Wilbanks, Eds., 2003). In this, the authors grappled with such difficult and controversial subjects as the "root causes" of terrorism. It is far beyond the scope of this chapter to discuss where various researchers and commentators would place such causes, which are tied to large political, economic, and cultural transformations, but it seems safe to say that even if a cause or causes could be accurately deduced, ameliorative steps might take years to show any effect. Thus, the qualities of surprise and ambiguity are likely to continue to be important for characterizing threats.

These main issues suggest a need for new kinds of research, especially research that cuts across several disciplines. In particular, there is a need for long-term studies within emergency management agencies and other relevant organizations, or at a minimum, studies with significant follow-up components over a period of years. There need to be studies of change of command processes and the management of information across successive governments. How is attention maintained across both time and space on particular issues? As an example of shifting attention and priorities, Project Impact, by most measures a successful mitigation program emphasizing outreach, public education, and multisectoral partnership building, was cut after the change in administrations in 2000.

We need renewed study of risk selection in particular places. How do we imagine risks, in particular the combination of old and new threats as well as complex emergencies? How is risk across space to be determined, and resources allocated? What happens to emergency managers and other public officials who are required to be on heightened alert for long periods of time? Is it really possible? Is it affordable? To what extent can heightened alerts be maintained? Geographers, sociologists, and psychologists have worked together on risk perception and communication models and guidelines. How can warnings and public information be delivered in a way that will be useful to people, given that some of the content of such messages, such as specificity of the nature of the risk, the location, and the action to take, is often not available? We need to know the parameters of capability, to ensure that planning and warnings are not built on mistaken, overoptimistic, or uninformed assumptions of capability.

Yet, right at the time that new information, new methods, and new knowledge is required, as well as new ways of looking at familiar hazards, the development of that knowledge may be impeded. Some researchers have had difficulty getting access to information, especially infrastructure data. In one instance, officials have declined interviews and a request to visit their facility, citing security concerns. And it is not just new data and new information that are suspect. Even old data can be spun in a sinister way. At the end of 2003, the FBI issued a bulletin to 18,000 law enforcement agencies, warning them to beware of people carrying almanacs and acting suspiciously, or making suspicious notations (Bridis, 2003). Unfortunately, no one looks more suspicious than a hazards researcher trained in observation who is writing notes, drawing maps, making sketches, and taking photographs. The danger, however, is that it is possible to paralyze knowledge, particularly if researchers continually have to struggle for access to sites, to facilities, and to data. Imagine the state of knowledge that would exist now if, forty years ago, Cold War security concerns had stifled hazard and disaster research, fixing us in a mid-century knowledge base. We would be much less able to confront present hazards, and even so there is a long way to go.

The turn of the century has brought with it a turn in the set of hazards that we face. Public officials must be tolerant of research and data gathering because energetic inquiry and the development of knowledge have always been the best ways of confronting new challenges. The individual parochial concerns of organizations, always concerned with appearance and legitimacy or the concerns of a narrow constituency, are at issue in this sphere as in any other, but we are at a point where organizations acting in their individual self interest jeopardize all of our collective security. Perhaps what is required as much as new knowledge is a refreshed recognition by public officials that disaster researchers have a calling and an ambition to reduce danger, and thus are acting in the public service.

REFERENCES

Alexander, D. (1991). Natural disasters: A Framework for Research and *Teaching. Disasters 15*(3): 209–226.

Alexander, D. (1993). *Natural Disasters.* Chapman and Hall, New York.

Amdahl, G. (2001). *Disaster Response: GIS for Public Safety.* ESRI Press, Redlands, CA.

Barrows, H.H. (1923). Geography as Human Ecology. *Annals of the Association of American Geographers 13*(1): 1–14.

Bridis, T. (2003). FBI Urges Police To Watch for People Carrying Almanacs. The Associated Press State and Local Wire. Last accessed via Lexis Nexis Academic, December 10, 2005.

Burton, I., Kates, R.W. and White, G.F.. (1993). *The Environment as Hazard.* Guilford, New York.

Cutter, S.L. (1993). *Living With Risk.* Edward Arnold, London.

Cutter, Susan L. (ed.) 1994. *Environental Risks and Hazards.* Prentice-Hall, Englewood Cliffs.

Cutter, S.L. (1996). Vulnerability to Environmental Hazards. *Progress in Human Geography 20*(4): 529–539.

Cutter, S.L., Richardson, D.R., and Wilbanks, T.J. (2003). *The Geographical Dimensions of Terrorism.* New York: Routledge.

Cutter, S.L. (2001a). The Changing Nature of Risks and Hazards. In Susan L. Cutter (ed.), *American Hazardscapes: The Regionalization of Hazards and Disasters.* Joseph Henry Press, Washington, D.C.. pp. 1–12.

Cutter, S.L. (2001b). Charting a Course for the Next Two Decades. In Susan L. Cutter (ed.), *American Hazardscapes: The Regionalization of Hazards and Disasters.* Joseph Henry Press, Washington, D.C.. pp. 157–165.

Cutter, S.L., Mitchell, J.T.,m and Scott, M.S. (2000). Revealing the Vulnerability of People and Places: A Case Study of Georgetown County, South Carolina. *Annals of the Association of American Geographers 90*(4): 713–737.

Cutter, S, Golledge, R., and Graf, W.L. (2002). The Big Questions in Geography. *Professional Geographer 54*(3): 305–317.

Dash, N., Peacock, W.G., and Morrow, B.H. (2000). And the Poor Get Poorer: A Neglected Black Community. In Peacock, W.G., Morrow, B.H.and Gladwin, H., eds. *Hurricane Andrew: Ethnicity, Gender, and the Sociology of Disasters.* Miami, FL: Florida International University. International Hurricane Center, Laboratory for Social and Behavioral Research. pp. 206–225.

Dear, M. (1988). The Postmodern Challenge: Reconstructing Human Geography. *Transactions of the Institute of British Geographers 13*: 262–274

Dow, K. (1999). The Extraordinary and the Everyday in Explanations of Vulnerability to an Oil Spill. *The Geographical Review 89*(1): 74–93.

Ellul, J. (1964). *The Technological Society.* Alfred A. Knopf, New York.

Fink, S. (1986). *Crisis Management: Planning for the Inevitable.* Saranac Lake: AMACOM.

Funtowicz, S.O., and Ravetz, J.R. (1990). *Uncertainty and Quality in Science for Policy.* Kluwer, Dordrecht.

Graham, J., and Kasperson, R.E. (1985). Television: A Social Hazard. In *Perilous Progress: Managing the Hazards of Technology,* Kates, R.W., Hohenemser, C., and Kasperson, J.X. eds. pp. 427–454. Westview, Boulder.

Harriss, R.C., Hohenemser, C., and Kates, R.W. (1978). Our Hazardous Environment. *Environment 20*: 6–15, 38–40.

Hewitt, K., ed. (1983). *Interpretations of Calamity.* Allen and Unwin, Winchester, MA.

Hohenemser, C., Kasperson, R.E., and Kates, R.W. (1985). Causal Structure. In *Perilous Progress: Managing the Hazards of Technology,* Kates, R.W., Hohenemser, C., and Kasperson, J.X., eds. pp 25–42. Westview, Boulder.

Hohenemser, C., Kates, R.W., and Slovic, P. (1985). A causal taxonomy. In *Perilous Progress: Managing the Hazards of Technology,* Robert W. Kates, Christoph Hohenemser, and Jeanne X. Kasperson, eds. pp. 67–89. Westview, Boulder.

Kasperson, R.E., and K.D. Pijawka. (1985). Societal Response to Hazards and Major Hazard Events: Comparing Natural and Technological Hazards. *Public Administration Review 45*: 7–18.

Kasperson, R.E., Renn, O., Slovic, P., Brown, H.S., Emel, J., Goble, R., Kasperson, J.X., Ratick, S. (1988). The Social Amplification of Risk: A Conceptual Framework. *Risk Analysis 8*(2): 177–187.

Kates, R.W. (1971). Natural Hazard in Human Ecological Perspective: Hypotheses and Models. *Economic Geography 47*: 438–451.

Kendra, J.M. and Wachtendorf, T.(2003). Creativity in Emergency Response after the World Trade Center Attack. In *Beyond September 11th: An Account of Post-Disaster Research.*

Natural Hazards Research and Applications Information Center, Public Entity Risk Institute, and Institute for Civil Infrastructure Systems. Special Publication No. 39. Natural Hazards Research and Applications Information Center, University of Colorado, Boulder, Co. pp. 121–146.

Middleton, N. (1999). *The Global Casino: An Introduction to Environmental Issues.* Second Edition. Arnold, London.

Mileti, D.S. (1999). *Disasters by Design: A Reassessment of Natural Hazards in the United States.* Joseph Henry Press, Washington, D.C.

Mitchell, J.K. (1990). Human Dimensions of Environmental Hazards: Complexity, Disparity, and the Search for Guidance. In *Nothing to Fear: Risk and Hazards in American Society*, Andrew Kirby (ed.), 131–175. University of Arizona Press, Tucson.

Mitchell, J.K. (1994). Natural Disasters in The Context of Megacities. Paper prepared for the Megacities and Disasters Conference. United Nations University, Tokyo. January 10–11, 1994.

Mitchell, J.K. (1999). Hazards and Culture: New Theoretical Perspectives. Paper presented at the 95th Annul Meeting of the Association of American Geographers. Honolulu, Hawaii, March 23–27, 1999.

Mitchell, J.K. (2001). Personal communication.

Mitchell, J.K., Devine, N., Jagger, K. (1989). A Contextual Model of Natural Hazard. *Geographical Review 79*(4): 391–409.

Otway, H. and Cannell, W. (1987). Review of *Perilous Progress: Managing the Hazards of Technology.* In *Economic Geography 63:* 76–78

Palm, R.I. (1981). Public Response to Earthquake Hazard Information. In *Annals of the Association of American Geographers 71*(3): 389–399.

Palm, R.I. (1990). *Natural Hazards: An Integrative Framework for Research and Planning.* Johns Hopkins University Press, Baltimore.

Peacock, W.G. and Ragsdale, A.K. (2000). Social Systems, Ecological Networks and Disasters: Toward a Socio-Political Ecology of Disasters. In Peacock, W.G., Morrow, B.H., and Gladwin, H. eds. *Hurricane Andrew: Ethnicity, Gender, and the Sociology of Disasters.* Florida International University, Miami, FL. International Hurricane Center, Laboratory for Social and Behavioral Research. pp. 20–35.

Peacock, W.G., Morrow, B.H., and Gladwin, H. (2000). *Hurricane Andrew: Ethnicity, Gender, and the Sociology of Disasters.* Florida International University. Miami, FL: International Hurricane Center, Laboratory for Social and Behavioral Research.

Pickles, J. (ed.) (1995). *Ground Truth: The Social Implications of Geographic Information Systems.* Guilford, New York.

Platt, R.H. (1999). *Disasters and Democracy: The Politics of Extreme Natural Events.* Island Press, Washington, D.C.

Platt, R.H. (2004). *Land Use and Society: Geography, Law, and Public Policy.* Revised Edition. Island Press, Washington, D.C.

Platt, R.H. (2004). Toward Ecological Cities: Adapting to the 21st Century Metropolis. *Environment 46*(5): 12–27.

Platt, R.H., Pelczarski, S.G., and Burbank, B.K.R. (1987). Cities on the Beach: Management Issues of Developed Coastal Barriers. Research Paper No. 224. Chicago: University of Chicago Department of Geography.

Platt, R.H., Miller, H.C., Beatley, T., Melville, J., and Mathedia, B.G. (1992). *Coastal Erosion: Has Retreat Sounded?* Program on Environment and Behavior, Monograph No. 53. Institute of Behavioral Science. University of Colorado, Boulder.

Postman, N. (1992). *Technopoly.* Vintage Books, New York.

Rodríguez, H, Wachtendorf, T., Trainor, J., and Kendra, J. (2005). The Great Sumatra Earthquake and Indian Ocean Tsunami of December 26, 2004: A Preliminary Assessment of Societal Impacts and Consequences. *Earthquake Engineering Research Institute Newsletter 39*(5): 1–7.

The 9/11 Commission Report: Final Report of the National Commission on Terrorist Attacks upon the United States. (2004). Norton, Norton.

Thomas, D.S.K. (2001). Data, Data Everywhere, But Can We Really Use Them? In Cutter, S.L. (ed.), *American Hazardscapes: The Regionalization of Hazards and Disasters.* Joseph Henry Press, Washington, D.C. pp.61–76.

Vargas, E. (2003). ABC News. Primetime Thursday: Blackout 2003, the Biggest in History. Last accessed via LexisNexis Academic December 10, 2005.

Wachtendorf, T., Connell, R., Monahan, B., and Tierney, K.J. (2002). Disaster Resistant Communities Initiative: Assessment of the Non-Pilot Communities. Final Report #48. Newark, DE, Disaster Research Center.

Weick, K.E. (1993). The Collapse of Sensemaking in Organizations: The Mann Gulch Disaster. *Administrative Science Quarterly 38*: 628–652.

White, G.F. (1973). Natural Hazards Research. In Richard J. Chorley (ed.) *Directions in Geography.* Methuen, London. pp. 193–216.

Winner, L. (1977). *Autonomous Technology: Technics Out-of-Control as a Theme in Political Thought.* MIT Press, Cambridge.

Zeigler, D.J., Johnson, Jr., J.H., and Brunn, S.D. (1983). *Technological Hazards.* Association of American Geographers, Washington, D.C.

Chapter 3

Meteorology, Weather and Emergency Management

Kent M. McGregor

ABSTRACT

The science of meteorology is deeply intertwined with the process of emergency management. Weather phenomena are the cause of many disaster events such as tornadoes and hurricanes and a factor in many others. Weather can also affect the way assistance is provided during or after an emergency. Since time to prepare is vital, much of meteorology is concerned with forecasting and issuing. This paper addresses the role of meteorology in tornadoes, hurricanes, floods, droughts, heat waves, wildfires and blizzards. The basic meteorological processes causing such disasters are discussed and selected examples are included from both the U.S. and other parts of the world. Finally, the future poses its own special brand of weather hazards due to the uncertainties and scale of global warming and consequent changes in global climate patterns.

INTRODUCTION

The relationship between weather and emergency management is fundamental yet complex. Weather causes many disasters that require an emergency response. Indeed meteorological processes determine the extent of the destruction to life and property. Meteorologists both forecast the impending event and survey the scene afterward to determine the magnitude of the atmospheric forces involved. This chapter is a survey of such relationships in the context of the most common types of disasters. This paper consists of five principal sections. The first section is a survey of disasters that are caused or influenced by meteorological processes. This includes the duration of the event, the duration of the consequences, and the scale of the impact. These are important considerations in determining the type of emergency response and the allocation of resources. The second section covers the process of developing a weather forecast and disseminating the result. Forecasting is the most common application of atmospheric science. Who gets the forecast and in what way are fundamental questions in the decision-making process. The third section is a primer on basic meteorology. To understand how extreme weather events develop, one must understand basic atmospheric processes. These include high and low pressure, winds, air masses, storms, cyclonic systems and related features on a weather map. The fourth section is the majority of the paper and reviews the major types of weather events that might require an emergency response. These are tornadoes, hurricanes, floods, droughts, heat waves, wildfires, and blizzards. It includes a discussion of the basic atmospheric processes causing each event with selected examples. The examples come from both the U.S. and countries around the world. The international perspective is required for a better understanding of what kind of emergency response is possible. Actions that could be taken easily in a modern country like the U.S. simply might not be possible in the developing nations. Finally, the fifth section is a discussion of current trends in atmospheric science that will continue into the future and have implications for the management of emergencies. These include continual development of models and supporting observation networks. Extreme weather events are increasingly viewed in the larger context of global atmospheric and oceanic forces. The best known of these is global warming. However, many regional climate cycles or oscillations have a pronounced affect on weather and extreme weather events. The El Niño phenomena is the best known of these oscillations. It affects not only the tropical Pacific, but places far away through what are called "teleconnections."

TYPES OF WEATHER-RELATED DISASTERS

Throughout history, weather events, of various kinds, have posed a hazard to human activities. Meteorological forces constitute both a direct hazard such as storms and consequent flooding, and indirect (associated) hazards such as the drift of smoke, ash and noxious fumes from an erupting volcano. Table 3.1 summarizes many of these weather related hazards. Of the twenty (20) items in this list, twelve (12) are caused directly by atmospheric forces, and weather is a factor in the remaining eight (8).

According to Burton, Kates and White (1993), approximately 90 percent of the world's natural disasters originate in four hazard types: floods (40%), hurricanes (20%), earthquakes (15%) and drought (15%). Floods are the most frequent and do the largest proportion of property damage. Droughts are the most difficult to measure in extent, property damage, and death toll.

IMPORTANT FACTORS TO CONSIDER

1. Time for event to develop and duration of occurrence. All of these events vary widely in time developing and length of time occurring. A tornado develops quickly and seldom lasts more than a few minutes. In contrast, droughts are the slowest developing weather hazard, but also the longest lasting. Flash floods can develop in a few minutes and be over in a few minutes, but the damage has been done.

2. Spatial extent or size of area impacted. Such events vary dramatically in their spatial extent. A microburst might be the most localized of weather related events while droughts, floods and pestilence can affect a large region of the globe. A lightning strike might be as localized as an event can get, and, yet set off wildfires destroying thousands of acres.

3. Potential number of people impacted. There are dramatic differences in the number of people that might be affected. A tornado may be a localized, short-lived event, but, it can affect thousands of people if it hits a city. A spill of hazardous materials might affect a few people in a nearby neighborhood, or in the case of the Bhopal, India, disaster, it can impact thousands. This disaster was instructive because it was fairly localized, yet, because of the dense population, it affected literally thousands of people.

4. When weather is not a direct cause, how might it impact or aggravate the event? Many types of disasters are not caused directly by weather; they are the result of human activity. Weather later becomes a factor after the disaster has occurred. A classic example is the melt down of the nuclear reactor in Chernobyl, Ukraine. Weather became a factor as radioactive gasses escaped into the atmosphere. These toxic gasses were carried by the winds and the rate of dispersal was determined by wind speed and direction and other atmospheric factors that determined the rate of mixing. As a result, Finland some 1000 miles away was heavily impacted.

5. The weather categories are not mutually exclusive. In fact, many types of emergencies will be accompanied or lead to others (like famine leads to disease). Some improbable combinations also can and do occur. During one of the worst floods in its history, the Red River flooded Fargo and Grand Forks, North Dakota. In Grand Forks, the natural gas lines broke; fires broke out and the downtown burned while still submerged in water.

Perhaps the slowest developing disasters are drought and famine. These are not typical emergency management situations initially because they develop slowly, perhaps over many months or even years, but they have the potential to impact the greatest area and the greatest number of people. As a result, they can require massive relief efforts. Indeed, mass starvation due to political strife is and continues to be one of the legacies of the twentieth century and continues today. The four horsemen of the apocalypse are still very much with us even in these post-modern times.

Table 3.1
WEATHER RELATED HAZARDS

	Time Developing	*Time Occurring*	*Spatial Extent*	*Number of People*	*Caused by Weather*
Tornado	fast	short	small	small	X
Hail	fast	short	small	small	X
Wind	fast	short	small to medium	small	X
Flood	slow to fast	short to long	medium to large	medium	X
Blizzard	medium	medium to long	large	medium to large	X
Hurricane	medium	medium to long	medium to large	medium to large	X
Air pollution	medium	medium to long	medium to large	medium to large	
Hazardous spills	fast	short to long	small to medium	small to medium	
Water pollution	slow to fast	medium to long	small to medium	medium	
Fire spread	fast	short to long	small	small to medium	
Disease	slow to fast	long	medium to large	large	
Heat wave	medium	medium to long	medium to large	large	X
Cold wave	medium	medium to long	medium to large	large	X
Drought	slow to fast	long	large	large	X
Volcano	medium to fast	short to medium	small to medium	medium	
Landslide	fast	short	small	small	
Transportation	fast	short	small	small	
Microburst.	fast	short	small	small	X
Fog	fast	short	small to medium	small to medium	X
Frost	fast	short	small to medium	small to medium	X

FORECASTING AND METEOROLOGICAL SCIENCE

Since so many disasters are caused by weather, probably the greatest contribution of atmospheric science is developing the weather forecast and issuing the warning. For example, the meteorologist is not only concerned with forecasting a developing severe weather situation, but also the location, size, and intensity of a tornadoes that might also form. He/she would also forecast the path the tornado might take given the parent thunderstorm characteristics and the prevailing steering winds. Could the tornado strike a heavily populated area? After the event, the meteorologist might look at additional data to determine the accuracy of previous estimates of wind speed for example.

Another important concern is simply gaining a better understanding of how the atmosphere works.

For example, there are still many questions about the exact environment in which a tornado develops (Hamill, et. al., 2005). Indeed, one of the mysteries in atmospheric science is why, given what seem to be two identical environments, one will develop a tornado and the other will not. Improving the basic understanding of atmospheric processes would improve not only the forecast lead-time but also the estimated impact of specific weather events. This is true for all events, drought or flood, hurricane or tornado, hail or fire. In the U.S. the various agencies in the National Oceanographic and Atmospheric Administration (NOAA) are responsible for both forecasts and basic research including the National Weather Service (NWS) and the National Hurricane Center (NHC). Private meteorological companies also provide specialized forecasts to their clients.

With any forecast or warning of an impending extreme weather event, there are always questions, of who gets the information, how quickly, and what is the best course of action to recommend. A good example is when to recommend evacuation in the face of an impending weather event. Generally, evacuation is more risky than seeking immediate shelter. However, in the case of the Oklahoma City tornado, the National Weather Service advised people to leave their homes and businesses to get out of the path of the oncoming tornado while there was still time. Such action undoubtedly saved many lives, however, there are uncertainties with this strategy. The tornado could change paths or speed of movement. Traffic or debris could slow or stop the evacuation.

The media play a critical role in transmitting such warnings and related information to the public. The National Weather Service can issue a perfect forecast but it must be successfully relayed to the individual citizen in time for them to decide on the best course of action in their individual case. There are a variety of ways in which this transmittal of warnings might be accomplished. The electronic media is perhaps the best example, but there are others. The inexpensive weather radios sound a special tone when activated by a signal on a special NWS frequency. Automated dialing systems for telephone notification are becoming more common. Internet notification is available as an option. As always, people will call friends and relatives who might be in jeopardy from severe weather.

Obviously, since weather is a cause or a factor in nearly all types of natural disasters, there is a tremendous amount of overlap with many other disciplines. Perhaps the strongest links are to government officials at all levels who must decide how best to respond to an emergency situation caused by or affected by weather. Links to the media are especially important in disseminating weather watches and warnings to the public. There are strong connections with civil engineers and hydrologists who design flood control works and predict how floods might affect a particular community. In the case of drought, there is interaction with agricultural specialists, and local water managers. In the case of hurricanes, there might be interaction with coastal geomorphologists.

METEOROLOGY: A PRIMER

Atmospheric pressure is the most fundamental concept in atmospheric science. A weather map is essentially a map of atmospheric pressure annotated with additional information. Small changes in atmospheric pressure cause large changes in the weather. If there is more air than usual at a given place, it is called high pressure. If there is less air than usual, it is called low pressure. At its simplest, air moves from high pressure areas to low pressure areas to equalize the pressure differences; these are called winds. Once winds start moving, they may be deflected from their original direction due to the earth's rotation. This is called the Coriolis force and is responsible for the pattern of rotation that winds develop around pressure cells. Winds move out of a high pressure cell and into a low pressure cell; however, because of the Coriolis force, they tend to spiral into a low and out of a high.

Pressure cells not only induce horizontal motions in air (winds), they also induce vertical motions. These vertical motions are critical in determining what the weather does. Low pressure causes upward (ascending) vertical motion and is associated with clouds, precipitation, and storms in general. High pressure causes downward (descending) vertical motion and is responsible for clear skies. High pressure is a bit difficult to understand because it can occur with both extremes of hot and cold temperatures, however the skies are clear in both cases.

Thus, storms are organized low pressure cells. Hurricanes, tornadoes, blizzards, heavy rainfall are all low pressure cells. The rising and cooling air causes the moisture to condense and fall to the surface. Storms are very effective at wringing moisture out of the atmosphere. In contrast, high pressure causes droughts and heat waves. As air descends toward the earth's surface, it heats up. When a large or strong high pressure cell becomes anchored in place during the summer, the combination of no rainfall, clear skies, descending and warming air can cause a heat wave. If this situation continues for weeks or months, it can cause a drought.

In the mid-latitudes, there is a special type of low pressure system called a cyclonic storm. Cyclones are displayed on the weather map with a large **L**. There is usually a cold front and a warm front connected to the center of low pressure. These fronts are the boundaries between tropical and polar air masses. Also in the mid-latitudes are areas of high pressure called anticyclones. These are displayed on the weather map with a large **H**. Both cyclones and anticyclones migrate across the U. S. from west to east pushed along by high altitude winds called the westerlies. The jet stream is the fastest part or core of the westerlies. The pattern or configuration of the westerlies and the jet stream determines the type of weather. Where the westerly winds make a northward bend, they create an area of high pressure aloft called a ridge. This ridge, in turn, makes an anticyclone at the surface. Where the westerly winds make a southward bend, they create an area of low pressure aloft called a trough. This trough, in turn, makes a cyclone at the surface. The alternating sequence of low pressure and high pressure, cyclone and anticyclone, establishes the changeable pattern of weather associated with mid-latitude locations.

In many parts of the world, the weather is heavily influenced by climatic cycles called oscillations. The best known of these is the El Niño/Southern Oscillation (ENSO) phenomena in the Pacific Ocean. The very intense 1997–98 ENSO event resulted in devastation around the world, and the resulting media coverage sharply focused public attention on the phenomenon. When sea surface temperatures (SSTs) are above normal in the eastern, equatorial Pacific, it is called an El Niño event. When sea surface temperatures (SSTs) are below normal in the eastern, equatorial Pacific, it is called an La Niña event. These events cause profound changes in the typical weather patterns around the tropical Pacific but their impact extends to many other parts of the world through what are termed "teleconnections." For example, El Niño events are associated with enhanced precipitation across the southern tier of the U.S. in spring and winter months. Other oscillations, such as the North American Oscillation (NAO) seem to have impacts more localized to a particular region of the planet. A better understanding of such oscillations will, hopefully, lead to better predictions of long-term climate variability. Glantz (2001) reviewed the ENSO phenomena including the history, growth in scientific understanding, monitoring activity and significance for the future.

TORNADOES

The central part of the United States has the highest incidence of tornadoes in the world. There, all of the ingredients are present like nowhere else in the world. Central Oklahoma is ground zero. At its simplest, tornadoes are created by the clash of air masses, but the pattern of upper air winds (westerlies) is equally important. In the central U.S., warm, humid tropical air is brought into contact with cool, dry polar air. These air masses with such vastly different characteristics are pulled together by the low pressure cells (cyclonic storm systems). Fronts are the boundaries between these air masses and thunderstorms often erupt along the fronts. Another important ingredient is called the "cap." This is a flow or layer of warmer, drier air pulled in at the mid-levels of the atmosphere from the southwest. This layer caps weaker convection cells and prevents the air from rising further. However, when a stronger convection manages to penetrate or break the cap, it can continue to rise very quickly. The analogy is to the hole in the dam. Once the dam has been breached, all of the water comes rushing through pushed by the pressure behind. Once the cap breaks, all of the heat and humidity rushes upward resulting in a monster thunderstorm. Lastly, the dynamics of the jet stream (the fastest part or core of the westerly winds) are important. The interaction of winds coming in from different directions and at different speeds creates shear forces in the atmosphere. This can, in turn, create a horizontal "tube" of air that rotates. For reasons that are not completely understood, upward convection can bend or tilt this tube to a vertical position. This is called the mesocyclone and, when the environment is just right, some of the rotation is translated into a smaller and much faster spinning vortex called a tornado funnel. The fastest wind speeds on earth occur in the strongest tornados probably a bit more than 300 mph. Table 3.2 shows the Fujita Scale of tornado winds and resulting damage.

Doswell, Moller and Brooks (1999) summarized the history and progress of storm spotters as part of the National Weather System procedures for forecasting tornadoes. They especially highlight the difficulty in disseminating warning information in a timely fashion so that the public has time to respond. They include an excellent review of the training that was offered to storm spotters over approximately a 50-year span. The authors claim the reduction in tornado fatalities is due, in part, to the efforts of the storm spotters.

Perhaps the most highly developed forecasting and warning system for tornadoes and related severe weather is in Oklahoma. Andra et al. (2002) evaluated the decision process and lead times in issuing the warnings for the strong tornadoes that developed on the 3rd of May, 1999. The lead-time for a warning issued by a human forecaster based on the mass of evidence was a median of 23 minutes. In contrast, the lead-time for a warning based on a tornado detection algorithm was 2 minutes for detection of the first tornado. While this might seem like an important difference in lead times, the algorithms did alert the meteorologist that a developing storm had potential to produce a tornado well before it actually did.

Table 3.2
FUJITA SCALE OF TORNADO WINDS AND DAMAGE

Fijita Scale	*Wind Speed mph (km/hr)*	*Damage*
F0	40–73 (68–118)	Light
F1	74–112 (119–181)	Moderate
F2	113–157 (182–253)	Considerable
F3	158–206 (254–332)	Severe
F4	207–260 (333–419)	Devastating
F5	261–318 (420–512)	Incredible

Morris et al. (2002) discussed the use of a system designed to get real time, detailed weather information to local emergency management authorities. The authors point out that, even in the information age, there is a big gap between what the National Weather Service does in issuing a warning and the ability of local authorities to access the detailed weather information necessary to implement their decisions. On May 3, 1999, the day of the massive Oklahoma City Tornado, over 25,000 files were shared. These were primarily real-time Weather Service Radar images that local managers used to make decisions affecting their jurisdiction. As a result local officials could be proactive rather than reactive in their approach to severe weather. A good example was what happened in Logan County during the outbreak. After one tornado destroyed the small town of Mulhall, rescue workers set up a command center to manage the emergency operations. Soon, these workers were advised to move their command center away from the path of additional oncoming tornadoes. In fact, they had to move their command center twice. The transfer of information made possible success stories that did not make the national news. The OK-FIRST system has won awards for technology (transfer) to local government. Perhaps, however, this could only be done in Oklahoma because of the very real concern the residents for severe weather, and the location of the National Weather Service facilities (Storm Prediction Center) in Norman Oklahoma. An additional factor is the success of the Oklahoma Mesoscale Network which gathers observations from every county in the state and makes them available in near real time through the internet.

Hammer (2002) evaluated the response to warnings during the Oklahoma City tornado and the resulting injury rates. Nearly half of the people fled their homes. One of the interesting findings was that no one was injured who fled either by foot or by vehicle. Most received a warning through the media although phones were also important. Golden (2000) reviewed the problem of public dissemination of tornado warnings and found that the area that needed improvement the most was not in the forecast but in communicating the warning effectively to the public so they had time to decide what action to take.

In contrast, during the 1987 Saragosa, Texas tornado, the warning system failed leaving the residents with little or no time to react (Aguirre, 1991). Saragosa is a small, remote, mostly Spanish-speaking community in west Texas. Many of the residents, if they were watching television, if they had a television, were tuned to a Spanish language cable channel. Typically, cable channels do not interrupt programming or scroll a weather warning across the screen. Since the tornado developed quickly, it was almost in the town before anyone received the warning.

In spite of the continued progress in the communication of weather warnings and the public's response, there is still room for improvement. Consequently, the National Weather Service (NWS) developed the StormReady program to help local communities develop preparedness plans for all types of severe weather. This is a grassroots program providing guidelines to help communities improve their emergency management operations. They are required to establish an emergency management center with 24-hour monitoring and has more than one way to receive severe weather warnings and also notify the public. They must have some way to monitor local weather conditions. They must increase public readiness through presentations to the community and training of storm spotters. Lastly, they must practice implementing their plans with periodic emergency exercises. Over a thousand communities nationwide have met these requirements and are active participants in the program.

HURRICANES

Hurricanes develop over warm tropical waters. Sea surface temperatures must be at least 27° C. or about 85° F. Indeed, the warm tropical waters are the principal source of energy for the hurricane. If the winds higher up (aloft) are light, the atmosphere above becomes saturated with humidity. All that is needed is a low pressure area called an easterly wave to initiate development and intensification of the storm. Easterly waves are pushed along from east to west by the tropical trade winds. The trade winds often curve northward (in the northern hemisphere), so Atlantic hurricanes have struck New England and Pacific typhoons have struck Japan.

Hurricanes have a unique combination of factors that make them especially destructive. The minimum wind speed for a hurricane is 74 mph. This is approximately the threshold for causing some minor damage. The very strongest hurricanes have wind speeds approaching 200 mph that will result in nearly total destruction of buildings. In addition, the torrential rains cause flooding and additional damage. As bad as the winds and rain are with hurricanes, they have one final especially devastating element called the storm surge. This is an artificial rise in sea level that increases the scale of flooding along the coasts. In 1969, Hurricane Camille hit the Mississippi coast with nearly 200 mph winds and a 28 ft. storm surge. Table 3.3 shows the Saffir-Simpson Scale of hurricane winds and damage.

Sheets and Williams (2001) provide a good overview of the history of Atlantic hurricanes including flying reconnaissance, attempted modification and modeling. Diaz and Pulwarty (1997) brought together experts from a wide rage of backgrounds to assess the socioeconomic impacts of hurricanes. These ranged form climatologists to representatives of the insurance industry.

Table 3.3
SAFFIR-SIMPSON SCALE OF HURRICANE DAMAGE

Category	*Wind Speed mph (km/hr)*	*Storm Surge*	*Damage*
1	74–95 (119–153)	4–5 ft	little structural
2	96–110 (154–177)	6–8 ft	minor structural
3	111–130 (178–209)	9–12 ft	some structural
4	131–155 (210–249)	13–18 ft	extensive structural
5	above 155 mph (249)	> 18 ft	some complete

Powell and Sim (2001) reviewed the accuracy of forecast on the timing and location of hurricane landfall. Their analysis showed that an early time bias of 1.5–2.5 hours for landfall of Atlantic Hurricanes. This has not improved much in recent years probably to the "least regret" strategy in the time prediction to account for unexpected storm acceleration. Thus, hurricane warnings could be issued 12 hours earlier (at 36 rather than 24 hours before landfall) without affecting the accuracy of the prediction. However, improving the accuracy of land-fall predictions has been difficult due to a number of related factors. For example, an important factor is the angle of the coast line relative to the projected path of the hurricane. Positional forecast errors were less for hurricanes in the Gulf (of Mexico) coast because they are moving perpendicular to the coast line. In contrast, hurricanes striking the Atlantic coast are generally moving more parallel to the coastline resulting in a diagonal path that results in larger positional errors. Position errors are 15–50 percent larger for parallel tracks than perpendicular tracks. There are additional problems in defining just what landfall is due to near misses and multiple strikes. Nevertheless, positional accuracy is important in the use of associated damage models like (storm inundation models). The errors in forecasting land-fall have to be low enough for their results to be usefull. Obviously, the timing and location of landfall are of paramount importance in evacuation planning. Finally, the predictions of models can be improved, not so much by improving the model *per se* but by gathering better observational data, and assimilating that data more effectively into the present model. Sorensen (2000) reported on the improvement forecasting and warning of natural hazards. The progress has been uneven, but hurricanes showed the most improvement.

Hurricane Andrew was the 3rd strongest hurricane ever to make landfall in the United States during the twentieth century. The result was one of the costliest natural disasters in U.S. history. Wakimoto and Black (1994) analyzed the relationship of the damage caused by Hurricane Andrew to the exact velocities in the eye wall. They concluded that the first period of highest winds stripped the surface of trees and other objects. This decreased the roughness of the surface and may have caused the second period of high winds to attain higher velocities than they would have obtained with a rougher surface to traverse. The winds reached a Fujita scale of F3, about 150 mph.

Watson and Johnson (2004) reviewed the current state-of-the-art in Hurricane loss estimation models. These models are very complicated because they link meteorology with everything that affects the dollar losses from hurricanes. Since these models are proprietary, the details of their assumptions and calculations are difficult to determine. However, these models suffer from any number of limitations common to all meteorological models. For example, it is difficult to determine exactly where wind speed was highest, how high it actually was, and how long it was sustained. It is also difficult to estimate dollar losses due to structural damage. It is also interesting to note that updated information on the meteorological specifics of a given hurricane, like Andrew, can noticeably change the damage estimates.

Pielke and Landsea (1999) explored the relationship of hurricane damages in the U.S. to the El Niño/Southern Oscillation (ENSO) phenomena in the Pacific Ocean. When sea surface temperatures are higher than normal in the eastern, tropical Pacific, it is called an El Niño event. When sea surface temperatures are lower than normal in the same region of the Pacific, it is called a La Niña event. La Niña years are also years when more hurricanes impact the U. S. In contrast fewer hurricanes occur during El Niño years. Such relationships provide some degree of predictability in the likelihood of a hurricane striking the U. S. in a given year.

Given the rapid development of the coastal areas of the U.S., the potential for hurricane damage increases each year; not because the frequency is increasing, but because there simply more people and structures along the coast each year. Having said that, given the four hurricanes and two tropical storms that impacted Florida in 2004 and the very active beginning to the season in 2005, the public seems to believe that the frequency is increasing and this is caused by global warming somehow. However, this is a short-term view, not a climatological fact.

Hurricanes have caused some of the worst natural disasters in history. One of the worst was the Indian Ocean hurricane that hit Bangladesh in 1970. Bangladesh is the low-lying delta of the Ganges-Brahmaputra River. It is an agricultural region with a very high population density. Consequently, there was no way to escape to higher ground even if there had been sufficient warning. Over 220,000 people died as well as an approximately equal number of large and small farm animals (Burton, Kates and White, 1993). While modern communications technology, like cell phones, would greatly speed the dissemination of a hurricane warning today, evacuation would still be a problem. The river delta environment is as much water as land, and roads are few and easily flooded.

FLOODS

Flooding can occur through a variety of meteorological processes resulting in excessive rainfall. The classic situation in the U. S. involves a winter with heavy snow accumulation that melts suddenly over soils that are already saturated with moisture and accompanied by persistent spring rains. The worst floods of the twentieth century on the Mississippi River occurred in 1927, 1973 and 1993. In all these cases, the meteorological causes were nearly identical especially the pattern of the upper level winds – the westerlies (Figure 3.1.). The core of the westerlies is the jet stream. It is not only the fastest part of the westerlies but its precise configuration determines the exact location of the boundaries between polar and tropical air masses. For example, in the spring and summer of 1993, there was a southward bend, or trough, in the jet stream over the Rocky Mountains and Great Plains with cold, Canadian air to the north. Meanwhile, the jet stream developed a northward bend or ridge over the Northeastern U. S. and Southeastern Canada. This allowed warm, humid tropical air

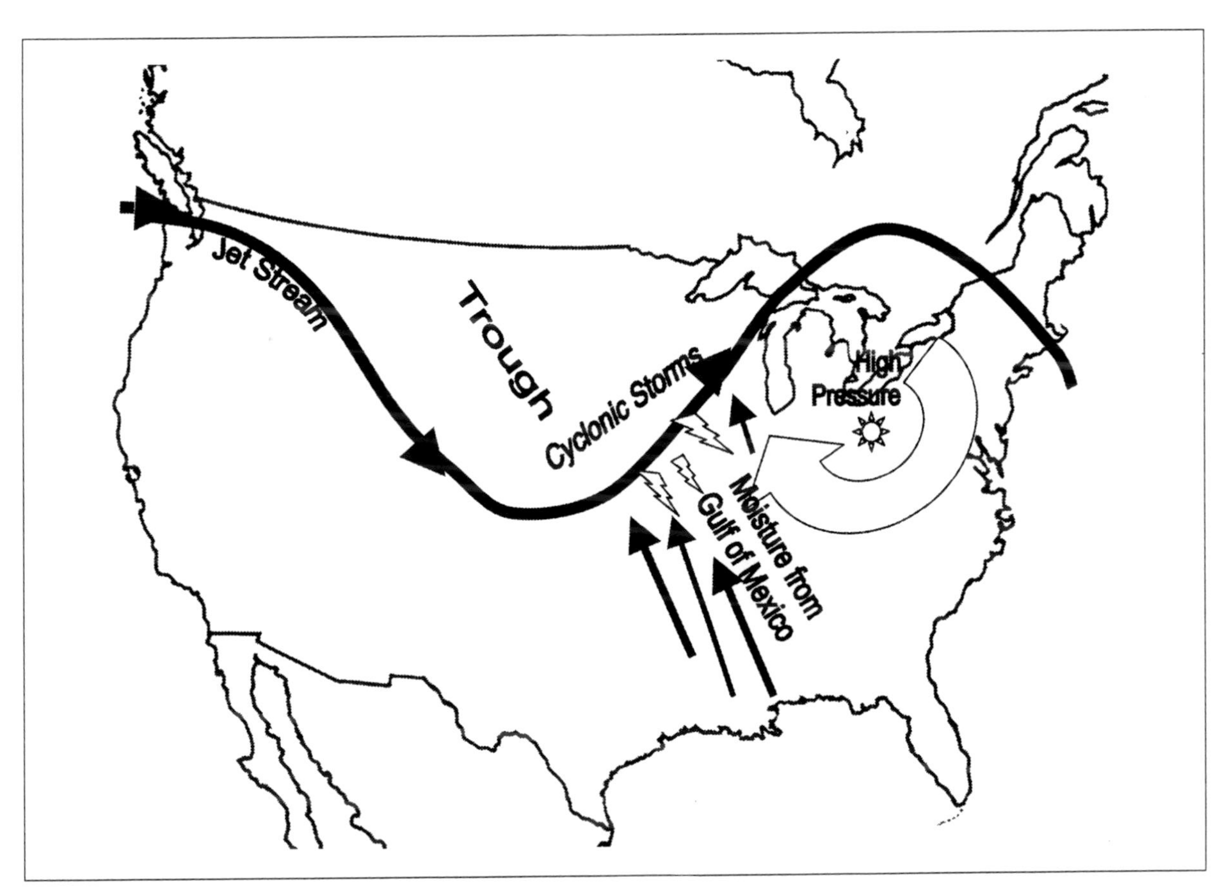

Figure 3.1. Jet Stream and other weather patterns causing the 1993 Midwest floods.

masses to penetrate northward as far as the Great Lakes. The pattern helped generate cyclonic system after cyclonic system that moved across the Midwestern states following the same path as the system before. The result was a situation called "training" where a series of thunderstorms follow the same track. The resulting rains just kept coming for months on end (Bell and Janowiak , 1995; NOAA, 1994; U.S.G.S., 1975).

The 1927 floods on the Mississippi River prompted the federal government to take action on flood disaster mitigation. Their approach was to build flood control structures miles away like dams, levee systems, diversion projects, etc. However, decades of subsequent flooding, especially 1973 and 1993, demonstrated that such measures were only partially successful. For example, the levees in St. Louis are relatively close to the river. During the 1973 flood, these levees did prevent flooding in St. Louis but constrained the flow so that the levees acted like a dam with a big hole in it. The water pooled behind this constriction and backed water up the river threatening to flood towns up stream like St. Charles. In 1993, the Corps of Engineers found themselves in the curious position of knocking holes in levees they had built to prevent flooding. However, such action was necessary. There was so much water that some additional lands had to be flooded to lower the water level in the river and, hopefully, mitigate flooding of nearby towns. Since structural controls did not completely attain the desired result, Congress passed a series of laws concerning flood insurance, control of development on flood plains and related land use planning and zoning.

The debate continues over just how all this planning, etc. should be done (Hayes, 2004). A group of government officials and academics reviewed the non-structural hazard mitigation plan developed by the Corps of Engineers and incorporated parts of it into their alternative plan. Part of the alternative plan involved development of a computer program to perform a cost/benefit analysis each individual structure to retrofit them to an acceptable flood-proof standard. This approach shows how detailed the planning and analysis process has become to mitigate the impacts of flooding in some locations.

Changnon and Kunkel (1995) showed the difficulty of determining whether future weather would be wetter or drier in the Midwest. During the period 1921–1985, floods increased in the northern Midwest, but not elsewhere in the study area. Cyclone frequencies, thunderstorm frequencies and heavy-precipitation events also increased. They concluded that increased future precipitation would lead to increased flooding and vice-versa, decreased precipitation would lead to more drought events. These conclusions may seem obvious to the layperson, but climate science does not necessarily require increased precipitation to translate into increased flooding. Many other factors determine whether it floods or not. This is especially true in urban environments.

Flash floods are an increasing threat in urban areas according to the American Meteorological Society (AMS) (2000). As a rural watershed becomes urbanized, floods will occur more frequently due to the increased amount of impervious surfaces. A stream channel that could carry all the runoff from a rural environment would flood dramatically after the watershed has become urbanized. Consequently, flash floods become "flashier." In their policy statement, the AMS points out that lead time for flash floods has increased to 50 minutes and much of the improvement is due to new technology and training. Radar technology now allows some reasonable estimate of rainfall rates and hydrometeorologists are now on staff at many National Weather Service forecast centers. Better linked meteorological – hydrological models continue to be developed supplemented by improved Geographic Information System (GIS) technology. In spite of these improvements, coordinated dissemination and preparedness programs by local governments are still necessary to mitigate the effects.

Weaver, Gruntfest and Levy (2000) reported on the flooding disaster that occurred July 28, 1997, in Fort Collins, Colorado, and the specific steps taken to mitigate the emergency management problems that occurred. Many of these modifications and improvements were in place when a second flood occurred April 30, 1999. This proved to be an excellent test of the new system. Many of the improved procedures involved better two-way communication between local authorities and the NWS including placement of automated rainfall and stream flow gauges. Much of this information could be transmitted directly to the NWS in real time. The procedures worked well demonstrating the old saying that it takes a disaster to prepare for a disaster.

The arrangements could be used as a model for the future. The city of Fort Collins, Colorado, created an Emergency Command Center staffed with just the right specialists, just the right communications, and ability to monitor events in real time with sensors at strategic locations in the field. The specialists also had a detailed knowledge of how similar events had impacted specific locations in the past. Unfortunately, such a model can only be implemented with modern information systems technology. Most of the rest of the world is simply at the mercy of the weather and its consequences.

The major rivers in China have a history of flooding and misery that are unequaled anywhere. The Huang (Yellow) River has flooded so many times in recorded history that it is called "China's Sorrow." The name refers not only to direct loss of life but also loss of crops and the resulting famine that follows. Similarly, the recurrent flooding of the Yangtze River prompted the Chinese government to take drastic action. The floods of 1998 were especially devastating and the relief efforts stretched the government's resources to the breaking point. Consequently, the government decided to build the Three Gorges Dam. This dam is a controversial project in several respects, but the sheer size of the project is impressive: some 600 feet tall, 1.5 miles long with a reservoir about 350 miles long. In additional to the obvious environmental destruction, there are immense potential social problems. Perhaps as many as 1.5 million people will have to be relocated. The whole project has been described as the biggest thing ever built. While this claim may not be exactly true, the project represents a tremendous investment and gamble by the Chinese government. Just to what extent it will have the intended benefits remains to be seen.

In China, modern technology has not only been used in the construction of the Three Gorges Dam, but also in the development of a modern system to better monitor and assess floods and other natural disasters (Zhang, et al, 2002). It is based on remote sensing, geographic information systems, and the Global Positioning System. The system illustrates the transfer of current technology to developing nations which greatly improves their ability to respond to an emergency.

Finally, it is important to remember that all floods are not caused by meteorological events. The 2004 Tsunami in the Indian Ocean will be remembered for a death toll of over 200,000. The shifting of the earth's crustal plates caused one of the very strongest earthquakes ever measured. The resulting tidal wave caused flooding on virtually all the coastlines surrounding the Indian Ocean. Within a few hours, thousands of coastal communities were utterly destroyed by flooding.

DROUGHTS

In the U.S., the cause of droughts is the configuration of the westerly winds. The driest years of the twenthiet century (1934, 1936, 1954, 1956, 1980) all have similar upper air patterns (Figure 3.2). The jet stream makes a large northern bend, called a ridge, across the middle of the country with smaller southern bends on each side across the west coast and east coast states. A large anticyclone (high pressure cell) forms below the ridge and begins to rotate. This pattern is very stable and is called an omega block, after the Greek letter omega, Ω. It can remain in place for the entire summer when its effects are most pronounced. The descending air in the anticyclone makes it nearly impossible for clouds to form or precipitation to fall. The clear skies and intense sunshine cause unusually high temperatures in the summer. However, there is an additional meteorological process contributing to the scorching temperatures which is often not fully appreciated. Basic atmospheric processes require rising air to cool off as it rises, and descending air to warm up as it descends. In a high pressure cell, the air descends and warms, resulting in even warmer temperatures than would occur due to the sunshine alone. The result can be 100° F. temperatures day after day.

Climatic factors also come into play. For example, the Great Plains of the U.S. is a place of climatic extremes. About one-third of the time, it is drier than normal, one-third of the time it is wetter than normal; so it is only within normal ranges the remaining one-third of the time. An important precursor of drought in the Great Plains region is a deficiency of soil moisture in the spring. As temperatures increase in the late spring and early summer, the ground temperatures

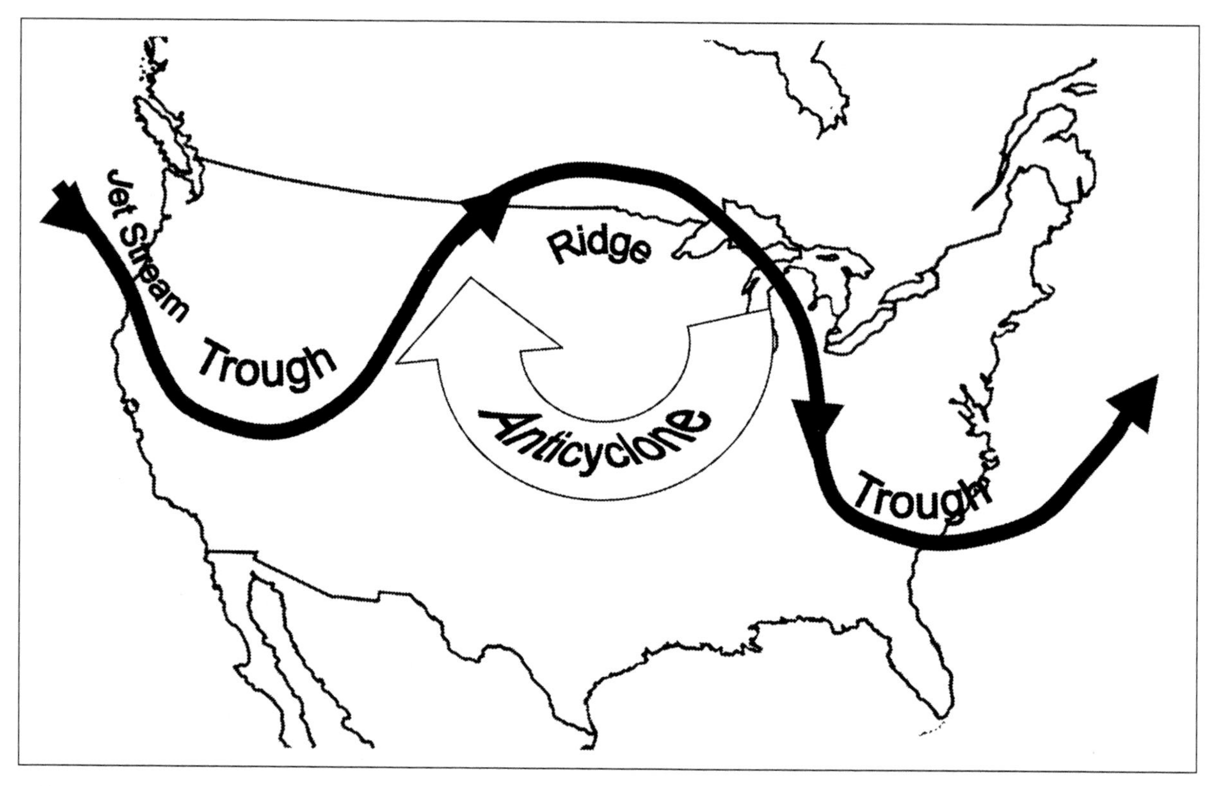

Figure 3.2. Jet stream and other weather patterns causing drought in the northern Great Plains and Midwest.

become hotter and hotter. This, in turn, sets up a positive feedback that helps to reinforce the strength of the anticyclone causing the drought. These forces were at work during summer drought of 1980 and the record high temperatures established at that time still stand in many locations.

The 1930s Dust Bowl Drought was the most severe drought to affect the U.S. during the twentieth century and the longest. McGregor (1986) showed that the 1950s drought was actually more intense, but simply did not last as long as the 1930s drought. The notoriety of the Dust Bowl was due as much to its social impact as its environmental catastrophe (Worster, 1979). Poor, destitute farmers migrated out of the region. The government developed relief programs that would have been unthinkable a decade earlier. By the 1950s, farming practices had changed, and a social safety net was in place that mitigated the impact of the 1950s drought. These included price supports, crop insurance, and improvements in land management techniques so the dust storms of thirties did not reoccur.

Recently the causes of the Dust Bowl drought has been linked to spatial pattern of Sea Surface Temperatures (SSTs) in the Pacific Ocean (Fye, Stahle and Cook, 2004). These included an anomalous pattern SST pattern in the north Pacific that endured for the entire eleven-year period of the drought. The patterns also included unusually cool temperatures in the eastern equatorial Pacific that today would be considered a La Niña pattern. Collectively, these results provide a clearer understanding of the atmospheric and oceanic conditions that caused this most infamous event and will lead to better forecasts of future droughts.

Droughts occur when there is a deficiency of precipitation usually over some extended period of time like several months or even years. In modern, developed countries, they have enormous economic consequences, but are not usually life threatening. In the developing world where a majority of the people are farmers and grow their own food, drought is equated with famine and may require massive relief efforts from other parts of the world. The 1968–75 drought in the Sahel of Africa is a case in point (Dalby, Church and Bezzaz, 1977; Glantz, 1986). The Sahel region is located south of the Sahara Desert and north of the forested lands of equatorial Africa. The people are primarily nomadic herders and subsistence farmers. When the rains failed, millions of animals died and the crops failed. An estimated 200,000 people died, and the migration of the remainder caused social chaos. The governments of these poor countries had little help to offer. Eventually aid poured into the region from wealthier nations. The drought did not actually end in 1975. After near-normal conditions in 1974 and 1975, the drought resumed again and lasted into the mid 1980s. The result has been continued famine and turmoil in parts of Africa (Glantz, 1987).

In a discussion of drought as a phenomenon, Hare (1987) makes an important distinction between drought and desiccation. In his view, drought is a temporary deficiency of precipitation and eventually the rains return. It is also more regional affecting, for example, part of the U.S. while another region might very well have more rain than usual. In contrast, desiccation is prolonged and gradually intensifying. It is also larger in scale. The 1970s and 1980s drought in Africa is a good example of such a desiccation where nearly the whole continent seemed to dry as a single coherent unit.

Wilhite (2002) points out that drought is often an indicator of unsustainable land and water management practices and humanitarian aid from outside only encourages these practices to continue. This, in turn contributes to the desertification process. The result is a vicious cycle. Wilhite argues for the development of a better system of preparedness, early warning systems, and mitigation strategies not only in the countries affected but by the international organizations that provide aid (Wilhite, Easterling and Wood, 1987).

What happens when a drought is forecast and then does not materialize? Changnon and Vonnahme (2003) reported on the consequences of such a failed forecast. In March of 2000, NOAA issued forecasts of spring and summer droughts for several states in the Midwest. The summer brought heavy rains instead. Various state and local water managers heeded the forecast and initiated actions such as authorizing water restrictions and/or holding meetings of drought response groups. For the most part the managers reported that such actions caused few problems. However, certain agricultural interests complained of large economic losses. The episode resulted in a loss of credibility and called into question the response of water managers to such a forecast in the future. Essentially, this entire episode is an example of the "cry wolf syndrome." It is an inevitable consequence of warnings that are issued for events that do not actually occur.

HEAT WAVES

Heat waves occur when a strong high pressure cell, an anticyclone, stalls over a particular place during the summer. The excessively high temperatures are caused by a combination of clear skies, intense sunshine, and descending (warming) air. These factors can also be aggravated by high humidity and the urban heat-island effect. Frequently, if winds are light, air pollutants can accumulate and make the situation even worse.

The heat wave has been called the silent killer. Perhaps as many as a 1,000 people a year die due to extreme heat. This is more than from any other type of weather related event in the U.S. The two most notable recent heat waves occurred in Chicago, Illinois in 1995 and in France in 2003. In both cases, there was a disproportionate number of deaths among the elderly and poor, and government officials were criticized for not responding more effectively.

The Chicago heat wave during the summer of 1995 has become legendary because it was such an extreme event and, as a result, at least 700 people died. Kleinenberg (2002) provided a detailed social history of the human impacts including a sympathetic analysis as to why the poor and elderly suffered a vast majority of the deaths. Many of these "incasts" lived in old tenements without air conditioning in high crime areas. Their windows are often nailed shut and in some cases their water and/or electricity had been cutoff for failure to pay their bills. Kleinenberg also condemed city officials for not doing more to aid the most vulnerable population.

Looking at the details of the meteorology, Karl and Knight (1997) conducted a statistical analysis of the Chicago heat wave and concluded that it was an extremely rare event with a probability of occurrence less than 0.1 percent. This probability was based on a 10,000 year simulation based on the mean and variance of temperatures. They also attempted to determine if such events might be more frequent in the future as a response to global warming, but had difficulty in accomplishing this goal.

The city of Philadelphia has created a heat watch/warning system. Ebi, et al. (2004) discussed the criteria for determining when a warning was needed and costs and benefits of issuing such a warning. They also demonstrated the statistical techniques used to estimate number of lives saved although there remain certain difficulties in accomplishing this goal. One of the most troublesome problems was determining the dollar value of a live saved as well as the costs of maintaining such a warning system.

During August, 2003, France was hit with an extraordinary heat wave that virtually paralyzed the country. During the first week of August, temperatures remained at 36°–37° degrees C. and some stations reported temperatures of 40° C. An estimated 11,400 people died and, again, most of them were elderly (Crabbe, 2003; Vandentorren, et al., 2004). The traditional August vacation season contributed to the problem. During this time, the country virtually shuts down as many people, including government officials, take their vacation. Many such officials were criticized for their reluctance to cut short their own holidays to deal with the crisis. Some hospital wards had even been closed during the August break. The end result was both a human catastrophe and a governmental crisis.

WILDFIRES

While forest fires, brush fires, and range fires are all natural phenomena, they have caused increasing dollar losses in the U.S. Much of this is due to the proliferation of suburban and low density rural development as people choose to live beyond the edge of the city in the so-called "exurbia" environment. Many of these homes are large and expensive so a single wildfire can cause millions of dollars in damage.

Wildfires are frequently aggravated by weather conditions. The potential for wildfires will be greater during a long, hot summer when high pressure is in control of the region. In the western U. S., such a high pressure can create winds that help fan the fires. In California, such winds are called Santa Ana winds and in the Rockies, they are called Chinook winds. In meteorological terms, both are katabatic winds, i.e., winds that move downslope and warm up as they do so. For example, in Colorado and Montana, Chinook winds move down the eastern slope of the Rocky Mountains. In California, Santa Ana winds move down the western slopes of the Sierra Nevada Mountains. These winds not only warm as they descend, their relative humidity decreases so they become desiccating winds absorbing moisture from everything they touch including the soils and vegetation. With a large anticyclone anchored over the western states, the clockwise pattern of rotation will cause Chinook winds to develop in the northern part and Santa Ana winds in the south. Both are associated with extreme fire danger.

One of the most notorious wildfires occurred near Los Alamos, New Mexico. Officially known as the Cerro Grande fire (Hill, 2000), it was one of the worst such incidents because the fire was set to burn off some excessive vegetation in about 900 acres. The fire got out of control and burned 48,000 acres including hundreds of homes. Damages were estimated at around a billion dollars. Over 18,000 citizens evacuated and 1,000 fire fighters eventually brought the fire under control.

The potential for wildfires is a function of accumulated vegetation, climate, moisture conditions, prevailing weather conditions, and human influence. Westerling, et al. (2003), analyzed the seasonal and interannual variability in wildfires in the western U.S. They discovered a strong relationship between previous moisture conditions and the incidence of wildfires. This relationship was so strong between that it was possible to forecast the severity of the upcoming fire season up to a year in advance.

Warner, et al. (2004) discussed the development and capabilities of a portable mesoscale model-based forecasting system for use by the U.S. Army and also for use in civilian emergency-response situations. While the system had obvious applications for operations in Afghanistan, it was also used during the 2002 Winter Olympics in Salt Lake City to predict the potential transport and dispersion of hazardous material. The system also has applications in wildfire monitoring and burn path prediction when meteorological conditions change rapidly.

BLIZZARDS

Blizzards are large, intense cyclonic storms during the winter season. They are justly famous for large accumulations of snow, high winds, and plunging temperatures. A strong blizzard can virtually paralyze an entire region of the country. Such storms cause power outages and bring transportation to a standstill. In early January, 1996, one of the strongest snowstorms of the century hit the East Coast dropping 17 to 30 inches of snow from Washington to Boston (Le Comte, 1997). Snow from two additional storms virtually paralyzed East Coast transportation and the federal government closed for three days. The total snowfall accumulated to between 20 and 50 inches. To make matters worse, the proverbial mid January thaw caused rivers to rise from the Ohio Valley tô New England and many areas flooded as far south as Washington, D.C.

DeGaetano (2000) summarized the meteorology and impacts of the ice storm that hit northern New York and New England in 1998. In spite of the fact that ice storms are regarded (SP) as relatively rare events, this storm was approximately comparable to at least three other similar events since 1948. Total economic impact was perhaps 2 billion dollars and direct impacts about 1 billion. At one time, nearly 600,000 customers were without electricity, and 1.4 million people lost electric power at some point. In addition to the usual impacts on utilities, other major losses occurred in the dairy and forest products (including maple sugar production). Over 300 people were admitted to hospitals and treated for carbon monoxide poisoning.

THE FUTURE

Several trends will continue into the future that are all intertwined. Forecasting and prediction will continue to be of paramount importance and will be done with increasingly complicated models. The observation networks that supply data to the models will become more elaborate and operate in near real time. The scale of the frame of reference will be larger, even global and include the oceans. Global warming will continue to influence everything in atmospheric science (Harvey, 2000).

Forecasting and prediction have been and will continue to be at the core of meteorological science. This will include both forecasting of immediate threats, like predicting the location and intensity of hurricane landfall, and longer range "seasonal" outlooks that will provide probabilities of some threatening weather event occurring like heavy rains or the number of Atlantic hurricanes. As atmospheric science progresses, the frame of reference will become larger, even global. For example, the development of extended droughts and the incidence of Atlantic hurricanes are influenced by oceanic conditions half a world away. A continued focus of attention will be the connections betweens conditions in the world's oceans and weather events elsewhere. As science progresses, and future ENSO events (and other oscillations) can be predicted with longer lead times, seasonal forecasts and perhaps even climatic forecasts become possible. The potential benefits for emergency management planning are immense.

A good example is Murnane's (2004) review of the impact that better climate forecasts would have on the reinsurance industry. Reinsurance is best described as insurance for insurance companies. It limits their losses in case of a major disaster in one place where they have an inordinate number of clients. Of all the potential disasters, reinsurance companies are most concerned about hurricanes since these, collectively, have the greatest impact on the global reinsurance business. One of the principal areas of research in current global climatology is focused on various oscillations or cycles in the earth's climate system. Such cycles seem to have a profound effect on the weather in various parts of the world including the incidence of hurricanes. Murnane described three atmospheric oscillations in detail: the Quasi-Binenial Oscillation (QBO), Arctic Oscillation (AO), and Madden-Jullian (MJO). The ability to predict these oscillations and their consequences (even interactions) would have a huge impact not only on atmospheric science but also on the reinsurance industry. Interesting, the models the industry uses are based on climatic probabilities of such events. However, they do not consider how an extreme rare event might alter the climatic probabilities. Michaels, et al., (1997) also noted that the models used by the insurance industry rely on historical data sets on storm frequency and assume that the probabilities will be the same in the future. Increasingly, the industry is questioning the wisdom of this traditional approach. The frequency of hurricanes may or may not increase in the future; but, either way, it is important for the insurance industry to incorporate better climate science into their models.

One of the more troubling trends as been the expansion of scale for atmospheric related phenomena from the regional to the global scale. Floods, droughts, air pollution emergencies are usually local or regional in scale. However, two types of air pollution, ozone depletion in the stratosphere and carbon dioxide enrichment of the atmosphere, are essentially global in their impact. Air pollution and ozone depletion may not pose immediate emergencies, but they are still of special importance because of the long-term impacts on human health. For the first time in history, it is clear the humans can and do impact the workings of the atmosphere at local, regional, and even global scales.

Global warming will continue to receive the most attention as a long-term threat. Global warming is especially troublesome because of the potential pervasive impact and the uncertainties associated with these impacts. A rise in sea level is perhaps the most obvious consequence, but there are many others like the supposed possibility of increased hurricane activity. Climate specialists do not all subscribe to the notion that global warming will result in increased hurricane activity. However, there is more general agreement that the climate variability will increase and this will cause more extreme weather events. If all this proves to be the case, the number of natural disasters will increase as well as the preparedness for emergency response.

SUMMARY

Weather extremes cause many different types of natural disasters requiring an emergency response. These could range from relatively local flash floods to drought, starvation and pestilence of Biblical proportions requiring an international response. The role of meteorology historically has been in forecasting the event, issuing the warnings and assessing the forces that caused the damage. Since there are so many different types of weather related disasters, meteorologists work with specialists from many different disciplines. These range from the media, to government officials to hydrologists to relief groups like the Red Cross. However, all share the common goal of protecting property and saving lives.

The meteorologist is responsible for forecasting an impending disaster. This traditional role is fundamental and will not change in the future. Today, the forecasts are based on models, and this trend will accelerate as more and more models will be linked together. The forecasts will become more refined with a better understanding of basic atmospheric processes and the collection of vast arrays of data through automated sensing systems. Modern communications are not only important in the transmission of these data but also in the rapid dissemination of the consequent forecasts, watches and warnings. Transmission and dissemination of warnings will also be improved by better organizational arrangements. A meteorologist and emergency manager will be on the same team similar to the Ft. Collins, Colorado arrangement.

The meteorological forces driving individual extreme weather events are increasingly understood in the context of larger regional or even global processes. Will global warming cause more variability in the weather at a particular place and hence lead to more extreme events? If so, the future for emergency managers will be a busy one.

REFERENCES

Aguirre, B. E. (1991). "Saragosa, Texas, Tornado, May 22, 1987: An Evaluation of the Warning System." *Natural Disaster Studies*, 3. National Academy Press, Washington, D.C.

American Meteorological Society (2000). "Prediction and Mitigation of Flash Floods." *Bulletin of the American Meteorological Society, 81*(6), 1338–1340.

Andra, D. L. Jr.; Quoetone, E. M. and Bunting, W. F. (2002). "Warning Decision Making: the Relative Roles of Conceptual Models, Technology, Strategy, and Forecaster Expertise on 3 May, 1999." *Weather and Forecasting, 17*(3), 559–517.

Bell, G.D. and Janowiak, J. E. (1995). "Atmospheric Circulation Associated with the Midwest Floods of 1993." *Bulletin of the American Meteorological Society, 76*, 681–695.

Burton, I., Kates, R.W., and White, G.F. (1993). *The Environment as Hazard*, 2nd. ed. The Guilford Press, New York.

Changnon, S. A. and Vonnahme, D. R. (2003). "Impact of Spring 2000 Droughts on Midwestern Water Management." *Journal of Water Resources Planning & Management, 129*(1), 18–26.

Changnon, S. A. and Kunkel, K. E. (1995). "Climate-Related Fluctuations in Midwestern Floods during 1921-1985." *Journal of Water Resources Planning & Management, 121*(4), 326–335.

Crabbe, C. (2003). "France Caught Cold by Heatwave." *Bulletin of the World Health Organization, 81*(10), 773–775.

Dalby, D., Harrison Church, R. J., and Bezzaz, F. (1977). *Drought in Africa.* Special Report 6. International African Institute, London.

DeGaetano, A.T. (2000). "Climatic Perspective and Impacts of the 1998 Northern New York and New England Ice Storm." *Bulletin of the American Meteorological Society, 81*(2), 237–254.

Diaz, H.F. and Pulwarty, R.S. (1997). *Hurricanes: Climate and Socioeconomic Impacts.* Springer-Verlag, London.

Doswell, C.A. III; Moller, A.R., and Brooks, H.R. (1999). "Storm Spotting and Public Awareness since the First Tornado Forecasts of 1948." *Weather and Forecasting, 14*, 544–447.

Ebi, K.L.; Teisberg, T.J., Kalkstein, L.S., Robinson, L., and Weiher, R.F. (2004). "Heat Watch/Warning Systems Save Lives." *Bulletin of the American Meteorological Society, 65*(8), 1067–1073.

Fye, F. K., Stahle, D.W., and Cook, V. (2004). "Twentieth-Century Sea Surface Temperature Patterns in the Pacific during Decadal Moisture Regimes over the United States." *Earth Interactions, 8*(1), 1–22.

Glantz, M.H. (2001). *Currents of Change: El Niño's Impact on Climate and Society*, 2nd ed. Cambridge University Press, Cambridge.

Glantz, M.H., ed.(1986). *Drought and Hunger in Africa.* Cambridge University Press, Cambridge.

Golden, J.H. and Adams, C.R. (2000). "The Tornado Problem: Forecast, Warning, and Response." *Natural Hazards Review, 1*(2), 107–118.

Hamill, T.M., Schneider, R.S., Broods, H.E. Forbes, G.S., Bluestein, H.B., Steinberg, M., Meléndez, D.L. and Randall, M. (2005). "The May 2003 Extended Tornado Outbreak." *Bulletin of the American Meteorological Society, 84*(4), 531–543.

Hammer, B. (2002). "Response to Warnings during the 3 May 1999 Oklahoma City Tornado: Reasons and Relative Injury Rates." *Weather & Forecasting, 17*(3), 577–582.

Hare, F.K. (1987). "Drought and Desiccation: Twin Hazards of a Variable Climate." In Wilhite, D.A., Easterling, W.E. and Wood, D.A. eds. (1987) *Planning for Drought.* Westview Press, Boulder, 3-9.

Harvey, L.D.D. (2000). *Climate and Global Environmental Change.* Prentice Hall, New York.

Hayes, B.D. (2004). "Interdisciplinary Planning of Non-structural Flood Hazard Mitigation." *Journal of Water Resources Planning & Management, 130*(1), 15–26.

Hill, B.T. (2000). "Fire Management: Lessons Learned from the Cerro Grande (Los Alamos) Fire and Actions Needed to Reduce Fire Risks." Associate Director, Energy, Resources and Science Issues, Resources, Community and Economic Development Division., testimony before Subcommittee of Forests and Forest Health, Committee on Resources, House of Representatives. Monday, Aug, 14. 2000. GAO/T-RCED-00-273.

Karl, T.R. and Knight, R.W. (1997). "The 1995 Chicago Heat Wave: How Likely Is a Recurrence?" *Bulletin of the American Meteorological Society, 78*(6), 1107–1119.

Kleinenberg , E. (2002). *Heat Wave: A Social Autopsy of Disaster in Chicago.* University of Chicago Press, Chicago.

Le Comte, D. (1997). "A Wet and Stormy Year." *Weatherwise, 50*(1), 14–23.

McGregor, K.M. (1986). "Drought During the 1930s and 1950s in the Central United States." *Physical Geography, 6*, 288–301.

Michaels, A., Malmquist, D., Knap, A. and Close, A. (1997). "Climate Science and Insurance Risk." *Nature, 389*, 225–227.

Morris, D.A., Crawford, K.C., Kloesel, K.A., and Gayland, K. (2002). "OK-FIRST: An Example of Successful Collaboration between the Meteorological and Emergency Response Communities of 3 May 1999." *Weather & Forecasting, 17*(3), 567–577.

Murnane, R.J. (2004). "Climate Research and Reinsurance." *Bulletin of the American Meteorological Society, 85*(5), 697–707.

National Oceanographic and Atmospheric Administration. (1994). *The Great Flood of 1993.* Silver Spring, MD.

Pielke, R.A. and Landsea, C.N. (1999). " La Niña, El Niño, and Atlantic Hurricane Damages in the United States." *Bulletin of the American Meteorological Society, 80*(10), 2027-2033.

Powell, M.D. and Aberson, S.D. (2001). "Accuracy of United States Tropical Cyclone Landfall Forecasts in the Atlantic Basin (1976–2000)." *Bulletin of the American Meteorological Society, 82*(12), 2749–2767.

Sheets, B. and Williams, J. (2001). *Hurricane Watch: Forecasting the Deadliest Storms on Earth.* Vintage Press, New York.

Sorensen, J. H. (2000). "Hazard Warning Systems: Review of 20 Years of Progress." *Natural Hazards Review, 1*(2), 119–125.

U.S. Geological Survey (1975). *The 1973 Mississippi River Basin Floods.* U.S.G.S. Professional Paper 937.

Vandentorren, S.; Suzan, F., Medina, S., Pascal, M., Maulpoix, A., Cohen, J. and Ledrans, M. (2004). "Mortality in 13 French Cities During the August 2003 Heat Wave." *American Journal of Public Health, 94*(9), 1518–1520.

Wakimoto, R. J. and Black, P. G. (1994). "Damage Survey of Hurricane Andrew and Its Relationship to the Eyewall." *Bulletin of the American Meteorological Society, 75*(2), 189–200.

Warner, T. J., Bowers, J.F., Swerdlin, S.P., and Beitler, B.A. (2004). "A Rapidly Deployable Operational Mesoscale Modeling System for Emergency-response Applications." *Bulletin of the American Meteorological Society, 85*(5), 708–17.

Watson, C.A. Jr. and Johnson, M.E. (2004). "Hurricane Loss Estimation Models." *Bulletin of the American Meteorological Society, 65*(11), 1713–26.

Weaver, J. C.; Gruntfest, E. and Levy, G.M. (2000). "Two Floods in Fort Collins, Colorado: Learning from a Natural Disaster." *Bulletin of the American Meteorological Society, 81*(10), 2359-2366.

Westerling, A.L., Gershunov, A., Brown, T.J., Cayan, D.R. and Dettinger, M.D. (2003). "Climate and Wildfire in the Western United States." *Bulletin of the American Meteorological Society, 81*(5), 595-605.

Wilhite, D.A. (2002). "Combating Drought through Preparedness." *Natural Resources Forum, 26*(4), 26–37.

Wilhite, D.A.; Easterling, W.E., and Wood, D.A., eds. (1987). *Planning for Drought.* Westview Press, Boulder.

Worster, D. (1979). *The Dust Bowl: The Southern Plains in the 1930s.* New York: Oxford University Press.

Zhang, J., Ahou, C., Xu, K., and Watanabe, M. (2002) "Flood Disaster Monitoring and Evaluation in China." *Global Environmental Change Part B: Environmental Hazards, 4*(2/3), 33–43.

Chapter 4

Engineering's Contributions to the Field of Emergency Management

Ana Maria Cruz

ABSTRACT

Engineering has contributed to the emergency management field in two important ways: in the setting of design and safety standards, and the actual design and construction of infrastructure used to prevent damage and losses caused by hazards. In this chapter the contribution of engineering to emergency management is presented. First, the evolution of the setting of engineering codes and standards in the United States is presented. Then, the contribution of engineering to hazard reduction is given by type of hazard and by type of infrastructure. Finally, conclusions and recommendations for future engineering research are presented.

INTRODUCTION

This chapter presents the main contribution of engineering to the emergency management field. Engineering encompasses several fields of study that have core engineering sciences in common (Channell, 1989). Rather than describe the contributions of each of the various engineering fields (e.g., civil and structural engineering, chemical engineering), the chapter will look at how engineering in general has contributed to hazard reduction.

Engineering has contributed to the emergency management field in several ways; helping communities reduce the risk from natural and technological hazards, but also in some cases contributing to the overall risk. This is illustrated in the following example. Important engineering flood control projects were implemented during the 1930s–1960s after the passing of the Flood Control Act in 1934. These flood control projects involved large and expensive construction ventures such as building of levees and floodwalls. An evaluation of a series of disasters in the late 1960s (particularly hurricanes Betsy and Camille) in the United States proved that engineered flood control measures alone, particularly structural mitigation measures, could often disrupt or destroy the natural environment, could be extremely costly, and could create a sense of false security (Godschalk, 1999). A need for non-structural mitigation alternatives for flood control (e.g., land use planning, relocation, protection of natural environmental features) was called for. Thus, protection of people and property from the impacts of natural and technological hazards requires a balanced use of engineered-measures as well as non-engineered ones.

Two of the most important contributions of engineering to emergency management have been in the setting of codes and standards, and the actual design and construction of infrastructure used to prevent damage and losses caused by hazards. The following sections describe the evolution of engineering codes and standards in the United States, and present the contribution of engineering to emergency management by type of hazard and by type of infrastructure.

THE EVOLUTION OF ENGINEERING CODES AND STANDARDS

Engineering design codes and safety standards in the United States have evolved over the years to incorporate lessons learned from past disasters or failures, as well as from research performed in the laboratory.

The task of setting engineering design and safety standards has often been the responsibility of engineering associations such as the American Society of Civil Engineers (ASCE), the American Society of Mechanical Engineers (ASME), or the American Petroleum Institute (API). The main purpose of setting design and safety standards has been for public safety and hazard reduction. However, the appropriate level of safety is not solely the decision of engineers or manufacturers but a societal choice. Thus, who participates in the decision making process and how these choices are made are crucial (Heaney et al., 2000). There is of course much debate on whether the public's interests are always fully considered (Keeney, 1983; Heaney et al., 2000). However, this discussion is outside of the scope of this chapter.

Until recently, design and safety standards have been set with primary regard to life safety issues. For example, buildings codes can assure that buildings are designed so that occupants can evacuate safely from a building, but they do not guarantee that the building itself will be inhabitable after a design level event. Recently, changes to codes and standards have been incorporated to minimize property damage, or to assure structural integrity of buildings. Heaney et al. (2000) observe that building regulations may now include other considerations such as accessibility for the disabled, historic preservation, and decrease of economic loss during design-level events.

Engineering codes and standards alone do not guarantee safety from natural and technological hazards. While engineering codes and standards have tried to address acceptable levels of risk and losses, they have been difficult to enforce and administer. The extensive damage to residential buildings (more than 215,000) during the Kocaeli earthquake in Turkey in 1999 resulted among other problems as a consequence of lack of oversight of building regulations (U.S. Geological Survey, 2000; Cruz, 2003). Many buildings were constructed without any engineering input, or often even if design plans had been approved by city engineering officials, poor quality materials and unqualified workers were used. Heaney et al. (2000) note similar problems during Hurricane Andrew in Florida in 1992. The authors report that lack of building code enforcement was in part blamed for the widespread building damage caused by the high winds during the hurricane.

Engineering codes and standards are dynamic, changing over time as new research findings and lessons from past disasters are incorporated, and as society's acceptable levels of risk change. Bringing buildings and other structures up to current buildings codes periodically is expensive and not always economically feasible. In addition, there is often an administrative lag time before new codes are actually incorporated. Thus, hazard reduction in buildings and structures cannot depend solely on engineering codes and standards, but must incorporate other hazard reduction strategies such as land use planning, insurance, education and awareness campaigns, and other non-structural hazard mitigation alternatives. Knowing how to combine various hazard reduction strategies and how to work with the various actors and stakeholders represents a challenge for the emergency manager.

ENGINEERING CONTRIBUTION TO HAZARD REDUCTION BY TYPE OF HAZARD

In this section the engineering contribution to hazard reduction is presented by type of hazard. This division is adopted from Heaney et al. (2000) because engineering practice has been organized by type of hazard, and by impact on type of infrastructure which will be described in a subsequent section.

Earthquakes

Earthquakes can have devastating effects on poorly constructed buildings and other infrastructure resulting in huge losses of life and property. Such is the case of the Kocaeli earthquake in Turkey in 1999 where more than 17,000 people lost their lives. Approximately 214,000 residential units and 30,000 business units sustained structural damage in the earthquake, and many complete or partial building collapses were reported (U.S. Geological Survey, 2000). Economic losses for the Turkey earthquake were estimated at US$16 billion (Tang, 2000). In the U.S., earthquakes have also taken their toll particularly in economic terms. Thousands of structures were damaged or destroyed by ground shaking during the Northridge earthquake in 1994, and total direct losses were estimated at more than US$20 billion (Tang, 2000).

Although the State of California has taken steps towards earthquake hazard reduction since the 1906 San Francisco earthquake, a national earthquake mitigation policy was not adopted until 1977 through the National Earthquake Hazards Reduction Program (NEHRP). The goals of the NEHRP include improving the understanding of earthquakes and their effects (e.g., predicting and forecasting), improving techniques to reduce seismic vulnerability of facilities and systems (e.g., through the adoption of updated seismic building codes and better construction practices), and improving seismic hazards identification (e.g., development of earthquake hazard maps) and risk-assessment methods and their use.

Earthquake engineering activities have mostly centered around impacts on buildings and lifeline systems (Heaney et al., 2000). The engineering design and construction of infrastructure (e.g., buildings, lifelines) to withstand earthquakes is vital, particularly in areas of high seismic risk. The adoption of appropriate seismic building codes for new structures and the retrofitting of older buildings to current engineering building codes can help minimize loss of life and property during earthquakes. In the United States adoption and enforcement of seismic building codes is left to the discretion of each state, with the exception of some seismic requirements for new and existing federal buildings (Executive Orders 12699 and 12941). The state of California, for example, has adopted the Unified Building Code (1997), which requires designing buildings for the 1 in 475-year earthquake event. However, some local communities in the state may choose to following stricter codes, such as the International Building Code (2000), which requires the design of new buildings for the 1 in 2475-year event. Each state has adopted various seismic construction standards for new buildings. The problem remains for older structures.

Earthquakes have the potential to disrupt lifeline systems. For example, electrical power outages were caused during the Taiwan (Kranz, 1999) and Turkey earthquakes in 1999 (Tang, 2000), the Kobe earthquake in Japan in 1995 (Erdik, 1998), and the Northridge earthquake in the United States in 1994 (Lau et al., 1995). Extensive damage to transportation routes was reported following the Kobe earthquake, which destroyed the city's main highway, several railroad tracks, and much of its port (Dawkins, 1995). Lau et al. (1995) reported extensive damage to gas distribution systems during the Northridge earthquake in California. This resulted in numerous fires, which consumed several single-story wood frame houses and over 70 mobile homes in the cities of San Fernando and Sylmar.

Damage to lifelines can have detrimental effects on emergency response activities. Loss of water due to multiple pipeline-breaks delayed emergency response to several of the gas-caused fires following the Northridge earthquake (City Administrative Officer, 1994). Steinberg and Cruz (2004) reported that loss of water and power outages following the Kocaeli earthquake hampered emergency response to earthquake-triggered hazmat releases.

Experience and observations during past urban earthquakes in the U.S. and around the world are used by engineers to assess vulnerability of lifelines to earthquakes (Erdik, 1998). After the 1971 San Fernando earthquake in California, which caused extensive damage to all types of lifelines, efforts were made to better understand the effects of earthquakes on lifelines and to advance the practice of lifeline earthquake engineering in the United States (Eguchi and Honegger, 2000). The Technical Council on Lifeline Earthquake Engineering, established after the San Fernando earthquake, develops guidelines and standards for the seismic design and construction of lifelines including electrical power and communications systems, gas and liquid fuel pipelines, transportation systems, and water and sewage systems (Heaney et al., 2000).

Lifeline vulnerability functions and estimates of time required to restore damaged facilities are provided in the ATC-25 report (ATC 1991) "Seismic Vulnerability and Impact of Disruption of Lifelines in the Conterminous United States" (Rojahn et al., 1992). The vulnerability functions are based on inventory hazard data and the elicitation of expert opinion methodology developed in the ATC-13 report (Erdik, 1998).

Floods

Floods account for about 80 percent of all declared disasters in the United States. Heaney et al. (2000)

observe that flood control in the United States represents a typical example of the long-term evolution of engineering design standards. The strong emphasis on mostly structural engineering flood control measures in the 1930–1960s proved unsustainable. The Midwest floods in 1993 led to a reevaluation and change in flood control policies in the nation, moving towards more integrated flood control management that combines structural and non-structural hazard reduction options (Godschalk et al., 1999).

Pilgrim (1999) notes the need for the engineering profession to broaden its scope from the merely technical aspects to those that directly affect communities and the environment. The author observes that the basic role of the engineer is shifting to provide more effective flood hazard reduction solutions. Thus, he adds, sociological, political, and environmental considerations are receiving increased recognition alongside technical aspects in order to provide effective flood mitigation alternatives for the community.

The United States Army Corps of Engineers (USACE) has done extensive work in flood mitigation control. The report "Flood proofing techniques, programs, and references," prepared for the USACE, presents a comprehensive review of flood proofing techniques (USACE, 1997).

Hurricanes

Hurricanes represent a major threat to areas in the United States along the east coast and the Gulf of Mexico. Although hurricanes are considered rare events (e.g., the probability that a particular 50-mile segment of coastal area along the U.S. Gulf Coast will be hit by a major hurricane in any given year is very low, ranging from close to 0.0% to 4.0%; Petak, 1982), nonetheless when a major hurricane makes landfall the results can be devastating. Hurricane Andrew, which impacted Florida and Louisiana in 1992, was considered one of the most costly disasters in U.S. history with economic losses estimated at almost $30 billion dollars (Jarell, 2001). In 2004, Florida and parts of the southeastern U.S. were impacted by four hurricanes in a period of six weeks. Hurricanes Charley, Frances, Ivan and Jeanne caused dozens of deaths, left thousands of people homeless, and knocked out power service for millions of people. These hurricanes were accompanied by high winds, torrential rainfall, storm surge and flooding, and hurricane-spawned tornadoes. According to the Federal Emergency Management Agency (FEMA, 2005) the 2004 Atlantic Hurricane Season was one of the busiest and most destructive in U.S. history (Petak and Atkisson,1982).

Engineering hazard reduction measures can be taken to minimize the loss of life and property during hurricanes. One of the main threats of a hurricane is the high wind speed. Engineering design codes are used to insure that buildings and structures are constructed to withstand particular wind speeds depending on the climatic characteristics of each region. The design wind speeds have been updated over the years, and in general the new codes require the use of higher design wind speeds (Cruz et al., 2001).

In the United States, ASCE provides the guidelines for the design and calculation of wind loads in the design standard ASCE 7 "Minimum Design Loads for Buildings and Other Structures" (ASCE 7, 1998). Theses provisions have been incorporated into other buildings codes (e.g., Uniform Building Code) (Cruz et al., 2001).

Heaney et al. (2000) point out that too often hurricane damage to residential structures is due to failure of roofing materials, doors and windows, and that these failures lead to weather penetration and damage. They note that during Hurricane Andrew, most damage losses in buildings were due to penetration of the weather envelope and not by the failure of major structural components. ASCE 7 has incorporated provisions for protecting the building envelope.

Droughts

Recent droughts in the United States since 1995 have demonstrated the vulnerability of the country to droughts despite improvements in weather forecasting and the development of new tools and technologies (Hayes et al., 2004). The U.S. Army Corps of Engineers evaluated the impacts of droughts in the United States (Dziegielewski et al., 1991). FEMA's "National Mitigation Strategy" report published in 1995 estimated drought losses as high as $8 billion (FEMA, 1995).

Droughts are slow-onset and relatively long-lasting events, sometimes making it difficult to determine when a drought begins. Droughts involve issues related to the supply and demand of water resources (Hayes et al., 2004). Therefore, drought hazard mitigation will involve both physical and social issues. Due to their complexity, there is no single universal remedy against these water-related extremes (Budhakooncharoen, 2003). Budhakooncharoen (2003) observes the need for a holistic approach involving applications of sustainable integrated water resources and comprehensive risk management. Therefore, both engineering and non-engineering approaches are needed for appropriate drought mitigation and risk management. Hayes et al. (2004) based on Wilhite (1997) and Wilhite and Vanyarkho (2000) present nine categories of state government actions in the U.S. for drought mitigation. Some of these include assessment programs, water supply augmentation and development of new supplies, technical assistance on water conservation and other water-related activities, and demand reduction/water conservation programs.

Landslides

Landslides and other ground-failure problems affect all 50 States and U.S. Territories. 36 States have moderate to highly severe landslide hazards (Spiker and Gori, 2003). Landslides are responsible for substantial human and economic losses in the United States. It is estimated that every year landslides cause 25 to 50 deaths, and cost between $1 and $3 billion in economic losses (National Research Council, 2004).

Landslides often accompany other natural hazards such as earthquakes, floods, hurricanes, and volcanic eruptions. The Northridge earthquake in 1994 triggered more than 11,000 landslides to the North and North West of the epicenter in the Santa Susana Mountains and the mountains north of the Santa Clara River valley (Jibson, 2002).

Hurricane Mitch in Central America in 1998 resulted in deadly landslides, which caused the majority of fatalities (Spiker and Gori, 2003).

The U.S. Geological Survey (USGS) has taken the lead in developing a national landslide hazard mitigation strategy. Spiker and Gori (2003) outlined the key elements of this strategy which include: (1) research; (2) hazard mapping; (3) real-time monitoring of active landslides; (4) loss assessment to determine potential impacts; (5) information collection, interpretation and dissemination; (6) guidelines and training of scientists, engineers, and decision makers; (7) public awareness and education; (8) implementation of loss reduction measures; and (9) emergency management.

As with other natural hazards, landslide hazard reduction includes both engineering and non-engineering measures. Engineering mitigation measures used include construction of earth-retaining structures, construction of surface water drainage systems, slope surface protection such as hydro-seeding, sprayed concrete and reinforced concrete grids, and recompaction of fill slopes (Kwong et al., 2004). Tunnels, although expensive, usually prove to be cost-effective in the long-term to avoid landslide hazard in transportation routes with slope problems (Bhasin et al., 2001). Spiker and Gori (2003) observe the need to establish standardized codes for excavation, construction and grading in landslide prone areas, as there is no nationwide standardization.

Fires Associated With Disasters

Natural and technological disasters can cause secondary events such as fires. Fires following earthquakes have caused the largest single losses due to earthquakes in the United States and Japan (Scawthorn et al., 1986; Della Corte et al., 2003). The California Seismic Safety Commission (CSSC) (ASCE-25 2002) reported approximately 110 earthquake-related fire ignitions due to gas pipeline breaks during the Northridge earthquake, and Menoni (2001) reported the destruction of almost 7000 buildings due to fire following the Kobe earthquake in Japan in 1995. The Turkey earthquake in August 1999 triggered multiple fires at one of Turkey's largest oil refineries (Steinberg and Cruz, 2004), and the recent Tokachi-oki earthquake in Japan in 2003 triggered a major fire in the oil storage farm of an oil refinery (Kurita, 2004).

Heaney at al. (2000) report that approximately 18 deaths and $180 million in economic damages are attributable annually to fires caused by natural disasters. A percentage of these losses result from damage to the natural and built environment caused by wildfires. Engineering approaches for fire hazard mitigation may include measures to prevent or delay fire ignition and fire spread, and to improve fire suppression (Zaghwl and Dong, 1994). Approaches for mitigation of losses caused by wildfires are presented in the Standard for Protection of Life and Property from Wildfire (National Fire Protection Association, 2002) and the Urban-Wildfire Interface Code (International Code Council, 2003).

ENGINEERING CONTRIBUTION TO HAZARD REDUCTION BY TYPE OF INFRASTRUCTURE

Natural and technological hazards often impact buildings and other structures. Typical engineering taxonomy divides infrastructure into buildings and lifeline systems (e.g., bridges, pipelines) (Heaney et al., 2000). In this section, the hazards and contribution of engineering to hazard reduction in these systems is presented. It is important to note that much of the research concerning natural hazard impacts on infrastructure has been in the area of earthquake hazard reduction particularly in California. Nevertheless, the lessons learned from earthquake hazard reduction can often be implemented for other types of natural hazards.

Buildings

Buildings are affected by floods, high winds, soil problems, snow, fires and earthquakes. In addition, Heaney et al. (2000) remark that because buildings are complex combinations of the foundation and structure, and the plumbing, electrical, heating, ventilation, air conditioning, and ancillary systems they may suffer damage when one or a combination of these systems fail. The following of appropriate buildings codes (e.g., Unified Building Code) can reduce the potential damage caused by natural hazards. However, there are instances when non-engineering hazard reduction alternatives can be more effective or less costly. The emergency manager must pay special attention to areas that are at greater risk (e.g., residences built in flood plains, constructions on or near known earthquake faults), and to older buildings, which have not been retrofitted to newer codes.

Bridges and Roadways

Bridges and roadways can be affected by floods, high winds, soil problems and earthquakes. Liquefaction, ground settlement, and slope instability can cause extensive damage to bridges and elevated highways during earthquakes, and other landmass movements.

Engineers work to improve the structural integrity and performance of bridges and roadways. There has been extensive research of potential damage to these lifeline systems during earthquakes. The Applied Technology Council (ATC) report "ATC-25" (ATC 1991) assigns 1, 2, 8, and 20 percent damage, for earthquake MMI[1] levels of VII, VIII, IX, and X, respectively, for non-upgraded major bridges in California. Damage to conventional bridges for the same MMI levels respectively, are 3, 10, 25, and 80 percent.

MMI – Modified Mercalli Intensity scale was developed in 1931 by the American seismologists Harry Wood and Frank Neumann. The scale is composed of 12 increasing levels of intensity that range from imperceptible shaking (I) to catastrophic destruction (XII) (Erdik, 1998). Erdik notes that the Northridge earthquake caused heavy damage to 10 viaducts and 157 overpasses. In addition, collapse and other damage (to bridges) resulted in the closing of 11 major roads in downtown Los Angeles.

Damage to transportation systems has often resulted in traffic congestion, and longer travel times delaying the arrival of emergency response teams and supplies. Damage to transportation systems has sometimes resulted in the complete isolation of whole communities. Identifying reliable transportation network designs that take into account accessibility/congestion, and dispersion/concentration of road networks is an area that is receiving increased attention (see Asakura and Kashiwadani, 2001; McFarland and Chang, 2001; Sakakibara, Kajitani, and Okada, 2001).

Underground Pipelines

Underground pipelines can be affected by earthquakes, poor ground conditions, liquefaction, flooding, storm surge, erosion and landslides. Experience

1. MMI – Modified Mercalli Intensity scale was developed in 1931 by the American seismologists Harry Wood and Frank Neumann. The scale is composed of 12 increasing levels of intensity that range from imperceptible shaking (I) to catastrophic destruction (XII).

from earthquakes around the world indicates that underground pipeline damage occurs in areas of fault rupture, liquefaction, and poor unstable ground (Erdik, 1998). Earthquakes have caused extensive damage to gas, water and wastewater and oil pipelines. Damage to gas pipelines can result in leaks, fires and explosions (Lau et al., 1995).

Engineers use field data from past disasters to estimate potential future damage to pipeline systems. Based on world wide data, Erdik reports that about 0.5–1 gas pipe breaks per one kilometer pipe occur during shaking intensity level VIII, depending on soil and pipe conditions. Rates can increase about 50 percent in shaking intensity level IX. The California Seismic Safety Commission (CSSC) (ASCE-25 2002) reported 35 gas system failures in older transmission lines, 123 failures of steel distribution mains, 117 failures in service lines, and 394 corrosion related leaks following the Northridge earthquake, an earthquake that has been considered mild with respect to future earthquakes that can be expected in the region.

It is estimated that in California water distribution lines can suffer 0.5, 1, 4, and 12 pipe breaks per kilometer, respectively, for MMI shaking values of VII, VIII, IX, and X according to ATC-25 (Erdik, 1998). Erdik observes that about half of these damage rates are applicable to gas lines, while about double these rates are applicable to sanitary sewer lines. Damage to water distribution lines was a major problem for fire protection following the Northridge earthquake (City Administrative Officer, 1994). Erdik (1998) reports over 2,000 water line breaks during the Kobe earthquake, having a negative effect on fire-fighting capabilities. Steinberg and Cruz (2004) reported that damage to the main water pipeline, which provided service to several industrial facilities in Korfez, severely hampered emergency response to the multiple earthquake-triggered fires at Turkey's largest oil refinery following the Kocaeli earthquake.

Ports and Marine Terminals

Ports and marine terminals are susceptible to hurricane winds and storm surge (Hanstrum and Holland, 1992). For example, several ports in Central America were severely affected by Hurricane Georges in 1998 (Beam et al., 1999). Protection of ports and harbors from wave action and storm surge may include natural or man-made breakwaters and surge barriers.

Ports and marine terminals are also affected by earthquakes and tsunamis, and liquefaction and soil stability problems during earthquakes (Tang, 2000; Erdik, 1998). Tang (2000) reported that ground shaking, settlement, and lateral displacement caused damage to port facilities on both the south and north shores of Izmit Bay following the Kocaeli earthquake. Erdik (1998) reported that widespread liquefaction and permanent ground deformation devastated the Port of Kobe, Japan, damaging more than 90 percent of the port's berths.

Damage to ports can have a severe economic effect on a region, as occurred following the Kobe earthquake, cutting Kobe off from the rest of Japan and the outside world (Cataldo, 1995). In addition, damage to port terminals of industrial facilities may result in spills at loading docks, such as occurred at Turkey's largest oil refinery. Several naphtha and LPG spills into Izmit Bay from broken loading arms at the oil refinery were reported following the Kocaeli earthquake (Steinberg and Cruz, 2004). The American Society of Civil Engineers' Ports and Harbors Committee has developed planning and design guidelines for small harbors (Sorensen et al., 1992), and the U.S. Army Corps of Engineers has done research concerning design and redevelopment of ports and harbors (Lillycrop et al., 1991).

Electrical Power Systems

Electrical power systems are highly susceptible to natural hazards. Damage to power systems can severely hamper emergency response capabilities. Power outages have been reported during most major hurricanes. Similarly, most major earthquakes have resulted in electrical power outages of varying lengths. Damage to electrical power systems during hurricanes is often caused by weather penetration of power stations and by toppling of transformers and electrical power lines and posts.

The most vulnerable components during earthquakes include generators and transformers, with damage often occurring due to improperly anchored equipment (Erdik, 1998). Indirect damage to electrical power lines and poles caused by building collapse can also be extensive, as was documented by Tang

(2000) following the Kocaeli earthquake. Potential damage values during earthquakes have been estimated. ATC-25 (ATC, 1991) assigns 16, 26, 42 and 70 percent damage values for non-upgraded electric transmission substations, respectively, for MMI levels of VII, VIII, IX, and X. For distribution substations the respective damage values are 8, 13, 25, and 52 percent (Erdik, 1998).

Engineers work to find ways to avoid or minimize disruption of electrical power systems during natural disasters. However, as the damage values above indicate, damage to these systems during a natural disaster event may be unavoidable. Thus, research efforts also involve developing methodologies and strategies to quickly repair and restore electrical power service.

ENGINEERING RESEARCH NEEDS

Heaney et al. (2000) present a comprehensive review of engineering research needs concerning codes and standards, and engineering research needs by type of hazard and type of infrastructure. The authors note that their results are highly influenced by the areas with current funding, which provide the resources to compile this information. The area with the highest research-funding budget is earthquakes ($13 million/year), followed by floods and hurricanes (each with less than $1 million/year).

Improvement, particularly when incorporating concerns beyond life safety issues, is needed in the area of codes and standards (Heaney et al., 2000). As society's perception of acceptable risk shifts, and as individuals and communities suffer ever-greater losses from disasters, the willingness to pay the price for stricter engineering codes and standards is also likely to increase.

Determining acceptable risk is an important issue in the setting of codes and standards. Derby and Keeney (1981) note that determining acceptable risk involves choosing the best combination of advantages and disadvantages from among several alternatives. What constitutes advantages and disadvantages will vary depending on the individuals or organizations involved in the decision-making process. Thus, to insure that codes and standards are in line with public values, those in charge of setting engineering codes and standards must work towards a more equitable participation of all sectors (actors and stakeholders) of the population. In this context, the need for better integration of scientific input from the various disciplines with the views of all stakeholders and actors involved is also essential (Heaney et al., 2000).

It was noted that there is still a large gap between the setting of codes and standards and their actual adoption and implementation. This represents a major challenge for local government officials and emergency managers, as well as for the scientific and engineering community. Development and evaluation of more cost-effective mitigation options, such as cheaper construction materials and innovative construction practices that still provide the desired levels of safety may encourage more businesses and home owners to adopt codes or retrofit older buildings. In developing countries, where economic resources are scarce, there have been several efforts to develop low cost, locally-based repair and retrofitting techniques for non-engineered, rural structures (see for example Asociacion Salvadoreña de Ingenieros y Arquitectos, http://www.asia.org.sv/).

The need for multihazard approaches to disaster management is increasingly called for. Heaney et al. (2000) note the need for more formal multihazard evaluation methodologies to assess the relative importance of the various hazards (e.g., earthquake, wind). To aid in this effort, the National Institute of Building Sciences (NIBS) established the Multihazard Mitigation Council (MMC) which works to reduce losses associated with natural and other hazards by promoting improved multihazard risk mitigation strategies, guidelines, and practices.

Further research is needed concerning the costs of disasters and the benefits of hazard mitigation (Heaney et al., 2000). There is relatively abundant data on deaths and injuries, and on losses caused by damage to buildings and infrastructure. However, there is limited data on the benefits obtained (cost not incurred) when a mitigation measure is effective. One effort currently underway at the MMC involves evaluating the data requirements and identifying possible methodologies to assess the benefits of hazard mitigation. Software programs such as FEMA's HAZUS-MH (http://www.fema.gov/hazus/) loss estimation methodology can be used to estimate potential damage and economic losses from earthquakes, floods and high winds. In addition to estimating economic losses due to natural disasters, HAZUS-MH can be used for emergency preparedness planning purposes as it provides estimates of number of possible deaths and injuries, as well as estimates of number of displaced households and shelter needs. Field et al. (2005) used HAZUS-MH to estimate potential losses caused by earthquakes of varying magnitudes along the Puente Hills blind-thrust fault beneath downtown Los Angeles. Their study points out the significant risk posed by this fault and other seismic sources in the region.

There is a need for more research concerning sustainable prevention measures and management of natural hazards. Disasters are complex events, which result from a combination of factors including urbanization, population growth and environmental degradation. Budhakooncharoen (2004) observes the need for more integrated disaster management that reduces human vulnerability to disasters, avoids past mistakes and satisfies a wide range of needs through sustainable hazard mitigation practices.

A research area that has received increased attention is the prevention of infrastructure failures and other technological disasters resulting as secondary effects of a natural or other large-scale disaster event. In 1996, the President's Commission on Critical Infrastructure Protection (PCCIP) was established to advance the understanding of the role of critical infrastructure systems in large-scale disasters. Infrastructure systems such as electric power, water, and telecommunications are becoming more and more interdependent. Thus, infrastructure failures in one system have the potential to "cascade" onto other systems, thereby severely compounding disruptions to society (McDaniels, 2005). The recent large-scale blackout in the Northeast of the United States in August 2003, and blackout that started in Switzerland and affected almost all of Italy in September 2003 are examples of how a single significant event can cause widespread disruption.

Natural disasters have the potential to cause other type of secondary disasters such as hazmat releases. Cruz et al. (2001) studied impacts of tropical cyclones on an oil refinery. The authors recommended identifying and evaluating methodologies to quantify the risks associated with natural hazard-triggered hazmat releases at these and other industrial plants. Steinberg and Cruz (2004) and Cruz and Steinberg (2005) have studied hazmat releases triggered by earthquakes. The authors provide a review of research needs concerning prevention of, preparedness for and response to these conjoint natural and technological (natech) disasters. Cruz et al. (2004) present the state of the art in risk management of conjoint natural and technological disasters in Europe. Future engineering research needs concerning natechs include assessment of the potential impacts of external hazards (e.g., earthquakes, floods) on both structural and non-structural components of industrial plants that use or handle hazardous chemicals, and the need to develop probabilistic hazard maps depicting areas where these conjoint events are most likely.

CONCLUSION

This chapter has reviewed the contribution of engineering to the field of emergency management. The contribution of engineering has been important in the setting of engineering codes and standards, and in the development of engineering resources, tools, and methodologies for use in mitigating the impacts of natural and technological hazards on the built environment. However, engineered-hazard mitigation options alone do not guarantee protection from natural and other hazards. Therefore, a holistic multihazard perspective that integrates social, economic, and environmental issues to hazard reduction is preferred. The engineering professionals, who contribute to hazard reduction, will be increasingly required to work across disciplines, and with many actors and stakeholders.

Engineering has contributed to our overall understanding of natural hazards and their impacts and the vulnerability of the built environment to these hazards. Improved understanding of natural hazards results in better forecasting of natural hazards, and more effective disaster prevention and mitigation practices and preparedness planning. One such example is HAZUS-MH, which not only provides estimates of potential damage and economic losses from natural disasters, but also provides useful data for emergency preparedness and response planning.

Engineering will continue to contribute to hazard reduction as cities become ever more complex and interdependent, and as new threats emerge (e.g., impacts of climate change, water scarcity, terrorism). Engineers will be required to apply new knowledge and skills to develop innovative and effective ways to prevent, prepare for, and respond to future disasters.

REFERENCES

Asakura, Y., and Kashiwadani, M. (2001). "Reliable Transport Network Design under Traffic Congestion." Proceedings of the Japan-US Workshop on Disaster Risk: Management for Urban Infrastructure Systems, Kyoto, Japan, 15–16 May.

ASCE-7 (1998). *Minimum Design Loads for Buildings and other Structures.* American Society of Civil Engineers.

ASCE-25 (2002). Improving Natural Gas Safety in Earthquakes. Draft Final Report. ASCE-25 Task Committee on Earthquake Safety Issues for Gas Systems, California Seismic Safety Commission, March.

ATC (1991). Seismic Vulnerability and Impact of Disruption of Lifelines in the Conterminous United States. Report ATC-25, Applied Technology Council, Redwood City, California.

Beam, A. R., de Caceres, L., and Moroney, M.J. Jr. (1999). "Restoration of maritime navigation systems in Central American Ports." *Oceans Conference Record* (IEEE), Vol. 3, p. 1317.

Bhasin, R., Grimstad, E., Larsen, J.O., Dhawan, A.K., Singh, R., Verma, S.K., and Venkatachalam, K. (2001). "Landslide Hazards and Mitigation Measures at Gangtok, Sikkim Himalaya." *Engineering Geology, 64,* 351–368.

BSSC (2003). "Research Needs Identified during Preparation of the 2003 Edition of the NEHRP Recommended Provisions for Seismic Regulations for New Buildings and Other Structures (Fema 450)." Building Seismic Safety Council, National Institute of Building Sciences, (May 2005) < http://www.bssconline.org/Research Needs.pdf>

Budhakooncharoen, S. (2003). "Floods and Droughts: Sustainable Prevention and Management." *Advances in Ecological Sciences, 19,* 1185–1194.

California Earthquake Project (2001). Scenario for a Catastrophic Earthquake on the Newport-Inglewood Fault. Joint Initiative of the Federal Emergency Management Agency and the California Office of Emergency Services, May.

Cataldo, A. (1995). "Japan Industry Weighing Kobe Earthquake Impact: Infrastructure Damage May Pose Problem." *Electronic News, 41* (2049), Jan. 23, p (2).

Channell, D.F. (1989). The History of Engineering Science: An Annotated Bibliography . Bibliographies in the History of Science and Technology. Garland Publishing, New York/London. 311 pp.

City Administrative Officer (1994). *City of Los Angeles Northridge Earthquake After-Action Report.* Report presented to the Emergency Operations Board, City of Los Angeles, CA, June 3.

Cruz, A. M., Steinberg, S.K., Vetere-Arellano, A.L., Nordvik, J.P., and Pisano, F. (2004). State of the Art in Natech (Natural Hazard Triggering Technological Disasters) *Risk Assessment in Europe.* Report EUR 21292 EN, DG Joint Research Centre, European Commission and United Nations International Strategy for Disaster Reduction, Ispra, Italy.

Cruz, A.M. and Steinberg, L.J. (2005). "Industry Preparedness for Earthquakes and Earthquake-Triggered Hazmat Accidents During the Kocaeli Earthquake in 1999: A Survey." *Earthquake Spectra,* May.

Cruz, A.M. (2003). Joint Natural and Technological Disasters: Assessment of Natural Disaster Impacts on Industrial Facilities in Highly Urbanized Areas. Dissertation. Tulane University, New Orleans, LA., p. 212. Pilgrim 1991 in manuscript – 1999 in references.

Cruz, A.M., Steinberg, L.J. and Luna, R. (2001). "Identifying Hurricane-Induced Hazardous Materials Release Scenarios in a Petroleum Refinery," *Natural Hazards Review, 2*(4), pp 203–210.

Dawkins, W. (1995). "Corporate Japan shakes in aftershock of quake – Some companies' losses from the Kobe tragedy may prove competitors' gains." *Financial Times* (London, England), International Company News, Feb. 24, p. 30

Della Corte, G., Landolfo, R., and Mazzolani, F.M. (2003). "Post-Earthquake Fire Resistance of Moment Resisting Steel Frames." *Fire Safety Journal, 38*(7), 593–612.

Derby, S. L., and Keeney, R. L. (1981). "Risk Analysis: Understanding 'How Safe is Safe Enough?'" *Risk Analysis, 1*(3), 217–224.

Dziegielewski, B. et al. (1991). "National Study of Water Management During Drought: A Research Assessment." IWR Rep. 91-NDS03, Institute of Water Resources, U.S. Army Corps of Engineers, Fort Belvoir, VA.

Eguchi, R.T., and Honegger, D.G. (2000). "Lifelines: Ensuring the Reliability of Water Supply Systems During Natural Disasters." *Standardization News, 30*(8), 24–27.

Erdik, M. (1998). "Seismic Vulnerability of Megacities." In *Seismic Design Practice into the Next Century: Research and Application.* Booth, E. (ed.). Rotterdam: Balkema.

FEMA (1995). *National Mitigation Strategy: Partnerships for Building Safer Communities.* Washington, D.C.

FEMA (2003). 2003 NEHRP Recommended Provisions for Seismic Regulations for New Buildings and Other

Structures. FEMA 450, Federal Emergency Management Agency, Washington, D.C.

FEMA (2005). 2004 Hurricane Season. May 6, 2005, URL: http://www.fema.gov/press/2004/hurricane_season.shtm#recovery_information

Field, E.H., Seligson, H.A., Gupta, N., Gupta, V., Jordan, T.H., and Campbell, K.W. (2005). "Loss Estimation for the Puente Hills Blind-Thrust Earthquake in Los Angeles, California." *Earthquake Spectra, 21*(2), 329–338.

Godschalk, D.R., Beatley, T., Berke, P., Brower, D.J. and Kaiser, E.J. (1999). *Natural Hazard Mitigation: Recasting Disaster Policy and Planning.* Island Press, Washington, D.C.

Hanstrum, B. N., and Holland, G.J. (1992). "Effects on Ports and Harbours of Tropical Cyclone Storm Surges: A Case Study at Port Hedland, Western Australia." National Conference Publication – Institution of Engineers, Australia, No. 82, Part 8, 201-203.

Hayes, M.J., Wilhelmi, O.V. and Knutson, C.L. (2004). "Reducing Drought Risk: Bridging Theory and Practice." *Natural Hazards Review, 5*(2), 106–113.

Heaney, J. P., Petarka, J. and Wright, L.T. (2000). "Research Needs for Engineering Aspects of Natural Disasters." *Journal of Infrastructure Systems, 6*(1), 4–14.

International Code Council (2003). *Urban-Wildfire Interface Code.* 2003 Edition. International Code Council, Whittier, California.

Jarrell, J.D., Mayfield, M., Rappaport, E.N., Landsea, C.W. (2001). The Deadliest, Costliest, and Most Intense United States Hurricanes from 1900 to 2000. NOAA Technical Memorandum NWS TPC-1, Miami, Florida.

Jibson, R.W. (2002). "A Public Health Issue Related to Collateral Seismic Hazards: The Valley Fever Outbreak Triggered by the 1994 Northridge, California Earthquake." *Surveys in Geophysics, 23*, 511–528.

Keeney, R.L. (1983). "Issues in Evaluating Standards." *Interfaces* (Providence, Rhode Island), *13*(2), 12–22.

Kranz, P. (1999). "Aftershocks in Taiwan." *News Week*, International Outlook, Global Wrap-up, Oct. 4, Number 3649, p. 62.

Kurita, T. (2004). "Observation of the Recent Earthquake Damage in Japan." In Vetere-Arellano, A.L., Cruz, A.M., Nordvik, J.P., and Pisano, F. (Eds.), Proceedings of the NEDIES International Workshop on Natech (Natural Disaster-Triggered Technological Disasters) Disaster Management, Report EUR 21054 EN, Ispra, Italy, 20–21 October.

Kwong, A.K.L., Wang, M., Lee, C.F. and Law, K.T. (2004). "A Review of Landslide Problems and Mitigation Measures in Chongqing and Hong Kong: Similarities and Differences." *Engineering Geology, 76*, 27–39.

Lau, D.L., Tang, A. and Pierre, J.R. (1995). "Performance of Lifelines During the 1994 Northridge Earthquake." *Canadian Journal of Civil Engineering, 22*(2), 438–451.

Lillycrop, W.J., Ippolito, M.A., Bottin, R.R. Jr. (1991). "Development of shallow draft harbor design criteria." Coastal Zone: Proceedings of the Symposium on Coastal and Ocean Management, Long Beach, California, Vol. 1, 584–591.

McDaniels, T. (2005). "Managing Infrastructure Interactions in Disasters." Presented at Kyoto University's Disaster Prevention Research Forum – 21st Century Center of Excellence Program, Kyoto, Japan, 9–13 May.

McFarland, K. and Chang, S.E. (2001). "A Methodology for Incorporating Urban Travel Demand Models into Mitigation Analysis." Proceedings of the Japan–US Workshop on Disaster Risk: Management for Urban Infrastructure Systems, Kyoto, Japan, 15-16 May.

Menoni, S. (2001). "Chains of Damages and Failures in a Metropolitan Environment: Some Observations on the Kobe Earthquake in 1995." *Journal of Hazardous Materials, 86*, 101–119.

National Research Council (2004). Parternerships for Reducing Landslide Risk: Assessment of the National Landslide Hazards Mitigation Strategy. Committee on the Review of the National Landslide Hazards Mitigation Strategy, Board on Earth Sciences and Resources, Division on Earth and Life Studies, National Research Council, The National Academies Press, Washington, D.C.

National Fire Protection Association (2002). NFPA 1144 Standard for Protection of Life and Property from Wildfire. 2002 Edition, National Fire Protection Association, 19 pp.

Olsen, J.R.; Lambert, J.H. and Haimes, Y.Y. (1998). "Risk of Extreme Events Under Nonstationary Conditions." *Risk Analysis, 18*(4), 497–510.

Petak, W.J. and Atkisson, A.A. (1982). *Natural Hazard Risk Assessment and Public Policy.* Springer Verlag, New York, pp. 322–353.

Pilgrim, D.H. (1999). "Flood Mitigation. Some Engineering Perspectives." National Conference Publication – Institution of Engineers, Australia, 1(91), Part 1, 205–217.

Rojahn, C.; Scawthorn, C. and Khater, M. (1992). "Transportation Lifeline Losses in Large Eastern Earthquakes." In Lifeline Earthquake Engineering in Central and Eastern U.S.: Proceedings of Three Sessions Sponsored by the Technical Council on Lifeline Earthquake Engineering, November, 87–101.

Sakakibara, H., Kajitani, Y., and Okada, N. (2001). "Comparative Study on the Structure of Highway Network by Use of Standardized Topological Index." Proceedings of the Japan–US Workshop on Disaster Risk: Management for Urban Infrastructure Systems, Kyoto, Japan, 15–16 May.

Scawthorn, C., Blackburn, F.T., and Seagraves, D.W. (1986). "Earthquakes, Fires and Urban Areas: All the Ingredients for a Disaster." *Fire Journal* (Boston), *82*(4), 5p.

Sorensen, P. H., Wortley, C.A., Hunt, F.G., Tobiasson, B.O., Childs, K.M., and Forster, C.G. (1992). "Planning and Design Guidelines for Small Craft Harbors." Ports'92, Jul 20–22, Seattle, WA.

Spiker, E.C., and P.L. Gori. (2003). "National Landslides Hazards Mitigation Strategy: A Framewok for Loss Reduction." *US Geological Survey Circular*, Issue *1244*, 1–54.

Steinberg, L.J., and Cruz, A.M. (2004). "When natural and technological disasters collide: lessons from the Turkey earthquake of August 17, 1999." *Natural Hazards Review, 5*(3), 121–130.

Tang, A.K. (Editor) (2000). "Izmit (Kocaeli) Earthquake of August 17, 1999 including Duzce Earthquake of November12, 1999: Lifeline Performance." Technical Council on Lifeline Earthquake Engineering, Monograph No. 17, ASCE, Virginia.

USACE (1997). "Flood proofing techniques, programs, and references" prepared by Dewberry and Davis with French & Associates, LTD., US Army Corps of Engineers National Flood Proofing Committee, Washington, D.C.

U.S. Geological Survey (2000). "Implications for Earthquake Risk Reduction in the United States from the Kocaeli, Turkey, Earthquake of August 17, 1999." U.S. Geological Survey Circular 1193. U.S. Geological Survey, United States Government Printing Office.

Wilhite, D.A. (1997). "State Actions to Mitigate Drought: Lessons Learned." *Journal of the American Water Resource Association, 33*(5), 951–959.

Wilhite, D.A., and Vanyarkho, O. (2000). "Drought: Pervasive Impacts of a Creeping Phenomena." In *Drought: A Global Assessment*, Vol. 1, D. A. Wilhite, ed., Routledge, New York, 245–255.

Zaghwl, A., and Dong, W.M. (1994). "A Neural Network Approach for Fire Following Earthquake Loss Estimation." IEEE International Conference on Neural Networks, Conference Proceedings, Vol. 4, 3271–3276.

Zimmerman, R., and Sparrow, R. (1997). "Workshop on Integrated Research for Civil Infrastructure." Final Report to the National Science Foundation, Washington, D.C.

Chapter 5

Sociology, Disasters and Emergency Management: History, Contributions, and Future Agenda[1]

Thomas E. Drabek

ABSTRACT

This chapter will summarize the contributions of sociologists to the study of disasters and the profession of emergency management. While some non-U.S.A. references will be made, most of the analysis will be limited to studies conducted within the U.S.A. by American scholars. The essay is divided into five sections: (1) history, including key literature reviews, definitions and issues of controversy; (2) major contributions to the knowledge base; (3) key points of overlap with other disciplines; (4) recommendations for emergency managers; and (5) future research agenda.

INTRODUCTION

Disasters have long been objects of study by sociologists. Indeed, prior to the 1980s the research literature was dominated by sociologically oriented analyses, followed by that of social geographers, e.g., Burton, Kates and White (1993). Given this rich and expansive legacy, this chapter will be limited to highlights, not detail. Readers desiring additional depth are advised to review the works referenced throughout. It should be noted that many important sociological contributions have been made by scholars researching disasters that occurred outside the U.S.A. Some of special importance are noted in this chapter, but most are not. The chapter is divided into five sections: (1) history, including key literature reviews, definitions and issues of controversy; (2) major contributions to the knowledge base; (3) key points of overlap with other disciplines; (4) recommendations for emergency managers; and (5) future research agenda.

HISTORY

While there are many definitions of sociology, most would agree that the focus of the discipline is the study of human interaction. Hence, when disaster strikes, sociologists have asked, "how do humans respond?" From the outset, starting with Prince's (1920) initial study of the collision of two ships in the Halifax harbor (December 6, 1917), this has been the key question that defined the sociological research agenda. The fundamental epistemological assumption was that while all disaster events were unique historic episodes, comparative analyses could identify elements of commonality, i.e., modal patterns of behavior. Literature reviews have summarized studies of individuals and their social units, ranging from families, to organizations to communities, e.g., Barton, 1969; Dynes, 1970; Quarantelli and Dynes, 1977; Kreps, 1984; Drabek, 1986). More recently, under the auspices of the FEMA Higher Education Project, Drabek (1996b, 2004) prepared detailed literature

1. I wish to thank Ruth A. Drabek for her assistance in the preparation of this chapter.

summaries for instructors of courses focused on the social dimensions of disaster. Collectively, these numerous synthesizing statements integrate the research conclusions from hundreds of post-disaster field studies. While preparedness and mitigation activities have been studied, the total aggregate of such inquires, like those examining "root causes" of disaster, pale in comparison to the number of post-event assessments (e.g., preparedness studies include Quarantelli, 1984; mitigation studies include Drabek et al., 1983; for assessments of "root causes" see Enarson et al., 2003).

Sociologists have argued that disasters may expose the key values and structures that define communities and the societies they comprise. Social factors that encourage both stability and change may thereby be documented. Thus, both core behavior patterns and the social factors that constrain them may be illuminated by the study of disaster. And while cultural differences may be associated with substantial variations in response, cultural similarities have been documented by those comparing the U.S.A. profile to responses by the British (e.g., Parker, 2000), Australians (e.g., Britton and Clapham, 1991) and others (e.g., Parr on New Zealand, 1997–1998; Domborsky and Schorr on Germany, 1986). In contrast, results from the former Soviet Union (Portiriev, 1998b), Japan (Yamamoto and Quarantelli, 1982), Italy (Quarantelli and Pelanda, 1989) and elsewhere (e.g., Bates and Peacock, 1992; Oliver-Smith and Hoffman, 1999) have documented the role of culture in pattern variation.

Typically, sociologists have differentiated *disasters* from *hazards*. Following most, for example, Drabek (2004) defined these terms as follows. A disaster is ". . . an event in which a community undergoes severe such losses to persons and/or property that the resources available within the community are severely taxed" (Drabek, 2004, Student Handout 2-1, p. 1). This conceptualization is consistent with these proposed or implied by the earliest research teams, e.g., Fritz, 1961; Dynes, 1970. In contrast, a hazard is ". . . a condition with the potential for harm to the community or environment." (Drabek, 2004, Student Handout 2-1, p. 1). For sociologists, the term *disasters* referred to specific events like Hurricane Jeanne (2004) whereas *hazards* define a class of threats like hurricanes, tornadoes, earthquakes, and so on. Thus, they refer to the hurricane hazard or the tornado hazard that reflects the risk, vulnerability, or exposure confronting families, communities or societies.

Flowing from these definitions, most sociologists view emergency management as ". . . the process by which the uncertainties that exist in potentially hazardous situations can be minimized and public safety maximized. The goal is to limit the costs of emergencies or disasters through the implementation of a series of strategies and tactics reflecting the full life cycle of disaster, i.e., preparedness, response, recovery, and mitigation" (Drabek, 2004, Student Handout 1-3, p. 1).

These terms have provided an important frame of reference for dozens of scholars who have sought to use the perspectives, concepts, and methods that define the broad field of sociology in their study of disaster. These applications have been nurtured by major research centers, most notably the Disaster Research Center. Since its founding at The Ohio State University in 1963, this unit has encouraged, integrated, and applied these tools to the study of disaster. After its relocation to the University of Delaware in 1985, the process of rapid arrival to disaster scenes continued. Implementation of a "quick response" funding process that was coordinated through the Natural Hazards Research and Applications Information Center at the University of Colorado has enabled dozens of scholars to gather perishable materials. At times these quick response field visits have facilitated larger and more focused studies. Important policy insights and recommendations have been proposed to emergency management professionals following such work.

Over time, however, key issues and concerns have precipitated much debate. Among these, two are most fundamental, and clearly are pushing alternative research agenda in very different directions. These issues reflect: (1) different definitions of the term "disaster"; and (2) degree of focus on vulnerability and/or risk-based paradigms.

Clearly there are basic and very real differences in viewpoints as to how the core concept of "disaster" ought to be defined. To some, like Murria (2004) the matter may best be pursued by an engineer or other non-sociologically oriented professional. So by comparing numerous dictionaries reflecting many different languages ranging from English, French, Spanish, Portuguese, and so on, the origins and nuances of the term "disaster" can be compared. Thus, within the Romance languages such as Spanish or French, "the noun *disaster* has magical, astral, supernatural and religious connotations . . ." (p. 127). For others, like the Poles and Czechs, ". . . the translation of the noun *disaster* comes from the translation of the English word of Greek origin *catastrophe*, i.e., catastrophe" (p. 127). In contrast, Dutch, Japanese, Arabs and others relate the term to such concepts as "great loss," "terrible happening," "big accident" or other such phrases that convey misfortune (p. 127).

Others, too, continue to wonder what the point of the question is. And so, even as recently as 2004, statements like the following characterize the literature. "When a hazard occurs, it exposes a large accumulation of risk, unleashing unexpected levels of impacts" (Briceño, 2004, p. 5). Despite the differentiations of many others continue to use the terms "disaster" and "hazard" interchangeably.

Starting with definitions that are event-based, many have proposed differentiations that reflect key analytical features of disasters. Kreps and Drabek (1996) proposed that some comparative analyses could be enhanced if disasters were viewed as a special type of social problem. Four defining features of such events, among others, are: (1) length of forewarning, (2) magnitude of impact, (3) scope of impact, and (4) duration of impact (p. 133). Reacting to criticisms from social constructionists (e.g., Stallings 1995) who emphasize the social processes whereby some events or threats are collectively defined as public concerns, while others are not, Kreps and Drabek (1996) emphasized that ". . . the essence of disaster is the **conjunction** of historical conditions and social definitions of physical harm and social disruption at the community or higher levels of analysis" (p. 142; for elaborations see Kreps, 1995a and 1995b).

Such a perspective has led some to propose elaborate typologies of differentiation whereby "levels" of disaster might be defined with precision. For example, by placing disaster within a framework of collective stress, Barton proposed that sources of threat (i.e., internal or external), system level impacted (i.e., family, organization, community), and other such features could differentiate natural disasters from riots, wars, revolutions and so on. More narrowly focused, Britton (1987) proposed a "continuum of collective stress" whereby classes of events could be grouped as either accidents, emergencies, or disasters (pp. 47–53). Reflective the thinking of his Russian colleagues and also the U.S.A. research base, Profiriev (1998a), proposed a typology that integrated numerous analytical criteria whereby different types of emergencies could be compared. These included such features as the "gravity of impact's effect" (i.e., emergencies vs. disasters vs. catastrophic situations); "conflict vs. non-conflict"; "predictability"; "rapidness of spreading" (p. 49). Most recently, Fischer (2003) has proposed a "disaster scale" that could facilitate comparative analyses by researchers *and* preparedness activities by practitioners (pp. 99–106). Drawing an analogy to the use of the Richter scale for easily communicating the severity of earthquakes, his ten "disaster categories" are ". . . based upon the degree of **disruption** and **adjustment** a community(s)/society experiences when we consider **scale**, **scope** and **duration of time**." Thus, "disaster category 1" is comprised of "everyday emergencies," whereas "disaster category 4" would be restricted to events of a major scale that impact small towns. Logically following them are such categories as "DC-8" (i.e., "massive large city"), DC-9 (i.e., "catastrophe") and DC-10 (i.e., "annihilation").

Reflecting his symbolic interactionist theoretical perspective Quarantelli (1987; 1998) has pressed scholars to retreat from frameworks focused exclusively on analytical features of crisis events or the "agents" that "cause" them. Rather, additional research questions ought to expand the agenda, e.g., what are the social processes whereby certain types of crisis situations become "legitimate" bases for social action? Why are there massive relief efforts following a tornado and yet many resist funding for programs assisting victims of the HIV-virus or famine?

Drabek (1970, 2000) has proposed that comparisons among disaster field studies could be integrated more effectively if this question was placed within a methodological framework. That is, the issue is viewed as one of "external validity." Researchers must answer "to what can we generalize?" By using a variety of event based criteria like "length of forewarning," he documented that the behavior of private business employees (1999), tourists (1996a) and others varied during evacuations triggered by hurricanes, floods, and tornadoes. Events reflecting different criteria were responded to somewhat differently.

Of course, such conclusions from a few field studies await the integrative efforts of others if disaster research is to be cumulative, and that is another reason why this key question of definition is so paramount. Implicit in the question, "What is a disaster?" is a fundamental question of strategy. That is, which approach will best permit the systematic accumulation of research findings flowing from separate disaster studies.

The second key issue confronting sociologists who are studying disasters pertains to the paradigms used. Most do not elaborate on the theoretical perspectives that might be guiding their field work although elements of functionalism, structuralism, symbolic interactionism, and other such frameworks can be identified. Many have built upon the "collective stress" framework first outlined by Barton (1969) although the nomenclature usually is modified. For example, Drabek elaborated on his "stress-strain perspective" (e.g., 1990, 1999, 2003) which had its origins in the early DRC studies (e.g., Haas and Drabek, 1970, 1973). Others have pursued the insights of social constructionists and moved into research agenda that usually are ignored by those rooted within a collective stress viewpoint. For example, Stallings (1995) carefully documented the "claims-making activities" of those who have "manufactured" the earthquake threat. This same perspective permitted Jenkins (2003) to document the shifting "ownership" of terrorism, both regarding the "guilty" and the "causes" being used to justify the killing of others.

In contrast, many (e.g., Mileti, 1999) have turned to environmental studies for help. By emphasizing the social desirability of "environmentally friendly" disaster mitigation policies, concepts of "sustainability," and "risk communication," "adoption of hazard adjustments" and others have redefined the research agenda (Mileti, 1980). Community education programs are designed and evaluated throughout the implementation process so as to guide emergency managers seeking to have community based disaster mitigation programs that will encourage development that may better "live with nature" rather than against. Mileti (1999), pp. 30–35) proposed that six core principles delineated this "Sustainable Hazards Mitigation Approach," e.g., "Maintain and, if possible, enhance environmental quality" (p. 31); "Foster local resilience to and responsibility for disasters" (p. 32); and "Adopt a consensus-building approach, starting at the local level" (p. 34).

Finally, some have proposed a paradigm shift reflecting a focus on the concept of vulnerability (e.g., Wisner, 2001). Citing such scholars as Mileti (1999) and Geis (2003), McEntire (2004) begins a recent article by stating that: "Scholars interested in disaster studies are calling for a paradigm shift" (p. 23). Among the reasons for such a shift, are "15 tenets" that include such observations as: "We have control over vulnerability, not natural hazards" (p. 23), "Vulnerability occurs at the intersection of the physical and social environments" (p. 24); "Variables of vulnerability exhibit distinct patterns" (p. 25). This last "tenet" was amplified significantly by Enarson et al. (2003) who designed an instructional guide for college and university professors entitled *A Social Vulnerability Approach to Disasters.* Building on the poignant criticisms of scholars like Hewitt (1983), this team nicely spelled out the basic elements of a social vulnerability paradigm and specified how it differs from "the dominate view" of disasters, e.g., focus on socio-economic and political factors rather than the physical processes of hazard; goal is to reduce vulnerability rather than damage. By documenting the differential and changing patterns of risk and vulnerability, long term levels injustice are highlighted. And so the "root causes" of disaster are exposed as are the policies and practices of those who benefit most by the existing social structure. Rather than accept differential exposures and losses by the politically weak, be they female, aged, or ethnic minorities, those adopting this paradigm question the status quo. They ask, "Why must the patterns of greed and financial corruption continue to perpetuate so-called disasters wherein those most vulnerable are disproportionately hurt?"

When one starts from a social vulnerability perspective, issues of disaster take on a very different look. For example, how did the attacks on the World Trade Center (2001) become defined as a "national" disaster? Oyola-Yemaiel and Wilson (2003) insightfully propose that ". . . we do not consider the terrorist attack itself as a disaster (system failure), we believe that the generalized conception of disaster as well as how the media and the authorities responded to the event illustrates *symptoms* of system failure" (p. 27). Hence, this perspective pushes researchers to examine the nature of vulnerability to terrorism in highly differentiated and interdependent societies. And in so doing, the nature of proposed solutions reflect root causes and basic societal processes that heretofore have rarely been the focus of disaster researchers. Oyola-Yemaiel and Wilson (2003), for example, offer the following.

> . . . rather than immense and impersonal business far away where the fate of the individual, the family, and the local community are in the hands of third parties, society should move forward to a social exchange that would enable local communities to have interdependence with the national system as well as independence of operation from it. At this point each community can sustain life independently outside the whole if needed. In so doing, the communities could become isolated from the threat of terrorism" (p. 37).

This case study underscores insightful conclusions proposed by Bankoff (2003). In contrast to western cultural norms, ". . . vulnerability has been proposed as the key to understanding a novel conceptualization of risk that attempts to break with the more causal, mechanistic attitudes that have characterized the relationship between human societies and their environments over past centuries . . ." (p. 6). Furthermore, "Social systems generate unequal exposure to risk by making some people more prone to disaster than others and that these inequalities in risk and opportunity are largely a function of the power relations operative in every society" (p. 6). Echoing the observations of Oyola-Yemaiel and Wilson (2003), Bankoff proposed that ". . . complexity may be just as much a source of vulnerability as it is an answer to risk" (p. 20). Thus, ". . . attempts to control the environment need to be replaced by approaches that emphasize ways of dealing with unexpected events, ones that stress flexibility, adaptability, resilience and capacity" (p. 20).

MAJOR CONTRIBUTIONS

Beyond the integrative reviews noted above, e.g., Dynes (1970), Barton (1969) and Drabek (1986), several collections summarize substantive contributions by sociologists to the study of disaster. Detailed statements are available in the collection edited by Dynes et al. (1987) that focus on such topics as: "Disaster Preparedness and Response Among Minority Citizens" (Perry); "Human Ecology" (Faupel); "Collective Behavior" (Wenger); "Organizational Change" (Stallings); "Emergent Structures" (Drabek) and "Social Change" (Bates and Peacock). Similarly, the collection of essays prepared in honor of E.L. Quarantelli that was edited by Dynes and Tierney (1994) also presents excellent summaries of both specific studies and broad perspectives such as "An Ecological Approach to Disasters" (Bates and Pelanda); "Public Risk Communication" (Fitzpatrick and Mileti); and "Post Disaster Sheltering and Housing" (Bolin).

As the diversity and depth of these topics indicates, a summary of contributions to the knowledge base is far beyond the limited space of this essay. But four broad topics stand out when a long-term view is applied: (1) disaster myths; (2) research methods; (3) theory; and (4) social criticism.

Disaster Myths

Historically, the most significant contribution of sociological research on disasters has been the correction of distorted images of human response (e.g., Quarantelli, 1960; Quarantelli and Dynes, 1972). Images of panic, looting, and other such anti-social behavior were debunked and properly labeled as myths. That is not to claim that such forms of anti-social behavior never occur. They do. But the image of such behavior as the prevailing response is an exaggeration that simply is wrong. Both the public and emergency officials were found to support such erroneous notions (Wenger et al., 1975; Wenger et al., 1980; Fischer, 1998). One of the most widely circulated documents among local emergency managers outlined these myths and the evidence that debunked them (Dynes et al., 1972). Today, many emergency management professionals point to disaster myths as the first item of substantive knowledge they associate with sociology.

Research Methods

Several excellent statements have been published that highlight unique contributions designed by sociologists studying disasters, e.g., Cisin and Clark, 1962, Drabek, 1970, Mileti, 1987, Stallings, 2002. Concerns raised by Killian (1956) in the 1950s (see the summary of his monograph in Stallings, 2002, pp. 49–93) are a sharp contrast to a range of more current issues such as those pertaining to electronic media raised by Dombrowsky (2002, pp. 305–319) or the uses of geographic information systems described by Nash (2002, pp. 320–333). Following the dictum that interesting questions should be pursued and appropriate methods designed, Drabek (2002) summarized numerous studies he directed that reflected varied types of methodological innovation. Some, like the analysis of police and fire department audio recordings built on unobtrusive data that many had not thought about collecting. Other innovations ranged from the construction of an elaborate police communications simulation to devising ways to track down tourists who were victims of Hurricane Andrew, Iniki, and other disasters.

Methodological innovations continue to be made as researchers seek to improve their understanding of disaster response and impact. Homan's (2003) recent explanation of the use of autobiography is a case illustration. Using materials at the Mass-Observation Archive at the University of Sussex, she demonstrated the utility of this approach and the range of new substantive research questions it permits. For example, "The 1989 Mass-Observation Directive sought to gauge, from personal perspectives, what people thought of the role of the media in disasters and the way in which they are reported, as well as issues apportioning blame and post-disaster relief work" (p. 64). If comparable materials were within the U.S.A. before and after the World Trade Center attacks in 1993 and 2001, important tracking of public perceptions could be available. Comparative analyses of shifts and continuities following earthquakes, hurricanes, and the like, could be most instructive in understanding the "manufacturing" processes being used by various groups within the society.

Theory

Evolving from years of analysis of interviews conducted by DRC staff, Kreps and his associates have moved toward a generalized theory of disaster response (e.g., Kreps, 1987; 1989; Kreps et al., 1994). When disaster strikes, emergent networks are born to handle the unique demands generated. Early on in the life of the DRC, a typology of organized disaster responses was formulated (Dynes, 1970). Many (e.g., Stallings, 1978) discovered that this typology helped make sense of the complex responses they observed in the field. The typology reflected two criteria: structure and tasks. Thus, established organized response units (Type I) reflected old structures being used to accomplish regular tasks. Conversely, emergent organized response units (Type II) reflected new structures being used to accomplish non-regular tasks. Expanding and extending units reflected non-regular tasks with old structures (Type II) or regular tasks with new structures (Type III). Kreps and his associates coded hundreds of DRC interviews so as to document the patterns of social structure that emerged during responses to disasters resulting from such agents as tornadoes, hurricanes, and the like. Their structural code reflected four key analytic qualities, i.e., domains, tasks, human and material resources, and activities. Their preliminary theoretical framework specified that various exogenous factors, e.g., event qualities were followed by social processes that defined the post-event organizing behavior which in turn produced various outcomes that could be assessed at both the individual and structural level. Their meticulous work lead them to conclude that the DRC typology was both an efficient and effective tool for understanding disaster behavior. Furthermore, if ". . . specifies nicely a micro-macro link between the individual and social structure" (Kreps et al., 1994, p. 191).

Building on the collective stress perspective noted above, Drabek (2003) formulated a model for predicting the relative effectiveness of disaster responses. This work paralleled the logic of the Kreps team, but introduced different concepts. Local emergency managers were viewed as being nested within state and federal systems that changed over time reflecting perceptions of threats, government policies, demographic trends, and other such factors (pp. 147–152). By implementing a series of managerial strategies, various forms of interagency networks are nurtured which spring into action when disaster threatens. Use or misuse of 26 specific coordination strategies predict the shape of the emergent response and its effectiveness. While far from complete, future comparative research along these lines will provide the foundation required for scientifically based theories of emergency management.

Finally, as Dynes (2002, 2003) has documented, social capital theory offers many important insights. This analysis was extended by Nakagawa and Shaw (2004) in their case study of reconstruction following the 1995 devastating earthquake in Kobe, Japan. Their results clearly documented that the high level of trust in local leaders by the community was the major factor that facilitated acceptance of the collective decisions made throughout the recovery process. They concluded that ". . . social capital and leadership in the community are the basic attributes, which are universal in nature, irrespective of the development stages of the country" (p. 29).

Social Criticism

A final area of contribution has taken the form of social criticism. Reflecting its historical roots, sociologists have offered "observations" about disaster responses that have highlighted fundamental flaws in both response and policy. This practice has reflected DRC publications since its origin. In its first publication, for example, Drabek, pinpointed "operational problems" stemming from inadequate inter-organizational coordination and communication (e.g., 1968, pp. 155–169). Years later (Drabek 1996a) reemphasized that business executives need to "resist threat denial" (1996a, p. 244), "do not overreact" (1996a, p. 245), and "debunk the panic myth" (1996a, p. 245). Dynes (1994, 1983) repeatedly has critiqued the planning and preparedness actions practiced by many who continue to fail because their top-down approach is fundamentally flawed, rooted in assumptions reflecting myth rather than research results.

Most recently, the Homeland Security Advisory System (HSAS) has been found lacking. For example, Major and Atwood (2004) documented that only 49 percent of U.S.A. citizens surveyed in April, 2003 (p. 82) responded that the system was "useful." Ambiguity in the announcements was the major complaint and it had real consequences. "The ambiguity of such announcements leaves the public with but one choice: not to prepare" (p. 97). Studies like these led Aguirre (2004) to a highly critical view.

"The current Homeland Security Advisory System does not draw from years of social science study and does not benefit the nation. It is not a warning system. At best, HSAS is a mitigation and anticipatory public relations tool" (p. 112).

DISCIPLINARY OVERLAPS

Sociologists studying disasters frequently have integrated both theory and methodological tools reflective of other disciplines into their work. Indeed, the first major textbook on emergency management (Drabek and Hoetmer, 1991), reflected a blending of concepts, conclusions, and analyses from sociology, public administration, and a wide variety of other disciplines. Such points of overlap within the literature at large are varied and numerous. The following *illustrations* document the point.

Response

A core theme in the analysis of disaster response is the concept of emergence, e.g., Drabek and McEntire 2002, 2003. Quarantelli (1996) summarized many of the key insights that had been accumulating over the years as scholars like Stallings (1978), Weller (1972), and Neal (1984) examined such dynamics. Most recently, Mendoça and Wallace (2004) have combined these insights with those from social psychologists like Weick (1993), and offered important new observations based on detailed examination of DRC interviews after Hurricane Camille (1969). In so doing they have developed a new methodology to specify the types of data required to document the "where, when and how" that improvisation occurs during disaster responses.

Recovery

Assessments of long-term impacts of disaster on individuals and communities illustrate, the close links between sociology, psychology, economics, and other disciplines. Drabek (2004) summarized numerous studies wherein various theoretical frameworks were used to document the lasting psychic pains following the *Exxon Valdez* oil spill (Arata et al., 2000), Hurricane Floyd (Willigen, 2001), the 1994 Northridge Earthquake (Siegel et al., 1999), the 1995 bombing of the Alfred P. Murrah Federal Building in Oklahoma City (Benight et al., 2000) and other disasters.

While controversy remains regarding the relative efficacy of alternative treatment modalities (NIMH, 2002, p. 9), the evidence is clear that most individuals cope well with even the worst of events. For some, however, the lingering pain of loss and fear continue although some interventions such as the "critical incident stress debrief" may offer promise (e.g., Mitchell and Everly, 2000). Economic and demographic shifts following such events as Hurricane Andrew (e.g., Peacock et al., 1987), have recast earlier "no-effect" conclusions reached by others (e.g., Rossi et al., 1978; Wright and Rossi, 1981). Short-term effects on such social phenomena as marriage rates (e.g., Cohan and Cole, 2002) regarding Hurricane Hugo), transportation patterns (e.g., Edwards et al., 2000 regarding Hurricane Floyd) and other social phenomena continue to be documented as does increased ethnic inequality such as that which occurred after Hurricane Andrew (Morrow and Peacock 1997). Future multidisciplinary collaboration will be required if the processes and outcomes disaster recovery are to be better understood.

Preparedness

Following extensive study of tourism evacuation behavior (e.g., Drabek, 1996a), FEMA supported Drabek's effort to team with faculty of the School of Travel Industry Management, University of Hawaii at Manoa (Drabek and Gee, 2000). An emergency management instructor guide was created for faculty within departments of tourism, hospitality, restaurant and travel management. This facilitated the diffusion of knowledge from the social sciences, especially sociology, into this professional area whose businesses reflect a catastrophic level of vulnerability (Drabek, 1994).

Diffusion of innovations has long been a focal point of sociologists and communications researchers (e.g., Rogers, 1962). Drabek (1991) documented the social history of the adoption and implementation of microcomputers into several local and state emergency management agencies during the late 1980s. Problematical aspects of such adoptions were specified by Quarantelli (1997). His observations contrast sharply with the advantages of such technology that are proposed by those coming from other disciplinary perspectives, e.g., Stephenson and Anderson, 1997; Gruntfest and Weber, 1998.

Mitigation

Learning from the wisdom of such social geographers as Gilbert White (1945), Mileti (1980) formulated a general paradigm for assessing human adjustments to the risks associated with environmental extremes. Over the years, his work matured so that by 1999 he was able to present a well developed framework of a "sustainable hazards mitigation approach" (Mileti, 1999, pp. 31–35). His approach has not been without criticism, however, and scholars like Aguirre (2002) have questioned both the content and direction. Others, like McEntire recognize both the strengths and weaknesses of applying sustainability to the study of disasters (McEntire and Floyd, 2003), but he is concerned that the perspective is not as holistic as implied by those who espouse it (McEntire et al., 2002, pp. 270–272).

EMERGENCY MANAGER RECOMMENDATIONS

As is evident from the above analysis, sociologists have offered recommendations to emergency managers for decades. Many of these were codified in the text edited by Drabek and Hoetmer (1991). These ideas built on the continuing stream of publications produced by DRC staff, graduates, and sociologists located at other universities. Most important among the recommendations are such principles as the following.

- An all-hazards approach is essential (Drabek and Hoetmer, 1991).
- Planning and preparedness activities are continuous processes, not goals to be accomplished and put aside (Dynes et al., 1972).
- Social science knowledge, not myths, should guide program activities, priorities, and implementation strategies (Quarantelli and Dynes, 1972).
- If disaster plans are to be relevant guides for the behavioral response, they must be developed by those who will implement them (Dynes and Drabek, 1994).
- Managing emergency responses requires the implementation of theoretical models that are resource based rather than authority based

(Dynes, 1994; Neal and Phillips, 1995; Drabek, 2003).

This last principle reflects another point of disciplinary overlap and points the way for a future research agenda. Writing from the perspective of a political scientist, Sylves (2004) stated the point with succinctness. "When it comes to the field of emergency management, the aim should be to develop new theory or adapt old theory to produce manageable policy. . . . the field must advance through the production of codified knowledge widely diffused to anyone who chooses to learn it" (p. 32).

These basic recommendations, and others like them, are being implemented in local emergency management agencies and related units of government more frequently than at any other time in the history of the nation. A recent write-up by a utility security manager in Bradenton, Florida is but one of dozens of illustrations that could be cited. In response to the federal mandates which amended the Safe Drinking Water Act, i.e., Public Health Security and Bioterrorism Preparedness Act of 2002, Brian Sharkey (2004) pressed for changes. Among the steps taken, all of which reflect the principles listed above, were these.

- "At the outset, the plan was developed with input from the department's senior staff. These are the people who are responsible for carrying out the plan, so they must have input and ownership." (p. 7)
- "Local emergency response agencies were involved in plan development. This ". . . allowed the emergency response agencies to integrate their plan with ours." (p. 7)
- "The plan is always considered an unfinished product. It has been made an active and evolving part of our working environment, and is not just another dust collector." (p. 7)

Unfortunately, vulnerabilities and risks are accelerating at rates that far exceed such increased capacities due to a whole host of social, demographic, technological, and political factors. So while much has been accomplished through applications of sociological research findings, the net result has been a society at increased risk, and globally, the situation is far worse.

FUTURE RESEARCH AGENDA

To celebrate the fortieth anniversary of the founding of the Disaster Research Center (DRC), numerous scholars gathered to reflect on the past and propose directions for the future (Rodríguez et al., 2004). At the end of their two day conference, they identified a list of research priorities. Among these was a vision of increased "globalization," more focus on vulnerability and development, increased multi and interdisciplinary research, emerging technologies, special population impacts especially children (and race, ethnic, gender, class and age inequalities), and new complex threats as represented by terrorist attacks (pp. 130–131). This listing and the elaborations provided are invaluable to any who might formulate their own research agenda. From this and other such efforts (e.g., Anderson and Mattingly, 1991; Simpson and Howard, 2001), two key themes merit priority.

1. *Alternative theoretical perspectives should be elaborated, encouraged, and compared.* Starting with the social problem orientation proposed by Kreps and Drabek (1996), disasters must be placed within the broader context of public policy, perception, and history. Similarly, analyses must be continued of the unique and continuing social injustices reflective in the inequalities of race, gender, age, etc. that are highlighted by those advocating social vulnerability perspectives (e.g., Enarson et al., 2003). So, too, must the insights from Mileti (1999) and others whose focus on mitigation led them to see the wisdom of the breadth of perspective inherent in sustainability theory. Different research questions may best be pursued through one of those perspectives or some other. The field will develop best through expansion, not premature closure.

2. *A global, rather than a national, focus must be developed.* There are many reasons why a global perspective must be nurtured. First, it is through cross-societal comparison that the issues of external validity can best be addressed (Drabek, 2000; Peacock, 1997). Second, as Dynes (2004) pointed out so effectively, the majority of disaster victims reside in underdeveloped countries where few research teams have ventured. Third, links between disaster consequences and other events, like resettlements caused by World Bank mitigation projects, should be assessed. "Without understanding the impoverishing

consequences of displacement, the inequalities between gainers and losers from such projects will be amplified and perpetuated: more than a few displaced people will end up worse off, poorer than before the project came into their midst" (Cernea, 2003, p. 37). Fourth, new threats, like terrorism, and the vulnerabilities they reflect must be viewed within an international context if preparedness, response, and mitigation policies are to be informed effectively. Dynes put it succinctly: "One of the other consequences of 9/11 was the effort to remove the burkas, which distorted the vision of those in Afghanistan. U.S. policy has insisted that we keep our burkas on, ignoring the lessons of Hamburg, Hiroshima, and New York" (Dynes, 2003, p. 21).

REFERENCES

Aguirre, B.E. (2004). "Homeland Security Warnings: Lessons Learned and Unlearned." *International Journal of Mass Emergencies and Disasters 22* (No. 2):103–115.

Anderson, W.A. and Mattingly, S. (1991). "Future Directions." pp. 311–335 in *Emergency Management: Principles and Practice for Local Government*, edited by Drabek, T.E., and Hoetmer, G.J. International City Management Association Washington, D.C.

Arata, C.M., Picou, J.S., Johnson, G.D., and McNally, T.S. (2000). "Coping with Technological Disaster: An Application of the Conservation of Resources Model to the *Exxon Valdez* Oil Spill." *Journal of Traumatic Stress 13*:23–39.

Bankoff, G. (2003). "Vulnerability as a Measure of Change in Society." *International Journal of Mass Emergencies and Disasters 21* (No. 2):5–30.

Barton, A.H. (1969). *Communities in Disaster: A Sociological Analysis of Collective Stress Situations.* Doubleday and Company, Inc., Garden City, New York.

Bates, F.L. and Peacock, W.L. (1987). "Disaster and Social Change." pp. 291–330 in *Sociology of Disasters: Contributions of Sociology to Disaster Research*, edited by Dynes, R,R, DeMarchi, B., and Pelanda, C. Franco Angeli, Milano, Italy.

Bates, F.L. and Peacock, W.G. (1992). "Measuring Disaster Impact on Household Living Conditions." *International Journal of Mass Emergencies and Disasters 10*: 133–160.

Bates, F.L. and Pelanda, C. (1994). "An Ecological Approach to Disasters." pp. 145–159 in *Disasters, Collective Behavior, and Social Organization*, edited by Dynes, R.R. and Tierney, K.J. Newark, Delaware: University of Delaware Press.

Benight, C.C., Freyaldenhoven, R.W., Hughes, J., Ruiz, J.M., Zoschke, T.A., and Lovallo, W.R., (2000). "Coping Self-Efficacy and Psychological Distress Following the Oklahoma City Bombing." *Journal of Applied Social Psychology 30*:1331–1344.

Bolin, R. (1994). "Postdisaster Sheltering and Housing: Social Processes in Response and Recovery." pp. 115–127 in *Disasters, Collective Behavior, and Social Organization,* edited by Dynes, R.R. and Tierney, K.J. University of Delaware Press, Newark, Delaware.

Briceño, S. (2004). "Global Challenges in Disaster Reduction." *Australian Journal of Emergency Management 19* (No. 1):3–5.

Britton, N.R. (1987). "Toward a Reconceptualization of Disaster for the Enhancement of Social Preparation." pp. 31–55 in *Sociology of Disasters: Contributions of Sociology to Disaster Research*, edited by Dynes, R.R. DeMarchi, B., and Pelanda, C. Franco Angeli: Milano, Italy.

Britton, N.R. and Clapham, K. (1991). *Annotated Bibliography of Australian Hazards and Disaster Literature, 1969–1989.* (Volume 1: Author Indexed; Volume 2: System Level Indexed). Armidale, New South Wales, Australia: Centre for Disaster Management, University of New England.

Burton, I., Kates, R.W., and White, G.F., (1993). *The Environment as Hazard.* 2nd ed. New York: Guilford Publishers, Inc.

Cernea, M.M. (2003). "For a New Economics of Resettlement: A Sociological Critique of the Compensation Principle." *International Social Science Journal 55*: 37–45.

Cisin, I. and Clark, W.B. (1962). "The Methodological Challenge of Disaster Research." pp. 23–47 in *Man and Society in Disaster*, edited by Baker, G.W. and Chapman, D.W. Basic Books, New York.

Cohan, C.L. and Cole, S.W. (2002). "Life Course Transitions and Natural Disaster: Marriage, Birth, and Divorce Following Hurricane Hugo." *Journal of Family Psychology 16*:14–25.

Dombrowsky, W.R. (2002). "Methodological Changes and Challenges in Disaster Research: Electronic Media and the Globalization of Data Collection." pp. 305–319 in *Methods of Disaster Research,* edited by Stallings, R.A. Xlibris Corporation, Philadelphia, Pennsylvania.

Dombrowsky, W.R. and Schorr, J.K. (1986). "Angst and the Masses: Collective Behavior Research in Germany." 1986. *International Journal of Mass Emergencies and Disasters 4*:61–89.

Drabek, T.E. (1968). *Disaster in Aisle 13.* College of Administrative Science. The Ohio State University: Columbus, Ohio.

_______. (1970). "Methodology of Studying Disasters: Past Patterns and Future Possibilities." *American Behavioral Scientist 13*:331–343.

_______. (1986). *Human System Responses to Disaster: An Inventory of Sociological Findings.* Springer-Verlag: New York.

_______. (1987). "Emergent Structures." pp. 190–259 in *Sociology of Disasters: Contribution of Sociology in Disaster Research*, edited by Dynes, R.R., DeMarchi, B. and Pelanda, C. Franco Angeli: Milano, Italy.

_______. (1990). *Emergency Management: Strategies for Maintaining Organizational Integrity.* Springer-Verlag: New York.

_______. (1991). *Microcomputers in Emergency Management: Implementation of Computer Technology.* Institute of Behavioral Science, University of Colorado.

_______. (1994). *Disaster Evacuation and the Tourist Industry.* Boulder, Colorado: Institute of Behavioral Science, University of Colorado.

_______. (1996a). *Disaster Evacuation and the Tourist Industry.* Boulder, Colorado: Institute of Behavioral Science, University of Colorado, Boulder, Colorado.

_______. (1996b). *Sociology of Disaster: Instructor Guide.* Emergency Management Institute, Federal Emergency Management Agency, Emmitsburg, Maryland.

_______. (1999). "Disaster Evacuation Response by Tourists and Other Types of Transients." *International Journal of Public Administration 22*:655–677.

_______. (2000). "The Social Factors that Constrain Human Responses to Flood Warnings." pp. 361–376 in *Floods*, (Vol. 1), edited by Parker, D.J. Routledge: New York.

_______. (2002). "Following Some Dreams: Recognizing Opportunities, Posing Interesting Questions, and Implementing Alternative Methods." pp. 127–153 in *Methods of Disaster Research*, edited by Stallings, R.A. Philadelphia, Pennsylvania: Xlibris Corporation.

_______. (2003). *Strategies For Coordinating Disaster Responses.* Institute of Behavioral Science, University of Colorado, Boulder, Colorado.

_______. (2004). *Social Dimensions of Disaster 2nd ed.: Instructor Guide.* Emergency Management Institute, Federal Emergency Management Agency, Emmitsburg, Maryland.

Drabek, T.E. and Gee, C.Y. (2000). *Emergency Management Principles and Application for Tourism, Hospitality, and Travel Management Industries: Instructor Guide.* Emergency Management Institute, Federal Emergency Management Agency: Emmitsburg, Maryland.

Drabek, T.E. and G.J. Hoetmer (eds.). (1991). *Emergency Management: Principles and Practice for Local Government.* International City Management Association: Washington, D.C.

Drabek, T.E. and McEntire, D.A. (2002). "Emergent Phenomena and Multiorganizational Coordination in Disasters: Lessons from the Research Literature." *International Journal of Mass Emergencies and Disasters 20*:197–224.

_______. (2003). "Emergent Phenomena and the Sociology of Disaster: Lessons, Trends and Opportunities from the Research Literature." *Disaster Prevention and Management 12*:97–112.

Drabek, T.E., Mushkatel, A.H. and Kilijanek, T.S. (1983). *Earthquake Mitigation Policy: The Experience of Two States.* Institute of Behavioral Science, The University of Colorado, Boulder, Colorado.

Dynes, R.R. (1970). *Organized Behavior in Disaster.* Lexington, Mass.: Health Lexington Books.

_______. (1983). "Problems in Emergency Planning." *Energy 8*:653–660.

_______. (1994). "Community Emergency Planning: False Assumptions and Inappropriate Analogies." *International Journal of Mass Emergencies and Disasters 12*:141–158.

_______. (2002). "The Importance of Social Capital in Disaster Response." Preliminary Paper #322. Newark, Delaware: Disaster Research Center, University of Delaware.

_______. (2003). "Finding Order in Disorder: Continuities in the 9-11 Response." *International Journal of Mass Emergencies and Disasters 21* (No. 3):9–23.

_______. (2004). "Expanding the Horizons of Disaster Research." *Natural Hazards Observer 28* (No. 4):1–2.

Dynes, R.R. and Drabek, T.E. (1994). "The Structure of Disaster Research: Its Policy and Disciplinary Implications." *International Journal of Mass Emergencies and Disasters 12*:5–23.

Dynes, R.R., De Marchi, B. and Pelanda, C. (eds.). (1987). *Sociology of Disasters: Contribution of Sociology to Disaster Research.* Franco Angeli, Milano, Italy.

Dynes, R.R., Quarantelli, E.L. and Kreps, G.A. (1972). *A Perspective on Disaster Planning.* Disaster Research Center, The Ohio State University, Columbus, Ohio.

Dynes, R.R. and Tierney, K.J. (eds.). (1994). *Disasters, Collective Behavior and Social Organization.* Newark, Delaware: University of Delaware Press.

Edwards, B., Van Willigen, M., Lormand, S., Currie, J., with Bye, K., Maiolo, J., and Wilson, K. (2000). "An Analysis of the Socioeconomic Impact of Hurricane Floyd and Related Flooding on Students at East Carolina University." (Quick Response Report #129). Natural Hazards Research and Applications Information Center, University of Colorado, Boulder, Colorado.

Enarson, E., Childers, C., Morrow, B.H., Thomas, D., and Wisner, B. (2003). *A Social Vulnerability Approach to Disasters.* Emergency Management Institute, Federal Emergency Management Agency, Emmitsburg, Maryland.

Faupel, C.E. (1987). "Human Ecology: Contributions to Research and Policy Formation." pp. 181–211 in *Sociology of Disasters: Contributions of Sociology to Disaster Research*, edited by Dynes, R.R., DeMarchi, B., and Pelanda, C. Franco Angeli, Milano, Italy.

Fischer, H.W. III. (1998). *Response to Disaster: Fact Versus Fiction and Its Perpetuation – The Sociology of Disaster.* Second Edition. University Press of America, Lanham, Maryland.

_______. 2003. "The Sociology of Disaster: Definitions, Research Questions, & Measurements. Continuation of the Discussion in a Post-September 11 Environment." *International Journal of Mass Emergencies and Disasters 21* (No. 1):91–107.

Fitzpatrick, C. and D.S. Mileti. (1994). "Public Risk Communication." pp. 71–84 in *Disasters, Collective Behavior, and Social Organization*, edited by Dynes, R.R. and Tierney, K.J. University of Delaware Press, Newark, Delaware.

Fritz, Charles E. 1961. "Disasters." pp. 651–694 in *Contemporary Social Problems*, edited by Robert K. Merton and Robert A. Nisbet. New York: Harcourt.

Geis, D.E. (2000). "By Design: The Disaster Resistant and Quality of Life Community." *Natural Hazards Review 1* (No. 3):151–160.

Gruntfest, E. and Weber, W. (1998). "Internet and Emergency Management: Prospects for the Future." *International Journal of Mass Emergencies and Disasters 16*:55–72.

Homan, J. (2003). "Writing Disaster: Autobiography as a Methodology in Disasters Research." *International Journal of Mass Emergencies and Disasters 21* (No. 2:51–80.

Haas, J.E. and Drabek, T.E. (1970). "Community Disaster and System Stress: A Sociological Perspective." p. 264–286 in *Social and Psychological Factors in Stress*, edited by McGrath, J.E. Holt, Rinehart and Winston, Inc, New York.

_______. (1973). *Complex Organizations: A Sociological Perspective.* The Macmillan Company, J.E.

Hewitt, K. (ed.). (1983). *Interpretations of Calamity.* Allen & Unwin Inc., Boston.

Jenkins, P. (2003). *Images of Terror: What We Can and Can't Know about Terrorism.* Aldine de Gruyter, New York.

Killian, L.M. (1956). *An Introduction to Methodological Problems of Field Studies in Disasters.* National Academy of Sciences/National Research Council, Washington, D.C.

Kreps, G.A. (1984). "Sociological Inquiry and Disaster Research." *Annual Review of Sociology 10*:309–330.

_______. (1987). "Classical Themes, Structural Sociology, and Disaster Research." pp. 357–401 in *Sociology of Disasters: Contribution of Sociology in Disaster Research*, edited by Dynes, R.R., DeMarchi, B., and Pelanda, C. Franco Angeli: Milano, Italy.

_______. (ed.). (1989). *Social Structure and Disaster.* Newark, Delaware: University of Delaware Press.

_______. (1995a). "Disaster as Systemic Event and Social Catalyst: A Clarification of Subject Matter." *International Journal of Mass Emergencies and Disasters 13*:255–284.

_______. (1995b). "Excluded Perspectives in the Social Construction of Disaster: A Response to Hewitt's Critique." *International Journal of Mass Emergencies and Disasters 13*:349–351.

Kreps, G.A. and Bosworth, S.L. with Mooney, J.A., Russell, S.T., and Myers, K.A. (1994). *Organizing, Role Enactment, and Disaster: A Structural Theory.* University of Delaware Press, Newark, Delaware.

Kreps, G.A. and Drabek, T.E. (1996). "Disasters Are Non-Routine Social Problems." *International Journal of Mass Emergencies and Disasters 14*:129–153.

Major, A.M. and L.E. Atwood. (2004). "Assessing the Usefulness of the U.S. Department of Homeland Security's Terrorism Advisory System." *International Journal of Mass Emergencies and Disasters 22* (No. 2):77–101.

McEntire, D.A. (2004). "Tenets of Vulnerability: An Assessment of a Fundamental Disaster Concept." *Journal of Emergency Management 2* (No. 2):23–29.

McEntire, D.A. and Floyd, D. (2003). "Applying Sustainability to the Study of Disasters: An Assessment of Strengths and Weaknesses." *Sustainable Communities Review 6* (Nos. 1 & 2): 14–21.

McEntire, D.A., Fuller, C., Johnson, C.W., and Weber, R. (2002). "A Comparison of Disaster Paradigms: The Search for a Holistic Policy Guide." *Public Administration Review 62* (No. 3):267–281.

Mendoça, D. and Wallace, W.A. (2004). "Studying Organizationally-situated Improvisation in Response to Extreme Events." *International Journal of Mass Emergencies and Disasters. 22* (No. 2):5–29.

Mileti, D.S. (1980). "Human Adjustment to the Risk of Environmental Extremes." *Sociology and Social Research 64*:327–347.

_______. (1987). "Sociological Methods and Disaster Research." pp. 57–69 in *Sociology of Disasters: Contributions of Sociology to Disaster Research*, edited by Russell R. Dynes, Bruna DeMarchi, and Carlo Pelanda. Milano, Italy: Franco Angeli.

_______. (1999). *Disasters by Design: A Reassessment of Natural Hazards in the United States.* Joseph Henry Press, Washington, D.C.

Mitchell, J.T. and Everly, G.S. (2000). "Critical Incident Stress Management and Critical Incident Stress Debriefings: Evolutions, Effects, and Outcomes." pp. 71–90 in *Psychological Debriefing: Theory, Practice, and Evidence*, edited by B. Raphael and J. Wilson. Cambridge, United Kingdom: Cambridge University Press.

Morrow, B.H. and Peacock, W.H. (1997). "Disasters and Social Change: Hurricane Andrew and the Reshaping of Miami?" pp. 226–242 in *Hurricane Andrew: Ethnicity, Gender and the Sociology of Disasters* edited by Peacock, W.H., Morrow, B.H., and Gladwin, H. Routledge, London.

Murria, J. (2004). "A Disaster, by Any Other Name." *International Journal of Mass Emergencies and Disasters 23* (No. 1):5–34.

Nakagawa, Y. and Shaw, R. (2004). "Social Capital: A Missing Link to Disaster Recovery." *International Journal of Mass Emergencies and Disasters. 22* (No. 1): 5–34.

Nash, N. (2002). "The Use of Geographic Information Systems in Disaster Research." pp. 320–333 in *Methods of Disaster Research*, edited by Robert A. Stallings. Philadelphia, Pennsylvania: Xlibris Corporation.

National Institute of Mental Health. (2002). *Mental Health and Mass Violence: Evidence-Based Early Psychological Intervention for Victims/Survivors of Mass Violence. A Workshop to Reach Consensus on Best Practices.* NIH Publication No. 02-5138. Washington, D.C.: U.S. Government Printing Office. (Accessed April, 2003 at: http://www.nimh.nih.gov/research/massviolence.pdf).

Neal, D.M. (1984). "Blame Assignment in a Diffuse Disaster Situation: A Case Example of the Role of an Emergent Citizen Group." *International Journal of Mass Emergencies and Disasters 2*:251–266.

Neal, D.M. and Phillips, B.D. (1995). "Effective Emergency Management: Reconsidering the Bureaucratic Approach." *Disasters: The Journal of Disaster Studies, Policy and Management 19*:327–337.

Oliver-Smith, A. and Hoffman, S.M. (eds.). 1999. *The Angry Earth: Disaster in Anthropological Perspective.* Routledge, New York.

Oyola-Yemaiel, A. and Wilson, J. 2003, "Terrorism and System Failure: A Revisited Perspective of Current Development Paradigms." *International Journal of Mass Emergencies and Disasters 21* (No. 3):25–40.

Parker, D.J. (ed.). 2000. *Floods.* London: Routledge.

Parr, A.R. (1997–98). "Disasters and Human Rights of Persons with Disabilities: A Case for an Ethical Disaster Mitigation Policy." *The Australian Journal of Emergency Management 12* (No. 4):2–4.

Peacock, W.G. (1997). "Cross-National and Comparative Disaster Research." *International Journal of Mass Emergencies and Disasters 15*:117–133.

Peacock, W.G., Killian, C.D. and Bates, F.L. (1987). "The Effects of Disaster Damage and Housing Aid on Household Recovery Following the 1976 Guatemalan Earthquake." *International Journal of Mass Emergencies and Disasters 5*:63–88.

Perry, R.W. (1987). "Disaster Preparedness and Response among Minority Citizens." pp. 135–151 in *Sociology of Disasters: Contributions of Sociology to Disaster Research*, edited by Dynes, R.R., DeMarchi, B., and Pelanda, C. Franco Angeli, Milano, Italy.

Porfiriev, B. (1998a). *Disaster Policy and Emergency Management in Russia.* Commack, New York: Nova Science Publishers, Inc.

_______. (1998b). "Issues in the Definition and Delineation of Disasters and Disaster Areas." pp. 56–72 in *What Is a Disaster?: Perspectives on the Question.* Routledge, New York.

Prince, S.H. (1920). "Catastrophe and Social Change, Based Upon a Sociological Study of the Halifax Disaster." Ph.D. thesis. Columbia University Department of Political Science: New York.

Quarantelli, E.L. (1960). "Images of Withdrawal Behavior in Disasters: Some Basic Misconceptions." *Social Problems 8*:68–79.

_______. (1984). *Organizational Behavior in Disasters and Implications For Disaster Planning.* National Emergency Training Center, Federal Emergency Management Agency, Emmitsburg, Maryland.

_______. (1987). "Presidential Address: What Should We Study? Questions About the Concepts of Disasters." *International Journal of Mass Emergencies and Disasters 5*:7–32.

_______. (ed.). (1998). *What Is a Disaster?: Perspectives on the Question.* Routledge, New York.

Quarantelli, E.L. and Dynes, R.R. (1972). "When Disaster Strikes (It Isn't Much Like What You've Heard and Read About)." *Psychology Today 5*:66–70.

_______. (1977). "Response to Social Crisis and Disaster." *Annual Review of Sociology 3*:23–49.

_______. (1996). "Emergent Behaviors and Groups in the Crisis Times of Disasters." pp. 47–68 in *Individuality and Social Control: Essays in Honor of Tamotsu Shibutani*, edited by Kwan, K.M.. Greenwich, CT: JAI Press.

_______. (1997). "Problematical Aspects of the Information/Communication Revolution for Disaster Planning and Research: Ten Non-Technical Issues and Questions." *Disaster Prevention and Management 5*:94–106.

Quarantelli, E.L. and Pelanda, C. (eds.). (1989). *Proceedings of the Italy-United States Seminar on Preparations for, Responses to and Recovery from Major Community Disasters, held October 5–10, 1986.* Disaster Research Center, University of Delaware, Newark, Delaware.

Rodrígiez. H., Wachtendorf, T., and Russell, C. (2004). "Disaster Research in the Social Sciences: Lessons Learned, Challenges, and Future Trajectories." *International Journal of Mass Emergencies and Disasters 22* (No. 2):117–136.

Rogers, E.M. (1962). *Diffusion of Innovations.* New York: The Free Press.

Rossi, P.H., Wright, J.D., Wright, S.R., and Weber-Burdin, E. (1978). "Are There Long Term Effects of American Natural Disasters?" *Mass Emergencies 3*:117–132.

Sharkey, B. (2004). "Drinking Water System Safety and Security Planning: Manatee County Utility Operations Department." *IAEM Bulletin 21* (No. 9):7.

Siegel, J.M., Bourque, L.B., and Shoaf, K.I. (1999). "Victimization after a Natural Disaster: Social Disorganization or Community Cohesion?" *International Journal of Mass Emergencies and Disasters 17*:265–294.

Simpson, D.M. and Howard. G.A. (2001). "Issues in the Profession: The Evolving Role of the Emergency Manager." *The Journal of the American Society of Professional Emergency Planners 8*:63–70.

Stallings, R.A. (1978). "The Structural Patterns of Four Types of Organizations in Disaster." pp. 87–103 in *Disasters: Theory and Research*, edited by E.L. Quarantelli. Beverly Hills, California: Sage.

Stallings, R.A. (1987). "Organizational Change and the Sociology of Disaster." pp. 239–257 in *Sociology of Disasters: Contributions of Sociology to Disaster Research*, edited by Dynes, R.R., deMarchi, B. and Pelanda, C. Franco Angeli, Milano, Italy.

_______. (1994). "Collective Behavior Theory and the Study of Mass Hysteria." pp. 207–228 in *Disasters, Collective Behavior, and Social Organization*, edited by Dynes, R.R. and Tierney, K.J. University of Delaware Press, Newark, Delaware.

_______. (1995). *Promoting Risk: Constructing the Earthquake Threat.* Aldine de Gruyter, Hawthorne, New York.

_______ (ed.). (2002). *Methods of Disaster Research.* Xlibris Corporation, Philadelphia, Pennsylvania.

Stephenson, R. and Anderson, P.S. (1997). "Disasters and the Information Technology Revolution." *Disasters: The Journal of Disaster Studies, Policy and Management 21*:305–334.

Sylves, R.T. (2004). "A Précis on Political Theory and Emergency Management." *Journal of Emergency Management 2* (No. 3):27–32.

Weick, K.E. (1993). "The Collapse of Sensemaking in Organizations: The Mann Gulch Disaster." *Administrative Science Quarterly 38*:629–652.

Weller, J.M. (1972). "Innovations in Anticipation of Crisis: Organizational Preparations for Natural Disasters and Civil Disturbances." Dissertation. Columbus, Ohio: The Ohio State University.

Wenger, D.E. (1987). "Collective Behavior and Disaster Research." pp. 213–237 in *Sociology of Disasters: Contributions of Sociology to Disaster Research*, edited by Dynes, RR, DeMarchi, B. and Pelanda, C. Franco Angeli, Milano, Italy.

Wenger, D.E., Dykes, J.D., Sebok, T.D., and Neff, J.L. (1975). "It's a Matter of Myths: An Empirical Examination of Individual Insight Into Disaster Response." *Mass Emergencies 1*:33–46.

Wenger, D.E., James, T.F., and Faupel, C.F. (1980). *Disaster Beliefs and Emergency Planning.* Disaster Research Project, University of Delaware, Newark, Delaware.

White, G.F. (1945). *Human Adjustment to Floods: A Geographical Approach to the Flood Problems in the United States.* (Research Paper No. 29). Chicago, Illinois: Department of Geography, University of Chicago.

Willigen, M.V. (2001). "Do Disasters Affect Individual's Psychological Well-Being? An Over-Time Analysis of the Effect of Hurricane Floyd on Men and Women in Eastern North Carolina." *International Journal of Mass Emergencies and Disasters 19*:59–83.

Wisner, B. (2001). "Capitalism and the Shifting Spatial and Social Distribution of Hazard and Vulnerability." *Australian Journal of Emergency Management 16*: 44–50.

Wright, J.D. and P.H. Rossi (eds.). (1981). *Social Science and Natural Hazards.* Cambridge, Massachusetts: Abt Books.

Yamamoto, Y. and Quarantelli, E.L. (1982). *Inventory of the Japanese Disaster Research Literature in the Social and Behavioral Sciences.* Disaster Research Center, The Ohio State University, Columbus, Ohio.

Chapter **6**

Research about the Mass Media and Disaster: Never (Well Hardly Ever) the Twain Shall Meet

Joseph Scanlon

ABSTRACT

A review of two areas of scholarship into the role of the mass media in crisis and/or disaster reveals a dichotomy. There is substantial research by scholars in a number of disciplines and by scholars in Journalism and Mass Communications. The two appear unaware of what each other is doing. Cross-referencing is rare. The scholarship shows that the media can play a critical role before, during and after such incidents. The media are essential, for example, for warnings to be effective and may be the single-most important source of public information in the wake of a disaster. The scholarship also shows that media reports that distort what happens in a disaster and lead to misunderstandings. Failure by officials to issue a warning, for example, may be a result of the myth that people panic, a myth perpetuated by the media. Media scholarship also shows, however, that in one area where the media are often criticized they are not guilty as charged: the limited research available suggests many victims and relatives of victims welcome the presence of the media and do not see journalists as intruders.

Oh, East is East and West is West, and never the twain shall meet.

– Rudyard Kipling, *Ballad of East and West*

Research about the role of Journalism in disaster has been done by disaster scholars from a number of areas of social science *and* by Mass Communications or Journalism scholars. The result is a dichotomy. The general social science literature on media and disaster rarely focuses on issues – such as ethical concerns – that dominate the Mass Communications and Journalism literature. The Journalism/Mass Com literature includes information that supports the findings from social science research but the authors do not make that connection. There is, in short, a great deal of information about the role the mass media play in crisis and disaster but it is found in two compartments.

When Tom Drabek reviewed the literature in the disaster field, he discovered a number of publications about mass media and disaster, but he also discovered that only a handful were published in Mass Communication or Journalism scholarly journals (Anderson, 1969; Drabek, 1986; Kueneman and Wright, 1975; Scanlon, Luukko and Morton, 1978; Waxman, 1973) or in monograph or book form (Singer and Green, 1972; Scanlon, 1976; Scanlon, Dixon and McClellan, 1982; Okabe, 1979). Similarly, when the author reviewed the main scholarly journals in the Mass Com/Journalism field – *Journalism and Mass Communications Quarterly*, *Journalism and Mass Communications Educator*, *Journalism Studies*, *Newspaper Research Journal*, *Quill*, *Mass Communications and Society*, *Public Relations Quarterly* and *Canadian Journal of Communication* – he discovered there were few articles about crises or disasters. When an article did appear even if it overlapped the disaster literature, the authors did not indicate that. Until September 11, 2001, that would have been the end of the story. However, since 9/11 the media have been giving massive attention to terrorism and to ethical issues related to terrorism – and the Mass Communications and Journalism literature has echoed that shift. But, once again, this new scholarship has not acknowledged the existing and relevant research.

This chapter reviews what is known about the media and crisis and/or disaster, whether this comes from the general social science literature or the Mass Communications and Journalism literature. It does not show – as the quote from Kipling implies – that the twain never meet. It does suggest a dichotomy. This is an important finding for, as E. L. Quarantelli has pointed out, practically everyone is willing to express views or opinions about what will happen in disasters, yet the great majority of people in Western society have only limited experience with disasters.

> So where do people get their images of disastrous phenomena if they do not base them on personal experiences? Some of the pictures they have undoubtedly come from deeply rooted cultural beliefs. . . . But we think a strong case can be made that what average citizens and officials expect about disasters, what they come to know on ongoing disasters, and what they learned from disasters that have occurred, are primarily if not exclusively learned from mass media accounts (Quarantelli, 1991, p. 2).

The social science literature has established that the media play a key role in many aspects of crisis and disasters. Mass media participation is critical, for example, for effective warning and the mass media may be the glue that binds societies in certain occasions. Yet, the media are also responsible for many of the misconceptions that exist about disaster, misconceptions that may lead to errors of judgment when disaster strikes. A review of texts suggests Journalism scholars are unaware of this. Strangely, the one area where media scholars have shown the most concern – the way journalists deal with survivors and relatives of victims – is the area where the limited available research suggests the media are not as guilty as painted.

MEDIA RESPONSE TO DISASTERS

It is now fairly well established what media does when disaster strikes. The media hear of the event, try to obtain more information, use their own files to add background to their stories, dispatch reporters and report anything they are told. Often they devote all their air time or much of the space available to that single story (Scanlon and Alldred, 1982). To gather material to fill this expanded news hole, the media draft anyone available. When two teenagers killed 15 students – including themselves – and wounded 13 others at Columbine High School in Colorado, KCNC-TV in Denver used every staff member available for its 13 hours of non-stop coverage:

> Well over 150 newsroom regulars and extras pitched in to make the extensive coverage possible. Off-duty employees came into the station without being summoned and took up posts. Newsroom hierarchies were discarded. Everyone, intern and news director alike, answered phones and responded when a need arose (Rotbart, 1999, p. 24).

On such occasions, the media will also use its technical resources and ingenuity to gather information. For example, when Mount St. Helen's erupted, NBC took a helicopter into the crater and persuaded a geologist to view and comment on the resulting tape. At Three Mile Island, staff from the *Philadelphia Inquirer* copied the license plates of all vehicles in the parking lot, traced the owners and started phoning them. Many were belligerent but 50 agreed to interviews (Sandman and Paden, 1979, p. 48).

All media monitor what their competitors are reporting and copy it if they think it is newsworthy. There are also many interconnections among the media. For example, almost all Canadian newspapers belong to the *Canadian Press* (CP) news agency. Everything is shared with CP which means any story produced by one paper is made available to every other paper. The electronic media have similar agreements. That's why visuals shot by one media outlet soon appear on stations around the world. These interconnections also mean that a false report can generate headlines around the world.

That, in fact, is exactly what happened in November, 1973, when Swedish radio broadcast a program about the nuclear power station at Barseback. The power station was still under construction but the program included dramatic fiction – set nine years in the future – about a radioactive release. That night and the next day all major Swedish media reported that the program led to widespread panic and that story was carried around the world by Reuters news agency. All those reports were based on an unsubstantiated report filed by one regional correspondent in Malmo:

> Panic was the main theme of his [report] panic in a whole country, perhaps two. [Malmo is just a short ferry ride from Denmark]. The telephone exchanges of the police stations, fire stations and mass media in two countries were reported to be jammed. People queuing before the civil shelters. Large crowds in the communities around Barseback taking to the roads. People in Malmo collecting their valuables and heading southward in their cars (Rosengren, Arvidson and Struesson, 1974, p. 12).

The story led to widespread comment and editorials, even questions in Parliament about how future similar panics could be avoided. The report in short was accepted as true because of the widespread belief among journalists that people do panic in crisis situations. But the researchers who interviewed 1,089 respondents found that while persons had reacted to the broadcast, there was not a single incident of flight or panic.

> The "behavioural" reactions to the programme as a rule consisted in contacting family members, relatives or neighbours, over the telephone or face-to-face. Other reactions were to close the windows, think over what to bring along in case of a possible evacuation, etc. No case of telephoning to the mass media, to the police or other authorities were found. . . . Nor did we find anyone having fled in panic (Rosengren, Arvidson and Struesson, 1974, p. 6).

The Barseback "panic" was a media invention that spread 'round the world. One reason why such a distorted account can be so readily accepted is that when major stories break, there is also widespread cooperation among reporters. That was true at Three Mile Island:

> From the moment the Harrisburg press corps heard about the accident [at Three Mile Island] . . . we all shared information. We got drawings and pierced together events. . . . We went out and got books on nuclear energy and compared them and discussed how a reactor works (Sandman and Paden, 1979, p. 16).

It was the same in Dallas, the day President Kennedy was assassinated.

> Throughout the day, every reporter on the scene seemed to do his best to help everyone else. Information came only in bits and pieces. Every one who picked up a bit or piece passed it on. I know no one who held anything out. Nobody thought about an exclusive. It didn't seem important. (Wicker, 1996, p. 28)

WARNINGS AND RUMOR CONTROL

In his review of the behavior of mass communications systems in disasters, Quarantelli concluded that passing on warnings is "Without doubt, the clearest and most consistent role [of mass media] in a disaster . . . (Quarantelli, 1991, p. 23). Warnings are effective only if they are specific about the threat, specific about who is affected and specific about what to do and – because persons hearing a warning from one source are inclined to check with another – they are effective only if they come from all possible sources. At Mount St. Helens, Perry and Green found that 80 percent of those who received a warning tried to confirm it with another source (Perry and Greene, 1983, p. 66). Since one source used to check is the media, an effective warning must come through the media as well as other channels.

When Peel Regional Police ordered an evacuation of Mississauga, just west of Toronto, they announced that there had been a train derailment and some cars were leaking chlorine and that there had been propane explosions and there could be more. The threat was clear. To make certain everyone knew if he or she was affected by this warning, they went door-to-door and were very clear about what residents should do – leave! Persons were told either to use their own vehicles or accept a ride on a Mississauga Transit bus. [Buses were coming along each street with police.] The warnings were reinforced by police cars using loud hailers alerting residents to the threat and the evacuation order.

Most important, instead of telling the media when they had ordered an evacuation, Peel Police told the media when they were about to order one – and provided maps so television could show precisely what area was to be evacuated next. Many residents first received the evacuation message over radio or television. Some heard first via a phone call from someone who had heard or seen a news reports (Scanlon and Padgham, 1980). They were ready to leave when the police arrived at their door.

The mass media can also play a vital role in keeping people informed after disaster strikes. When ice jams blocked the river and water poured over the dykes in Peace River, Alberta, officials ordered evacuations of several residential areas, schools and part of the downtown business section – almost at the same time. Families were separated – some were still at home but some were already at work – and some agencies had trouble locating their staff. [Their employees had been forced out of their homes and their offices had also been evacuated.] Everyone tuned in to local radio: 100 percent of a sample reported that was how they kept informed. Many said the only time they were really worried was when the local station temporarily went off the air. It was one of the businesses evacuated and had to reestablish in a building above the flood plain (Scanlon, Osborne and McClellan, 1996).

The media can also be critical in putting down rumours. When a severe windstorm hit Nova Scotia, there was a rumour that the ferry between North Sydney and Port aux Basques, Newfoundland, had sunk. The rumour stopped when the mayor sent a reporter from the one radio station still on the air to interview the ferry captain at the docks. The captain said that the voyage had been a rough one but his ship was fine (Scanlon, 1977). That killed the rumor.

CONVERGENCE

The media not only cover dramatic events, they cover them in a massive way. Within 24 hours, there were 325 media personnel in the isolated Newfoundland community of Gander after an air crash involving the 101st Airborne, several thousand media in Lockerbie, Scotland after the crash of Pam Am 103. There were media-created helicopter traffic jams over Coalinga, California after the earthquake and a media city with its own mayor and Saturday evening entertainment near the Branch Davidian compound during the stand-off at Waco. John Hansen, the assistant fire chief, handled media relations after the bombing at Oklahoma City:

> By the second day, we had nicknamed the media area "satellite city" as there was almost a two square block area of nothing but satellite trucks and live trucks lined up side by side. Several prestigious network television journalists told me they had never seen that many media trucks covering any single incident, including the O. J. Simpson trial. . . . As more and more reporters arrived from all across the country, I admit that I was in awe. On the other side of the microphones and tape recorders were the voices and faces we all know from "Nightline," "20/20," "Dateline," "48 Hours" and other shows (Hansen, 1998, pp: 56–57).

In 1957, Charles Fritz and J. H. Mathewson labeled this type of massive response to disaster as, "convergence." They said that in the wake of a disaster there are three types of convergence: personal convergence – the actual physical movement of persons on foot, by automobile or in other vehicles; informational convergence – the movement or transmission of messages; and material convergence – the physical movement of supplies and equipment (Fritz and Mathewson, 1957). They said all these forms of convergence cause problems – for example informational convergence jams telephones making emergency communications difficult.

> [In Lockerbie] massive congestion to the public telephone network . . . brought normal telecommunications almost to a standstill [because of] an insatiable demand for telephone lines for emergency and support services and for voluntary agencies and the media (McIntosh, 1989).

Fritz and Mathewson said convergence is a direct result of media reports partly because early media reports are not specific enough to satisfy the needs and curiosity of those hearing them.

> One of the most effective ways of securing such lead time would be to delay public announcements of disaster until the organized units would have had an opportunity to arrive on the scene. . . . The possibility of this type of coordination between the broadcast media and official disaster agencies should receive further consideration (Fritz and Mathewson, 1957, p. 75).

This conclusion was largely accepted for nearly 40 years; but it is flawed. For one thing, in a disaster the initial response is not by emergency personnel but by survivors and in a real disaster with widespread damage and destruction, there is no scene. But the major weakness with this conclusion is that convergence is not triggered solely by the media. In a study of a tire fire – 14 million used rubber tires burned for 18 days – Scanlon identified hundreds of responders, all legitimate: for example, 12 police detachments and three police forces, 26 fire departments, 27 federal government agencies, 60 voluntary agencies. None of that was triggered by the media. In fact media reports were, in the initial stages, quite limited (Scanlon and Prawzick, 1991).

Similarly, when a downtown office building filled with gas and exploded in North Bay, Ontario in 1975, there was no news coverage until 19 minutes after the explosion. In those 19 minutes, news spread by word of mouth so quickly that 80 percent of those interviewed by students belonging to Carleton University's Emergency Communications Research Unit (ECRU) reported they had first learned of the explosion by word of mouth. Only 20 percent first learned through radio or television.

> Asked if they had seen the disaster site, roughly half the people in the sample said, "Yes." A great many of them also said they got there very quickly. Eight point two per cent . . . said they had seen it within half an hour. Assuming that the sample was reasonably accurate, this means somewhere between 3,000 and 4,000 persons were at the site within the first hour. . . . Of those who went, about 45 per cent said they went from simple curiosity. . . . Only a small percentage – eight per cent – said they went because their jobs took them there (Scanlon and Taylor, 1975).

Since most of those persons learned through interpersonal sources, convergence was not solely or even mainly a result of news reports. Incidentally, those high-speed informal networks have their usages. For example, the passengers on the hijacked aircraft that crashed in Pennsylvania on 9/11 learned what was happening through calls on their cell phones. And it was informal rather than formal networks that led to such a quick response from neighbouring communities after North America's worst catastrophe, the 1917 Halifax, Nova Scotia explosion. (Approximately one-fifth of the residents were killed or injured when a French ship carrying munitions caught fire, then exploded in the city's harbour.) Within hours relief trains were en route from nearby centers. Convergence is not just a short-term problem. At Halifax, it was a problem for weeks and, at one point, passengers on all incoming trains were screened to block all but authorized arrivals. After 9/11 emergency services became almost frantic trying to stop volunteers – many of them emergency professionals – from flocking to the scene. Though the media are not responsible for much convergence, they can add to the problem by making unwarranted assumptions. It is not uncommon for media to say that nurses and physicians are desperately needed or that blood donors are wanted even though no such requests have come from official sources. The result is further convergence. If the media are their first source of information, people turn to other sources. A study of how persons learned about two hurricanes showed that more than 60 percent first saw both warnings on television, 17 and 25 percent heard first on radio.

> Apparently the warning messages [the ones seen on television or heard on radio] triggered the formation of a kind of hurricane culture. . . . Residents turned from the media to more personal communications channels, while maintaining environmental surveillance through the media. . . . Residents acted in accordance with their own perceptions of the situation, and those perceptions drove, and were affected by, all that they saw and heard (Ledingham and Masel Walters, 1989, p. 43).

Similarly, if other sources come first, people turn to the media (Kanihan and Gale, 2003, p. 89). On 9/11 when persons were informed by word of mouth about the attacks, they turned immediately to the mass media, especially television.

> Technically any single communication channel can not meet the information demands. . . . Our data on citizen preference suggest two important conclusions. First, a mix of channels should be used to send messages. Second, the news media need to be systematically incorporated into this mix (Perry and Lindell, 1989, p. 62).

Media, Victims and Relatives

While the media perform a number of useful roles in crisis and disaster, there is one thing they do that arouses considerable criticism – and that is the way they treat victims and their relatives. When Pam Am 103 went missing over Lockerbie, Scotland, journalists waiting for information about the flight were cordoned off near the first class lounge at New York's Kennedy Airport. Seeing them, a woman asked what the fuss was about. An official said a Pam Am plane had crashed. She asked the flight number. He replied, "1-0-3." She collapsed on the floor, screaming, "Not my baby. Not my baby." While her husband tried to shield her, photographers and television crews recorded her grief.

> All I remember is losing control. . . . I remember lights all over. I felt like I was being raped by the media. I am usually a woman who is very much in control. I'll have to say that was one of the few moments in my life where I was out of control. And I felt the media chose

> that moment. I felt violated. I felt exploited. And there was no one there to protect me (Deppa, 1994, p. 29).

When she finally left Kennedy airport, she noticed something on her taxi's front seat:

> I saw a newspaper – I can't remember what the headlines were but it had to do with Pam Am – so I asked the driver, "Can I see that newspaper?" It was the Daily News. And there on the front page was a picture of myself on the floor of the airport, and I was actually appalled. I just couldn't believe it (Deppa, 1994, p. 33).

Incidents like that made Everett Parker of the United Church of Christ openly critical of the mass media during meetings of the Committee on Disasters and the Mass Media:

> Day in and day out, we see reporters bullying statements out of stricken people; they take pride in their ability to do so. . . . It is dehumanizing to stick a camera and a microphone in the face of an injured or bereaved person and demand a statement. It is unconscionable for reporters and editors to use the human elements in disaster to feed the morbid curiosity of viewers, listeners and readers (Parker, 1980, p. 238).

Yet the media are not as guilty Parker charges. Although there is a widespread perception that in the wake of incidents the media act as ghouls, harassing victims and the relatives of victims, showing no sensitivity, this perception is misleading. Both anecdotal and research data suggests some victims and relatives welcome a chance to talk to reporters. After the 1985 Gander air crash – a crash that took the lives of U.S. soldiers – an officer was assigned to media relations at the soldiers' home base, Fort Campbell, Kentucky. He told the media that the military intended to protect the privacy of the soldiers' families. They would have access to families only if the families requested it. To his surprise, a number of families did ask to speak to the media. This same approach was used by Oklahoma State University after basketball players and athletics staff was killed in a plane crash:

> While the media were given new information whenever it became available, they were also asked to respect the privacy of families involved in the tragedy. During the university's memorial service on Jan. 31, 2001, PIO staff members ensured that the media were restricted to a specific designated area. Media were asked not to harass family members; however, family members who felt comfortable talking to the media were not discouraged (Wigley, 2003).

Although they are pressured by editors to do such interviews, reporters find approaching such persons distasteful.[1] Kim Brunhuber recalled shooting visuals of relatives of the victims of the Swissair crash off Nova Scotia. Brunhuber was outside the Lord Nelson Hotel, where the relatives were staying:

> She catches sight of our camera 20 feet away, lowers her head, pulls back part of her black dress to hide her face. When we put our report together we stay with the shot until the moment she shields her face. Saving us the public acknowledgment of our grim voyeurism. Days later, what I suspect becomes clear. I can edit the shot, but I can't edit my guilt (Brunhuber, 1998).

Though many may share Brunhuber's guilt, reporters often discover they are made welcome when they approach relatives of those who died. These relatives are anxious to talk to someone and the reporter is anxious to listen. The result can be a relationship satisfying to both parties. When the Broadcast Standards Committee of the United Kingdom interviewed 210 victims of violence and disaster, including 54 who had been interviewed by reporters, three-quarters said they were not offended. That was especially true of those involved in a disaster. Most who complained were upset with newspapers, especially tabloid reporters, not with broadcast journalists. Survivors said they were prepared to be interviewed if the stories had a purpose, for example, "exposed the human frailties and negligences that had contributed to major disasters and so help to minimize the danger of such disasters happening again" (Shearer, 1991).

INTRUSION RESENTED

There have, however, been cases where the intrusion was obvious and journalists and journalism

1. After 16 children and a teacher were killed at a school in Dunblane, Scotland, some reporters who had been ordered to interview victims' families made sure their approaches were noticed by police officers who ordered them to leave. This allowed them to explain their failure to their editors.

organizations are becoming wary of this. After the Pam Am 103 crash, for example, there was a vigil in the Hendricks chapel at Syracuse University because a group of Syracuse students was among the victims:

> As the chapel filled, the media were asked to stay away from the area in front of the raised platform, where chaplains and representatives of the various faiths would lead that service. Photographers were asked not to use flash. But the emotion generated by the event, especially in the moments of meditation between scriptures and sacred music created compelling pictures, and the whir or the automatic levers advancing film echoed from both sides of the sanctuary. Soon flashes began going off. Upstairs, at the back of the balcony, a local television reporter "went live" over protests of students in the area (Deppa, 1994, p. 51).

It is because of incidents like that two widely shared codes of ethics now caution against insensitive approaches.

> Professional electronic journalists should treat all subjects of news coverage with respect and dignity, showing particular compassion to victims of crime or tragedy [and that] Professional electronic journalists should refrain from contacting participants in violent situations while the situation is in progress (Radio Television News Directors Association Code, 2000).

> Show compassion for those who may be affected adversely by news coverage. Use special sensitivity when dealing with children and inexperienced sources or subjects. Be sensitive when seeking or using interviews or photographs of those affected by tragedy or grief (Society of Professional Journalists).

However, reporting texts have described approaching survivors and their relatives as difficult but necessary:

> One of the toughest things that a reporter has to do while covering a disaster is to interview the families of the victims. At no other time does the public's right to know seem to come into direct conflict with people's right to privacy. Professionals realize that if they handle the interviews with a great deal of sensitivity, they can offer survivors an opportunity to grieve openly and to eulogize a loved one (Itule and Anderson, 1984, p. 348).

One journalism publication – *Nieman Reports* – produced some guidelines for such approaches. It suggested reporters ask for permission to do such interviews and that they should stop taking notes or recording if the interviewee asked them to. It also suggests the journalist state precisely what the interview will be about before starting. These things, it says, will give the interviewee a sense of power and reduce their uneasiness (Cote and Bucqueroux, p. 27). Interestingly, the woman whose photo was taken at Kennedy airport agreed to talk to reporters in Syracuse after she returned home:

> I think it was the way the media approached me on the phone. . . . They were not pushy. They asked permission. . . . They knew it was a difficult time and they would accept the fact if I chose not to. . . . And soon after that there was some information that Pam Am had received a warning about this, that Pan Am had received notification. At that point . . . I had a sense of anger that needed to be acknowledged. So I think that was another factor that influenced my decision (Deppa, 1994, pp. 33–34).

HUMAN INTEREST STORIES

By making such approaches and using the information they acquire to write about the victims as individuals, the media play another role that is largely ignored in the literature. They do what might be called "humanize" events.[2] They do this by not just providing a broad overview of what has happened but by focusing on the individuals involved. *Life Magazine* did that on the Viet Nam war when it ran the photo of every service person killed in a single week. The *New York Times* did that after 9/11 when day after day it ran photos and brief articles about those who died as a result of the terrorist attack, material later incorporated into a book (*New York Times*, 2003).

However, this "humanization" process – especially when it is done immediately after an incident – can have a down side. It can lead to a distorted impression of the impact of an event. Noting that "human interest" stories are staple items in disaster coverage, Wenger, James and Faupel suggests they tend to focus on those who were most severely impacted:

2. I am indebted to a former researcher, David Tait, for this idea: Tait is now a member of faculty at Carleton University.

> Such stories detail the plight of the individual who has been "wiped out" by the disaster, who has lost their family, or suffered great misfortune. Of course, such individuals are covered by the media because they "stand out" from the other victims; they are sought by the media. However, these atypical cases are often presented as if they were typical. . . . Death, economic loss, human suffering, and social disruption are the standard themes in the media's portrayal of disaster. For the audience, the apparent image is one of total destruction (Wenger, James and Faupel, 1980, p. 40).

Another aspect of this humanization process is the attempt to link an event elsewhere to the publication's perceived audience. Journalists call this searching for a "local angle." This means events are more likely to be reported if they occur close to the place of publication and more likely to be reported by media in a specific country if that country's nationals are involved. For example, when Gladys and Kurt Engel Lang reviewed the 139 disasters included in a book illustrating front pages from the *New York Times*, they found:

> Of the 18 really big stories . . . those for which coverage ran over four different pages – 5 occurred within the New York area. Because only 7 of the 139 disasters were in the New York area, it seems evident that the local ones get special treatment in the *Times* (Lang and Lang, 1980, p. 217–272).

Anyone watching American, Canadian and British television after the tsunami hit Asia would have been acutely aware of this phenomenon. The media in all three countries tended to focus on stories about victims from their own countries and about response activity by their own personnel including military personnel.

HUMAN BEHAVIOUR

Disaster research has shown that victims are not dazed and confused and in shock but instead do most if not all of the initial search and rescue. It has shown that panic is so rare it is difficult to study and that the real problem is not panic but an unwillingness to believe the clearest possible warnings (Quarantelli and Dynes, 1972)..For example, when the freight train derailed in Mississauga, even though they could smell chlorine and could see and hear propane tanks exploding and flying through the air like flaming missiles, a few still refused to leave. Research has also shown that looting in the wake of disasters does happen – it did after Hurricane David – but it is extremely rare. Usually, crime rates fall (Scanlon, 1992). When hundreds of passengers were diverted to Gander, Newfoundland as a result of the attacks on the United States on September 11, 2001, there was not a single crime reported while the passengers were in that community (Scanlon, 2002).

The few reporting texts that have touched on disaster coverage appear unaware of this research – they assume the myths are true:

> The enterprising reporter covering a disaster can often add color to a story, using quotes from survivors telling of their escapes, tales of courage or cowardice, descriptions of carnage and panic (Metz, 1991, p. 297).
>
> Another difficulty is the emotional nature of the event: tragedy, destruction, pestilence and death are emotion-packed news events. In the midst of fear, panic and loss, sources become confused and antagonistic (Stone, 1992, p. 144).

Kueneman and Wright found reporters felt it was their duty to shape their stories to avoid panic:

> The following comments from interviews are characteristic of their orientation. "You must be very careful that you don't over-emphasize what is taking place." "I think you can create a good deal of panic if you're not very careful on the air; you can scare people out of their wits." "We are caught in a dilemma: we try not to minimize the danger, yet try not to create panic" (Kueneman and Wright, 1973, pp. 671–72).

Others have suggested reporters should be sympathetic to officials with the same goal:

> Because official sources are often worried that the press is going to distort the story, they may sanitize information before releasing it. At the same time, reporters must realize that public officials are trying to avoid unnecessary panic (Itule and Anderson, 1984, p. 97).

The media may downplay negative stories, especially in their own communities. For example, in the wake of Hurricane David in Dominica, there was substantial looting; but journalists covering the hurricane tended to ignore that and, even when they did not ignore it, their reports tended not to be broadcast:

> The rampant looting behavior during the hurricane's strike on Dominica in Roseau [Dominica's capital], and later the looting of stored relief supplies in both Roseau and Melville-Hall Airport, was common knowledge among local officials and residents. It was even observed first hand by several reporters. However, this . . . received minimal attention in most news reports (Rogers and Sood, 1981, p. 65).

That type of caution showed up on television networks in the wake of the terrorist attacks on September 11, 2001:

> There is no point in allowing this thing to appear worse than it is, it is already horrendous, and we don't need to make it worse by misstating numbers and we want you to keep that in mind – CNN anchor Aaron Brown.

> Tom as you point out we try not to exaggerate very much in this circumstance, and yet in many ways it's hard not to exaggerate just the things we have been seeing and the things we are told – NBC reporter Pat Dawson talking to Tom Brokaw (Reynolds and Barnett, 2003, p. 698).

None seemed aware that people find it easier to cope with the truth, with clear factual accounts of what is known about what is happening. It is lack of clarity and confusion, not accuracy, that makes persons uneasy. Yet, this same misunderstanding showed up in an article published in *Journalism Quarterly*, the leading scholarly journal:

> At Three Mile Island, reporters faced a pressure that was new to science reporting. Residents of the area monitored news reports for hints of whether to flee. Overly alarming coverage could have spread panic; overly reassuring coverage could have risked lives (Stephens and Edison, 1982, p. 199).

Scanlon concluded:

> A review of Journalism text books suggests that the authors who deal with disaster coverage often state as fact what disaster scholars have shown to be inaccurate. Perhaps that explains why the myths about disaster are perpetuated in the media. Most likely, the students who used these texts were influenced by the inaccurate representations . . . when they became reporters (Scanlon, 1998, p. 45).

Wenger found that those who learned about disasters mainly through the media were more likely than others to believe the myths (Wenger, 1985) though this does not seem to affect what disaster-stricken individuals do. While they believe the myths they do not act as if they do. They believe there will be panic but do not panic. They believe looting will occur but do not loot. Unfortunately, the same is not true for organizations. For example, emergency agencies hold back warnings for fear of panic. This happened in China:

> Officials in Amuer were frightened that a false warning [of a forest fire] might panic local residents. Unfortunately, district fire officials did not recognize the seriousness of the threat and no warning was issued. When the fire reached Amuer at 11 p.m., nearly four hours after it struck Xilinji, many persons had gone to bed. That made it much more difficult to alert everyone and organize an evacuation. There were 25 deaths (Xuewen, 1999).

COMMAND POST VIEW

Quarantelli concluded that media ignorance may lead to what he calls the "command post" view of disaster. By that he means that since journalists don't know the important role that survivors play in search and rescue and initial transport to hospital – and are unaware how limited a role emergency personnel play in early response – they will assume that emergency officials know what is going on. Their reports will reflect the official view of what has happened and is happening. Journalism texts seem knowledgeable of this problem but not how to solve it:

> Writing the first story of a major disaster such as an earthquake or a tornado presents a particular challenge. Officials are often unsure and can only guess (Harriss, Leiter and Johnson, 1992, p. 293).

> . . . when tornadoes slice through cities, efforts may be concentrated on finding those trapped in the rubble of buildings. No one may be certain who is trapped or where they are or, for that matter, whether anyone is trapped at all (The Missouri Group).

> . . . the safest thing for any reporter . . . is to say quite frankly what people can see for themselves that no one has any accurate casualty figures and that it may take some time to arrive at an accurate count (Hohenberg, 1983).

> The enormity of destruction is so vast that no one source can accurately assess the toll in human lives until

the disaster begins to abate and energy can be devoted to gauging human and property losses (Mencher, 1981).

Rich agrees much early information is unreliable, but suggests reporting it anyway:

> Estimated costs of damages and property loss: Initially these accounts – from insurance agents, fire departments, police officials or state officials – are often inaccurate, but they add an essential element to the story (Rich, 1997, p. 489).

She also suggests reporters should check into looting: "Check with the police to find out about looting or other post-disaster crimes. . . ." (Rich, 1997, p. 489).

Most authors either imply or state that eventually there will be accurate figures on injury and death, which is not correct. UPI correspondent Jack Virtue reported in the aftermath of an earthquake in Guatemala that "The number of lives lost may never be accurately tallied. Many bodies will never be found." Similarly, Janet Kitz, author of Shattered City, a book on the catastrophic 1917 Halifax explosion, wrote:

> I am frequently asked how many people died in the explosion, but I am reluctant to give a definite answer. I have come across so many different figures; for example, 1,635 or 1,963. No list I have seen has ever included all the people I know to have died. I believe the figure was higher than 2000 (Kitz, 1989, p. 15).

If the death toll cannot be calculated with precision, it would be harder still to calculate the injury toll because in the wake of disaster many victims decline to go for medical help for what they see as minor injuries. Even those who do go for help are often not recorded accurately. In disasters, record keeping is one of the first casualties (Scanlon, 1996).

Another problem with the "command post" approach to coverage of disasters is that it tends to ignore non-traditional activities such as search and rescue, conducted mainly by volunteers working in emergent groups. Wenger and Quarantelli found that only 8.6 percent of newspaper articles on disaster and 8.4 percent of electronic media reports on disaster mention search and rescue. When search and rescue was mentioned, those stories inevitably relied to some extent on non-traditional sources. In other words, to cover search and rescue activity, reporters would have been forced to use non-traditional sources for their information. These sources were often missed so an important activity was given rather slight attention (Wenger and Quarantelli, 1989, p. 62).

The lack of understanding of disaster was reflected in another way. Most media did not have disaster plans for their own organizations – no plans as to how they would continue to operate in such conditions, no plans as to how they would deal with the demands of disaster coverage:

> Even in the minority of those outlets that had engaged in prior planning, it was generally of inadequate quality. . . . Furthermore, those plans were usually outdated, never exercised and often could not be located by the staff (Wenger and Quarantelli, 1989, p. 33).

On 9/11 the *Wall Street Journal* proved an exception. The *Journal* had a back-up facility with equipment installed and the decision to get it up and running was made as soon as the first plane hit the first tower.

> The facility . . . had been outfitted in the past 18 months with a couple of classroom-sized spaces full of computer work stations. And in recent months, Journal editors under Pensiero's direction [Jim Pensiero is the Journal's assistant managing editor] had spent a couple of Saturday mornings making up the paper there, in case of emergency.
>
> The South Brunswick offices seemed from another world – a comfortable, modern, suburban campus with expansive green lawns. The two "emergency" newsrooms were ready to go, and staff had prepared additional ones, so that 55 workstations were operational – most with Hermes pagination and edited software. . . . The Journal's copy chief, Jesse Lewis, was on the premises (Baker, 2001, p. 13).

The move was handled so well that the Journal managed to deliver to their subscribers all but 180,000 of its normal 1.8 million copies. The paper was somewhat smaller than usual – two sections instead of three – and had one other unusual characteristic. For the second time in the paper's history it had a banner headline:

TERRORISTS DESTROY WORLD TRADE CENTER, HIT PENTAGON IN RAID WITH HIJACKED JETS

The only previous banner headline was for Pearl Harbor.

It might be assumed that when disaster occurs, all media would cover it, especially if it occurred in their coverage area. That is not the case – at least for radio. Wenger and Quarantelli found that while newspapers and television provided extensive coverage of disaster – 83.3 percent of television stations pre-empted regular programming – many radio stations ignored disasters, even local ones:

> A total of 18.6 per cent of the radio stations [examined for the study] did not cover the disaster in their community at all. Three of these were small stations with no news department; they continued with their normal programming. . . . Thirty per cent of the stations who covered the disaster in their area never pre-empted local programming, and 28.3 per cent did not increase their normal time allocated for news (Wenger and Quarantelli, 1989, p. 39).

While all television and print media tended to expand news coverage – television replaced regular programs, newspapers enlarged the space for news – the two did so differently. The electronic media tended to reduce the normal gate-keeping function by which editors control what goes over the air. Reports are broadcast live and interviews with those calling in or reached by phone are run without editing (Waxman, 1973; Sood, Stockdale and Rogers, 1987).

> Video tape was not edited as carefully as usual and significantly more live coverage was aired. One station we studied, for example, devoted hours to live coverage of a major toxic spill from its own helicopter. Raw tape brought to the station was aired in unedited form. Another station in the immediate aftermath of a tornado aired live footage shot out of the station's back door, and also placed raw tape taken by a citizen with a home videocamera on the air (Wenger and Quarantelli, 1989, p. 14).

In contrast, print media assign the task of "rewrite" to reporters who normally would not perform that function. Those rewrite persons take material filed by reporters in the field and shape it. Then the stories pass through the normal gate-keepers, editors and copy editors. Editors' perceptions about disaster still controls what appears in the paper. Stories that challenge editors' misconceptions will not be published (see Breed, 1955).

JOURNALISM LITERATURE

Very little of the material cited so far comes from Mass Communication or Journalism publications or from authors located in Schools of Journalism and/or Mass Communications published before 9/11. The exceptions were three books – *Bad Tidings Communications and Catastrophe*, a book by Lynne Masel Walters, Lee Wilkins and Tim Walters (Walters, Wilkins and Walters), *The Media and Disasters Pam Am 103* by Joan Deppa and others at Syracuse University (Deppa, 1994) and *Media Ethics Issues* by Philip Patterson and Lee Wilkins Philip (Patterson and Wilkins, 1998) – and a few articles: Scanlon's article critiquing reporting texts, Shearer's review of how survivors felt about the media, Wigley's report on how Oklahoma State University dealt with the media after a fatal plane crash, and Sood, Stockdale and Rogers' articles about how the news media operate in natural disasters.

There were a few other articles in the major Journalism journals prior to 9/11 – and some add a little to our knowledge of the media in disaster, usually in relation to ethics. These include: a study of unethical use of visuals in television coverage of crises (Smith, 1998); a review of media coverage of an earthquake prediction (Showalter, 1995); and a review of media coverage of two mass fatalities, one at the Hillsborough football grounds in Sheffield, the other at an elementary school in Dunblane, Scotland (Jemphrey and Berrington, 2000).

Smith reviewed coverage of several incidents to see if television used archival visuals that portrayed an inaccurate image of current conditions without identifying the fact the visuals were dated. This is against network news guidelines. He found more than 1,000 visuals used more than once; only seven of the thousand properly identified:

> Among stories about the Exxon Valdez oil spill, 23 separate video clips of oiled shorelines were recycled more than once, including a scene of oily rocks used 16 times by CBS and a helicopter shot of oiled shoreline used 10 times by NBC. None were labeled as file footage though it was sometimes apparent from the reporter's narration that the video did not represent current conditions (Smith, 1998, pp: 252–253).

Showalter described what happened when Iben Browning claimed conditions were ripe for an earthquake in the New Madrid Earthquake Zone – on December 3, 1990.

> . . . Browning was not a geologist or seismologist, he had no formal training in climatology, his doctorate was in zoology not physiology, he had not predicted the Loma Prieta earthquake, and what he called his projection was based on a widely discredited theory (Showalter, 1995, p. 2).

Nevertheless the Browning projection received widespread coverage, partly because – though all stories included someone challenging him – the challenge and his "prediction" were given equal play.

> . . . it appears that the different ways journalists and scientists define balanced coverage will remain a problem. For journalists, it is sufficient to present two opposing viewpoints. For scientists, such a practice represents biased reporting because it places a single individual on one side of an issue on equal footing with hundreds if not thousands of scientists on the other side of the issue (Showalter, 1995, p. 10).

It's an issue that also shows up in coverage of terrorist activity. A statement by previously unknown persons is matched by a statement from an authority, perhaps even someone as important as a White House spokesperson or the President. This raises the status of the hostage takers. It is possible to argue that the best approach might be for the media to ignore something but this, too, raises issues. However there may be a "catch 22." Ralph Turner found that when the media disregarded rumours about earthquakes, this might have been counterproductive:

> A substantial minority of the population believes that the scientists, public officials, and news people know more about the prospect of earthquake than they are willing to tell the public – and that responsible public leaders are withholding information indicating that awful things are going to happen. . . . By ignoring rumors rather than airing them and presenting authoritative contradiction, the media may have fostered the conviction that valid information was being withheld (Turner, 1980, p. 283).

Another article concluded that perceptions influence journalistic behaviour and news reports. After the Hillsborough soccer crowd crush incident (96 persons died as a result of overcrowding at one end of the field), the media were aggressive in going after survivors and the relatives of victims because they saw the deaths as a result of hooliganism and alcohol. After a massacre at an elementary school in Dunblane, Scotland, the media were far more sensitize about grief, even agreed to leave the community before the funerals.

> Pre-existing negative impressions of Liverpool combined with journalistic selectivity were crucial in shaping press coverage. . . . Initial accounts focused on football hooliganism (an important political issue at the time) and alcohol as primary causal factors, therefore established those involved as less-than-innocent victims. . . . Early reports . . . stated unequivocally that Liverpool supporters had "forced a gate" leading to the crush inside the ground. In cross-examination during the Home Office Inquiry . . . Chief Superintendent Duckenfield, the senior officer in charge, admitted he had lied about supporters forcing the gates and "apologized for blaming the Liverpool fans for causing the deaths." Despite this denial Duckenfield's initial comment established an international reported myth which still persists (Jemphrey and Berrington, 2000, p. 473).

This was in sharp contrast to Dunblane where a man shot and killed 16 children and a teacher and shot and injured 13 other children and three adults:

> The positive and sympathetic portrayal of the community had an effect on the behaviour of journalists, particularly the British press. The agreement to leave before the funerals took place was described by journalists as 'unprecedented'. . . . Such a decision by the national press is unusual, though the local press may be more sensitive to community feeling. . . (Jemphrey and Berrington, 2000, p. 481).

POST 9/11

The morning of September 11, 2001, changed the amount of attention in the media and the media literature given to untoward events. Jack Lule found that *New York Times* editorials were consumed by the attack for more than a month afterwards (Lule, 2002).

> From 12 September to 12 October the paper published eighty-four editorials, usually three a day. Of these eighty-four, fifty-eight (69 percent) were directly related to the consequences and aftermath of the terrorist attacks. For the first eight days after the attack every editorial confronted some feature of September 11. No other issue was worth of consideration. And for seventeen

days, the only other issue that merited attention was the New York mayoral race to decide the successor to Rudy Giuliana, a race ultimately shaped by the attacks (Lule, 2002, p. 280).

Another scholar found the same concentration in the issues of three news magazines and found that the coverage followed a pattern:

> During the month following the attacks, these three magazines [*Time*, *Newsweek* and *U.S. News & World Report*] told a cohesive story of the tragedy and its aftermath, a story that moved from shock and fear to inspiration and pride. They did so by using testimony from readers and mourners across the country, as well as from victims and witnesses of the attacks. These actors participated, along with the journalists themselves, in the performance of a ritual with symbolic visual representations of candles, portraits of the dead and the American flag. . . . Overall, this coverage corresponded with the stages of a funeral ceremony. In that sense, it provided evidence that journalism plays an important role in – and can in certain circumstances be a form of – civil religion (Kitch, 2003, p. 222) [There was a similar approach after the assassination of President Kennedy.]

There was continued attention to ethical issues. The Kratzers interviewed editors to determine how they decided it was appropriate to use photos of persons trapped in the upper floors of the twin towers or photos or photos of individuals jumping to their deaths:

> The results reveal that many of the editors . . . engaged in debate about running the photographs and the main issues that emerged were reader response, the victims' privacy, and the ability of the photographs to communicate the story. Although many editors found the photographs disturbing, the overwhelming reason for publishing them was that they added to the visual storytelling of what happened. Many editors believed that readers needed to be exposed to the disturbing images in order to fully comprehend the story of the day (Kratzer and Kratzer, 2003, p. 46).

This was in line with what Deppa and others found in their study of coverage of Pam Am 103 at Lockerbie, specifically when a body was brought down from a roof:

> The day they brought the body down, the photographers were running around stupid, a neighbourhood resident recalled. They were running through my garden, up onto my step to get as near as they could to a photo of it being brought down. That was really ghastly and I thought they were pigs at the time.

Three print publications – *Time*, *Newsweek* and the *Washington Post* – used those photos. Scottish television was more discreet: ". . . my cameraman actually got a very close-up shot of it [the body]. I thought we can't use this. I said, "Can you imagine how the relatives of this particular person would feel if they saw that?" The cameraman . . . agreed. He said not to show the close-up of the body."

Television was equally discreet in its coverage of the Columbine school incident in Littleton, Colorado.

> Nor did the station opt to show gore. KCNC editors had plenty of film to exploit had they wanted. In particular, cameramen captured one police SWAT team dragging two of the victims' bodies across the school lawn – images that never once aired. In the heat of the story chase, newsroom editors talked about their responsibilities to decency and community values. No one dissented (Rotbart, 1999, p. 24).

Because print photographers are unable to match the immediacy of radio or the drama of movement conveyed by television, they tend to be aggressive in trying to get visuals others don't have. Pijnenburg and Van Duin noticed that in Belgium in the wake of the Zeebrugge ferry accident:

> Some journalists behaved also rather badly when a funeral chapel was installed in Zeebrugge's sports centre. They had to be dissuaded to enter the building "manu militari" by the police forces and emergency services' personnel. But it was impossible to prevent aggressive photographers from pursuing and harassing completely distressed relatives of victims on their way to and entering the funeral chapel (Pijnenburg and Van Duin, p. 342).

RUMOURS NOT REPORTED

Although after 9/11 print editors were willing to print photos that seem marginal, they did not, according to Lasora, publish many rumors. He scoured the web looking for post 9/11 rumors.

> . . . someone rode a piece of the World Trade Center to safety, that gasoline prices would soar, that terrorists would attack a major shopping mall on Halloween, that

additional terrorists were thwarted at the New York area airports, that Jews working in the tower were warned ahead of the attacks, that the hijacked jet that crashed in Pennsylvania was shot down by U.S. forces, that videos of Palestinians celebrating the attacks was fake, and that Nostradamus predicted the attacks (Lasora, 2003).

He also found stories that persons – in one case fire fighters, in another police – were found alive in the rubble of the World Trade Center, days after the collapse.

> In the aftermath of the terrorist strikes, major newsmagazines, newspapers, broadcast news stations and cable news stations reported scores of stories related to the attacks. Yet despite unusually difficult reporting circumstances these media did a remarkably good job of separating out false rumors. This study found only four cases where the mainstream news media carried false reports. Furthermore, while they were disseminated widely, these stories were in most cases corrected quickly, once the truth was uncovered (Lasora, 2003, p. 14).

The rumors that were published were all about persons found alive in the rubble. Even though every report was wrong, there were "good news stories" editors could not resist.

Lasora notes that the media usually corrected the rumours as more information became available. This is in line with disaster research which suggests rumours spread in the wake of a disaster may persist until they are contradicted. And, though he did not mention this, his findings also fit with disaster research that shows that, in time of disaster, print media tend to maintain the traditional gate-keeping functions[3] but electronic media do not. And it was television that captured the bulk of the audience after 9/11. Wilson found that young and old alike, no matter what their previous media habits, turned to television – and others found that it was there they heard the rumours:

> Frequency of TV use before the attacks was not a significant predictor of the degree of dependency on TV after the attacks. Apparently, individuals who used radio, print media and the web during normal times relied on TV to a greater degree in the months following the crisis, and the leading force in this change was the perception of threat. It is clear that TV is the medium of choice in a national crisis, and this preference is not simply the result of habit (Wilson, 2004, p. 354).

Those ties to television developed very quickly, often within minutes of the attacks.

> Half of our respondents first learned of the attacks from the broadcast media (28 percent from television and 16 percent from radio). Interestingly, we found that 6 percent of our respondents found out from a mix of broadcast and interpersonal channels: These respondents indicated that someone (often a parent) telephoned them and simply told them to "turn on the TV." The magnitude of the events was so large, incomprehensible, and, at first unclear, that some people alerted others interpersonally but quickly instructed them to see the images on television to explain the catastrophe. Almost half (48 percent) of our respondents learned about the tragedies from another person (Kanihan and Gale, 2003, pp: 82–83).

And – just as disaster scholars would have predicted – the electronic media allowed rumors and commentary to be broadcast:

> . . . journalists who covered the breaking news of the September 11 terrorist attacks used multiple roles to deliver information including that of expert and social commentator; they reported rumors, used anonymous sources, and frequently included personal references in their reporting regardless of which role they assumed. . . . The content of breaking news reported live is fundamentally different than the content of news stories that are produced with more time to check for violations of journalistic conventions. Further the role of the journalist is less clear during breaking news (Reynolds and Barnett, p. 669).

TERROR

Recently, U.S. media have had to deal with the fact that many destructive events are not caused by nature or human frailty but are deliberate acts. Despite Pam Am 103 – the plane that crashed at Lockerbie – and the bomb at the World Trade Center and

3. The concept of gate keeping was first elaborated roughly 50 years ago. See, for example, David Manning White (1950) "The 'Gatekeeper': A Case Study in the Selection of News" *Journalism Quarterly* Vol. 27 No. 4 pp: 383–390.

Oklahoma City and other incidents elsewhere, that message was been slow to sink in. However, the aerial attacks on the twin towers of the World Trade Center drove it home. Destructive events are now covered in two ways. There is coverage of the aftermath of the incident, and there is coverage of those who state openly that they caused it and the government reaction to that. In the past, when incidents involving human error occurred, the media often searched for a scapegoat, for someone to blame (Bucher, 1957). They no longer have to look. Terrorists are not only willing to admit responsibility for their acts; they plan them for maximum attention.

The reason Black September took Israeli athletes hostage at the Munich Olympics was because the media were on hand. One reason why Al-Qaida flew aircraft into buildings in New York City is because New York City is a media centre. Massive coverage was guaranteed. The Palestinian Liberation Organization's (PLO) representative at the United Nations said that the first PLO aircraft hijacking "aroused the consciousness of the world to our cause and awakened the media and world opinion much more and more effectively than 20 years of pleading at the United Nations" (Hickey, 1976, p. 12).

How rapidly the media are taken over by even minor incidents – and how quickly they allow those involved to set the agenda – was shown in March, 1977, when Hanafi Muslims occupied three Washington, D. C., buildings, killing one man and taking hostages. CBS in Chicago got a call from a man stating he was an Hanafi Muslim at the Chicago Muslim temple. Without checking, CBS allowed him live on air:

> The young man who could have been Santa Claus, for all the reporters knew . . . was addressing nearly two million people. . . . As it turned out he had much to say but it did not pertain to the siege in Washington (Jaehnig, 1978, p. 719).

The fact so much attention is given to these incidents raises the question of whether this leaves the impression that these persons are much more powerful than they really are. Al Qaida' s attack on 9/11 was successful in the sense it caused massive damage in New York City and led the U.S. government to take actions which disrupted North American and trans-Atlantic air travel. Yet, other terrorist groups have managed to get a very high profile though their numbers were very small indeed:

> . . . anxious newspaper readers . . . were led to believe that the German Baader-Meinhof group, the Japanese Red Army, the Symbionese Liberation Army . . . were mass movements that ought to be taken seriously indeed. . . . Yet these were groups of between five and 50 members. Their only victories were in the area of publicity (Lacquer, 1976, p. 102).

Since 9/11, many terrorist groups have used a similar approach to Al Qaida. They have selected a high-profile target and they have gone after a Western country, often a country, like Spain with a connection to the United States. There was the bomb that killed 202 persons, many of them Australians in Bali on October 12, 2002. There were the incidents at the theatre in Moscow and the school in Beslan, and the hijacking of two Russian aircraft, all the work of Chechens. There were the bombings at Luxor and Taba in Egypt, the second linked to the first incident at the World Trade Center. [A leaflet left at the scene demanded the release of Umar Abd al-Rahman, who was imprisoned by life after being convicted in connection with it.] There was the prolonged hostage-taking in Lima by Peru's Marxist-Leninist Tupac Amaru Revolutionary Movement. There was the attack on the Central Bank in Colombo, Sri Lanka, which was tied to the Liberation Tigers of Tamil Eelan, and, of course, there was the attack in Madrid, which was tied to Al Qaida.

The ability of terrorists to create an event which catches the media's complete attention indicates another significant role the media play in disasters – and in disaster research – agenda setting. As Bernard Cohen pointed out in The Press and Foreign Policy:

> It [the press] may not be successful much of the time in telling people what to think, but it is stunningly successful in telling its readers what to think about. And it follows from this that the world looks different to different people, depending not only on their personal interests, but on the map that is drawn for them by the writers, editors and publishers of the papers that they read (Cohen, 1963, p. 13).

In fact when it comes to disasters, the agenda-setting function is greater than this. To a large extent – as Scanlon pointed out in the foreword to a forthcoming book: *What is a Disaster?* Rogers and Sood pointed out much earlier – the agenda-setting power of the media determines which events come to public attention and which do not:

> The media have the ability to tell us that some issue of topic is news today, and by their silence, that millions of others are not. Certain media like the New York Times set the agenda not only for their own readers, but for many other of the mass media. By their very decision to cover (or not to cover) a disaster, or some aspect of a disaster, and by the prominence (or lack of prominence) given such coverage, the media wield great influence on authorities' decisions to seek (or not to seek) more information concerning that disaster (Rogers and Sood, 1980, p. 2).

This is not true for just the authorities or the public. Those who study disasters are also influenced by the attention paid by the media. That is why events in countries like the Soviet Union did not influence disaster scholarship because they were never reported. Chernobyl, for example, became important because the increased radioactivity it caused was noticed in Sweden.

The most serious ethical issue raised by 9/11, however, is probably the one that showed up only in the *Columbia Journalism Review.* The *Review* reported, for example, that Condoleezza Rice had convinced editors not to broadcast in full tapes released by Osama bin Ladin or his associates. She told TV executives that those tapes might contain coded messages and she added they that could increase anti-American sentiments among Muslims in the United States and elsewhere. The executives went along with her request. The *Review* also raised the issue as to whether the so-called "war on terrorism" meant that reporters writing about domestic issues had to consider whether their stories would give aid and comfort to the enemy.

> As veteran war correspondents already know, information is a weapon of war. One has to assume that terrorists have constant access to the Internet and CNN. Premature disclosure of a U.S. operation . . . could cost the lives of American combat troops. . . . It is now clear that reporting risks are no less serious on the domestic front. . . . U.S.-based journalists – whose first impulse has always been getting out the news fast – now need to pause and filter it like any other war correspondent. No matter what the topic, they must ask: Does the public's need to know outweigh the harm it might cause . . .? This question might well influence how much detail to include when news outlets break stories about, say, oil tanker construction, Amtrak procedures, building ventilation, pesticide factories (25).

The *Wall Street Journal* . . . ran a massive piece on September 28 detailing inconsistencies in security precautions at airports across the country. . . . Many editors say the *Journal* performed a public service. The story certainly could have put useful pressure on the FAA and airport authorities to make the security more stringent and consistent. The problem is, of course, that one man's public service article is another man's tip sheet for murder (Hansen, 2001, p. 25).

INFORMATION CRITICAL

There are, however, two other aspects to reporting terrorism. First, in some cases, high-speed mass communications may be critical to public safety. That's because incidents such as chemical contamination change the nature of the threat. In normal mass casualty incidents, most initial search and rescue and transport to hospital is done by the survivors. There is some risk to victims in being handled by unskilled persons. However, on balance, the victims are more likely to survive if they reach the hospital quickly. During an incident involving chemical contaminants, this situation changes dramatically. Now every person who comes into contact with a contaminated victim risks becoming another victim. That's exactly what happened during the Sarin gas attack in Tokyo: the victims included thousands of passers-by who tried to help those who were attacked, scores of firefighters and paramedics and even more than a dozen emergency physicians at the closest hospital.

> When the first call reached the Tokyo Fire Department, it dispatched all available equipment (Pangi, 2002, p. 17). On arrival, firefighters, despite seeing victims gasping for breath, rushed into the station without taking precautions. Of the 1,364 firefighting personnel dispatched, 135 – 10 percent – became affected by direct or indirect exposure. In addition, 135 (9.9 percent) of EMTs showed acute symptoms and had to be treated (Okumura et al., 1998; Nosaki et al., 1995).

When a TV crew started shooting visuals of victims at one transit station some persons shouted at them that they should be assisting rather than reporting. The crew ended up loading some victims into their van and transporting them to hospital. When they arrived at the hospital, they discovered that no one there had been informed.

A review of past incidents suggests that the first emergency agency to identify the problem – and the threat of widespread contamination of civilian and other responders – may be a hospital emergency ward. It will then be up to that hospital to advise other hospitals, other emergency agencies and the public. If that is to happen, the warning message will need to be transmitted accurately and quickly over all possible channels – partly because many responders will be transporting victims in private vehicles.

> It is inevitable in such situations that Good Samaritans will assist, and will drive some victims to hospital, unaware that they, themselves are in danger. They must be warned. To do this may require some sort of warning over car radios plus the AMBER Alert now used in criminal cases (Scanlon, 2004, p. 33).

This will work only if there is careful planning and if the media understand what needs to be done and why and why it must be done so quickly. Given the fact the media adopt a command post approach to coverage of such incidents this should not be difficult.

> Informants from radio and television were both willing to accept a partial responsibility to serve as a communication link from emergency officials to the general population. They acknowledged that the nature of their technology allowed for the rather immediate transmission of emergency messages to citizens (Wenger and Quarantelli, 1989).

The second problem is that terrorist incidents may also involve journalists and their sources as victims and this may complicate the information gathering and sharing process. During the anthrax incidents on Capital Hill in Washington, both journalists and their usual sources – Congressional staff – were worried about their own safety. Inevitably they shared rumours with each other and, in the absence of credible information those rumours were reported (Bullock, Haddow and Bell, 2004, p. 7). There was a closer connection between journalists and terrorists at NBC in New York City when a woman working for Tom Brokaw opened an envelope containing a white powder that turned out to be anthrax. Robert Windrem sat at the next desk:

> . . . the day careened from one development to another. Press conferences were held; reporters and satellites trucks gathered outside our windows; studios were shut down; CDC epidemiologists armed with clipboards and swabs walked through the newsroom; hundreds of NBC employees were herded onto a floor below for interviews with police detectives, testing with nasal swabs, and dispensation of Cipro. Just after 6 p.m., as I watched, the FBI formally taped off the news desk where the envelope had been opened three weeks before. It was now a crime scene (Windrem, 2001, p. 19).

Windrem said the event finally struck home when he realized he could just as easily have been the victim.

SUMMARY AND CONCLUSIONS

As shown in this book, scholars in many disciplines – Sociology, Geography, Political Science, Law, Public Administration, Economics to name just a few – have discovered a great deal about human and organizational behavior in crisis and disaster. Some of that scholarship has focused on the role of the mass media and, as a result, we know a great deal about the roles media can and do perform before, during and after disasters. This includes warning, keeping people informed in the aftermath of disaster, correcting rumours. Journalism scholars have added to that knowledge, pointing out, for example, that journalists are often troubled by ethical issues, that television is becoming the source of first choice in crises and that there is a great deal of interaction between media and interpersonal sources. Scholars in more than one discipline have shown that electronic media perform differently than print media in crises: the latter have more time to shape the news using established "gatekeeping" procedures.

Yet, much of the knowledge that scholars in other disciplines have acquired about the media and disaster has not made it into the reporting texts and much of the scholarship by media scholars has not been integrated into the disaster literature. Thus, while news stories all too often reflect these myths both in what they include and what they omit, others still misunderstand the effects of journalistic behavior and the way it impacts on victims of disaster. A review of the role of the media in disasters suggests that the media and disasters are inevitably intertwined but in many ways they are still strangers.

These findings do suggest two lessons for those in positions of responsibility during crises or disasters. The first one is that the media can play a critical role before, during and after such incidents. The media are essential, for example, for warnings to be effective and may be the single most important source of public information in the wake of a disaster. The second lesson is that the media have to be monitored and handled with care because it is media reports that distort what happens in a disaster and lead to misunderstandings. Failure by officials to issue a warning, for example, may be a result of myths created by the media.

REFERENCES

Anderson, W.A. (1969). "Disaster Warning and Communication Processes in Two Communities." *Journal of Communication 19*: 92–104.

Baker, R. (2001). "The Journal on the Run" *Columbia Journalism Review*, November/December: pp. 16–17.

Breed, W.B. (1955). "Social Control in the Newsroom." *Social Forces* Vol. 33 pp. 326–335.

Brooks, B., Kennedy, G., Moon, D.R. and Ranly, R. [The Missouri Group] (1992). *News Reporting & Writing*, St. Martin's Press, New York. p. 228.

Brunhuber, K. (1998). "The real story at Peggy's Cove," *The Sunday Herald* September 13 [Quoted from a clipping, no page number available].

Bucher, R. (1957). "Blame and hostility in disaster." *American Journal of Sociology* Vol. 62 pp. 467–475.

Bullock, J.A., Haddow, G.D., and Bell, R. (2004). "Communicating During Emergencies in the United States." *The Australian Journal of Emergency Management* Vol. 19 No. 2 p. 7.

Cohen, B. (1963). *The Press and Foreign Policy*. Princeton University Press, Princeton.

Cote, W. and Bucqueroux, B. (1996). "Tips on Interviewing Victims." Nieman Reports.

Deppa, J. with Russell, M., Hayes, D., and Flocke, E.L. (1994). *The Media and Disasters: Pam Am 103*. New York University Press, New York. p. 29.

Drabek, Thomas E. (1986). *Human System Responses to Disaster: An Inventory of Sociological Findings*. New York: Springer-Verlag.

Fritz, C.E. and Mathewson, J. H. (1957). *Convergence Behavior in Disasters: A Problem in Social Control*. National Academy of Sciences, National Research Council, Washington, D.C.

Hansen, J. (1998). "Handling the Media in Times of Crisis: Lessons From the Oklahoma City Bombing." Patterson, P., and Wilkins, L. *Media Ethics Issues Cases*. McGraw Hill, Boston. pp: 56-57.

Hansen, C. (2001). "Over Here We're All War Correspondents Now." *Columbia Journalism Review*. November/December pp. 25–28.

Harriss, J., Leiter, K. and Johnson, S. (1992). *The Complete Reporter: Fundamentals of News Gathering, Writing and Editing*. Macmillan Publishing Company, New York.

Hickey, H. (1976). "Terrorism and Television." *TV Guide* July 31–August 7.

Hohenberg, J. (1983). *The Professional Journalist*. Holt, Rinehart and Winston, New York. p. 142.

Itule, B. and Anderson, D. (1984). *Contemporary News Reporting*. Random House, New York.

Jaehnig, W. (1978). "Journalists and terrorism: Captives of the Libertarian Tradition." *Indiana Law Journal* Vol. 53 p. 719.

Jemphrey, A. and Berrington, E. (2000). "Surviving the Media: Hillsborough, Dunblane and the press." *Journalism Studies 1*(3): 469–484.

Kanihan, S.F., and Kendra L.G. (2003). "Within 3 Hours, 97 Percent Learn About 9/11 Attacks." *Newspaper Research Journal 24*(1): 89.

Kitch, C. (2003). "'Mourning in America': Ritual, redemption and recovery in news narrative after September 11." *Journalism Studies 4* (2).

Kitz, J. (1989). *Shattered City: The Halifax Explosion and the Road to Recovery*. Nimbus Press, Halifax.

Kratzer, R.M. and Kratzer, B. (2003). "How Newspapers Decided To Run Disturbing 9/11 Photos." *Newspaper Research Journal 24* (1).

Kueneman, R.M. and Wright, J.E. (1975). "New Policies of Broadcast Stations for Civil Disturbances and Disasters." *Journalism Quarterly 52*(4): 670–677.

Lacquer, W. (1976). "The Futility of Terrorism." *Harper's*. March p. 102.

Lang, G.E. and Lang, K. (1980). "Newspaper and TV Archives: Some Thoughts About Research on Disaster News." *Disasters and the Mass Media*. The National Research Council, Washington, D.C., pp. 269–280.

Lasora, D. (2003). "News media Perpetuate Few Rumors About 9/11 Crisis." *Newspaper Research Journal 24*(1): 10–21.

Ledingham, J.A. and Walters, L.M. (1989). "The Sound and Fury: Mass Media and Hurricanes." Walters, L.M., Wilkins, L. and Walters, T. eds. *Bad Tidings Communications and Catastrophe*. Lawrence Erlbaum and Associates, Hillsdale. p. 43.

Lule, J. (2002). "Myth and Terror on the Editorial Page: The New York Times Responds to September 11, 2001." *Journalism & Mass Communication Quarterly 79*(2) pp. 275–309.

McIntosh, N. (1989). *Lockerbie: A Local Authority Responds to Disaster.* Dumfries and Galloway Regional Council, Dumfries.

Mencher, M. (1981). *News Reporting and Writing.* William C. Brown Company, Publishers, Dubuque. p. 491.

Metz, W. (1991). *Newswriting From Lead to "30."* Prentice Hall, Englewood Cliffs, p. 297.

Mileti, D.S. and Beck, E.M. (1975). "Communication in Crisis: Explaining Evacuation Symbolically." *Communication Research* 2 January pp. 24–29.

New York Times (2003). *Portraits: 9/111/01: The Collected 'Portraits of Grief' from the New York Times.* Henry Holt and Company, New York.

Nosaki, H., Hori, S., Shinozawa, Y., Fujishima, S., Takuma, K., Sagoh, M., Kimura, H., Ohki, T., Suzuki, M., and Aikawa, N. (1995). "Secondary exposure of medical staff to sarin vapor in the emergency room." *Intensive Care Medicine 21* p. 1032 [1032–1035].

Okabe, K. et al. (1979). *A Survey Research on People's Responses to an Earthquake Prediction Warning.* Institute of Journalism and Communication, Tokyo.

Okumura, T., Suzuki, K., Fukada, A., Kohama, A., Takasu, N., Ishimatsu, N., and Hinohara, S. (1998). "The Tokyo Subway Sarin Attack: Disaster Management Part I: Community Emergency Response." *Academic Emergency Medicine 6*(8) 615 [613–617].

Pangi, R. (2002). "Consequence Management in the 1995 SARIN Attacks on the Japanese Subway System." Belfer Center for Science and International Affairs Discussion Paper 2002-4 Harvard: John F. Kennedy School of Government, p. 17.

Parker, E.C. (1980). "What is Right and Wrong with Media Coverage of Disaster?" *Disasters and the Mass Media.* The National Research Council, Washington, D.C. pp. 237–240.

Paul, M.J. (2001). "Interactive Disaster Communication on the Internet: A Content Analysis of Sixty-Four Disaster Relief Home Pages." *Journalism & Mass Communication Quarterly 78*(4) pp. 739–753.

Perry, R. and Greene, M. (1983). *Citizen Response to Volcanic Eruptions: The Case of Mount St. Helens.* Irvington Publishers: New York.

Perry, R.W. and Lindell, M. (1989). "Communicating Threat Information for Volcano Hazards." Walters, L.M., Wilkins, L., and Walters, T., eds. *Bad Tidings Communications and Catastrophe.* Lawrence Erlbaum and Associates, Hillsdale, p. 62.

Pijnenburg, B. and Van Duin, M.J. (1990). "The Zeebrugge Ferry Disaster: Elements of a communication and information processes scenario." *Contemporary Crises 14*: 321–349.

Prince, S.H. (1920). *Catastrophe and Social Change.* Columbia University: New York.

Quarantelli, E. L. (1991). *Lessons From Research: Findings on Mass Communications System Behavior in the Pre, Trans and Postimpact Periods.* Disaster Research Center: Newark.

Quarantelli, E. L. and Russell Dynes (1972). "When Disaster Strikes (It Isn't Much Like What You've Heard and Read About)." *Psychology Today 5*(9) pp. 66–70.

Reynolds, A. and Barnett, B. (2003). "This just in . . . How National TV News Handled the Breaking 'Live" Coverage of September 11. *Journalism & Mass Communication Quarterly 80*(3): 698.

Rich, C. (1997). *Writing and Reporting News: A Coaching Method.* Wadsworth Publishing Company, New York.

Rogers, E.M. and Sood, R. (1981). *Mass Media Operations in a Quick-Onset Natural Disaster: Hurricane David in Dominica.* Natural Hazards Research and Applications Information Center, Boulder.

Rogers, E.M. and Sood, R. (1980). "Mass Media Communication and Disasters: A Content Analysis of Media Coverage of the Andhra Pradesh Cyclone and the Sahel Drought." *Disasters and the Mass Media.* The National Research Council, New York. pp. 139–157.

Rosengren, K.E., Arvidson, P., and Struesson, D. (1974). *The Barseback Panic.* University of Lund, Lund.

Rotbart, D. (1999) "An Intimate Look at Covering Littleton." *Columbia Journalism Review* May/June pp: 24–35.

2000 RTNDA Code of Ethics and Professional Conduct (adopted September 14, 2000).

Salwen, M.J. (1995). "News of Hurricane Andrew: The Agenda of Sources and the Sources' Agendas." *Journalism & Mass Communication Quarterly 72*(4): 826–840.

Sandman, P.M. and Paden, M. (1979). "At Three Mile Island." *Columbia Journalism Review 18* (July–August) p. 48.

Scanlon, J. (2002). "Helping the Other Victims of September 11: Gander Uses Multiple EOCs to Handle 38 Diverted Flights." *Mass Emergencies and Disasters 20*(3) pp. 369–398.

Scanlon, J. (2004). "High Alert Chemical terrorism and the safety of first responders." *Royal Canadian Mounted Police Gazette 66*(2) p. 33.

Scanlon, J. (1996). "Not on the Record: Disasters, Records and Disaster Research." *International Journal of Mass Emergencies and Disasters 14*(3) November pp. 265–280.

Scanlon, J. (1998). "The Search for Non-Existent Facts in the Reporting of Disaster." *Journalism and Mass Communication Educator 53*(2) p. 45.

Scanlon, J. and Prawzick, A. (1991). "Not Just a Big Fire: Emergency Response to an Environmental Disaster." *Canadian Police College Journal 13*(4): 229–259.

Scanlon, J. and Taylor, B. (1975). *The Warning Smell of Gas.* Emergency Preparedness Canada, Ottowa.

Scanlon, J. and Hiscott R. (1985). "Not Just the Facts: How Radio Assumes Influence in an Emergency." *Canadian Journal of Communication 11*(4) pp. 391–404.

Scanlon, J., Osborne, G., and McClellan, S. (1996). *The Peace River Ice Jam and Evacuation: An Alberta Town Adapts to a Sudden Emergency.* Emergency Communications Research Unit, Ottawa.

Scanlon, J. with Padgham, M. (1980). *The Peel Regional Police Force & The Mississauga Evacuation.* Canadian Police College, Ottowa.

Scanlon, T. J. (1992). *Disaster Preparedness: Some Myths and Misconceptions.* The Emergency Planning College, Easingwold.

Scanlon, T.J. (1977). "Post-Disaster Rumor Chains: A Case Study." *International Journal of Mass Emergencies and Disasters 2*: 121–126.

Scanlon, T.J. (1976). "The Not So Mass Media: The Role of Individuals in Mass Communication." Adam, G.S. ed. *Journalism, Communication and the Law.* Prentice-Hall of Canada, Scarborough. p. 104–119.

Scanlon, T.J. and Alldred, S. (1982). "Media Coverage of Disasters: The Same Old Story." Jones, B. and Tomazevic, M. eds. *Social and Economic Aspects of Earthquakes.* Institute for Testing and Research in Materials and Structures, Cornell University and Ljubljana, Ithaca, pp. 363–375.

Scanlon, T.J., Luukko, J. and Morton, G. (1978). "Media Coverage of Crises: Better Than Reported, Worse Than Necessary." *Journalism Quarterly 55*(1), pp. 68–72.

Scanlon, T.J. with Dixon, K. and McClellan, K. (1982). *The Miramichi Earthquakes: The Media Respond to an Invisible Emergency.* Emergency Communications Research Unit, Ottowa.

Shearer, A. (1991). *Survivors and the Media.* John Libbey & Company Limited, London.

Showalter, P. (1995). "One newspaper's coverage of the 1990 earthquake prediction." *Newspaper Research Journal 16*(2), pp. 2–13.

Singer, B.D. and Green, L. (1972). *The Social Functions of Radio in a Community Emergency.* Copp Clark, Toronto.

Smith, C. (1998). "Visual Evidence in Environmental Catastrophe TV Stories." *Journal of Mass Media Ethics 13*(4), pp. 246–257.

Society of Professional Journalists, *Code of Ethics.*

Sood, R., Stockdale, G., and Rogers, E.M. (1987). "How the News Media Operate in Natural Disasters." *Journal of Communication 37*, 27–41.

Stone, G. (1992). *Newswriting.* Harper Collins, New York, p. 144.

Stephens, M. and Edison, N. (1982). "News coverage of the incident at Three Mile Island." *Journalism Quarterly 59*, p. 199.

Turner, R. (1980). "The Mass Media and Preparations for Natural Disaster." *Disasters and the Mass Media.* The National Research Council, Washington, D.C., pp. 281–292.

Walters, L.M., Wilkins, L. and Walters, T., eds. (1989). *Bad Tidings Communications and Catastrophe.* Lawrence Erlbaum and Associates, Hillsdale.

Waxman, J. (1973). "Local Broadcast Gatekeeping During natural disaster." *Journalism Quarterly 50*, pp. 751–758.

Wenger, D. (1985). "Mass Media and Disasters." *Preliminary Paper #98.* Disaster Research Center, Newark, N.J.

Wenger, D., James, T., and Faupel, C. (1980). *Disaster Beliefs and Emergency Planning.* Disaster Research Center, Newark, N.J.

Wenger, D. and Quarantelli, E.L. (1989). *Local Mass Media Operations, Problems and Products* in Disasters Report Series #19. Disaster Research Center, Newark, N.J. p. 62.

Wicker, T. (1966). "The Assassination." Ruth Adler, ed., *The Working Press.* G. P. Putman and Sons, New York.

Wigley, S. (2003). "Relationship Maintenance in a Time of Crisis: The 2001 Oklahoma State University Plane Crash." *Public Relations Quarterly 48*(2), p. 40.

Wilson, L. (2004). "Media Dependency During a Large-Scale Social Disruption: The Case of September 11." *Mass Communication and Society 7*(3), p. 354.

Windrem, R. (2001). "They Are Trying to Kill Us." *Columbia Journalism Review* November/December, pp. 18–19.

Xuewen, S. (1999). "The 1987 Daxinganling Forest Fire." Paper presented in Amsterdam at the International Conference of Local Authorities Confronting Disasters and Emergencies, April 22–24.

Chapter 7

Disasters: A Psychological Perspective

Margaret Gibbs and Kim Montagnino

ABSTRACT

Psychological research has shown that disasters can cause serious mental health consequences for victims. These consequences take the form of Posttraumatic Stress Disorder and a variety of other disorders and symptoms which have been less investigated. The more stress, defined in a variety of ways, within the disaster, the more likely there are to be emotional consequences. Vulnerability factors within the victim operate in complex ways, but seem related to the extent of stress experienced by the victim and the available resources, broadly defined, with which to deal with it. The mental health profession has developed a variety of strategies with which to ameliorate the effect of disaster. Although recent research on single session debriefing has produced disappointing results, many techniques and therapies have been validated as successful interventions for disaster victims.

INTRODUCTION

Unlike other disciplines, which have come more recently to the study of disasters, psychology has concerned itself with disasters' impacts on victims for much of its own short history. As long ago as 1944, Lindemann published an observation of the psychological aftermath of the Coconut Grove nightclub fire in Boston. Besides the obvious involvement psychologists have in attempting to relieve distress of victims, disasters have a relationship to several important psychological constructs. Disasters allow psychologists to perceive the operation of trauma on emotional functioning, an operation which mental health practitioners as far back as Freud have been interested in understanding. Stress research is a central and crucial explanatory factor in many fields of psychology, especially community psychology, which considers stress the central ingredient to the formation of psychopathology (e.g., Albee, 1997; Dohrenwend, 1998). There is an ethical limit to the extent that stress can be manipulated in the laboratory, and disasters allow psychologists the opportunity to observe how extreme stress impacts individuals and groups.

Because of psychology's interest in trauma and stress, its definition of disaster has differed somewhat from that employed in other fields. In the 1970s, after the Vietnam War and the discovery of its impact on veterans, and after the discovery of the long-term effects of child sexual abuse, the mental health field conceptualized a disorder specifically related to the consequences of trauma, Post-Traumatic Stress Disorder (PTSD) (American Psychiatric Association, 2000). We will define PTSD later in the chapter. Here we are making the point that because of the interest by psychologists in PTSD, there has been some blurring between the concepts of victimization from any source and victimization from disaster. For instance, vulnerability factors to PTSD in victims of an earthquake may be similar to vulnerability factors to PTSD in victims of rape, and effective treatments may also be similar, so that studying a broader group of victims may be useful in understanding disasters.

In spite of many differences in opinion (e.g., Quarantelli, 1998), the definition of disaster in use in this chapter agrees with that of most psychologists (e.g., Barton, 1969; Norris, Friedman, Watson, Byrne, Diaz and Kaniaty, 2002) who regard disasters as involving an unexpected or uncontrollable event rather than a long-term experience. That is, a disaster is something that could happen within a war (e.g., My Lai, or many other less well-known examples) rather than the war itself, or Three Mile Island rather than Love Canal. These examples illustrate the difficulty with the distinction, and some researchers think that our concept of disaster should include chronic disaster (Couch and Kroll-Smith, 1985). Dynes (2004) has argued that social scientists need to expand their definition of disaster to encompass events like war, genocide, and refugee experiences that are critical in third world countries.

Disasters are also usually viewed as a collective experience, excluding personal disasters like sexual abuse or automobile accidents, unless these involve a large number of people. Again, the dividing line can be unclear. The type of event, with its various dimensions, can affect our perceptions. We might not consider an automobile accident that killed 13 people to be a disaster, even if many others were involved or witnessed it, but the killing of 13 in the shootings at Columbine certainly qualifies.

With the passage of time, study of disasters has become less descriptive and more quantitative, attempting to resolve some of the methodological problems of this research. The focus has moved from the question of whether there are significant long-term psychological impacts of disasters, to studying the types of impact that occur and what factors in the disaster and in the individual increase the likelihood of emotional damage. Interventions to assist victims have been developed. Most recently, there has been more focus on the effectiveness of these interventions. This chapter will explore in turn each of these areas: methodology of disaster research; extent of psychological impact of disasters; types of psychological sequelae; damaging aspects of disaster; vulnerability factors; psychological interventions for victims; and the effectiveness of these interventions.

METHODOLOGY OF DISASTER RESEARCH

Early studies of disaster tended to be descriptive. Lifton (1967) described the emotional impacts of Hiroshima, Coles (1967) portrayed the effect of political disaster on children (1967), and Erikson (1976) painted the picture of the aftermath of the Buffalo Creek floods in West Virginia. While some researchers (e.g., Edelstein, 2004) still favor a qualitative approach, most psychological disaster research today tends to be quantitative.

Problems exist for the social scientist who wishes to study disaster. Experimental design requires random assignment of participants to experimental and control conditions. Even if a mad scientist wanted to conduct such an experiment, controlling a disaster is an oxymoron. Disaster research can only attain the status of quasi-experimental design, with comparison groups, not controls. Since disasters occur unpredictably, pre-test data on victims are usually not available. Psychologists called into a disaster are usually there to provide help. Researchers can seldom obtain access to the disaster at its onset, and if they do find access, the exigencies of the situation usually preclude administration of standard instruments in a standardized fashion. Victims usually have no motive to participate in research, and follow-up studies are often difficult to arrange. Samples of victims vary from those directly impacted, to rescue workers, to the families of the bereaved. It is difficult to compare Western victims to those from third world countries, as their circumstances and resources are so different, and for the same reasons it is difficult to compare victims from different ethnic groups within a culture.

Disasters also vary widely in the amount and the nature of the stress they involve: duration; loss of life; personal injury, or injury to loved ones; property damage; terror; helplessness; gruesome sights, sounds and smells; dislocation from one's home; availability of social support – all these factors may differ in a flood as contrasted to an earthquake, or between one flood and another, or between one victim's and another's experience of the same flood. One special differentiation between types of disaster is the natural vs. the technological, or human-caused, disaster. Natural disasters tend to involve lack of control over natural forces, like wind, that we expect to be uncontrollable, while technological disasters can be less defined, especially if they include toxic exposure, and can involve a loss of control over an area of life in which we expect control, like drinking water (Baum, Fleming, and Davidson, 1983). Terrorism is a special form of technological disaster, and the most recent addition to the typology of disaster (Ursano, Fullerton, and Norwood, 2003).

As psychologists conducted more disaster research, they began to develop standardized measures, beginning with the Impact of Events Scale (Horowitz, Wilner and Alvarez, 1979). Many measures have been devised to diagnose post-traumatic stress disorder (PTSD), one of a number of psychological consequences of disaster. The National Center for PTSD (2003) currently lists 15 adult PTSD self-report measures, 4 interview measures, and 9 measures for children. Obviously, many other standardized measures of other types of psychopathology have been administered. Measures have also been developed to identify vulnerability factors and intervening variables, e.g., the Peritraumatic Dissociation Experiences Questionnaire (Marmar, Weiss, Schlenger et al., 1994), and World Assumptions Scale (Janoff-Bulman, 1985).

Comparison groups, if not actual control groups, were introduced early into the research. A step forward in the confusing array of studies on different disaster's with different samples and different methods came with the meta-analysis of Rubonis and Bickman (1991), which found small but consistent post-disaster effects on levels of psychopathology across different types of study and types of disaster.

Robins, Fischbach, Smith, Cottler, Solomon, and Goldring (1986) seized upon a fortuitous (or infortuitous) series of events at the Times Beach site in Missouri. Interviews had taken place in that area for the Epidemiological Catchment Area study which documented the prevalence of psychiatric problems in the country. Then the area was struck both by floods and the discovery of dioxin. This allowed the comparison of the effects of a natural and a technological disaster, with pre-test information available for a sample of the victims, and documented that change had occurred.

When Norris et al. reviewed the disaster literature in 2002, she found six other studies that were able to obtain true pre-test measures for their samples. Comparison of types of disaster exposure for the same sample remains rare. Norris et al. also report that many recent disaster studies used follow-up formats and probability sampling methods. In short, methodology has improved dramatically, and conclusions can comfortably be drawn about the psychological impact of disasters.

EXTENT OF PSYCHOLOGICAL IMPACT OF DISASTERS

As noted, Rubonis and Bickman found in their 1991 meta-analysis consistent but small post-disaster effects upon psychopathology. Many other review chapters and articles have been written (e.g., Gibbs, 1989, 1992; Green and Solomon, 1995; Katz, Pellegrino, Pandya, Ng, and DeList, 2002; Sundin and Horowitz, 2003), concluding that post-disaster effects are greater and more pervasive than Rubonis and Bickman's inferences. The most thorough recent review is that of Norris et al. (2002a). The authors analyzed 160 different disaster studies, with a total of over 60,000 participants, and did not conduct a meta-analysis because of the difficulty in deriving effect sizes from descriptive studies. Using a rough four-point scale to rate level of pathology, they found that only about 10 of studies found minimal impairment, about half the studies found moderate impairment, and the remaining 40 percent found severe or very severe impairment. Severe impairment was equivalent to rates of psychopathology in the participants of between 25 and 50 percent.

For many years there has been a debate over whether the effect of disaster on mental health was important. One side of the debate came from the sociological point of view (e.g., Quarantelli and Dynes, 1985), which focused on the adaptive nature of community response, both in the immediate aftermath of a disaster and in most people's long-term response. The majority of people function adaptively during and after a disaster, and the old notion (Kinston and Rosser, 1974) that individuals will experience panic, wander aimlessly and be dependent has been shown to be untrue (Wenger, Dykes, Sebok, and Neff, 1975). But that is a different matter from focusing on the toll that the disaster takes on some individuals. There is so much evidence now of the damage to individuals, that to our minds, the debate has been resolved. Norris, Friedman, and Watson (2002b) conclude that the field does not need new studies indicating that disaster causes serious psychopathology; we know this to be the case. Instead, we should focus on understanding what aspects of disaster are most devastating, and what characteristics of individuals make them vulnerable, issues we will address later in the chapter.

FORMS OF PSYCHOPATHOLOGY RESULTING FROM DISASTERS

If it is clear that disasters cause psychopathology, it is less clear what form that psychopathology takes. Since the mental health profession developed the PTSD diagnosis, PTSD has been the main focus of research on the aftermath of disaster. The criteria for PTSD include (APA, 2000): (1) having been exposed to a traumatic and fearful event; (2) re-experiencing the traumatic event, usually in flashbacks or nightmares; (3) avoidance of situations and stimuli that could reawaken the trauma, for example, numbing one's feelings or withdrawing from others; and (4) increased level of arousal, for instance, sleep difficulties, irritability, and concentration problems.

Norris et al. (2002a) reported that 68 percent of their research samples assessed for and found PTSD in disaster victims. The second most common psychiatric problem was depression, found in 36 percent of the samples. Anxiety in various forms was shown in 32 percent of the samples. Health concerns were also often present (23% of the samples). It was not usually clear whether victims' health concerns were realistic, or were based on somaticizing the stress of the experience (North, 2002). Alcoholism and drug abuse were not often investigated but when they were, levels of abuse have been found to rise after disasters.

What is not clear from the above figures is what the actual rate of various psychopathologies might be if each study had assessed for all of them. Norris et al. (2002b) recommend that all disaster researchers use a standard measure of psychopathology so that it can be more clearly determined which disorders are linked to undergoing disaster.

Victimization, primarily child physical and sexual abuse, has been shown to lead to other diagnoses beyond the ones investigated in disasters. These include schizophrenia and other psychoses (Neria, Bromet, Sievers, Lavelle, and Fochtmann, 2002), dissociative disorders (Coons and Milstein, 1986) and borderline personality disorder (Herman, Perry and Van der Kolk, 1989). None of these diagnoses has been investigated to see if higher rates result after disaster, although dissociative symptoms have been reported during and after some disasters (Marmar, C. R., Weiss, D. S., Metzler, and DeLucchi, 1996; Weiss, Marmar, Metzler, and Weiss et al., 1995) and can be part of the avoidance criterion of PTSD (APA, 2000). It would be valuable to look at long-term vulnerabilities of childhood victims of disaster to these disorders. Little research of any kind has been conducted looking at long-term consequences of disasters for children.

An issue that has been discussed in the literature is whether symptoms of other disorders found after disasters are part of the PTSD syndrome or whether they are independent consequences. There are several possible explanations for the overlap that often is observed. Symptoms within diagnoses do overlap, symptoms of other diagnoses could be sub-clinical cases of PTSD, PTSD could increase vulnerability to other diagnoses, and other diagnoses could increase vulnerability to PTSD (McMillen, North, Mosley and Smith, 2002). In particular, the fact that depression and PTSD are both common consequences of disaster is of interest. Greening, Stoppelbein, and Docter (2002) conducted an interesting study in which they looked at attributions for the negative outcomes of the Northridge earthquake. Victims who made what have been labelled depressogenic attributions, seeing negative outcomes as related to internal, stable and global causes (Abramson, Seligman, and Teasdale, 1978), were more likely to develop depressive symptoms, but not PTSD symptoms. Livanou, Basoglu, Salcioglu and Kalender (2002) looked at PTSD and depression as outcomes of the Turkish 1999 earthquake, and found that there were different predictors for each. Research into the relationship between different outcomes of disaster is continuing, but the lack of solid findings points out that we know little about the actual mechanism of how symptoms are caused by disaster stress.

Aspects of Disaster Which Contribute to Psychopathology

In general, the nature of the disaster and the extent of the trauma it wreaks are more predictive of the extent of psychopathology that follows than are characteristics of the victims (Sundin and Horowitz, 2003). The more stressful the disaster experience, it appears, the more negative the consequences, but it is not always possible to identify which of the many factors within a disaster make it more stressful. Theorists have identified the following as important characteristics: mass violence (Norris et al., 2002); the experience of terror and horror (Bolin, 1985); duration of the disaster (Baum and Davidson, 1985, Bolin, 1985); and the amount of unpredictability and lack of control (Baum and Davidson, 1985; Thoits, 1983).

First responders and disaster workers are at special risk for PTSD and other negative emotional consequences of disaster (Gibbs, Lachenmeyer, Broska, and Deucher, 1996; Norris, 2002a) . This vulnerability has usually been perceived to be related to the experience of the work rather than to any inherent vulnerability factors, as often people choosing these professions have high levels of emotional hardiness. Looking at disaster workers who dealt with the aftermath of the World Trade Center disaster of September 11th provides an example. Working with dead bodies and body parts after a major disaster is something that almost everyone finds extraordinarily stressful, and perhaps the experience could be said to define horror. Disaster workers' experience of the disaster is often more long-term than that of other victims, as for instance the long-term digging out after September 11th. In addition, the experience of helplessness and lack of control is prevalent, as workers searched for but were unable to find identifiable bodies.

Psychologists have many theories about what causes the disorders of PTSD, depression, anxiety reactions, etc. (e.g., Barlow, 2000), but little conclusive about what it is exactly about a disaster that leads to emotional damage.

As we have mentioned, most psychologists identify stress as a leading cause of psychopathology, but theories as to how stress affects its victims are varied. Some focus on the physiological overload of stress (e.g., Selye, 1976), some on the unpredictability and uncontrollability of stress (e.g., Kelly, 1955) and some on the conditioning that takes place between a frightening stressor and other aspects of life, with a resulting avoidance of stimuli that are reminders (Mowrer, 1960). Losses in a disaster, of other people, of material goods, of one's own health and security, are also critical (Nolen-Hoeksema, 1990). Some theorists focus on the shift in cognitions that take place after a disaster. Janoff-Bulman and Frieze (1983) speculated that cognitions shift after a disaster. The individual asks "Why me?" and the answer involves a change in one's sense of invulnerability, in the world's predictability, and in one's own worth.

VULNERABILITY FACTORS

Research has identified a number of characteristics of victims that make them more vulnerable to disaster effects. Vulnerability factors include, but are not limited to, socioeconomic status (SES), available resources, previous level of psychopathology, age, social/family factors, gender and ethnicity.

Regarding SES, Norris et al. (2002a) found that 13 of 14 samples which investigated socioeconomic status and disaster outcome found lower socioeconomic status to be associated with increased post-disaster distress. Studies included a wide range of disasters: an air disaster (Epstein, Fullerton, and Ursano, 1998), an industrial disaster (Vila, Witowski, and Tondini, 2001), floods (Ginexi, Weihs, Simmens, 2000), and an earthquake (Lewin, Carr, and Webster, 1998). Individuals who live in poverty tend to have fewer resources available to them to attenuate the effects of disaster.

Pre-existing psychopathology is a risk factor for developing psychopathology related to a trauma (Norris et al., 2002a) in that individuals who suffer from a psychological disorder are more susceptible to further distress in the aftermath of a disaster. For example, pre-disaster anxiety disorders (Asarnow et al., 1999), depression (Knight, Gatz and Heller, 2000), and suicidal ideation (Warheit, Zimmerman and Khoury, 1996) were found to increase the likelihood of post-disaster psychopathology.

In terms of age, Norris et al. (2002a) noted that middle-aged adults appear to be the group most affected by disasters. This age group may have more burdens and stresses (Thompson, Norris and Hanacek, 1993), such as caring and providing support for a family, that may be amplified in the aftermath of a disaster.

Social network characteristics influence vulnerability. For example, a lack of perceived (Bromet, 1982; Dougall, Hyman and Hayward, 2001) or received (Sanchez, Korbin and Viscarra, 1995; Udwin, Boyle and Yule, 2000) social support may lead to greater post-disaster distress.

These risk factors do not operate in isolation. Any single factor is often interrelated with others. We will illustrate this complex interaction within the context of two variables: gender and minority or third world ethnicity, both of which Norris et al. (2002a) in their review cite as among the most robust of vulnerability factors.

GENDER. Norris et al. (2002a) stated that in 94 percent of 49 studies which investigated the issue, female survivors of disaster were more seriously affected than were males. There are several possible explanations for this difference. For example, as mentioned in the previous paragraph, low socioeconomic status is a risk factor for post-disaster psychopathology, and women more often live in poverty than men (Belle, 2000).

The gender difference may be in part explained by differences we often observe between the genders in the way psychological distress is expressed. In general, women are more likely than men to acknowledge psychological symptoms and to report them (Nolen-Hoeksema, 1990). After a disaster, males may suppress feelings of psychological distress because of the expectation that men must be strong and capable (Wolfe and Kimerling, 1997). As discussed in a previous section, the most commonly investigated post-disaster reactions are PTSD, depression, and other forms of anxiety. Substance abuse and other acting out behaviors, such as interpersonal violence, are seldom assessed. Men are more likely to express psychological distress through these kinds of behaviors, rather than reporting neurotic-type symptoms like depression and anxiety (Myers, Weissman, Tischler, et al., 1984).

Women have higher pre-disaster rates of depression and most anxiety disorders than men (Myers et al., 1984), putting them at risk for disaster-related distress. Furthermore, there may be some experiences that women are more likely to have that may contribute to the development of PTSD post-disaster. The experience of rape and sexual assault is higher among women than men (Kessler, Sonnega, Bromet, 1995), and it has been shown that when compared with other forms of trauma, unwanted sexual contact is more likely to result in PTSD (Breslau, Davis and Andreski, 1997; Kessler et al., 1995). Pulcino et al. (2003) found that the experience of previous unwanted sexual contact increased a woman's likelihood of endorsing PTSD symptoms after the September 11th attacks by 33 percent.

The interaction of gender and various social/family factors highlights the interconnectedness of vulnerability factors. While men typically cope using individual and immediate decision-making, women use their social network to process and work through problems (Kawachi and Berkman, 2000; Taylor, Klein and Lewis, 2000). After a disaster, changes often occur in one's social network (Kaniasty and Norris, 1997). In a study with victims of Hurricane Andrew, Norris, Perilla, Riad, Kaniasty and Lavizzo (1999) noted that nearly all of the events that were experienced in common by the sample were related to changes in the social environment. Womens' PTSD symptoms have been shown to increase as their available social supports decrease, a finding that was not true for men (Pulcino et al., 2003). Change in the social network, which may involve a decrease in available social support, may be more devastating for women than for men due to its negative effect on their coping ability.

Traditionally, women have been assigned the role of caregiver, a role that may lead to increased stress levels in the aftermath of a disaster. First, for women who are primary caretakers, the extra stress of caring for children and the home may fall disproportionately on them. Norris et al. (2002a) noted in their review of disaster studies that being a parent, especially a mother, was associated with higher disaster-related distress. In a study with survivors of the 1999 earthquake in Turkey, a higher percentage of women than men reported that their first thoughts were of their family (Yilmaz, 2004). Second, women may be more likely to provide care for others affected by disaster (Kaniasty and Norris, 1995; Solomon, Smith, Robins, 1987). In a study with vicarious victims of the September 11th attacks, more than twice as many women than men reported engaging in collective helping behavior (Wayment, 2004). When women offer support to other people, not only can they be further exposed to the trauma through contact with others, but they also may be burdened by the stress of providing support in times of need (Solomon et al., 1987). A particularly devastating situation may be the one in which a woman provides support services to others in the aftermath of a disaster, but does not receive an equal amount of social support back, especially in light of our previous discussion on coping styles.

There may be something about the traditional caregiving role that leads to vulnerability. A brief investigation of gender, ethnicity and this role will again highlight the complexity of the interaction between vulnerability factors. Studies with members of varying cultural groups have suggested that the gap between PTSD symptoms in men and women is higher in societies that are more traditional (Norris et al., 2002a). Norris et al. (2001) conducted a study using a sample of non-Hispanic White and Black Americans affected by Hurricane Andrew and Mexicans affected by Hurricane Paulina. In all cultural groups, women reported more PTSD symptomology than men. However, this gap was widest in the Mexican sample and smallest in the Black sample. Since, when compared to non-Hispanic White American culture, Mexican culture is understood to be more traditional in its adherence to gender roles (Chia, Wuensch and Childers, 1994; Davenport and Yurich, 1991), and Black American culture is understood to be more egalitarian in its gender role definitions (Davenport and Yurich, 1991; McAdoo, 1988), the results suggest that women who assume the traditional female role are most vulnerable to post-disaster psychopathology.

MINORITY OR THIRD WORLD ETHNICITY. Norris et al.'s study with Americans and Mexicans brings us to our consideration of a second vulnerability factor, ethnicity. Post-disaster effects in developing countries tend to be greater than in the U.S. (Norris et al., 2002a), and within the U.S., adult members of ethnic minority groups are more negatively affected by disasters (Norris et al., 2002a; Perilla, Norris and Lavizzo, 2002). Differential exposure to disasters may account for some of these differences. For example, in the U.S., ethnic minority members are often concentrated in the lower income strata and are more likely to live in less safe homes and at risk areas (Quarantelli,1994), increasing their trauma exposure.

Factors beyond the amount of exposure to disaster-related trauma are likely in operation as well. Again, poverty leads to lower access to post-disaster resources for minorities (Kaniasty and Norris, 1995). Also related to low socioeconomic status is a higher pre-disaster exposure to community violence. Similarly, immigrant members of minority groups or individuals who live in developing nations may live or have lived in cultures where they are likely to have experienced trauma. This could include the community or personal violence that is common in countries characterized by political or social unrest. Previous exposure to community or personal trauma increases the risk of post-disaster psychopathology. For example, Perilla et al. (2002) found that the incidence of neighborhood and personal trauma was higher among the Black and Latino participants in their study, and that the severity of their exposure accounted for much of their higher rates of PTSD post-Hurricane Andrew.

There may also be culturally-influenced ways of interpreting or expressing distress that account for the vulnerability of minority groups. Members of an ethnic minority group may have experienced prejudice, discrimination or oppression. These experiences can result in psychological vulnerability in general, but could also be related to the way trauma is expressed. African-Americans, for instance, may, because of experiences of oppression, become hypervigilant to perceived threats and this in turn could result in the expression of certain post-traumatic symptoms (Allen, 1996). The Latino concept of susto, which refers to an experience of fright, is often to what Spanish-speaking individuals attribute any symptoms they experience (Hough et al., 1996; Kirmayer, 1996). The incidence of a disaster is consistent with this cultural concept, as it represents a singular traumatic event to which one can attribute distress. In this way, the expression of PTSD in response to a disaster is quite culturally consistent.

In certain cultures, such as African-American and Latino ones, family ties are emphasized and there is a strong reliance on the family for social support (Chia et al., 1994; Hatchett and Jackson, 1993; Sabogal, Marin and Otero, 1987). As in the discussion of women, disruption to the family or social network that can occur post-disaster can lead both to a loss of available support for minority group members (Kaniasty and Norris,1997) and to increased stress that comes with the obligation of tending to others' needs. In addition, such a family-orientation can also result in less receipt of outside sources of support (Kaniasty and Norris, 2000).

Fatalism, the tendency to attribute the causes for things to a higher power, such as nature or God, is associated with Latino culture and sometimes with African-American culture (Pepitone and Triandis, 1987). Such a worldview can lead to poor psychological outcomes in response to distress because one's personal power is perceived as minimal (Mirowsky and Ross, 1984; Wheaton, 1982). In an interesting study with children affected by Hurricane Andrew, Lee (1999) found that African-American and Hispanic students often received information about the cause of the hurricane that was inconsistent with Western science, and sometimes consistent with fatalism.

To summarize these vulnerability factors, like features of disasters that contribute to psychopathology, they seem primarily related to the extent of stress experienced, before, during, and after the disaster, and the available resources to deal with it. We have cited, for instance, the findings that both women (Pulcino, et al., 2003) and minority victims (Perilla, et al., 2002; Quarantelli, 1994) may have experienced more trauma before or during the disaster than white males. Resources include material resources, like money and infrastructure of a western vs. third world culture, social resources, like social networks and the way these may impact males and females differently, and coping style resources, which may vary by gender and culture. Understanding risk factors can assist us in designing interventions, both at the individual and community level, for survivors of a disaster.

PSYCHOLOGICAL INTERVENTIONS FOR VICTIMS

Numerous individuals and organizations have written about disaster planning and interventions from a psychological perspective (e.g., Ehrenreich, 2001; Jacobs, 1995; Roberts, 2000; SAMHSA, 2000). In the panoply of ideas and techniques put forward, a valuable model for looking at psychological interventions for disaster victims is that provided by Caplan (1964), the father of community psychology, who developed the model of prevention of mental disorder. If, as the community psychology model posits, stress is the major cause of psychopathology, the best way of preventing psychopathology is to reduce the stress of the environment. This is *primary prevention*, and as it applies to disasters, primary prevention places psychology squarely in the process of emergency preparedness. Psychologists, might for instance, help develop campaigns to persuade the public not to build houses in a flood plain, or find ways to increase the public's emergency preparedness through education, or influence legislation that requires insurers to provide disaster insurance or prompt payment of benefits after a disaster. Because psychology has so much to contribute to education and policy development, it is important for emergency managers to involve psychology in all their planning efforts.

Secondary prevention in the Caplan model involves identifying people at risk, and intervening to assist them. As applied to disasters, secondary prevention requires psychologists to conduct rapid screening after disasters and to begin interventions as soon as possible. Again, emergency managers need to include psychologists in the immediate aftermath of a disaster.

This type of prevention is often labeled crisis intervention, an attempt to reduce the stress of a crisis at the time it occurs. Lindemann's (1944) groundbreaking research at the Cocoanut Grove nightclub fire, mentioned earlier, involved helping survivors and the bereaved express their grief, in the belief that this would reduce their later symptoms. Caplan (1964) proposed that a crisis is a turning point, and that individuals in crisis can either cope successfully and thereby enhance their ability to cope, or they can make maladaptive attempts to cope, and thereby decline in their psychological functioning.

As we have noted in the section on vulnerability, the availability of resources is critical to postdisaster adjustment, and Caplan idenified the providing of resources as a major form of crisis intervention. Resources include material resources (for instance, helping victims locate temporary housing after a flood, or locate missing family members) and social resources (for instance, providing emotional support to an individual who lost a family member in the flood, and locating other individuals who can provide support). Social resources may be especially critical for female victims, as we have mentioned. Psychologists should be involved in the allocating of resources after a disaster by emergency managers.

Helping deal with coping resources is another form of crisis intervention. While many models of crisis counseling have been proposed and discussed (e.g., McGee, 1992; Roberts, 2000), most tend to be solution focused, with an emphasis on the victim's strengths and finding appropriate solutions to the problems they face. In general, active problem solving strategies are more effective than passive ones (e.g., Lazarus and Folkman, 1980). One issue for psychologists applying crisis intervention to disasters is that often there are not good solutions to the crisis, regardless of the individual's coping strengths.

One type of crisis intervention, critical incident stress debriefing (CISD), has received a great deal of attention of late. Developed by Jeffrey Mitchell in the early eighties, the model has a strict format and is applied to victims, family members, and especially rescue workers, including fire and police personnel. It is conducted in groups, and includes seven phases: (1) introduction; (2) facts about what happened in the crisis; (3) thoughts about what happened; (4) feelings about what happened; (5) symptoms; (6) teaching/information about stress and stress management; and (7) re-entry (Mitchell and Everly, 2000). In the next section, we will discuss the effectiveness of CISD.

Traditional psychotherapy falls into the category of secondary prevention. A number of interventions have been developed for victims with PTSD. Similar to strategies for other anxiety disorders, therapists use exposure (e.g., Foa and Kozak, 1986) to require clients tò revisit the trauma of the disaster experience. The theory is that in dealing with a traumatic event, we use avoidance strategies to reduce the pain, and these avoidance strategies are part of the symptom picture. More psychodynamic therapists may work to have disaster victims confront their feelings about their experience, using different labels from the behaviorist, but doing similar work.

Usually, cognitive restructuring is also a part of therapy for individuals with PTSD. We have mentioned that disasters lead to a shift in cognitions (Janoff-Bulman and Frieze, 1983) and victims of disaster often have distorted beliefs regarding their safety, the likelihood of another disaster, their personal worth, and so forth.

Many forms of therapy, too numerous to list, have been developed for other disorders, such as depression and anxiety, which may result from disasters. These therapies are not specific to the treatment of postdisaster survivors. It is important that emergency managers be able to provide some forms of therapeutic intervention to victims and responders after a disaster.

Tertiary prevention in the Caplan model involves preventing further deterioration of those already emotionally disturbed, and is less relevant to disaster work. It might apply to long-term victims of disaster, like Vietnam veterans, whose problems persist, and who may need new and as yet undeveloped forms of treatment.

EVALUATION OF PSYCHOLOGICAL INTERVENTIONS FOR DISASTER VICTIMS

Psychology has a long history of evaluation research, again making it an important partner for emergency managers who need to assess the effectiveness of their planning and interventions. Psychologists have investigated the effectiveness of therapeutic interventions for disaster victims with mixed results.

Using the Caplan model for looking at psychological interventions is useful because it provides perspective on the multitude of interventions that are included. Recent focus on the efficacy of Critical Incident Stress Debriefing has taken emphasis away from the many efficacious types of intervention that mental health fields have developed, and perhaps represents a backlash against an overly enthusiastic application of the CISD model. There was little empirical investigation of the efficacy of CISD in its early days. More recently, CISD and other debriefing approaches have been scrutinized intensely. Here again, the approaches are under investigation for their efficacy with many types of victimization, not just disasters. Mitchell and Everly (2000) argue that the findings are mixed because of the variability of the training and skill of the provider. Many studies, however, have found no positive results beyond that of a placebo condition (Humphries and Carr, 2001; Rose, Brewin, Andrews and Kirk, 1999) or no treatment (Conlon, Fahy, and Conroy, 1998; Kenardy, Webster, Lewin, Carr, Hazell, and Carter, 1996). Some studies with randomized assignment to groups have actually found that trauma victims who underwent debriefing showed higher levels of symptoms than those who did not (Bisson, Jenkins, Alexander, and Bannister, 1997; Mayou, Ehlers, and Hobbs, 2000). This issue has even reached the popular press: an article in the New Yorker focused on the lack of benefit of debriefing for individuals suffering from reactions to the September 11 attack on the World Trade Center (Groopman, 2004), and the New York Times featured an article about the inappropriateness of psychological help for non-Western victims of disasters (Satel, 2005). A number of reviews (Arendt and Elklit, 2001; Ehlers and Clark, 2003; Van Emmerik, Kamphuis, Hulsbosch, and Emmelkamp, 2002; Litz, Gray, and Bryant, 2002; Raphael, 2000) conclude that the lack of benefit for debriefing after disasters means it should be used cautiously, never be compulsory, and that further research is necessary.

Since CISD has been the primary technique used post-disaster, these findings have thrown the whole issue of psychology's ability to understand and help disaster victims and responders into question. It is important to note that psychotherapy itself is effective (Lambert and Ogles, 2004), and has been effective in treating PTSD (Marks, Lovell, Noshirvani, Lavanou, and Thrasher, 1998). There are several possible explanations for this disparity between the effectiveness of CISD and psychotherapy in treating PTSD: (1) CISD may be too short-term and unfocused to have enough of an impact; (2) Psychotherapy may need to be adapted to the particular situation of disaster victims; and (3) There may be characteristics of disaster that are different from other traumatic stress, making intervention more difficult. We will take each of these explanations in order.

CISD is an extremely short-term form of treatment. Reviews which compare debriefing with cognitive behavioral therapy (CBT) (Litz, et al., 2002; Ehlers and Clark, 2003) show that CBT is more effective in ameliorating trauma symptoms, perhaps because it is longer term and more focused on symptoms.

CISD is also a treatment provided mainly by other disaster workers, trained in the process, rather than by professional psychologists, although professionals certainly sometimes provide CISD. Barker and Pistrang (2002) argue convincingly that the processes of social support and psychotherapy are overlapping and should be conceptualized in similar ways. For instance, the outcome of professional helping seems to be no more helpful than paraprofessional helping (e.g., Faust and Zlotnick, 1995). Hogan, Linden and Najarian (2002) review 100 studies on social support intervention and conclude that they in general are helpful, although we do not know which kinds of interventions work best for which problems. It seems logical to suppose that there are disaster interventions that would be helpful when administered at the time of the crisis by paraprofessionals, although further refining of approaches is obviously necessary.

New approaches to disaster are being developed. A relatively new and controversial therapy for PTSD is Eye Movement Desensitization (EMDR), which involves controlled eye movements back and forth while the client is thinking about the trauma which occurred. Empirical findings are mixed (e.g., Taylor, Thordarson, Maxfield, Federoff, Lovell, and Ogrodniczuk, 2003). The explanatory mechanism for why the technique should work is involved and many psychologists find it unconvincing. For victims of fire, Krakow, Melendrez, Johnston, et al. (2002) described a sleep dynamic therapy, involving psychoeducational approaches about sleep, and found that both sleep disturbances and other anxiety and depressive symptoms lessened. Basoglu, Livanou and Salcioglu (2003) report that a single session with an earthquake simulator diminished symptoms of traumatic stress in earthquake victims. Smyth, Hockemeyer, Anderson, et al. (2002) administered the task of writing about victimization experiences in Hurricane Floyd, and found that it reduced the relationship between intrusive thoughts and symptoms, not as dramatic a finding as that of Pennebacker and Harber (1993) who had earlier reported that writing down one's feelings about a disaster can ameliorate symptoms. Lange, Rietdijk, Hudcovicova, van de Ven, Schrieken, and Emmelkamp (2003) have incorporated writing tasks into an Internet treatment for posttraumatic stress, which they report as successful. Neuner, Schauer, Klaschik, Karunakara and Elbert (2004) describe an effective narrative exposure therapy for PTSD in Sudanese refugees, in which participants replayed the events of their life until they formed a coherent narrative. Pitman, Sanders, Zusman, et al. (2002) report that propranolol administered to victims of trauma interferes with memory of the event and ameliorates the potential for PTSD.

The issue of special characteristics of disaster which make psychological interventions more problematic should be addressed. Individuals in a disaster are more likely to see their needs as physical and real rather than as emotional, especially in a non-Western culture (Satel, 2005). Emotional problems may only emerge years later, as with many Vietnam veterans. It may be that psychologists, in their work with other emergency managers, need to educate individuals about possible emotional reactions, rather than stepping in to try to intervene too quickly with those who are not in search of services.

We have already noted that most psychological efforts are directed to helping individuals develop active coping strategies, rather than passive, fatalistic ones. It is sometimes the case in disaster, however, that there are no active strategies to take. One issue which has not been sufficiently discussed is that of individual styles and needs. Fullerton, Ursano, Vance, and Wang (2000) reported that female emergency workers were three times more likely than males to seek out debriefing. Our previous discussion of gender differences in vulnerability would suggest that women may be in special need of social support services postdisaster. Roth and Cohen (1986) discuss the fact that individuals seem to have preferred styles for either avoiding or approaching stress, and that these styles are difficult to change. Both avoidance and confrontation can be helpful depending on the circumstances. Most psychologists, going back as far as Lindemann, assumed that individuals need to confront the trauma of a disaster. It may be that enabling individuals to avoid effectively is just as useful, especially when the trauma is severe and there is little that can be done to change the situation.

Another issue is the perception that needing and taking help from a psychologist is stigmatizing. Jenkins (1998) reports, for instance, that co-workers were the most frequently sought out resource (by 94% of emergency workers dealing with a mass shooting) and the most consistently useful source of emotional support. Although counselors were equally effective, only 50 percent of victims sought them out. Again, education from psychologists about the possible emotional consequences of disaster could normalize this process, and make it easier for victims to seek help. It may also be that forms of paraprofessional intervention, other than CISD, need to be developed.

Gray, Maguen and Lidz (2004) point out that current crisis interventions focus on PTSD and its prevention, and that the wide range of victim responses, which we have reviewed earlier, demands a more nuanced and individuated range of treatments. Few interventions have been tailored to the needs of children (Wooding and Raphael, 2004), and it is possible that many interventions for children need to be addressed to their parents (Norris, 2001).

In summarizing psychology's achievements in understanding and dealing with disaster, the following seem clear. Disasters can cause severe psychological disturbance, with many victims experiencing PTSD, depression and anxiety. More research is needed to determine the entire range of disorders and the frequency of their occurrence after disaster. The severity and duration of the disaster will predict much of the extent of the reaction. Vulnerability factors in the individual do play a part, with gender, age, previous level of psychopathology, poverty, ethnicity and social support correlating with extent of post-disaster psychopathology in victims. These variables interact in complex ways. Mental health fields intervene both pre-disaster in emergency planning and post-disaster in crisis intervention, debriefings, psychotherapy, and evaluation of emergency management efforts. Recent research has questioned the usefulness of single session debriefings, but there is support for longer-term interventions and there is the promise of new types of interventions for disaster victims.

REFERENCES

Abramson, L. Y., Seligman, M. E. P., & Teasdale, J. D. (1978). Learned helplessness in humans: Critiques and reformulation. *Journal of Abnormal Psychology, 87*: 49–74.

Albee, G. P. (1997). Speak no evil. *American Psychologist, 52*: 1143–1144.

Allen, I. (1996). PTSD among African Americans. In Marsella, A. & Friedman, M. (Eds.) *Ethnocultural aspects of posttraumatic stress disorder: Issues, research, and clinical applications* (pp. 209–238). Washington, DC: American Psychological Association.

American Psychiatric Association. (2000). *Diagnostic and Statistical Manual of Mental Disorders*, Fourth Edition, Text Revision. Washington, DC: American Psychiatric Association.

Arendt, M., & Elklit, A. (2001). Effectiveness of psychological debriefing. *Acta Psychiatrica Scandinavica, 104*: 423–437.

Asarnow, J., Glynn, S. & Pynoos, R. (1999). When the earth stops shaking: Earthquake sequelae among children diagnosed for pre-earthquake psychopathology. *Journal of the American Academy of Child & Adolescent Psychiatry, 38*: 1016–1023.

Barker, C. & Pistrang, N. (2002). Psychotherapy and social support: Integrating research on psychological helping. *Clinical Psychology Review 22*: 361–379.

Barlow, D. (2000). *Clinical handbook of psychological disorders: A step-by-step treatment manual.* (2nd ed.). New York: Guilford.

Barton, A. H. (1969). *Communities in disaster: A sociological analysis of collective stress situations.* New York: Doubleday.

Basoglu, M., Livanou, M., & Salcioglu, E. (2003). A single session with an earthquake simulator for traumatic stress in earthquake survivors. *American Journal of Psychiatry, 160*: 788–790.

Baum, A., & Davidson, L.M. (1985). A suggested framework for studying factors that contribute to trauma in disaster. In B.J. sowder (Ed.), *Disasters and mental health: Selected contemporary perspectives* (pp. 29–40). Rockville, MD: National Institute of Mental Health.

Baum, A., Fleming, R. & Davidson, L. M. (1983). Natural disaster and technological catastrophe. *Environment and Behavior, 15*: 333–354.

Belle, D. & Doucet, J. (2003). Poverty, inequality, and discrimination as sources of depression among U.S. women. *Psychology of Women Quarterly, 27*(2): 101–113.

Bisson, J. I., Jenkins, P. L., Alexander, J. & Bannister, C. (1997). Randomised controlled trial of psychological debriefing for victims of acute burn trauma. *British Journal of Psychiatry, 171*: 78–81.

Bolin, R. (1985). Disaster characteristics and psychosocial impacts. In B. J. Sowder (Ed.), *Disasters and mental health: Selected contemporary perspectives* (pp. 3–28). Rockville, MD: National Institute of Mental Health.

Breslau, N., Davis, G. & Andreski, P. (1997). Sex differences in posttraumatic stress disorder. *Archives of General Psychiatry, 54*: 1044–1048.

Bromet, E. (1982). Mental health of residents near the Three Mile Island reactor: A comparative study of selected groups. *Journal of Preventive Psychiatry, 1*: 225–276.

Caplan, G. (1964). *Principles of preventive psychiatry.* New York: Basic Books.

Chia, R., Wuensch, K. & Childers, J. (1994). A comparison of family values among Chinese, Mexican, and American college students. *Journal of Social Behavior & Personality, 9*: 249–258.

Coles, R. (1967). *Children of crisis.* Boston: Little Brown.

Conlon, L., Fahy, T. J., & Conroy, R. (1998). PTSD in ambulant RTA victims: A randomised controlled trial of debriefing. *Journal of Psychosomatic Research, 46*: 37–44.

Coons, P. M. & Milstein, V. (1986). Psychosexual disturbances in multiple personality: Characteristics, etiology, and treatment. *Journal of Clinical Psychiatry, 47*: 106–110.

Couch, S. & Kroll-Smith, J. S. (1985). The chronic technical disaster: Toward a social scientific perspective. *Social Science Quarterly, 66*: 564-575.

Davenport, D. & Yurich, J. (1991). Multicultural Gender Issues. *Journal of Counseling & Development, 70*: 64–71.

Dohrenwend, B. P. (1998). *Adversity, stress and psychopathology.* Oxford, England: Oxford University Press.

Dougall, A., Hyman, K. & Hayward, M. (2001). Optimism and traumatic stress: The importance of social support and coping. *Journal of Applied Social Psychology, 31*: 223–245.

Dynes, R. (2004). Expanding the horizons of disaster research. *National Hazards Observer, 28*(4): 1.

Edelstein, Michael R. (2004). *Contaminated communities: Coping with residential toxic exposure.* Boulder, CO: Westview Press.

Ehlers, A., & Clark, D. M. (2003). Early psychological interventions for adult survivors of trauma: A review. *Biological Psychiatry, 53*: 817–826.

Ehrenreich, J. H. (2001). Disasters: A guidebook to psychosocial intervention. http://www.massey.ac.nz/~trauma/issues/2002-1/smyth.htm.

Epstein, R., Fullerton, C. & Ursano, R. (1998). Posttraumatic stress disorder following an air disaster: A prospective study. *American Journal of Psychiatry, 155*: 934–938.

Erikson, K. (1976). *Everything in its path.* New York: Simon and Schuster.

Faust, D. & Zlotnick, C. (1995). Another Dodo Bird verdict: Revisiting the comparative effectiveness of professional and paraprofessional therapists. *Clinical Psychology and Psychotherapy, 2*: 157–167.

Foa, E. B. & Kozak, M. J. (1986). Emotional processing of fear: Exposure to corrective information. *Psychological Bulletin, 99*: 20–35.

Fullerton, C. S., Ursano, R. J., Vance, K., & Wang, L. (2000). Debriefing following trauma. *Psychiatric Quarterly, 71*(3): 259–276.

Gibbs, M. S. (1989). Factors in the victim that mediate between disaster and psychopathology: A review. *Journal of Traumatic Stress, 2*: 489–513.

Gibbs, M. S. (1992). Disasters: Their impact on psychological functioning, and mediating variables. In Gibbs, M. S., Lachenmeyer, J. R., & Sigal, J. (Eds.) *Community Psychology and Mental Health* (pp.195–213). New York: Gardner.

Gibbs, M, Lachenmeyer, J. R., Broska, A. and Deucher, R. (1996). Effects of the AVIANCA Aircrash on Disaster Workers. *International Journal of Mass Emergencies and Disasters 14*(1): 23–32.

Ginexi, E., Weihs, K. & Simmens, S. (2000). Natural disaster and depression: A prospective investigation of reactions to the 1993 Midwest floods. *American Journal of Community Psychology, 28*: 495–518.

Gray, M. J., Maguen, S. Lidz, B. T. (2004). Acute psychological impact of disaster and large-scale trauma: Limitations of traditional interventions and future practice recommendations. *Prehospital and Disaster Medicine, 19*(1) http://pdm.medicine.wisc.edu.

Green, B. & Solomon, S. (1995). The mental health impact of natural and technological disasters. In Freedy, J. & Hobfoll, S. (Eds.), *Traumatic Stress: From Theory to Practice* (pp. 163–180). New York: Plenum.

Greening, L., Stoppelbein, L. & Docter, R. (2002). The mediating effects of attributional style and event-specifid attributions on postdisaster adjustment. *Cognitive Therapy and Research, 26*: 261–274.

Groopman, J. (2004). The grief industry: How much does crisis counseling help – or hurt? *New Yorker*, January 26, 30–38.

Hatchett, S. & Jackson, J. (1993). African American extended kin systems: An assessment. In McAdoo, H. (Ed). *Family ethnicity: Strength in diversity* (pp. 90–108). Thousand Oaks, CA: Sage Publications, Inc.

Herman, J. L., Perry, J. C. & Van der Kolk, B. A. (1989). Childhood trauma in borderline personality disorder. *American Journal of Psychiatry, 146*: 490–495.

Hogan, B. E., Linden, W., & Najarian, B. (2002). Social support interventions: Do they work? *Clinical Psychology Review 22*: 381–440.

Horowitz, M., Wilner, M. & Alvarez, W. (1979). Impact of Event Scale: A measure of subjective stress. *Psychosomatic Medicine, 41*: 209–218.

Hough, R., Canino, G. & Abueg, F. (1996). PTSD and related stress disorders among Hispanics. In Marsella, A. & Friedman, M. (Eds.). *Ethnocultural aspects of posttraumatic stress disorder: Issues, research, and clinical applications* (pp. 301–338). Washington, DC: American Psychological Association.

Humphries, C. & Carr, A. (2001). The short term effectiveness of critical incident stress debriefing. *Irish Journal of Psychology, 22*: 188–197.

Jacobs, G. A. (1995). The development of a national plan for disaster mental health. *Professional Psychology: Research and Practice, 26*: 543–549.

Janoff-Bulman, R. (1985). The aftermath of victimization: Rebuilding shattered assumptions. In C.R. Figley (Ed.), *Trauma and its wake: The study and treatment of post-traumatic stress disorder.* New York: Brunner/Mazel.

Janoff-Bulman, R. & Frieze, I. H. (1983). A theoretical perspective for understanding reactions to victimization. *Journal of Social Issues, 39*: 1–17.

Jenkins, S. R. (1998). Emergency medical workers, mass shooting incident stress and psychological recovery. *International Journal of Mass Emergencies and Disasters, 16*: 119–143.

Kaniasty, K. & Norris, F. (1995). In search of altruistic community: Patterns of social support mobilization following Hurricane Hugo. *American Journal of Community Psychology, 23*: 447–477.

Kaniasty, K. & Norris, F. (1997). Social support dynamics in adjustment to disasters. In Duck, S. (Ed). *Handbook of personal relationships: Theory, research and interventions* (2nd ed.) (pp. 595–619). New York, NY: John Wiley & Sons, Inc.

Kaniasty, K. & Norris, F. (2000). Help-seeking comfort and receiving social support: The role of ethnicity and context of need. *American Journal of Community Psychology, 28*: 545–581.

Katz, C. L., Pellegrino, L, Pandya, A, Ng, A. & DeLisi, L. E. (2002). Research on psychiatric outcomes and interventions subsequent to disasters: A review of the literature. *Psychiatric Research, 110*: 201–217.

Kawachi, I. & Berkman, L. (2000). Social ties and mental health. *Journal of Urban Health, 78*: 458–467.

Kelly, G. A. (1955). *The psychology of personal constructs.* New York: Norton.

Kenardy, J. A., Webster, R. A., Lewin, T. J., Carr, V. J., Hazell, P. L., & Carter, G. L. (1996). Stress debriefing and patterns of recovery following a natural disaster. *Journal of Traumatic Stress, 9*: 37–49.

Kessler, R., Sonnega, A. & Bromet, E. (1995). Posttraumatic stress disorder in the National Comorbidity Survey. *Archives of General Psychiatry, 52*: 1048–1060.

Kinston, W. & Rosser, R. (1974) Disaster: Effects on mental and physical state. *Journal of Psychosomatic Research, 18*: 437–456.

Kirmayer, L. (1996). Confusion of the senses: Implications of ethnocultural variations in somatoform and dissociative disorders for PTSD. In Marsella A. & Friedman, M. (Eds.). *Ethnocultural aspects of posttraumatic stress disorder: Issues, research, and clinical applications* (pp. 131–163). Washington, DC: American Psychological Association.

Knight, B., Gatz, M. & Heller, K. (2000). Age and emotional response to the Northridge earthquake: A longitudinal analysis. *Psychology & Aging, 15*: 627–634.

Krakow, B. J., Melendrez, D. C., Johnston, L. G., Clark, J. O., Santana, E. M., Warner, T. D., Hollifield, M. A., Schrader, R., Sisley, B. N., & Lee, S. A. (2002). Sleep dynamic therapy for Cerro Grande fire evacuees with posttraumatic stress symptoms: A preliminary report. *Journal of Clinical Psychiatry 63*: 673–683.

Lambert, M. J. & Ogles, B. M. (2004). The efficacy and effectiveness of psychotherapy. In Lambert, M. J. (Ed). *Bergen and Garfield's Handbook of psychotherapy and behavior change* (5th Edition) (pp. 139–193). New York: Wiley.

Lange, A., Rietdijk, D., Hudcovicova, M., van de Ven, J-P., Schrieken, B., & Emmelkamp, P. M. G. (2003). Interapy: A controlled randomized trial of the standardized treatment of posttraumatic stress through the Internet. *Journal of Consulting and Clinical Psychology, 71*: 901–909.

Lazarus, S. and Folkman, R. S. (1980). An analysis of coping in a middle-aged community sample. *Journal of Health and Social Behavior*, 21: 219–239.

Lee, O. (1999). Science knowledge, world views, and information sources in social and cultural contexts: Making sense after a natural disaster. *American Educational Research Journal, 36*: 187–219.

Lewin, T., Carr, V. & Webster, R. (1998). Recovery from post-earthquake psychological morbidity: Who suffers and who recovers? *Australian & New Zealand Journal of Psychiatry, 32*: 15–20.

Lifton, R. (1967). *Death in life: Survivors of Hiroshima.* New York: Random House.

Lindemann, E. (1944). Symptomatology and management of acute grief. *American Journal of Psychiatry, 101*: 141–148.

Litz, B. T., Gray, M. J., Bryant, R. A (2002). Early interventions for trauma: Current status and future directions. *Clinical Psychology: Science and Practice, 9*(2): 112–134.

Livanou, M., Basoglu, M., Salcioglu, E. and Kalender, D. (2002). "Traumatic stress responses in treatment-seeking earthquake survivors in Turkey." *Journal of Nervous and Mental Disease, 190*: 816–823.

Marks, I., Lovell, K., Noshirvani, H., Livanou, M., & Thrasher, S. (1998). Treatment of post-traumatic stress disorder by exposure and/or cognitive restructuring: A controlled study. *Archives of General Psychiatry, 55*: 317–325.

Marmar, C. R., Weiss, D. S., Metzler, T. J. & DeLucchi, K. (1996). Characteristics of emergency service personnel related to peritraumatic dissociation during critical incident exposure. *American Journal of Psychiatry, 153*, Festschrift Supplement: 94–102.

Marmar, C. R., Weiss, D. S., Schlenger, W. E., Fairbank, J. A., Jordan, B. K., Kulka, R. A. & Hough, R. L. (1994). Peritraumatic dissociation and post-traumatic stress in male Vietnam theater veterans. *American Journal of Psychiatry, 151*: 902–907.

Mayou, R. A., Ehlers, A., & Hobbs, M. (2000). Psychological debriefing for road traffic accident victims: Three year follow-up of a randomised controlled trial. *British Journal of Psychiatry, 176*: 589–593.

McAdoo, H. (1988). *Black Families* (2nd Ed.). Thousand Oaks, CA: Sage Publications, Inc.

McGee, T. F. (1992). Crisis intervention. In Gibbs, Lachenmeyer, and Sigal, (Eds.). Community psychology and mental health (pp. 139–156). New York: Gardner Press.

McMillen, C., North, C., Mosley, M., & Smith, E. (2002). Untangling the psychiatric comorbidity of posttraumatic stess disorder in a sample of flood survivors. *Comprehensive Psychiatry, 43*: 478–485.

Mirowsky, J. & Ross, C. (1984). Mexican culture and its emotional contradictions. *Journal of Health & Social Behavior, 25*: 2–13.

Mitchell, J. T. (1983). When disaster strikes . . . The critical incident stress debriefing process. *Journal of Emergency Medical Services, 8*: 36–9.

Mitchell, J. T. & Everly, Jr. G. S., (2000). CISM and CISD: Evolutions, effects and outcomes. In Raphael, B. & Wilson, J. P. (Eds.). *Psychological debriefing: Theory, practice and evidence.* Cambridge, UK: Cambridge University Press.

Mowrer, O. H. (1960). *Learning theory and behavior.* New York: Wiley.

Myers, J. K., Weissman, M. M., Tischler, G. L., Holzer, C. E., Leaf, P. J., Orvaschel, H., Anthony, J. C., Boyd, J. H., Burke, J. D., Kramer, M. and others. (1984). Six-month prevalence of psychiatric disorders in three communities 1980–1982. *Archives of General Psychiatry, 41*: 959–967.

National Center for PTSD (2003). Assessment instruments. http://www.ncptsd.org//publications/assessment/

Neria, Y., Bromet, E. J., Sievers, S., Lavelle, J. and Fochtmann, L. J. (2002). "Trauma exposure and posttraumatic stress disorder in psychosis: Findings from a first-admission cohort." *Journal of Consulting and Clinical Psychology, 70*: 246–251.

Neuner, F., Schauer, M., Klaschik, C., Karunakara, U., & Elbert, T. (2004). A comparison of narrative exposure therapy, supportive counseling, and psychoeducation for treating Posttraumatic Stress Disorder in an African refugee settlement. *Journal of Consulting and Clinical Psychology, 72*: 579–587.

Nolen-Hoeksema, S. (1990). *Sex differences in depression.* Stanford, CA: Stanford University Press.

Norris, F. H., Friedman, M. J., Watson, P. J., Byrne, C. M., Diaz, E., & Kaniasty, K. (2002a). 60,000 disaster victims speak: Part 1. An empirical review of the empirical literature, 1981-2001. Psychiatry, 65: 207–239.

Norris, F. H., Friedman, M. J., & Watson, P. J. (2002b). 60,000 disaster victims speak: Part II. Summary and implications of the disaster mental health research. *Psychiatry, 65*: 240–260.

Norris, F., Perilla, J., Ibanez, G., & Murphy, A. (2001). Sex differences in symptoms of posttraumatic stress: Does culture play a role? *Journal of Traumatic Stress, 14*: 7–28.

Norris, F., Perilla, J., Riad, J., Kaniasty, K. & Lavizzo, E. (1999). Stability and change in stress, resources, and psychological distress following natural disaster: Findings from Hurricane Andrew. *Anxiety, Stress & Coping: An International Journal, 12*: 363–396.

North, C. S. (2002). Somatization in survivors of catastrophic trauma: A methodological review. *Environmental Health Perspectives, 110*: 636–640.

Pennebaker, J. W. & Harber, K. (1993). A social stage model of collective coping: The Loma Prieta Earthquake and the Persian Gulf War. *Journal of Social Issues, 49*(4): 125–146.

Pepitone, A. & Triandis, H. (1987). On the universality of social psychological theories. *Journal of Cross-Cultural Psychology, 18*: 471–498.

Perilla, J., Norris, F. & Lavizzo, E. (2002). Ethnicity, culture, and disaster response: Identifying and explaining ethnic differences in PTSD six months after Hurricane Andrew. *Journal of Social & Clinical Psychology, 12*: 20–45.

Pitman, R. K., Sanders, K. M., Zusman, R. M. et al. (2002) Pilot study of secondary prevention of posttraumatic stress disorder with propranolol. *Biological Psychiatry, 51*: 189–192.

Pulcino, T., Galea, S., Ahern, J., Resnick, H., Foley, M. & Vlahov, D. (2003). Posttraumatic stress in women after the September 11 terrorist attacks in New York City. *Journal of Women's Health, 12*: 809–820.

Quarantelli, E. L. (1994). *Future disaster trends and policy implications for developing countries.* Newark, DE: Disaster Research Center.

Quarantelli, E. L. (1998). (Ed.) *What is a disaster? Perspectives on the question.* London: Routledge.

Quarantelli, E. L. & Dynes, R. A. (1985) Community responses to disasters. In B. J. Sowder, (Ed.). *Disasters and mental health: Selected contemporary perspectives* (pp. 158–168). Rockville, MD: National Institute for Mental Health.

Raphael, B. (2000). Conclusion: Debriefing - science, belief and wisdom. In Raphael, B & Wilson, J. (Eds). *Psychological debriefing: Theory, practice and evidence.* Cambridge, UK: Cambridge University Press.

Roberts, A. R. (2000). *Crisis intervention handbook: Assessment, treatment, and research* (2nd edition). Oxford, U.K.: Oxford University Press.

Robins, L., Fischbach, R., Smith, E., Cottler, L., Solomon, S., & Goldring, E. (1986). Impact of disaster on previously assessed mental health. In Shore, J. (Ed.). *Disaster Stress Studies: New Methods and Findings* (pp. 21–48). Washington, DC: American Psychiatric Press.

Rose, S., Brewin, C., Andrews, B., & Kirk, M. (1999). A randomized controlled trial of individual psychological debriefing for victims of violent crime. *Psychological Medicine, 29*: 793–799.

Roth, S. & Cohen, L. J. (1986). Approach, avoidance, and coping with stress. *American Psychologist, 41*: 813–819.

Rubonis, A. V. & Bickman, L. (1991). Psychological impairment in the wake of disaster: The disaster-psychopathology relationship. *Psychological Bulletin, 109*: 384–399.

Sabogal, F., Marín, G. & Otero-Sabogal, R. (1987). Hispanic familism and acculturation: What changes and what doesn't? *Hispanic Journal of Behavioral Sciences, 9*: 397–412.

SAMHSA (2000). *Field Manual for Mental Health and Human Service Workers in Major Disasters.* Washington, DC: U.S. Department of Health and Human Services.

Sánchez, J., Korbin, W. & Viscarra, D. (1995). Corporate support in the aftermath of a natural disaster: Effects on employee strains. *Academy of Management Journal, 38*: 504–521.

Satel, S. (2005) Bread and Shelter, Yes, Psychiatrists, No. *The New York Times*, Tuesday, March 29, p. F5.

Selye, H. (1976). *The stress of life.* New York: McGraw Hill.

Smyth, J. M., Hockemeyer, J., Anderson, C., Strandberg, K., Koch, M., O'Neill, H. K., & McCammon, S., (2002). Structured writing about a natural disaster buffers the effects of intrusive thoughts on negative affect and physical symptoms. *The Australasian Journal of Disaster and Trauma Studies*, 2002-1. http://www.massey.ac.nz/~trauma/issues/2002-1/smyth.htm.

Solomon, S. Smith, E. & Robins, L. (1987). Social involvement as a mediator of disaster-induced stress. *Journal of Applied Social Psychology, 17*: 1092–1112.

Sundin, E. C. & Horowitz, M. J. (2003). Horowitz's Impact of Events Scale: Evaluation of 20 years of use. *Psychosomatic Medicine, 65*: 870–876.

Taylor, S., Thordarson, D. S., Maxfield, L., Federoff, I. C. Lovell, K., & Ogrodniczuk, J. (2003). Comparative efficacy, speed, and adverse effects of three PTSD treatments: Exposure therapy, EMDR, and relaxation training. *Journal of Consulting and Clinical Psychology, 71*: 330–338.

Taylor, S., Klein, L. & Lewis, B. (2000). Biobehavioral responses to stress in females: Tend-and-befriend, not fight-or-flight. *Psychological Review, 107*: 411–429.

Thompson, M. P., Norris, F.H., & Hanacek, B. (1993). Age differences in the psychological consequences of Hurricane Hugo. *Psychology & Aging, 8*: 606–616.

Udwin, O., Boyle, S. & Yule, W. (2000). Risk factors for long-term psychological effects of a disaster experienced in adolescence: Predictors of Post Traumatic Stress Disorder. *Journal of Child Psychology & Psychiatry, 41*: 969–979.

Ursano, R. J., Fullerton, C. S., & Norwood, A. E. (2003). *Terrorism and disaster: Individual and community mental health interventions.* Cambridge, UK: Cambridge University Press.

Van Emmerik, A. A. P., Kamphuis, J. H., Hulsbosch, A. M., & Emmelkamp, P. M. G. (2002). Single session debriefing after psychological trauma: A meta-analysis. *Lancet, 360*: 766–771.

Vila, G., Witowski, P. & Tondini, M. (2001). A study of posttraumatic disorders in children who experienced an industrial disaster in the Briey region. *European Child & Adolescent Psychiatry, 10*: 10-18.

Warheit, G., Zimmerman, R. & Khoury, E. (1996). Disaster related stresses, depressive signs and symptoms, and suicidal ideation among a multi-racial/ethnic sample of adolescents: A longitudinal analysis. *Journal of Child Psychology & Psychiatry, 37*: 435–444.

Wayment, H. (2004). It could have been me: Vicarious victims and disaster-focused distress. *Personality & Social Psychology Bulletin, 30*: 515–528.

Weiss, D. S., Marmar, C. R., Metzler, T. J., & Ronsfeldt, H. M. (1995). Predicting symptomatic distress in emergency services personnel. *Journal of Consulting and Clinical Psychology, 63*: 361–368.

Wenger, D. E., Dykes, J. D., Sebok, T. D., Neff, J. L. (1975) It's a matter of myths: An empirical examination of individual insight into disaster response." *Mass Emergencies, 1*: 33–45.

Wheaton, B. (1982). A comparison of the moderating effects of personal coping resources in the impact of exposure to stress in two groups. *Journal of Community Psychology, 10*: 293–311.

Wolfe, J. & Kimerling, R. (1997). Gender issues in the assessment of posttraumatic stress disorder. In Wilson, J. & Keane, T. *Assessing psychological trauma and PTSD* (pp. 192–238). New York, NY: Guilford Press.

Wooding, & Raphael, B. (2004). Psychological impact of disasters and terrorism on children and adolescents: Experiences from Australia. *Prehospital and Disaster Medicine, 19*, http://pdm.medicine.wisc.edu.

Yilmaz, V. (2004). A statistical analysis of the effects on survivors of the 1999 earthquake in Turkey. *Social Behavior & Personality, 32*: 551–558.

Chapter 8

Anthropological Contributions to the Study of Disasters[1]

Doug Henry

ABSTRACT

This chapter addresses the contributions of anthropology towards the field of disaster studies and emergency management. Anthropology's concern with the holistic study of humanity in relation to social, political, cultural, and economic contexts, as well as the breadth of its studies done internationally, seem to make it well-positioned to answer calls from within the field of disaster studies for an "expanded horizon." This article examines contemporary contributions and investigations, following the life cycle of a disaster event, from pre-disaster vulnerability, conceptions of risk, individual and social responses and coping strategies, and relief management. It concludes by providing recommendations for future research.

INTRODUCTION

Anthropology attempts to engage its subjects holistically and comparatively, placing its focus on the broader context of human interactions in contemporary, historical, and prehistorical time, as well as the interrelationships between cultural, social, political, economic, and environmental domains. In its approach to studying disasters, this has meant calling attention to how risks and disasters both influence and are products of human systems, rather than representing simply isolated, spontaneous, or unpredictable events. There is especial concern with how cultural systems (the beliefs, behaviors, and institutions characteristic of a particular society or group) figure at the center of that society's disaster vulnerability, preparedness, mobilization, and prevention. Understanding these cultural systems, then, figures at the center of understanding both the contributing causes to disasters as well as the collective responses to them.[2] A holistic approach examines the complex interrelationships between humans, culture, and their environment, from the human actions that may cause or influence the severity of disaster, to the position of social vulnerability that defines disaster impact, to the range of socio-cultural adaptations and responses, including the impact of aid and the infusion of donor money. The comparative, relativistic approach of the discipline has often given it a critical stance, privileging local knowledge and local ways of management, while problematizing the dominant models of relief.

Given calls within disaster studies for an "expanded horizon" more inclusive than the current domestic, natural hazards focus, anthropology seems ideally situated to make a contribution to the field. Its own broad perspective includes what Dynes (2004) calls "slow-onset" disasters, public health epidemics, and complex emergencies. Because anthropologists often

1. Special thanks to the editors, and to Anthony Oliver-Smith, Peter Van Arsdale, and Linda Whiteford for their helpful comments on earlier drafts.

2. Besides the focus on "systems" I have taken here, two other anthropological approaches to disasters bear note: (1) a typological approach, categorizing disasters by their logical type, such as drought, flood, cyclone, earthquake, chemical disaster, etc. (Franke, 2004), and (2) a "processual" approach, which highlights that pre-disasters, disasters, and relief are continuous events which serve as instigators of social interactions, transformations, and reorganization (Hoffman and Lubkemann, 2005).

work in the developing world, where vulnerability to disasters is the highest, they have been positioned to comment on issues like risk, change, management, and assistance. Some of the most complete reviews of the field have been done by Anthony Oliver-Smith (see, for instance, his 1996 *Anthropological Research on Hazards and Disasters*, and 1999, *The Angry Earth: Disaster in Anthropological Perspective* (edited with Susanna Hoffman), which provide much of the basis and inspiration for this current review); many of the conceptual categories that follow are his. The chapter is organized to follow anthropology's contributions to the complete life cycle of disaster, from issues of vulnerable and perceived risk, to individual and social responses and coping strategies, to relief and recovery efforts.

PRE-DISASTER RISK AND VULNERABILITY

Culture influences that some people within the social system are more vulnerable to disasters than others. Ethnic minorities, disempowered castes or classes, religious groups, or occupations may live or work in physical areas that are relatively disaster-prone (Torry, 1979; Zaman, 1989; Haque and Zaman, 1993; Bankoff, 2003). For example, the mortality from the 1976 earthquake in Guatemala so disproportionately impacted the poor (unable to afford standard construction, and forced to live in landslide-susceptible ravines and gorges) that the disaster was called a "classquake" (Blaikie et al., 1994). In addition, cultural ideas about gender occupations and gender roles may predispose women (and often, by association, children) to be disproportionately represented among groups whom disasters strike, or who are most vulnerable to its effects (Agarwal, 1992; Shaw, 1992; Fothergill, 1996; Bari, 1998). Studies of vulnerability and risk have thus focused largely on environmental and technological susceptibility, such as at living near waste disposal sites (Johnston, 1994; Pellow, 2002), water contamination (Fitchen, 1988), workplace contact with toxic chemicals or dust (Sharp, 1968; Michaels, 1988; Petterson, 1988), and industrial accidents (Wallace, 1987). Pre-disaster inequalities within social relationships have also been shown to exacerbate tensions and discrimination during times of crisis or relief (Jackson, 2003). Torry (1986), for instance, showed how pre-disaster religiously sanctioned inequality existing in India structured the provision of relief during famine in such ways that reinforced the cultural model of customary discrimination. He notes that social adjustments during crisis "are not radical, abnormal breaks with customary behavior; rather they extend ordinary conventions" (1986: 126). Working with Bangladeshi communities resettled from erosion prone riverine areas, Haque and Zaman (1993) suggest that relief efforts that ignore broader cultural institutions like religious and sociopolitical organization, may do so at their own peril, in that they ignore factors that influence or limit how communities are able to organize and respond to their own situation.

In parallel with work in other disciplines, anthropology has sought ways to call attention to (and alleviate) structural conditions of predisaster vulnerability that predispose some communities to experience disaster or that increase the severity of disaster impact. Such conditions include gender inequality, global inequities, endemic poverty, racism, a history of colonial exploitation, imbalances of trade, and underdevelopment. Poor or ethnic minority groups may have little choice but to live in sub-standard housing on or near unstable land prone to flooding, drought, disease, or environmental pollution (Bodley, 1982; Johnston, 2001). The developing world experiences three times the disaster-induced death rates of the developed world (UNDRO 1984). Paul Farmer, a medical anthropologist, takes stock of the profound and spreading social disaster within the poorest countries of the world that HIV/AIDS and tuberculosis infection represent (1999, 2004). With millions dead and tens of millions of children left orphaned in Africa alone, Farmer places the blame for the epidemic squarely on structural forces: the poverty and racism that heighten vulnerability by preventing the poor from receiving education and health care access, the multinational greed that prevents life-prolonging treatment drugs from reaching the poor, and neo-liberal economic policies that force governments to slash safety nets and reduce spending on crucial social services (see also Schoepf et al., 2000).

Research from the African Sahel has shown that economic pressures associated with colonialism and global trade induced unsustainable practices that increased the local vulnerability to desertification, famine, and starvation (Turton, 1977; Fagan, 1999). Oliver-Smith notes the "socially created pattern of vulnerability" that Spanish-induced changes in building materials, design, and settlement patterns induced in Andean cultures, that contributed to higher mortality during a 1970 earthquake in Peru (1994). The pressure for economic development, modernization, and growth through means such as mining, deforestation, urbanization, and hydroelectric dams, can lead to dramatic environmental degradation, loss of food security, and increasing disease vectors, thus elevating vulnerability to natural and infectious hazards (Scudder and Colson, 1982; Simonelli, 1987; Cernea, 1990; Shipton, 1990; Hunter, 1992; Lerer and Scudder, 1999).

There is also research on how various actors involved in pre-disaster situations assess and define risk and vulnerability. Anthropologists have emphasized local models of risk construction, and stressed the importance of understanding the sociocultural context of judgments and indigenous linguistic categories and behaviors about what is dangerous and what is not. They note that public perceptions about risk and acceptability are shared constructs; therefore, understanding how people think about and choose between risks must be based on the study of culturally-informed values as well as their social context of poverty or power (Wolfe, 1988; Cernea, 2000). Douglas and Wildavsky (1982) note, for instance, that scientific ratios that assess levels of risk are incomplete measures of the human approach to danger, since they explicitly try to exclude culturally constructed ideas about living "the good life." Risky habits or dangerous behaviors are conformations to lifestyles, and thus become evaluated within other socially and culturally evaluated phenomenon. Food, money, or lifestyle may outweigh perceived vulnerability. People live in Los Angeles, for example, not because they like breathing smog, but to take advantage of job opportunities, or because they value natural beauty, a warm climate, and so forth. Altering risk selection and risk perception, then, depends on changing the social order. From the point of view of sociology, Mileti (1999) similarly argues that any shift in vulnerability-preparedness must include a shift in cultural premises that privilege technological solutions, consumerism, and short-term, non-sustainable development. He notes that in the U.S., centralized attempts to guard against natural disasters, especially those that employ technological means to control nature, may ultimately create a false sense of security that can exacerbate the risk of even more damage occurring. For instance, dams and levees meant to protect communities from flooding along the Mississippi River basin actually encouraged denser settlement patterns and industrial development in flood-prone areas, which inflicted much greater losses during a large flood that caused the levees to fail. Paine (2002), in writing about Israeli citizen responses to violence from the Palestinian uprising, notes that consciousness of risk can actually be socially negated. Particularly for Zionist Israelis, the acceptance of religious identity and collective mission supersedes any rational calculation of vulnerability. Finally, Stephens (2002) writes how political culture can shape risk assessment. In Europe in the years following the Chernobyl disaster, risk assessment has been effectively delegated away from individual or personal level to the realm of scientific "authoritative experts." Stephens' work shows the pressure among these experts to both inform an anxious public about the levels of risk surrounding nuclear energy, nuclear accidents, and radiation danger, and simultaneously assuage the public that everything is "normal" and "under control."

RESPONSES TO DISASTER

Individual and Organizational Responses

As Oliver-Smith notes, hazards and disasters challenge the structure and organization of society. Much anthropology, therefore, examines the behaviors of individual actors and groups within the events surrounding a disaster. The anthropology of disaster response has focused on changes occurring within cultural institutions like religion, ritual, economic organization, and politics, especially concerning the relative degrees of local cooperation or conflict, the ability of local institutions to mitigate the impact of a disaster, and the differential capabilities of response due to ethnicity, gender, age, and socioeconomic status (Das, 1997). Pannell (1999), for instance, notes that inland resettlement of a coastal community because of volcanic activity involved dramatic and destabilizing changes in subsistence, organization, and identity. Research has also focused on how vulnerable populations variously respond to both the crisis and the provision of aid, in particular the aged (Guillette, 1993), women (Vaughan, 1987; Shaw, 1992; Alexander, 1995; Bari, 1998), and children (Gordon et al., 1996; Tobin and Whiteford, 2001; Shepler, 2003). Each of these populations may have different coping mechanisms, different vulnerabilities, and different capabilities (Anderson, 1994; Nordstrom, 1998; Skelton, 1999). Research has also focused on the interactions and interrelationships between donors, providers and recipients of aid (Oliver-Smith, 1979).

With the rise in occurrence and severity of technological disasters such as oil spills and chemical explosions have come anthropological studies of community and corporate responses. Research into the Exxon-Valdez Alaska oil spill uncovered how communities recovered from the stress and impact of the spill. Some of these have shown that disasters can stimulate a range of social responses, from initial anger and denial to social integration and cohesiveness, as new groups form to initiate bargaining for responsibility and obligation (Button, 1992). Loughlin's work on responses to the Bhopal, India, chemical explosion shows how corporate and community definitions as to disaster, culpability, and accountability can be at odds, and that disasters may stimulate new forms of local activism and social consciousness (Loughlin, 1996). Such research provides grounding for the concept of "environmental justice" (Johnston, 1994), which attempts to define rights for those communities whose subsistence is primarily dependent on an ecological relationship with their surrounding natural resources.

Though disaster-literature typically focuses on the population-level, disaster-related trauma may have individual effects that become expressed in culturally informed ways, in response to fire (Maida et al., 1989), earthquake (Bode, 1989; Oliver-Smith, 1992), technological disaster (Palinkas, et al. 1993), or complex emergencies (Jenkins, 1996; Caruth, 1996; Young, 1997; Henry, 2000a). Anthropologists have come to use the analytical term "embodiment" to focus on the complex meanings of disaster-related trauma that become manifest in individuals, as the lived experiences of disaster, and the creative ways that survivors use to comprehend the trauma done to their lives, and attempt to move on (Kleinman et al., 1997; Green, 1999; Anderson, 2004; Scheper-Hughes and Bourgois, 2004). Henry, for instance, notes how *haypatensi* (from the English medical condition "hypertension") evolved in the war-torn areas of Sierra Leone as a new kind of local sickness experienced by refugees and internally displaced persons in response to their experiences of violence, displacement, and the provision of relief aid (Henry, 2000b). Cathy Caruth (1996) notes that when traumatic experience is remembered, a "historical narrative" is created in which the events become restructured and resituated in ways that help the survivors understand and make sense of what happened, and move forward (see also Malkki, 1990). The analyses of trauma narratives have been recognized as valuable in helping illuminate how people come to make sense of the violence done within disaster (Poniatowska, 1995; Jenkins, 1996; Coker, 2004). As Coker points out, the references within that narrative need not be straightforward, but become locally understood and expressed indirectly in culturally defined idioms (somatization, a new kind of sickness, Divine punishment, supernatural wrath, spirit possession, etc.).

Anthropologists have often been critical of the dominant Western classification-diagnoses "PTSD," or Post-Traumatic-Stress-Disorder, especially its claim to represent universal "human" responses to extremely traumatic situations. They trace the diagnosis through the historical construction of "trauma," especially in its transformations by nineteen and twentieth century scientists, and the study of war-traumatized WWII soldiers and Vietnam veterans (Hermann, 1992; Young, 1997; Petty and Bracken, 1998). Bracken notes that PTSD was created through Western categories, and that the therapy it entails is shaped by Western ideas of cognitivism, in which trauma is located as an event inside a person's head, rather than representing a social phenomenon, where recovery might be bound up with the recovery of the wider community. Instead, in line with the Western model, children and adults are universally encouraged to talk about traumatic experiences, or draw, paint, or use storytelling, in effect provoking them to relive the trauma. Not surprisingly, anthropologists have been critical of this kind of approach; it conflicts with the holistic and relativistic approaches of anthropology described above – "Talking cures" or counseling that ignores other family or community members may be cross-culturally inappropriate, especially in other parts of the world, where conceptions of individuality and person may be much more connected to the social context than in the Western world (Young, 1995; Brett, 1996). Parker (1992) further notes that any implication that traumatic disaster is "temporary" ignores the fact that many people live with chronic insecurity, economic frailty, and extended states of trauma.

Responsive Belief Systems and Coping Strategies

Since the beginning of the discipline, anthropologists have been interested in how people draw upon and alter their belief systems in efforts to come to terms with events of catastrophic change, violence, loss, resettlement, and even humanitarian relief (Lindstrom, 1993; Maida, 1996). These events can involve changes in social institutions like religious beliefs or customs (Stewart and Harding, 1999), social organization (Colson, 1973; Oliver-Smith, 1977), attitudes and values (Bode, 1977; Oliver-Smith, 1992), even marriage institutions (Loizos, 1977).

Anthropologists have shown some of the adaptive coping strategies that even relatively isolated world populations have traditionally used to respond and cope with disasters from the environment, such as flood, drought, conflict, earthquake, volcanic explosion, and disease (Turton, 1977; Torry, 1978a; Zaman, 1989; Tobin and Whiteford, 2002). Archaeology, for instance, has used the material record to provide long-term depth for understanding the human-environment relationship in both historical and pre-historical time. This has involved using flora, fauna, and material remains to examine the relationship between contextual variables like the magnitude or speed of a disaster with social variables such as population density, wealth distribution, and political complexity, in order to assess how disasters have impacted human response and social adaptation over time (McGuire et al., 2000; Bawden and Reycraft, 2001). Some of the work here notes how disasters can instigate cultural evolution (Minnis, 1985; Mosely and Richardson, 1992); others note the disastrous consequences of unsustainable environmental practices that human behavior can cause (Fagan, 1999; Redman, 1999; Dods, 2002). In contemporary time, Elizabeth Colson has pointed out the creative coping mechanisms that can occur within social systems as a result of the upheaval of forced relocation, such as flexible forms of social organization, familial obligations, occupations, and belief systems (1973, 2003). Monica Wilson notes how the cultural norms of hospitality in southern Africa enabled shipwrecked explorers and traders to be welcomed and integrated into the social order of local communities (1979). Davis echoes this, noting that the suffering involved in traumatic experience is social – "the experience of war, famine, and plague is continuous with ordinary social experience; people place it in social memory and incorporate it with their accumulated culture (1992: 152). For Davis, suffering results not so much from a "breakdown" in the proper functioning of the social order, but rather is itself a painful part of the social organization. This includes the culturally diverse ways that people mourn, and how they draw upon culturally and religiously defined symbols to find strength (Bode, 1989; Hoffman, 1995).

In some areas of the world, people have long had to deal with social disruption, such as areas in the African Sahel, where drought, famine, and political insecurity have become somewhat common, if not always anticipated, events. In Sudan, for example, Van Arsdale (1989) coins the term "adaptive flux" to refer to the indigenous self-help tactics and long-term coping strategies that have evolved to enable people to survive under fluctuating, harsh, and erratic conditions in what is a socio-economically and geographically peripheral area. People may activate migration networks that send some family members to urban areas, farmers may enact systems of crop rotations or sharing of draft animals to increase the chance of a successful harvest, or they may rely on grass-roots political councils to mobilize food resources or security during scarcity or political instability. In Ethiopia, for example, Hailu et al. (1994) note that these kind of local council decisions were able to mobilize 6,000 peasants to build a dry-weather road to eastern Sudan in a short time. This later enabled relief-assistance to reach the area during famine.

Adaptive strategies can, however, become strained under the larger-scale of vulnerability that has frequently accompanied the transformations inflicted on indigenous societies since Western contact, colonialism, industrialization, and incorporation into the world market. Already mentioned was how British colonialism and economic pressures in the East African Sahel eroded (and in some cases, outlawed) preexisting indigenous methods of drought survival, and increased the local vulnerability to desertification, famine, and starvation (Turton, 1977; Fagan, 1999).

Responses Within Political Organization

Anthropologists have noted how disasters can alter political organizations and power relations between individuals, the state, and international actors. Disasters may provide a kind of structuring idiom that allows people to more clearly apprehend their own political situation and their own position of power (or marginality) relative to that of the state (Chairetakis, 1991; Button, 1992). Chairetakis notes that where states or political parties are able to exploit the situation by being seen as a major player in relief, relief efforts can bolster the dominant political interests of those already in power (see also Blaikie et al., 1994). Davis, writing about the consequences of earthquake and tsunami in Alaska, notes that disaster assistance functioned to increase the integration of native groups into the state (Davis, 1986). Alternately, disaster and relief can stimulate the development of subaltern means, identities, or interests. Robinson et al. (1986), for instance, writing about local responses following the 1985 Mexico City Earthquake, note how neighborhood and student organizations recovering from the quake felt empowered to mobilize and demand more accountability from the political party in power.

Responses Within Economic Systems

Anthropologists have always been interested in the material and economic exchange of peoples, especially in terms of production, distribution, consumption, the allocation of scarce resources, and the cultural rules for the distribution of commodities. Because disasters and disaster relief can so dramatically impact material subsistence and exchange, anthropologists have looked at the changes that disasters can bring to economic systems and related mechanisms like employment, sharing, egalitarianism, and morality (Dirks, 1980). Torry, for example, studying Hindu responses to famine, notes that social inequalities situated within caste or other.sanctioned structures can produce marked inequalities in access to resources, and the unequal distribution of relief items (Torry, 1986). Oliver-Smith, writing about immediate responses to avalanche and earthquake in Peru, notes that previously existing stratifications like class and ethnicity can temporarily disappear in a short-lived wave of altruism. Once national and international aid appears, however, old divisions can reemerge, and conflicts over access to resources begin again (1979, 1992).

Providing Relief: Development and Power

As mentioned above, anthropology has sought ways to alleviate the structural conditions of predisaster vulnerability that predispose individuals, groups, or societies to experience disaster, or that increase the severity of disaster's impact. Targeting these structural conditions, then, has often involved a search for ways to incorporate the goals and mechanisms of "sustainable development" into the paradigm of "relief" (Cuny, 1983; Kibreab, 1987; Slim and Mitchell, 1992; Zetter, 2003; Anderson and Woodrow, 1998). The relief paradigm is criticized for being externally managed and non-participatory, or for failing to recognize and affirm local institutions or skills with which communities might be involved in the management of their own disasters. Critics note the singular tendency of the relief model to implement top-down strategies which preclude situational flexibility or genuine local participation, or for biases which pathologize the victims or survivors and encourage aggressive, external interventions, or for the "restricting logic" that relief bureaucracies impose on the recipients of aid, thus creating dependent, helpless, powerless populations (Harrell-Bond, 1993; Adams and Bradbury, 1995; De Waal, 1997; Platt, 2000). They posit that a more developmental approach is ultimately more beneficial in helping prevent future disasters, in that development is more likely to target the structural forces attributed to be at the root causes of vulnerability. Developmentalists assert that emergency relief should be temporary, and that any aid should be quickly followed by rehabilitation, focusing on "capacity building" and "supporting local structures" (see Boutros-Ghali, 1992). Critics of this counter that these words can be merely excuses for reducing food and medical entitlements, which then shift the burden to local communities without properly assessing their capacities to manage it (Macrae et al., 1997; Bradbury, 1998; Macrae, 1998).

The international system of relief can dramatically impact previously remote or marginal areas, and create new and previously inconceivable kinds of employment, education, opportunity, even aspirations, for people. As noted, however, new opportunities tend to fall along preexisting restrictions of gender roles and expectations, class, nationality, or religion (Ferguson and Byrne, 1994; Anderson and Woodrow, 1998; Sommers, 2001; Shepler, 2002), and can even result in heightened tension or conflict (Jackson, 2003). As mentioned above, the comparative and relativistic stance of the discipline has given it an often critical stance towards dominant Western models of relief, often giving voice instead to local knowledge and local ways of management (Harrell-Bond, 1993; De Waal, 1997). Others have analyzed the media, and how those affected by disaster are portrayed in popular print. This includes a critique of the media for appropriating images and stories of others' experiences of pain and suffering as a commodity to be bought, sold, manipulated, or marketed in order to attract more donations (Feldman, 1995; Kleinman and Kleinman, 1997; Gourevitch, 1998).

Some have directly confronted the structural imbalances embedded in the relationships between refugees and the humanitarian community. This calls attention to the fact that the very field of "emergency management" often involves an a priori assumption that local people are in need of external managers, and are unable to provide for themselves (Torry, 1978b, see also Mileti, 1999). Though not an anthropologist, Platt (2000) argues that U.S. disaster policy since 1950 has supplanted moral and community concern with government subsidies and financially-expressed compassion that fosters co-dependency, effectively providing disincentives to local governments in their own attempts to create disaster-resistant communities. Ino Rossi, in studying the long-term reconstruction following an earthquake in Italy, notes that local priorities can be overlooked when they differ from those of donors, and relief agents, and governments (Rossi, 1993). The control of information by donors may be linked to anxiety, frustration, and feelings of powerlessness among recipients (Button, 1995; Henry, 2000a). Malkki, in her work with Rwandan refugees in the Congo, notes that humanitarian knowledge is discursively powerful, and may operate to silence local agendas that run contrary to its own (1996). The recipients of aid are not completely powerless, however; Henry (2002) notes how refugees living in remote, marginal, border areas learned to adapt to the system providing relief aid by interchanging identities between "citizens" and "displaced" in order to maximize benefits and empower themselves on an international stage dominated by foreign relief efforts.

Because one of the most common social reactions to a crisis is flight, problems associated with the management of post disaster population upheaval and resettlement have been examined in considerable detail. One avenue of productive exploration has been with populations fleeing complex emergencies, obtaining shelter in camps set up for refugees and internally displaced people (IDPs) (Colson, 2003); this includes the effects of camp policies on the displaced themselves. A growing body of research questions the international community's motivations in persistently encouraging the placement of refugees in separate, demarcated camps (Harrell-Bond, 1986, 1994; Van Damme, 1995), as opposed to self-settlement. Infectious disease rates may be higher in camps, despite aggressive, centralized public health interventions; nutritional problems may be higher, especially where there is no individual access to means of subsistence, and environmental damage is greater. Morbidity and mortality may be underreported, as camp dwellers have an interest in concealing any drop in their numbers in order to maintain relief-supply entitlements. There may be further "invisible" damages from introducing a foreign aid system, which undermines local values of sharing, cooperation, or hospitality, that hold society together. Yet despite this research, local and international agencies, usually under UN auspices, use relief supplies to encourage the settlement of displaced people into camps, with the rationale that centralized groups of displaced people are easier to distinguish from the general population and manage.

Starting from Foucault's *Discipline and Punish* (1979), Malkki notes how camps for displaced people can be seen as discrete loci of asymmetric power – set apart, clearly bounded, and with formalized, hierarchical structures. Almost as in a hospital, mobility in the camps may be restricted by numerous identification stations and check points. The "sick" are the displaced, uncertain of the necessary course of events to get them back to a desired state, at the mercy of the authoritative knowledge of camp administrators, and secure in their residence only by maintaining a demonstration of helplessness (Hitchcox, 1990; Malkki, 1995; Muecke, 1992). Harrell-Bond (1986, 1993), Mazur (1988), and Kibreab (1993) document how the paradigms in use by the UNHCR tend to characterize the displaced as "helpless," despite abundant evidence to the contrary. This bureaucratic conceptualization may be reinforced by an ideological belief that refugees lack the motivation or capacity to work out solutions for their own self-sufficiency (Van Arsdale, 1993; Gibbs, 1994). Such characterization has severe implications for how refugees are treated: camp authorities may react negatively when refugees demonstrate their own competence, as personal initiative is seen as interfering with the "smooth functioning" of the camps (Williams, 1990). Camps may thus impose a kind of "restricting logic" on its members; powerless to take charge of events affecting their lives, the displaced often become dependent on aid agencies for their basic subsistence (Marchal, 1987).

The history of the modern system of international humanitarian relief has received recent attention, as has the ambivalent nature of its entry into disaster affected areas (Crew, 1998; Middleton and O'Keefe, 1998). De Waal (1993, 1997), for example, in his critique of the self-serving nature of humanitarian interventions in Africa and Asia, implicates the "relief industry" as perpetuating (and exacerbating) the very famines and conflict they purportedly try to alleviate (see also Jackson, 2003). Also noted is the inappropriateness of some aid, especially food relief. Henry (2000a) notes that supplying cornmeal and boxed breakfast cereal to West African refugees whose staple is rice may have satisfied regional political and economic pressures within the World Food Program, but did little to alleviate local hunger. Having no idea how to prepare cornmeal into recognizable "food," Sierra Leonean refugees were forced to sell the cornmeal to local traders in exchange for bags of rice. Unfortunately, this was a bad deal for those in need, as the poor exchange rate in a cash-poor environment meant worsened malnutrition and hunger. Finally, a growing literature looks at the impact of disaster resettlement on host country populations. Gebre (2003) notes that while displaced populations receive aid, research coverage, and policy attention, those hosting the displaced can themselves undergo extreme strain and upheaval, though their plight remains largely unnoticed. Similarly, Leach (1992) and Henry (2002) note that relief efforts for Liberian and Sierra Leonean refugees could upset and sour traditionally cordial host-guest and extended family relationships and obligations.

ASSESSMENT

In summary, anthropology offers the field of disaster studies broad comparative, contextual, and cross-cultural perspectives, particularly from its extensive work in the developing world. Its holistic approach frames disasters within their social, cultural, political, economic, and environmental relationships, from the human behaviors that can cause or influence the severity of disaster, to culturally informed adaptations and responses, to the relative social vulnerabilities that mitigate or magnify a disaster's impact.

The anthropology of disasters works under the assumption that those suffering under crisis are not empty vessels stripped bare of their cultural makeup; on the contrary, cultural institutions figure at the center of a society's disaster vulnerability, preparedness, mobilization, and prevention. It follows then, that disaster preparedness as well as relief and reconstruction aid could be more appropriate, efficient, and economical if an understanding of the experiences and perspectives of local communities and institutions were taken into account. This includes understanding the larger social and organizational cultures that may interfere with practices of sustainable, long-term development. Given the top-down biases of emergency relief, anthropology needs to continue to seek practical ways to incorporate local technical knowledge, insight, skills, desires, and needs into the management of disaster situations, so that local people and institutions might be affirmed in identifying problems and offering solutions towards the management of their own situation, and that local capacities may be strengthened to resist future emergencies. Morren (1983), for instance, notes that Kalahari Bushmen in Africa, on the front lines of disaster as first responders to drought, can be remarkably effective in limiting loss and facilitating relief (see also Torry, 1988).

In addition, more ethnographic research is needed on the organizational cultures and constraints of relief agencies themselves, along the lines of Kent's *Anatomy of Disaster Relief* (1987). This should move beyond the merely critical, to offer practical solutions as to how to address the gap between research and practice, perhaps focusing on how bureaucratic barriers might be transcended in order to encourage situational flexibility and generate genuine grass-roots participation. This should include the moral, social, political, and economic values that the recipients of aid attach to the items being provided in relief (Prendergast, 1996).

Finally, there are methodological concerns that need to be addressed. More critical research is needed into how social scientists can professionally yet ethically conduct research during and in the midst of disasters. The professional concerns include the identification of methodological biases, such that our work can remain both academically sound and yet policy relevant. There are also ethical concerns that arise from the researcher's position of relative privilege and power (Nordstrom and Robben, 1995; Greenhouse et al., 2002; Jacobsen and Landau, 2003). This is particularly true for international disasters that occur in developing countries, whose people experience more extreme forms of vulnerability and stress. Through its concern for local sensitivities, anthropology needs to ask how it may better structure questions and better seek information in ways that inflict the least harm from people under situations of severe duress.

REFERENCES

Adams, M. and Bradbury, M. (1995). *Conflict and Development: Organizational Adaptation in Conflict Situations.* Oxfam Discussion Paper 4. Oxford: Oxfam.

Agarwal, B. (1992). "Gender Relations and Food Security: Coping with Seasonality, Drought, and Famine in South Asia." In Beneria, L. and Feldman, S. (ed.). *Unequal Burden: Economic Crises, Persistent Poverty and Women's Work.* Boulder: Westview Press. pp. 181–218.

Alexander, E. (1995). *Gender and Emergency Issues: A Synthesis of Four Case Studies: Malawi, Mozambique, Angola, and Zaire.* Report prepared for the World Food Programme.

Anderson, M. (1994). "Understanding the Disaster-Development Continuum: Gender Analysis is the Essential Tool." *Focus on Gender 2*(1): 7–10.

Anderson, M. (2004). *Cultural Shaping of Violence: Victimization, Escalation, Response.* West Lafayette: IN: Purdue University Press.

Anderson, M. and P. Woodrow. (1998). *Rising from the Ashes: Development Strategies in Times of Disaster.* Boulder: Lynne Rienner.

Bankoff, G. (2003). "Constructing Vulnerability: the Historical, Natural, and Social Generation of Flooding in Metropolitan Manila." *Disasters 27*(3): 224–238.

Bari, F. (1998). "Gender, Disaster, and Empowerment: A Case Study from Pakistan." in Enarson, E. and Hearn-Morrow, B. (Ed.). *The Gendered Terrain of Disaster: Through Women's Eyes.* London: Praeger. pp. 1–8.

Bawden, G. and Reycraft, R. (2001). *Environmental Disaster and the Archaeology of Human Response Anthropological Papers No. 7.* Albuquerque: Maxwell Museum of Anthropology and University of New Mexico.

Blaikie, P., Cannon, T., Davis, I., and Wisner, B. (1994). *At Risk: Natural Hazards, People's Vulnerability, and Disasters.* London: Routledge.

Bode, B. (1977). "Disaster, Social Structure, and Myth in the Peruvian Andes: the Genesis of an Explanation." *Annals of the New York Academy of Sciences 293*: 246–274.

Bode, B. (1989). *No Bells To Toll: Destruction and Creation in the Andes.* New York: Scribners.

Bodley, J. (1982). *Victims of Progress* (second ed.). Menlo Park, CA: The Benjamin/Cummings Publishing Company, Inc.

Boutros-Ghali, B. (1992). *An Agenda for Peace.* New York: United Nations.

Bradbury, M. (1998). Normalising the Crisis in Africa. *Disasters 22*(4): 328–338.

Brett, E. (1996). "The Classification of Posttraumatic Stress Disorder." In van der Kolk, B., McFarlane, A., and Weisath, L. (Eds.). *Traumatic Stress: the Effects of Overwhelming Experience on Mind, Body, and Society.* New York: Guilford Press. pp. 117–128

Button, G. (1992). *Social Conflict and Emergent Groups in a Technological Disaster: the Homer Area Community and the Exxon-Valdez Oil Spill.* Unpublished Ph.D. thesis. Brandeis University.

Button, G. (1995). "What You Don't Know Can't Hurt You: The Right to Know and the Shetland Islands Oil Spill." *Human Ecology 23*: 241–257.

Caruth, C. (1996). *Unclaimed Experience: Trauma, Narrative, and History.* Baltimore: Johns Hopkins University Press.

Cernea, M. (2000). "Risks, Safeguards, and Reconstruction: A Model for Population Displacement and Resettlement."

In Cernea, M. and McDowell, C. *Risk and Reconstruction Experiences of Settlers and Refugees*, edited by. Washington, D.C.: The World Bank. pp. 11–55

Cernea, M. (1990). "Internal Refugee Flows and Development Induced Population Displacement." *Journal of Refugee Studies 3*: 320–329.

Chairetakis , A. (1991). *The Past in the Present: Community Variation and Earthquake Recovery in the Sele Valley, Southern Italy, 1980–1989.* Unpublished Ph.D. thesis presented to Columbia University.

Coker, E. (2004). "Traveling Pains: Embodied Metaphors of Suffering Among Southern Sudanese Refugees in Cairo." *Culture, Medicine, and Psychiatry 28*: 15–39.

Colson, E. (1973). *The Social Consequences of Resettlement*, Kariba Studies IV. Manchester: Manchester University Press.

Colson, E. (2003). "Forced Migration and the Anthropological Response." *Journal of Refugee Studies 16*(1): 1–18.

Crew, E. (1998). *Whose Development? An Ethnography of Aid.* New York: Zed Books.

Cuny, F. (1983). *Disasters and Development.* New York: Oxford University Press.

Das, V. (1997). *Social Suffering.* Berkeley: University of California Press.

Davis, N. (1986). "Earthquake, Tsunami, Resettlement, and Survival in Two North Pacific Alaskan Native Villages." In Oliver-Smith, A. (ed.) *Natural Disasters and Cultural Responses.* Williamsburg: College of William and Mary. pp. 123–154

Davis, J. (1992). "The Anthropology of Suffering." *Journal of Refugee Studies 5*(2): 149–161.

De Waal, A. (1993). "In the Disaster Zone: Anthropologists and the Ambiguity of Aid." *Times Literary Supplement 4711*(5): 5–6.

De Waal, A. (1997). *Famine Crimes: Politics and the Disaster Relief Industry in Africa.* Oxford: African Rights and James Currey.

Dirks, R. (1980). "Social Responses During Severe Food Shortages and Famine." *Current Anthropology 21*: 21–44.

Dods, R. (2002). "The death of Smokey Bear: the Ecodisaster Myth and Forest Management Practices in Prehistoric North America." *World Archaeology 33*(3): 475–487.

Douglas, M. and Wildavsky, A. (1982). *Risk and Culture: An Essay on the Selection of Technical and Environmental Dangers.* Berkeley: University of California Press.

Dynes, R. (2004). "Expanding the Horizon of Disaster Research." *Natural Hazards Observer 28*(4): 1–2.

Fagan, B. (1999). *Flood, Famines, and Emperors: El Niño and the Fate of Civilizations.* New York: Basic Books.

Farmer, P. (1999). *Infections and Inequalities: The Modern Plagues.* Berkeley: University of California Press.

Farmer, P. (2004). *Pathologies of Power: Health, Human Rights, and the New War on the Poor.* California Series in Public Anthropology. Berkeley: University of California Press.

Feldman, A. (1995). "Ethnographic States of Emergency." In Robben, A., and Nordstrom, C. (Eds.). *Fieldwork Under Fire.* Berkeley: University of California Press. pp. 224–253

Ferguson, P. and Byrne, B. (1994). *Gender and Humanitarian Assistance: A Select Annotated Bibliography.* Bibliography compiled for the Office of Women in Development, U.S. Agency for International Development.

Fitchen, J. (1988). "Anthropology and Environmental Problems in the U.S.: the Case of Groundwater Contamination." *Practicing Anthropology 10*(5): 18–20.

Fothergill, A. (1996). "Gender, Risk and Disaster." *International Journal of Mass Emergencies and Disasters 14*(1): 33–56.

Franke, R. (2004). "Review of Catastrophe and Culture: The Anthropology of Disaster." *American Anthropologist 106*(4): 765–766.

Gebre, Y. (2003). "Resettlement and the Unnoticed Losers: Impoverishment Disasters among the Gumz in Ethiopia." *Human Organization 62*(1): 50–61.

Gibbs, S. (1994). "Post-war Social Reconstruction in Mozambique: Re-framing Children's Experience of Trauma and Healing." *Disasters 18*(3): 268–276.

Gordon, N., Farberow, N., and Maida, C. (1996). *Children and Disasters.* New York: Brunner/Mazel.

Gourevitch, P. (1998). *We Wish to Inform You That Tomorrow We Will Be Killed With Our Families: Stories From Rwanda.* New York: Farrar, Straus, and Giroux.

Green, L. (1999). Lived Lives and Social Suffering: Problems and Concerns in Medical Anthropology. *Medical Anthropology Quarterly 12*(1): 3–7.

Greenhouse, C., Mertz, E., and Warren, K. (Eds.). 2002. *Ethnography in Unstable Places: Everyday Lives in Contexts of Dramatic Political Change.* Durham: Duke University Press.

Guillette, E. (1993). *The Role of the Aged in Community Recovery Following Hurricane Andrew.* Boulder, CO: Natural Hazards Research and Applications Information Center Quick Response Program.

Hailu, T., Wolde-Georgis, T., and Van Arsdale, P. (1994). Resource Depletion, Famine, and Refugees in Tigray. In Adelman, H. and Sorenson, J. (Eds.). *African Refugees: Development Aid and Repatriation.* Boulder: Westview Press: pp. 21–42.

Haque, C. and Zaman, M. (1993). "Human Responses To Riverine Hazards in Bangladesh: A Proposal for Sustainable Development." *World Development 21*: 93–107.

Harrell-Bond, B. (1986). *Imposing Aid: Emergency Assistance to Refugees.* Oxford: Oxford University Press.

Harrell-Bond, B. (1993). "Creating Marginalised Dependent Minorities: Relief Programs for Refugees in Europe," *Refugee Studies Program Newsletter 15*: 14–17.

Harrell-Bond, B. (1994). "Pitch the Tents: an Alternative to Refugee Camps." *The New Republic.* September 19th and 26th.

Henry, D. (2000a). *Embodied Violence: War and Relief Along the Sierra Leone-Guinea Border.* Unpublished Ph.D. thesis presented to Southern Methodist University. Dallas, Tx.

Henry, D. (2000b). *Hypertension: Somatic Expressions of Trauma and Vulnerability During Conflict.* Paper presented at the annual meetings of the American Anthropological Association, 99th. San Francisco, CA.

Henry, D. (2002). Réfugiés Sierra-léonais, et aide humanitaire en Guinée: la réinvention d'une "citoyenneté de frontière." *Politique Africaine.* 85 (mars 2002): 56–63.

Herman, J. (1992). *Trauma and Recovery.* New York: Basic Books.

Hitchcox, L. (1990). *Vietnamese Refugees in Southeast Asian Camps.* Hampshire: St. Anthony's/Macmillan Press.

Hoffman, D., and Lubkemann, S. (2005). Warscape Ethnography in West Africa and the Anthropology of "Events." *Anthropological Quarterly 78*(2): 315–327.

Hoffman, S. (1995). *Culture Deep and Custom Old: the Reappearance of a Traditional Cultural Grammar in the Aftermath of the Oakland-Bekeley Firestorm.* Paper presented at the annual meetings of the American Anthropological Association, 94th. Washington, DC.

Hunter, J. (1992). "Elephantiasis: A Disease of Development in North-East Ghana." *Social Science and Medicine 35*(5):627–649.

Jackson, S. (2003). Freeze/Thaw: Aid, Intervention, Uncertainty, and Violence in the Kivus, Eastern DR Congo. Paper presented at the annual meetings of the American Anthropological Society, 102nd. November 19, 2003. Chicago, IL.

Jacobsen, K. and L. Landau. (2003). "The Dual Imperative in Refugee Research: Some Methodological and Ethical Considerations in Social Science Research on Forced Migration." *Disasters 27*(3): 185–206.

Jenkins, J. (1996). "The Impress of Extremity: Women's Experience of Trauma and Political Violence." In Brettell, C. and Sargent, C. *Gender and Health, an International Perspective.* New York: Prentice Hall, pp. 278–291.

Johnston, B. (1994). *Who Pays the Price? The Sociocultural Context of Environmental Crisis.* Washington, DC: Island Press.

Johnston, B. (2001). "Function and Dysfunction: Human Environmental Crisis and the Response Continuum." In Messer, E. and Lambeck, M. (Ed.) *Thinking and Engaging the Whole: Essays on Roy Rappaport's Anthropology,* edited by. Ann Arbor: University of Michigan Press, pp. 99–121

Kent, R. (1987). *Anatomy of Disaster Relief: The International Network in Action.* New York: Pinter Publishers.

Kibreab, G. (1987). *Refugees and Development in Africa: The Case of Eritrea.* Trenton, NJ: Red Sea Press.

Kibreab, G. (1993). "The Myth of Dependency Among Camp Refugees in Somalia: 1979-1989." *Journal of Refugee Studies 6*(4): 321–349.

Kleinman, A. and Kleinman, J. (1997). "The Appeal of Experience; the Dismay of Images: Cultural Appropriations of Suffering in Out Times." In Kleinman, A., Das, V., and Lock, M. (Eds.). *Social Suffering.* Berkeley: University of California Press. pp. 1–24

Kleinman, A., Das, V. and Lock. M., eds. *Social Suffering.* Berkeley: University of California Press.

Leach, M. (1992). *Dealing With Displacement. IDS Research Reports 22.* Sussex, England: Institute of Development Studies.

Lerer, L., and Scudder, T. (1999). "Health Impacts of Large Dams." *Environmental Impact Assessment Reviews 19*: 113–123.

Lindstrom, L. (1993). *Cargo Cult: Strange Stories of Desire from Melanesia and Beyond.* Honolulu: University of Hawaii Press.

Loizos, P. (1977). "A Struggle for Meaning: Greek Cypriots." *Disasters 1*: 231–239.

Loughlin, K. (1996). "Representing Bhopal." In Margus, G. (Ed.). *Late Editions 3: The Net, News, and Videotape.* Chicago: University of Chicago Press.

Macrae, J. (1998). "The Death of Humanitarianism?: An Anatomy of the Attack." *Disasters 22*(4): 309–317.

Macrae, J., Jaspars, S., Duffield, M., Bradbury, M. and Johnson, D. (1997). "Conflict, the Continuum, and Chronic Emergencies: A Critical Analysis of the Scope for Linking Relief, Rehabilitation, and Development Planning in Sudan." *Disasters 21*(3): 223–243.

Maida, C., Gordon, N., Srauss, G. (1989). "Psychosocial Impact of Disasters: Victims of the Baldwin Hills Fire." *Journal of Traumatic Stress 2*: 37–48.

Maida, C. (1996). *Crisis and Compassion in a World of Strangers.* New Brunswick: Rutgers University Press.

Malkki, L. (1990). "Context and Consciousness: Local Conditions for the Production of Historical and National Thought Among Hutu Refugees in Tanzania." In Fox, R., (Ed.). *Nationalist Ideologies and the Production of National Cultures.* Washington, DC: The American Anthropological Association. pp. 32–62.

Malkki, L. (1995). "Refugees and Exile: From 'Refugee Studies' to the National Order of Things." *Annual Review of Anthropology 24*: 495–523.

Malkki, L. (1996). "Speechless Emissaries: Refugees, Humanitarianism, and Dehistoricization." *Cultural Anthropology 11*(3): 377–404.

Marchal, R. (1987). "Production sociale et recomposition dans l'éxile: le cas Érythréen." *Cahiers d'études africaines 27*, 3/4: 393–410;

Mazur, R. (1988). "Refugees in Africa: The Role of Sociological Analysis and Praxis," *Current Sociology 36*: 43–60.

McGuire, W., Griffiths, D., Hancock, P., and Stewart. I. (2000). Archaeology of Geological Catastrophes Geological Society Special Publication, No. 171. London: Geological Society of London.

Michaels, D. (1988). "Waiting for the Body Count: Corporate Decision-Making and Bladder Cancer in the U.S. Dye Industry." *Medical Anthropology Quarterly 2*(3): 215–232.

Middleton, N. and O'Keefe, P.. (1998). *Disaster and Development: The Politics of Humanitarian Aid.* Chicago: Pluto Press.

Mileti, D. (1999). *Disasters by Design: A Reassessment of Natural Hazards in the United States.* Washington, DC: Joseph Henry Press.

Minnis, P. (1985). *Social Adaptation to Food Stress: A Prehistoric Southwestern Example.* Chicago: University of Chicago Press.

Morren G. (1983). "A General Approach to the Identification of Hazards and Responses." In Hewitt, K. (Ed.). *Interpretations of Calamity.* Edited by K. Hewitt. New York: Allen & Unwin. pp. 284–297.

Moseley, M, and Richardson, J. (1992). Doomed by Disaster. *Archaeology 45*: 44–45.

Muecke, M. (1992). New Paradigms for Refugee Health Problems. *Social Science and Medicine 35*(4): 515–523.

Nordstrom, C. (1998). Girls Behind the (Front) Lines. In Lorentzen, L. and Turpen, J. *The Women and War Reader.* New York: New York University Press. pp. 80–89.

Nordstrom, C., and Robben, A. (eds). (1995). *Fieldwork Under Fire: Contemporary Studies of Violence and Survival.* Berkeley: University of California Press.

Oliver-Smith, A. (1977). "Disaster Rehabilitation and Social Change in Yungay, Peru." *Human Organization 36*: 491–509.

Oliver-Smith, A. (1979). "Post Disaster consensus and Conflict in a Traditional Society: the Avalanche of Yungay, Peru." *Mass Emergencies 4*: 39–52.

Oliver-Smith, A. (1992). *The Martyred City: Death and Rebirth in the Peruvian Andes.* Prospect Heights, IL: Waveland Press (2nd edition).

Oliver-Smith, A. (1994). "Peru's Five-Hundred Year Earthquake: Vulnerability in Historical Context." pp. 3–48 in *Disasters, Development, and Environment.* Edited by A. Varley. London: Wiley.

Oliver-Smith, A. (1996). "Anthropological Research on Hazards and Disasters." *Annual Review of Anthropology 25*: 303–328.

Oliver-Smith, A., and Hoffman, S. (Eds.). (1999). *The Angry Earth: Disaster in Anthropological Perspective.* New York: Routledge.

Paine, R. (2002). "Danger and the No-Risk Thesis." pp. 67–90 in *Culture and Catastrophe: The Anthropology of Disaster.* Edited by S. Hoffman and A. Oliver-Smith. Santa Fe: School of American Research Press.

Palinkas, L., Downs, M., Petterson, J., and Russell, J. (1993). "Social, Cultural, and Psychological Impacts of the Exxon-Valdez Oil Spill." *Human Organization 52*: 1–13.

Pannell, S. (1999). "Did the Earth Move for You? The Social Seismology of a Natural Disaster in Maluku, Eastern Indonesia." *The Australian Journal of Anthropology 10*(2): 129–144.

Parker, M. (1992). Social Devastation & Mental Health in Northeast Africa: Some Reflections on an Absent Literature. In Allen, T. (Ed.). *In Search of Cool Ground: War, Flight and Homecoming in Northeast Africa.* London: James Currey. pp. 262–273.

Pellow, D. (2002). *Garbage Wars: The Struggle for Environmental Justice in Chicago.* Cambridge, MA: MIT Press.

Petterson, J. (1988). "The Reality of Perception: Demonstrable Effects of Perceived Risk in Goinia, Brazil." *Practicing Anthropology 10*: 8–12.

Petty, J. and Bracken, P. (1998). *Rethinking the Trauma of War.* London: Free Association Books

Platt, R., (Ed.). (2000). *Disasters and Democracy: The Politics of Extreme Natural Events.* Washington, D.C.: Island Press.

Poniatowska, E. (1995). *Nothing, Nobody: The Voices of the Mexico City Earthquake.* Translated by A. Camacho de Schmidt and A. Schmidt. Philadelphia: Temple University Press.

Prendergast, J. (1996). *Frontline Diplomacy: Humanitarian Aid and Conflict in Africa.* Boulder, CO: Lynne Rienner.

Redman, C. (1999). *Human Impacts on Ancient Environments.* Tucson: University of Arizona Press.

Robinson, S., Hernandez Franco, Y. F., Mata Catrejon, R., Bernard, H. (1986). ""It Shook Again – The Mexico City Earthquake of 1985." In Oliver-Smith, A. *Natural Disasters and Cultural Responses.* Williamsburg, VA: College of William and Mary. pp. 81–123.

Rossi, I. (1993). *Community Reconstruction After an Earthquake.* Westport: Preager.

Scheper-Hughes, N. and Bourgois, P. eds. (2004). *Violence in War and Peace: An Anthology.* Oxford: Blackwell Publishing.

Schoepf, B., Schoepf, C., and Millen, J. (2000). "Theoretical Therapies, Remote Remedies: SAPs and the Political Ecology of Poverty and Health in Africa." In Kim, J., Millen, J., Irvin, A., and Gershman, J. (Eds.). *Dying for Growth: Structural Adjustment and the Health of the Poor.* Monroe, ME: Common Courage Press. pp. 91–125.

Scudder, T., and Colson, E. (1982). "From Welfare to Development: a Conceptual Framework for the Analysis of Dislocated People." In Hansen, A. and Oliver-Smith, A. (Eds.). *Involuntary Migration and Resettlement: The Problems and Responses of Dislocated People.* Boulder, CO: Westview Press. pp. 267–287

Sharp, G. (1968). *Dust Monitoring and Control in the Underground Coal Mines of Eastern Kentucky.* Master's thesis submitted to University of Kentucky Department of Anthropology.

Shaw, R. (1992). 'Nature,' culture, and disasters: floods and gender in Bangladesh." In Cross, E. and Parkin, D. *Bush Base: Forest Farm: Culture, Environment, and Development.* London: Routledge. pp. 200–217.

Shepler, S. (2002). "Les Filles-Soldats: Trajectoires d'apres-guerre en Sierra Leone." *Politique Africaine 88*: 49–62.

Shepler, S. (2003). "Educated in War: The Rehabilitation of Child Soldiers in Sierra Leone." pp. 57–76 in *Conflict Resolution and Peace Education in Africa.* Edited by E. Uwazie. Lanham, MD: Lexington Books.

Shipton, P. (1990). "African Famines and Food Security: Anthropological Perspectives." *Annual Review of Anthropology 19*: 353–394.

Simonelli, J. (1987). "Defective Modernization and Health in Mexico." *Social Science and Medicine 24*(1): 23–36.

Skelton, T. (1999). "Evacuation, Relocation, and Migration: Monserratian Women's Experiences of the Volcanic Disaster." *Anthropology in Action 6*(2): 6–13.

Slim, H. and Mitchell, J. (1992). "Towards Community Managed Relief: A Case Study from Southern Sudan." *Disasters 14*(3): 265–269.

Sommers, M. (2001). *Fear in Bongoland: Burundi Refugees in Urban Tanzania.* New York: Berghahn Books.

Stephens, S. (2002). "Bounding Uncertainty: the Post-Chernobyl Culture of Radiation Protection Experts." In Hoffman, S. and Oliver-Smith, A. *Culture and Catastrophe: The Anthropology of Disaster.* Santa Fe: School of American Research Press. pp. 91–112 .

Stewart, K, and Harding, S. (1999). "Bad Endings: American Apocalypsis." *Annual Review of Anthropology 28*: 285–310

Tobin, G., and Whiteford, L. (2001). "Children's Health Characteristics under Different Evacuation Strategies: The Eruption Of Tungurahua Volcano, Ecuador," *Papers of the Applied Geography Conferences,* Vol. 24, pp. 183–191.

Tobin, G., and Whiteford, L. (2002). "Community Resilience and the Volcano Hazard: The Eruption of Tungurahua and the Evacuation of the Faldas, Ecuador." *Disasters 26*(1): 28–48.

Torry, W. (1978a). "Natural Disasters, Social Structure, and Change in Traditional Societies." *Journal of Asian and African Studies 13*: 167–183.

Torry, W. (1978b). "Bureaucracy, Community, and Natural Disasters." *Human Organization 37*: 302–308.

Torry, W. (1979). "Anthropological studies in Hazardous Environments: Past Trends and New Horizons." *Current Anthropology 20*: 517–541.

Torry, W. (1986). "Morality and Harm: Hindu Peasant Adjustments to Famines." *Social Science Information 25*: 125–160.

Torry, W. (1988). "Famine Early Warning Systems: the Need for an Anthropological Dimension." *Human Organization 47*: 273–281.

Turton, D. (1977). "Response to Drought: the Mursi of Southwestern Ethiopia." *Disasters 1*: 275–287.

United Nations Disaster-Relief Coordinator (UNDRO). (1984). Disaster Prevention and Mitigation, Vol. II, Preparedness Aspects. New York: United Nations.

Van Arsdale, P. (1989). "The Ecology of Survival in Sudan's Periphery: Short Term Tactics and Long-term Strategies." *Africa Today 36*(3/4): 65–78.

Van Arsdale, P., (Ed.). (1993). *Refugee Empowerment and Organizational Change: A Systems Perspective.* Arlington, Va: American Anthropological Association Committee on Refugee Issues.

Van Damme, W. (1995). "Do Refugees Really Belong in Camps? Experiences from Goma and Guinea." *Lancet 346*: 360–362.

Vaughan, M. (1987). *The Story of an African Famine: Gender and Famine in Twentieth-Century Malawi.* Cambridge: Cambridge University Press.

Wallace, A. (1987). *St. Clair: A Nineteenth-Century Coal Town's Experience with a Disaster Prone Industry.* New York: Knopf.

Williams, H. (1990). "Families in Refugee Camps." *Human Organization 49*(2): 100–109.

Wilson, M. (1979). "Strangers in Africa: Reflections on Nyakyusa, Nguni, and Sotho Evidence. In Shack, A. and Skinner, E. (Eds.). *Strangers in African Societies.* Berkeley: University of California Press. pp. 51–66.

Wolfe, A. (1988). "Environmental Risk and Anthropology." *Practicing Anthropology 10*: 1.

Young, A. (1995). *The Harmony of Illusions. Inventing Post-Traumatic Stress Disorder.* New Jersey: Princeton University Press.

Young, A. (1997). "Suffering and the Origins of Traumatic Memory." In Kleinman, A., Das, V., and Lock, M. (Eds). *Social Suffering.* Berkeley: University of California Press. pp. 245–260.

Zaman, M. (1989). "The Social and Political Context of Adjustment to Riverbank Erosion Hazard and Population Resettlement in Bangladesh." *Human Organization 48*: 196–205.

Zetter R. (2003). *Shelter Provision and Settlement Policies for Refugees: A State of the Art Review* Studies Nordiska Afrikainstitutet (The Nordic Africa Institute)/Swedish International Development Agency, Studies on Emergencies and Disaster Relief Report No. 2, Uppsala, Sweden.

Chapter 9

Social Work and Disasters

Michael J. Zakour

ABSTRACT

This article reviews the contributions of the social work profession to disaster research, with an emphasis on contributions in the last two decades. Social workers have been active in disaster relief since the U.S. Civil War and the Settlement House Movement of the late nineteenth century. Social workers have defined disasters primarily in terms of the social and psychological impact of natural and technological hazards. The social work profession has been largely concerned with disaster-related issues such as prevention of severe disruption during disaster, impacts on systems at multiple levels of analysis, and availability of services to high-risk populations such as children and low-income persons. Social workers have contributed research findings on traumatic stress, disaster volunteers, vulnerable populations, organizations and interorganizational networks, environmental disasters, cross-cultural and international issues in disasters, and improved measurement and theory. Disaster research in social work is largely based in research from sociology and psychology. Gaps in social work knowledge, and suggestions for future research, are discussed. Finally, substantive suggestions for emergency management and social work are offered.

HISTORICAL OVERVIEW

The social work profession has long been involved with disaster relief, both through the profession's roots in the provision of wartime relief, and its concern with the physical environment of people. Beginning with the Civil War and continuing with the formal role of social workers in Vet Centers, social workers have helped treat the trauma resulting from wartime deployment (Pryce and Pryce, 2000). In the social work perspective, the environment is included among the physical, biological and social factors influencing the welfare of individuals, groups, and populations. Since the late nineteenth century social workers have intervened in the microenvironments of people to improve their health status, residential living environment, workplace conditions, and social and psychological functioning (Zakour, 1996a). An important focus for these interventions has been the urban environment of immigrants to the United States before 1900. Crowded and unhealthy tenement living, poor public health, and elevated morbidity and mortality in these urban settings led to collaborative efforts by social workers and public health workers to seek to reform urban systems and conditions.

These early urban reform efforts were closely related to the Settlement House movement led by Jane Addams at Hull House, and to the Charity Organization Societies. The settlement house workers lobbied for public health reforms which resulted in a sharp decrease in morbidity and mortality from epidemic disease in urban areas (Zakour, 1996a). Settlement workers in Chicago provided disaster relief and services to victims of the Chicago Fire in 1871. Charity Organization Societies (COS) responded to the San Francisco Earthquake of 1906, using their tradition of interagency coordination to improve disaster response. Both the COS and the settlement house workers represent early movements within social work emphasizing both community mobilization and services coordination. Community mobilization to improve environmental conditions for individuals, households, and populations provided the roots for environmental concerns in social work today. More effective coordination continues to be a focal point for improvements in disaster response within social work. Coordination promises to make services accessible for vulnerable populations, as well as link services together to provide for improved continuity of care for victims of disaster (Zakour and Harrell, 2003).

This article describes how social workers currently define disaster, vulnerability, and emergency management. The central concerns of social work disaster research are discussed. The contributions of social workers to disaster research are reviewed, with an emphasis on the research findings of the last two decades. The relationship of social work disaster research to research in other disciplines and professions is summarized, and gaps in social work disaster knowledge are described. Finally, suggestions for emergency managers, and for future research in social work and related disciplines, are offered.

DEFINING DISASTER, VULNERABILITY, AND EMERGENCY MANAGEMENT

Social work disaster researchers define disaster primarily through social disruption and collective stress, though physical hazards are an important part of the definition of disaster. Vulnerability in disasters refers to social structural factors leaving populations such as low-income groups, children, and older individuals disproportionately at risk for loss during disaster. Communities are vulnerable because of their demographic, cultural, historical, or ecological characteristics. Emergency management in the social work perspective is the management and coordination of the disaster social services delivery system so that important resources are redistributed to vulnerable populations heavily impacted by disaster.

Definition of Disaster

In social work research, disasters are seen as a type of collective stress situation, in which many individuals fail to have their needs met through societal processes (Barton, 1969). Disasters are distinguished from other types of collective stress because, first of all, disasters are crisis situations (Quarantelli, 1998). This approach is consistent with the use of crisis intervention frameworks in social work disaster research (Miller, 2003). Furthermore, conflict situations such as riots and wars are generally not defined as disasters in social work research. Conflict situations, as compared to natural and technological disasters, involve very different responses of organizations and other social systems. However, conflict situations are related to disasters both because competition is present in disasters, and because disasters and conflict crises often lead to high levels of collective stress and traumatic stress.

Disasters are often defined in social work research using a stress framework, with a focus on the stressor and the impacted system. This conceptualization of disaster allows for the examination of disaster impacts at micro, mezzo, and macro levels of analysis. Stress theory classifies disaster impacts according to type, demands on the impacted system, and duration (Dodds and Nuehring, 1996). These properties of disaster are consistent with Barton's (1969) typology of collective stress situations. Social work disaster researchers also focus on the mitigation, preparedness, response and recovery periods of disasters. The stress framework supports the generation of different research questions at each of these disaster phases, and for disaster impacts from individual to societal levels.

This definition of disaster is consistent with the social work definition of disasters as events causing human loss and suffering sufficient to create social disruption. Disasters warrant an extraordinary response from outside the immediately impacted area or community. Though the response required may be extraordinary, disasters may usefully be viewed as an extension of everyday events (Streeter, 1991). This definition permits a long-term developmental orientation, such that political, social, economic, and environmental forces work together to undermine a system's ability to cope with new stresses. The definition also shares some similarities with those of scholars who conceptualize disasters as socially defined occasions leading to radically changed behaviors to meet the crisis. According to these scholars, disasters are conceptualized within a social change perspective and are viewed as multidimensional (Quarantelli, 1998).

Though in social work research disasters are defined with an emphasis on social disruption, the environmental aspect of hazards is not excluded. The use of systems theory in social work is based partly on Duncan's POET framework which points to the interaction of variables related to population, organization, environment, and technology (Norlin and Chess, 1997; Quarantelli, 1998). In ecological theory, a type of systems theory, the physical and social environments of individuals and collectives are of equal importance in shaping human welfare. The physical nature of hazards, in an ecological perspective, is an important aspect of the definition of disasters in social work disaster research. In the POET framework, the natural and built environments interact with societal variables and may lead to disasters. Environment and society mutually affect one other.

Vulnerability

Social workers define vulnerability in reference to both individuals and communities. Vulnerability at the individual level refers to social structural factors which increase individuals' probability of suffering long-term and serious social, psychological, and health problems after a disaster (Thomas and Soliman, 2002). The primary theoretical foundation for vulnerability is distributive justice (Soliman and Rogge, 2002). In this formulation, the market value of individuals and populations is inversely related to the level of risk from natural and technological hazards that people are exposed to. Social vulnerability is therefore a continuum in which lower levels of socioeconomic status are associated with greater social vulnerability (Rogge, 2003).

Two of the most important social structural factors are poverty and social isolation. Poverty and lack of household wealth means that individuals will be less likely to recover from the material and health impacts of disasters. These individuals are less likely to have insurance to cover disaster losses, they may be on fixed or very limited incomes, and will have difficulty repaying low-interest disaster loans. Health conditions existing before disaster will likely be exacerbated by disaster, and lack of disposable income will make it very difficult for these low-income populations to afford health and mental health care. Older individuals, people of color, recent immigrants, and children are disproportionately represented among low-income populations (Sanders, Bowie, and Bowie, 2003). Children are especially vulnerable because they are dependent on adult caregivers for survival and recovery in disasters. Because they are developing physiologically, children are also highly vulnerable to environmental and technological disasters. Asthma and other respiratory problems, endocrine and immune system damage, and loss of IQ are among the documented or suspected consequences of chemical exposure in children (Rogge, 2003).

Often income level is related to social isolation. Social isolation from neighbors, kin, and formal organizations means that individuals and households will be unable to mobilize social capital to recover after a disaster. Isolated individuals will have difficulty obtaining information to help them make evacuation decisions, and to obtain relief services from formal organizations. These individuals suffer from a lack of social support and network ties, either to core networks of kin and neighbors, or to geographically dispersed networks which include aid organizations. Older individuals, and households consisting only of older individuals, tend to be especially socially isolated (Sanders et al., 2003).

Vulnerability is defined at the community level by the community's demographic, historical, cultural, and ecological characteristics. Poverty rate is a demographic variable negatively associated with community survival and recovery during major, long-term disasters (Sherraden and Fox, 1997; Sundet and Mermelstein, 1996). The level of functioning of local governments also predicts survival during community disasters. Communities are vulnerable when they contain few disaster social services organizations, and when these organizations and their programs are poorly coordinated. The lack of a developed disaster relief network of organizations also makes it difficult for community members to access services after a disaster. Disaster response and mitigation programs may be lacking in vulnerable communities. Vulnerable individuals and households tend to reside in communities whose other residents have similar social and demographic characteristics. Partly because low-income communities tend to have a poor tax base, the degree of vulnerability of communities coincides with the vulnerability of populations (Zakour, 1996b; Zakour and Harrell, 2003).

Emergency Management

Emergency management in social work disaster research is defined as management of the disaster social service system, which includes disaster organizations as well as the mass assault after a disaster. Emergency management focuses on preparedness for disasters, and planning for coordination of community resources during disasters (Gillespie, 1991). In the social work perspective, an important goal for emergency managers is inclusion of diverse organizational representatives and community leaders in overall disaster planning. Participants in the planning process should include representatives of informal community organizations serving vulnerable populations such as children, single-parent families, low-income individuals, and members of ethnic minorities (Harrell and Zakour, 2000).

CENTRAL DISASTER ISSUES AND CONCERNS

Social work research in disasters is consistent with the profession's concern with prevention, a generalist approach to social problems, and the equitable distribution of resources. Research on prevention focuses on understanding intervention in the social and physical environments of individuals in order to mitigate or ameliorate serious psychosocial problems. Generalist approaches in social work research and practice in disasters examine interventions in systems at different levels of abstraction, to respond to disaster impacts on a large number of societal systems. Research on access to resources by vulnerable populations seeks to improve the equality of service delivery to vulnerable populations.

Prevention

Just as prevention is part of the mission of the social work profession, disaster social work is concerned with intervention in the social and physical environments of individuals and groups as a means of preventing serious long-term social, health, and mental health problems after disaster (Rogge, 2003). The immediate social environment of individuals consists of their social support networks, including family, friends, and formal social services organizations. These networks are often disrupted by disasters of regional scope. Disaster relief programs using volunteers may seek to reconstitute these support networks to minimize disruption of social functioning and to facilitate recovery. Disaster social work involves not only expertise in service provision, but also interorganizational practice to improve coordination. An effective and coordinated network of disaster services organizations helps individuals, households, and communities recover and avoid long-term psychological and social problems (Zakour, 1996b).

In addition to reconstituting social support networks, restoration to pre-disaster levels of functioning depends on reconstruction of the physical environment. Housing and other infrastructure make up an important part of the physical environment of individuals and households, and these may be damaged or destroyed by disasters. Social work disaster services include helping people qualify for aid for home reconstruction and for replacement of other material losses. Volunteer programs managed by social workers also provide skills and personnel for rebuilding and for management of temporary shelters. Intervention in the physical environment represents a type of secondary prevention limiting disruption in systems after a disaster.

Prevention is most embodied in community disaster mitigation. This may involve rapid dissemination of information in a public education format to induce vulnerable populations to evacuate in the face of disaster warning. It also involves mobilizing community groups to support mitigation projects such as building codes to increase the built environment's resilience to earthquakes, floods, or high winds associated with tornadoes or tropical systems. Primary prevention is viewed as the most effective means of lessening traumatic events in refugee camps (Drumm, Pittman, and Perry, 2003). With highly vulnerable populations such as children, prevention can take place through ensuring that children are not exposed to chemicals and other substances released during environmental and technological disasters. By avoiding exposure of people at an early age to harmful substances, it is possible to limit or prevent long-term damage to children's health and cognitive functioning (Rogge, 2003).

Social Problems at Multiple Levels

Consistent with the multidimensional nature of disaster, social work disaster research is concerned with social problems at multiple levels of analysis.

These include the individual, family, group, organizational, community, and societal levels (Streeter and Murty, 1996). Intervening at multiple levels is part of the historical mission of the profession, and it includes prevention through services at the organizational, community, and societal levels to improve individual wellbeing. Research at the individual, family, and household levels has examined the effectiveness of interventions to restore people and small groups to pre-disaster levels of social and psychological functioning. Groups, organizations, and service delivery systems have been studied to improve coordination and effectiveness of the disaster-relevant interorganizational network (Gillespie, Colignon, Banerjee, Murty, and Rogge, 1993).

Research at the community level seeks to understand community characteristics which increase vulnerability to disaster. Research at the societal level analyzes policy from governmental and other large organizations contributing to or mitigating vulnerability. For example Soliman (1996), in a study of chronic technological disaster in a rural community, used a social process model which focuses on the social and psychological impact of toxic exposure through complex interactions among various levels of analysis. Research has also taken place from the micro to macro levels simultaneously, such as research utilizing an ecosystems approach in refugee camps (Drumm et al., 2003).

Access to Disaster Services

A central concern for social work is facilitating access to needed services. The mission of the social work profession includes creating linkages between vulnerable populations and service systems, and creating linkages among service systems to make resources more accessible to people (Minahan and Pincus, 1977). This is of particular importance to disaster services in social work because the populations most vulnerable to disaster are often protected by fewer mitigation projects and served by fewer disaster relief organizations with relatively low service capacities (Zakour and Harrell, 2003). The focus on access to services is important to social work partly because of the profession's historic concern with poverty and urban ecology. Many vulnerable populations are concentrated in urban jurisdictions with high rates of poverty, a lower tax base, and low rates of volunteering in emergency services organizations. These social and demographic conditions are associated with lessened access to disaster services.

Research in social work has focused on a variety of social work practice methods which aim to improve access to disaster services. These methods include case management, case finding, outreach, advocacy, brokering, information and referral, and helping clients apply for or qualify for services. Case management involves finding programs, services, and resources for a client, and ensuring provision of a meaningful combination of services. Case finding after a disaster is important because many disaster victims are unaware of available disaster services, or fear stigmatization resulting from receiving social services. Outreach means increasing access to services by creating satellite locations for programs, to make services more geographically and socially accessible. Advocacy involves using professional contacts within organizations to persuade program intake workers that a particular client is qualified for a service. Brokering involves exchanging clients among programs or organizations to meet client needs for multiple services, and to ensure the movement of clients through various programs in service systems (Harrell and Zakour, 2000; Zakour and Harrell, 2003).

CONTRIBUTIONS OF SOCIAL WORK TO THE DISASTER KNOWLEDGE BASE

Social work disaster research has contributed new findings on the uses and effectiveness of debriefings, disaster volunteers and service delivery, vulnerable populations in disasters, organizational and interorganizational behavior in disasters, environmental disasters, cross-cultural and international aspects of disaster response, and improved measurement and theory. The scope of these contributions reveals a generalist approach to disaster response, such that social work research and intervention focuses on systems of various sizes and levels of abstraction. Improved measurement and theory has resulted from research on volunteerism, organizations, interorganizational networks, communities, and vulnerable populations.

Traumatic Stress Interventions

Psychological debriefing is highly consistent with social work's orientation. Debriefing emphasizes coping mechanisms, community social support, and social connections through networking (Miller, 2003). Because disasters are conceptualized as impacting more than one person, debriefings take place in a group format. Debriefings are led by one or more facilitators, and involve a typical format. Victim accounts of what occurred are reviewed, and there is reflection about cognitive, emotional and physical reactions.

Psycho-educational teaching about typical stress responses and useful coping mechanisms provides a framework for understanding the traumatic event. Ideas and plans for healing, self-care, and mutual aid and support are then elicited. Participants in debriefings, including professional facilitators and disaster victims, generally report debriefing to be helpful (Miller, 2003). Debriefing has been found to be effective for first responders in disasters, such as police and fire personnel in the Oklahoma City Bombing (Callahan, 2000). However, debriefings may not be effective for disaster victims other than first responders. Also, group debriefing formats are reported to be less helpful than individual counseling in eastern European (Kosovar) refugee camps, based on the self-reports of refugees (Drumm et al., 2003).

Volunteers

The use of volunteers for service delivery has historically been an important issue in social work. Social work research on disaster volunteerism represents a significant contribution to the disaster literature (Zakour, Gillespie, Sherraden, and Streeter, 1991). Social workers are prominent among disaster volunteers, making up a large percentage of trained disaster volunteers (Cosgrove, 2000). In a study of disaster organizations providing relief services to people, Zakour and Harrell (2003) found that smaller, less formal organizations, often excluded from the officially comprised disaster interorganizational network, could be an important source of trained volunteer personnel in a disaster. An argument is made for inclusion of these informal organizations, which are often smaller mental health, religious, or minority-run organizations. Both understanding their disaster roles, and inclusion of these informal organizations, may be facilitated through use of more refined measures of the range of types of organizations an organization has a cooperative link with in disasters (Harrell and Zakour, 2000).

Volunteers are often engaged in multiple roles in disaster social services. One important role is facilitating psychological debriefings using a social work strengths-based perspective. Though professional social workers have strong psychotherapeutic skills in group settings, volunteers can help facilitate group processes while reducing the social distance between facilitators and debriefing group members (Miller, 2003). Another important role of volunteers and volunteerism in disaster is through the integration of for-profit and voluntary sectors and organizations. Because paid workers often volunteer with career development as a goal, the for-profit and volunteer sectors provide resources for each other. Volunteers can play an important role in coordination of the disaster social service system through their multiple affiliations in voluntary and for-profit organizations. Training volunteers not only increases the capacity of disaster services organizations, but also develops human capital valuable for career advancement of individuals (Zakour, 1994).

The level of volunteerism of disaster organizations is operationalized as the percentage of volunteers among an organization's total staff, and the types of appreciation shown to volunteers by the organization. Volunteerism been has shown to increase the capacity of organizations to provide disaster services, which in turn increases an organization's geographic range of service delivery. Increased capacity and geographic range helps organizations to overcome distance barriers in disaster services delivery, as well as to insure that services are provided to vulnerable populations residing in isolated geographic segments of communities (Zakour and Gillespie, 1998).

The relative lack of volunteerism and volunteer resources accounts for much of the explained variance in reduced access to disaster social services in some populations. The relative lack of disaster volunteers and programs of effective volunteer management was found to lead to lessened access to resources for African Americans, households led by those 75 years or older, and for female-headed households with small children (Zakour and Harrell, 2003). Municipalities with the greatest need for disaster services were shown to have the lowest levels of volunteer resources during disasters.

Vulnerable Populations

Social work disaster research has made a number of important contributions in the area of vulnerable populations. Individuals may be vulnerable because of their low socioeconomic status and social isolation, or because of their residence in communities vulnerable to disaster. This research has been consistent with sociological research pointing to organizational, communication, and physical barriers which limit access to services for vulnerable populations (Phillips, 1993; Phillips, Garza, and Neal, 1994). Outreach counselors in disasters report the usefulness of a training program for identifying and treating vulnerable populations such as children and the frail elderly (Soliman, Raymond, and Lingle, 1996). Children represent an important special needs population in disasters, and have been extensively studied by social workers. Prevention research in social work has focused on children because they are more susceptible to a variety of environmental stressors, and children are likely to develop long-term problems if not treated in a timely fashion. Krueger and Stretch (2003) in a study of 3,876 children in grades K-12 impacted by flooding in 1993, found that about 9 percent (n=366) required follow-up psychosocial interventions because of the flooding. For these children, predictors of risk as evidenced by elevated anxiety or depression were (a) presence of disaster impact, (b) harm to the family, (c) a need to evacuate the residence, and (d) lack of household disaster recovery. Floodwater in the residence was not a statistically significant predictor of risk. Residential evacuation led to the need for long-term services because of elevated anxiety which involved physiological manifestations of anxiety rather than worry or social concerns. Evacuation was not related to elevated depression (Krueger and Stretch, 2003). Rogge (2000) shows that, at the international level, socioeconomic inequalities interact with poor environmental conditions to amplify the risks of poor children. Environmental degradation resulted from the interaction of socioeconomic inequality with (a) human population growth and consumption; (b) reduction in per capita water supply and water quality; (c) loss of renewable natural resources; (d) loss of biodiversity; (e) environmental damage due to the application of industrial, commercial, and household chemicals; and (e) global warming.

Several other populations which are vulnerable to disasters are African-Americans, female-headed households with small children, and those 75 years and older (Sanders et al., 2003; Zakour and Harrell, 2003). African American and female-headed households are disproportionately low income. Older individuals tend to have sparse social support networks of family and friends, and are often have reduced access to formal disaster services (Thomas and Soliman, 2002). Each of these groups was found to be more vulnerable in a disaster because of the geographic areas in which they reside, the smaller number of organizations providing disaster services in these areas, and the lower capacities and network interaction of community organizations serving these populations. Geographic barriers were also found to increase the vulnerability of these populations because of their geographic isolation from the disaster interorganizational network. Geographic distance is a barrier to interorganizational links promoting resource sharing after a disaster. Additionally, when vulnerable populations are relocated after disaster, they lose support networks of family, friends, and voluntary associations because of the greater geographic distances of relocation centers to these networks.

Cherry and Cherry (1996) found social action research to be effective in influencing disaster organizations such as FEMA to extend the timeframe for additional disaster aid, which was of benefit for populations made homeless by Hurricane Andrew. Action research in this study differed from traditional research on vulnerable populations because the investigators operationalized variables to be highly sensitive to the needs of participants and to highlight the plight of vulnerable disaster victims (Cherry and Cherry, 1997). The research was also designed to withstand critique by others during legal challenges.

Organizational and Interorganizational Research

Research on organizations and interorganizational networks has been an important agenda for social work since its initial involvement in disaster studies (Gillespie, Sherraden, Streeter, and Zakour, 1986). Much of this research examines the behavior of organizations in disaster settings, and compares behavior by organizational type (Zakour et al., 1991). Typologies distinguishing among established, extending, expanding, and emergent organizations have been used in this research, which has compared established and extending organizations such as fire and police departments to expanding and emergent organizations which are often volunteer and smaller organizations.

Because organizations active in disaster may have roles related to their organizational type, a number of studies have examined interorganizational linkages among types of organizations. Zakour and Gillespie (1999) discuss the advantages of linking the governmental and nonprofit sectors in the context of the disaster-relevant interorganizational network. One of these advantages is the integration of the significant resources of governmental agencies with the greater knowledge of community needs of local nonprofits. Other advantages include increasing the capacity of nonprofit organizations active in serving geographically isolated neighborhoods, and making the greater service delivery efficiency of governmental organizations more accessible to community nonprofits and their constituents during a disaster. Harrell and Zakour (2000) found that the total number of different types of organizations which a given disaster organization is linked to is significantly and positively associated with an organization's preparedness for disaster. Range of types is also positively related to the number of special populations an organization is able to serve during disasters, and the organization's total disaster social services capacity. They suggest that each type of organization, including smaller, informal organizations, have fundamentally different resources in a disaster, so that linkage to a larger range of types of different organizations translates into an organization's access to different types of disaster resources.

Coordination of the disaster interorganizational network, particularly among public and private disaster organizations, has been shown to improve the effectiveness of the disaster social services system (Liu, Gillespie, and Murty, 2000; Robards, Gillespie, and Murty, 2000). Interorganizational coordination is conceptually defined as the deliberate interdependence of autonomous units for a common purpose. This definition of coordination is broadly consistent with sociological definitions of multiorganizational coordination, though substantial differences exists among different researchers' concepts of coordination (Drabek and McEntire, 2002). One aspect of interorganizational networks which promotes resource flow and coordination in uncertain environments is systems redundancy (Streeter, 1992). If too little redundancy exists in network roles or linkages, then the system is more vulnerable to complete breakdown if one or more actors is damaged by a disaster. Joint disaster drills also increase interorganizational links along with disaster capacity. Murty (2000) found that rural disaster management organizations developed greater awareness of disasters as well as higher capacities after the false prediction of an earthquake by Browning. This awareness diminished when the earthquake did not occur, but more regional organizations, and those organizations participating in a large-scale earthquake drill well after the predicted date, showed the least decline in awareness of disasters and in interorganizational relationships.

Zakour (1996b) developed a theoretical path model of interorganizational links using data from 52 local disaster services organizations in a midwest metropolitan area. In the sociological literature on disasters, both cooperative links and the resulting network of disaster organizations make up the foundation for multiorganizational coordination in disaster (Drabek and McEntire, 2002). In the model, geographic distance has a direct effect on cooperative links, and was negatively related to links. However, organizational variables including percentage of volunteers in an organization, type of organization (social service versus emergency management), geographic range of service delivery, and number of different means of showing appreciation to volunteers were more influential than geographic distance in determining cooperative links among volunteer organizations. All of these organizational variables were positively related to links, except for type, which was negatively related. The use of organizational variables as control variables reduced the magnitude of the relationship between geographic distance and links among organizations. Furthermore, for the 29 organizations with the greatest percentage of volunteers among total staff, both volunteer and paid, the relationship between distance and links was not significant. The relationships in this path model suggest that emergency management organizations in metropolitan areas tend to have low organizational permeability (Zakour and Gillespie, 1998). Low permeability is associated with few interorganizational links, quasi-military organizational characteristics, and low ability to recruit and retain disaster volunteers.

Environmental Issues

Social work research in environmental issues in disaster shows how environmental exposure affects systems at a variety of levels (Soliman, 1996), from individual stress responses to community social organization. Similar to studies of urban ecology and vulnerability, a number of social work studies examine the geographic distribution of risk and preparedness (Rogge, 2000). Risk of exposure to technological hazards is most highly related to the population density of geographic areas. As the population density increases, there is a decline in voting and social participation, and this reduces a population's ability to take legal and political action to oppose the location of industrial sites proximate to the population. These highly populated urban areas also have large percentages of working-class individuals who place industrial development at a higher priority than avoiding fugitive chemical releases. Though household income, level of education, and a higher percentage of non-elderly adults in a geographic area are related to population density, these three socioeconomic factors are negatively related to level of toxic risk (Rogge, 1996).

Rogge (1998) found that locality development and social planning models of community organization were effective in cleaning up long-term river pollution in Chattanooga-Hamilton County, but only for middle-class neighborhoods and jurisdictions. In low-income and African-American neighborhoods, which tended to be located on more marginal land next to the polluted waterway, social action models involving adversarial interactions were effective. Community-based development was successful in this setting because of (a) extensive citizen participation; (b) a broadly and flexibly defined vision of sustainability focusing on interactive systems of environment, economy, and social equity; (c) an appropriate balance between long-term vision and current action; (d) coalition building across public and private organizations, private and not-for-profit enterprises, business entrepreneurs, and environmentalists; and (e) optimism about locating solutions.

Cross-Cultural and International Research

Recently, social work researchers have begun exploring the effectiveness of disaster interventions developed in the United States and delivered in cross-cultural or international settings (Soliman and Silver, 2003). Most disaster research in the United States has studied middle-class populations, and it has not been clear to what extent research findings transfer to either cross-cultural or international settings. This lack of attention to developing nations may result from a view of disasters as random, isolated acts of God, so that the disproportionate risk of low-income nations and communities has been ignored (Streeter, 1991). However, disaster planning should be incorporated into social development planning, and should focus on mitigation, response, and recovery. Disasters may lead to significant loss of social and economic infrastructure, which damages development efforts. Rogge (2000) shows that, at the international level, socioeconomic inequalities and poor environmental conditions place children at higher levels of risk in disasters. Long-term environmental degradation and population growth in developing nations are further increasing the vulnerability of children.

Many effective methods of helping disaster victims through social services are not feasible in cross-cultural and international settings. Cultural norms about helping, who is to be helped first, and which groups are eligible for aid may prevent the provision of disaster relief professionals and volunteers from the United States and other neo-European societies (Puig and Glynn, 2003). For example, though everyone affected by a disaster is initially viewed as deserving of aid in the U.S., in some developing nations very low status populations may not be viewed as appropriate recipients of aid. Bell and associates (2000) in a study of an airline disaster in South Korea found that victim's families as well as governmental and nonprofit agencies to be antagonistic toward disaster volunteers from rural areas. The lower social status of the rural volunteers led to mistrust between volunteers active in the mass assault, and organized disaster workers as well as victims' families. This resulted in blame being assigned to volunteers for any shortage of aid in the immediate aftermath of the plane disaster.

A few studies have demonstrated the applicability of social work interventions in international or cross-cultural settings. Despite the presence of significant survival needs, Kosovar refugees reported emotional traumatization which could be treated by mental health professionals (Drumm et al., 2003). In addition to emotional traumatization, refugees reported a lack of knowledge concerning family members and relatives, as well as a lack of a normalized routines, meaningful activities, and self-determination in the camps. After an air-traffic disaster in Egypt, a social worker used a human-relations model sensitive to culturally-defined needs of victims, their families, and relatives (Soliman, in press). This culturally sensitive approach was found to be more effective than a command and control model for managing response to this disaster.

Debriefing models developed for mainstream populations in the United States have been extended to groups such as Native Americans. Social work researchers have generally found group models to be highly effective, though diagnostic categories developed for debriefing models in the United States have been shown to be only partially applicable to Native Americans. For example, traditional behaviors associated with spiritualism are incorrectly diagnosed as hallucinations accompanying post-traumatic stress disorder. A combination of debriefing and traditional healing approaches seems to provide for the most improvement of individuals suffering from long-term traumatic stress (Brave Heart, 2000).

Improved Theory and Measurement

THEORY. A number of research studies in social work have contributed to theory regarding communities in disaster, as well as interorganizational cooperation in local communities. In an application of the Social Process Model (Edelstein, 1988) to toxic exposure in a rural community, Soliman (1996) found evidence to support four postulates of this model. For the first postulate, that toxic exposure involves complex interactions among various levels of society, it was found that inadequate policies and regulations regarding river pollution have caused residents to lose trust in every system they have relied on to solve the problem, including the Environmental Protection Agency. Postulate 2 states that toxic exposure influences victim behavior and how they perceive and comprehend their lives. In this study, 86 percent of the residents believed river pollution is a very serious matter, and 14 percent also believed the pollution will have long-term effects on their lives. Postulate 3 states that toxic exposure is stressful, and forces victims to adopt a coping response. In this study of the Pigeon River, over 80 percent of respondents exhibited stress responses, and most community members participated either in individual efforts, a social movement organization, or a class-action suit to end the pollution. Postulate 4 of the social process model states that toxic contamination is inherently stigmatizing. Evidence to support this includes the refusal of agricultural and industrial interests to invest in the small rural community because of fears that toxic wastes from the Champion Plant have infiltrated the region's water, soil, and other resources.

Sundet and Mermelstein (1996) in a study of the 1993 Mississippi flood used a systems approach to understand predictors of community survival. Demographic, historical, ecological, cultural, and organizational factors were examined. Among demographic variables, only poverty rate was a strong predictor of community survival. Paradoxically, high poverty rates predicted survival because residents were unable to afford relocation, even with FEMA loans. Historical dimensions predicting survival were harmonious inter-governmental relations, lack of internal community conflict, and horizontal integration of community subsystems. Ecological factors predicting survival were communication capacity (e.g. marketing tourism), and proximity to resources external to the community. Strong local government leadership to respond to the flood disaster was also associated with community survival, though other organizational behaviors were not.

Zakour and Harrell (2003) found that urban geography and human ecology mirror a social structure of stratification related to vulnerability among communities. For geographic areas with high percentages of African Americans, female-headed households, and older individuals, community organizations have lower capacities for service provision, and have fewer links to the officially-comprised disaster services network of organizations. The geographic distribution of disaster organizations and cooperative links among these organizations coincides with urban geography and ecology.

Zakour (1996b) developed a theoretical path model of interorganizational links with high goodness-of-fit. This model extends Barton's (1969) work to the organizational level. This model shows that organizational characteristics determine both an organization's distance to all other disaster organizations, as well as its total cooperative links to these organizations. However, distance is shown to have a negative relationship with cooperative links. These theoretical relationships support the distinction between localistic and cosmopolitan organizations (Zakour and Gillespie, 1998). Compared to cosmopolitan organizations, localistic organizations have a higher mean distance from all other organizations, have fewer cooperative links, lower levels of appreciation shown to volunteers, and a lower percentage of trained volunteers among total staff during disaster conditions. Localistic organizations are often emergency management organizations in suburban municipalities. Contrary to intuitive expectations, local organizations are less volunteer oriented, and may function to block resource distribution throughout a metropolitan area. Cosmopolitan organizations have a larger geographic range of service delivery, are geographically centrally located, and are higher in volunteerism.

MEASUREMENT. Several studies have produced improved measurement of disaster volunteerism, disaster's impact on children, interorganizational networks, and sociodemographic variables related to disasters and disaster relief. Some important improvements in measurement have come from volunteerism research. By asking organizational representatives about the number of trained volunteers they can count on during a disaster, a more reliable measurement of volunteer capacity and the volunteerism of organizations is produced (Gillespie et al., 1986; Zakour et al., 1991). This measure has also been used to determine the percentage of volunteers among the total staff of disaster organizations, paid and volunteer, forming a measure of the level of volunteerism of organizations (Zakour, 1996b; Zakour and Gillespie, 1998). The volunteerism measure is a more refined measure of organizational volunteerism, and is positively associated with organizational capacity in disasters. A related advance is in the measurement of career-development and training of volunteers. Zakour (1994) developed a Guttman Scale using data from a survey of 1349 American Red Cross Volunteers. This scale reveals a temporal dimension to skills development among volunteers. The scale has a reproducibility coefficient of .93 and displays a high level of reliability. It conforms to the requirements of a scaling model, with a scalability coefficient of .72.

Harrell and Zakour (2000) developed a measure to assess the range of different types of organizations that a particular organization has a cooperative link with during disaster conditions. A larger value on this measure was found to be predictive of the variety of needed relief resources that both disaster organizations and the constituents of organizations are able to access.

Krueger and Stretch (2003) developed and tested the Children and Adolescents Protocol on Flood Impact (CAPFI). This self-report instrument contains 10 disaster impact items assessing magnitude of disaster; questions about harm to self, family, friends, and neighbors; amount of recovery; and demographic items. The instrument gathered information in a rapid fashion, and measured subjective feelings. The CAPFI proved relatively inexpensive to administer. A study of the Mississippi Flood of 1993 provided validating information for the CAPFI, as well as the Revised Children's Manifest Anxiety Scale (RCMAS), and the Children's Depression Inventory (CDI). The RCMAS has three subscales measuring worry, physiological anxiety, and concerns about social occasions.

Structural equivalence analysis, a type of social network analysis, has been used to understand changes in the nature of positions between a predisaster and postdisaster network of organizations. Suggestions for the use of structural equivalence analysis to complement standard modes of analysis for disaster organizations were offered (Streeter and Gillespie, 1992). This simultaneous use of network and more standard methods of analysis was shown to have great potential to advance disaster social work research and theory. Gillespie and Murty (1994) used structural equivalence analysis to identify peripheral and isolate positions in a disaster service delivery network. Peripheral positions contained organizations with only a single link to the main network of disaster organizations, while isolate positions had no links to the larger network. The organizations occupying the peripheral and isolate positions have substantial experience in responding to disasters, but the resources of these organizations could not be easily mobilized during a disaster.

Structural equation modeling, which simultaneously assesses the reliability of measures and tests relationships among variables, has been used to examine the relationship among sociodemographic variables and fugitive toxic emissions. This was done using the 1990 United States Census and geographic information systems. Toxic risk was operationalized as the density (pounds/square mile) of fugitive chemical releases (Rogge, 1996). In qualitative research in cross-cultural as well as international settings, investigators discovered the importance of matching socio-demographic characteristics of interviewers and interpreters with those of subjects. In societies segregated by gender, age, or ethnicity, the presence of an interviewer or interpreter who differs socially or demographically from the interviewees may cause subjects to be less forthcoming or honest (Drumm et al., 2003). Additionally, during interviews of Albanian Kosovars, it was found that members of the same ethnic group who are geographically dispersed have markedly different dialects from each other.

RELATIONSHIP OF SOCIAL WORK DISASTER RESEARCH TO OTHER DISCIPLINES

Social work research has historically drawn on research in psychology and sociology, and these two disciplines have contributed much to social work research on disasters. The sociology of disasters has been a foundation for social work research into interorganizational networks, coordination of the disaster system, and human ecological theories of disaster vulnerability. There was little research within social work on disasters until the last two decades (Dodds and Nuehring, 1996). David Gillespie promoted disaster research in social work through work with students and associates at Washington University in St. Louis (e.g., Gillespie et al., 1993; Gillespie et al., 1986). In 1995, the Disaster and Traumatic Stress Symposium, founded and originally chaired by Zakour, was held at the Council on Social Work Education's Annual Program Meeting in San Diego, and has been a part of the Annual Program Meeting since 1995. In the last decade a significant amount of disaster research has been conducted by social work researchers and educators (Padgett, 2002).

The sociology of disasters has been a major foundation on which social work research in disasters is based. Sociological investigators have provided an important knowledge base on vulnerable groups, such as the elderly, children, families, and racial minorities (Cherry and Cherry, 1996). Barton's (1969) theory of collective stress and related concept of the altruistic community forms a theoretical foundation for the study of disaster social services, and the creation of theoretical models of disaster relief in social work (Zakour et al., 1991; Zakour and Harrell, 2003). The study of organized volunteers, organizations, and interorganizational networks has been strongly influenced by sociological research (Drabek, 1970; Drabek and McIntire, 2002; Dynes, 1970) and by research with a sociological perspective conducted by social work educators (Gillespie et al., 1993; Gillespie et al., 1986).

Current social work research on vulnerable populations and environmental factors continues to draw on the work of sociological investigators focusing on societal factors in disaster causation (Mileti, 1999; Peacock, Morrow, and Gladwin, 1997). Social work disaster research at the community level has also been based on socio-political ecological models from sociology (Peacock and Ragsdale, 1997). In this perspective, communities are defined as an ecological network of actors which are usually social systems such as organizations and large groups. This perspective focuses on heterogeneity and social inequality in disaster impacts and outcomes. An important issue in ecological networks is the promotion of coordination of the network during a disaster, as well as the mitigation of destructive competition among systems. Research within this perspective also focuses on the flow of resources in a disaster, including people, information, services, and capital. Social work research with a network, geographic, and community focus, as well as research on social inequality, is inspired by the socio-political ecological approach (Harrell and

Zakour, 2000; Rogge, 1996, 1998; Soliman, 1996; Zakour and Gillespie, 1998; Zakour and Harrell, 2003).

Psychology has been a foundation for social work research on trauma and traumatic stress, as well as debriefing interventions during crisis (Streeter and Murty, 1996). Psychological debriefings are conducted by practitioners in psychology, psychiatry, as well as social work. Mitchell originated the use of debriefings in disasters and other crises (Mitchell, 1983), and this technique has been adapted by social workers who use the Mitchell model of debriefing. Social workers have also sought to use debriefing as a means to help prevent severe post-traumatic stress and post-traumatic stress disorder (Bell, 1995; Callahan, 2000). Investigators in each of these disciplines have sought to evaluate the efficacy of debriefings (Miller, 2003), and social work researchers have been influenced by researchers in psychology and psychiatry is assessing the usefulness of debriefings.

KNOWLEDGE GAPS AND FURTHER RESEARCH

An important gap in social work knowledge in disasters is the effectiveness of debriefing for different populations, especially the effectiveness of debriefing for prevention of post-traumatic stress. Most social work disaster research and intervention have not been studied cross-culturally, and further research in needed on vulnerable populations and communities. These include adolescent populations subject to chronic community violence, low-income and older populations at-risk for natural disasters, and small and rural communities facing environmental hazards.

Psychological Debriefing

The effectiveness of debriefing, a widely used intervention in social work, remains unclear. There appears to be evidence that debriefings are unable to prevent future PTSD for those exposed to critical incidents, and that those who receive debriefings report them to be helpful. However, methodological problems in assessing the effectiveness of debriefing include studying different types of critical incidents, using different definitions of debriefing, offering debriefings at different time intervals after the critical incident, and comparing varied recipient groups. Though both practitioners who conduct debriefings and disaster victims who receive them report that debriefings are helpful, it has been difficult to demonstrate that debriefings lead to measurable positive outcomes (Miller, 2003). Also, knowledge of the percentage of victims who experience psychopathology after a disaster, and who may benefit from debriefings, is limited (Padgett, 2002).

Though there is evidence that structural inequality influences provision of social services after disaster (Zakour and Harrell, 2003), there has been little research on inequality or discrimination in the area of debriefings. It is not known if racism or ageism, for example, have a significant impact either on access to debriefing interventions, or on the outcome of the debriefing process (Miller, 2003). As with other social work interventions in disaster, controlled outcome studies on debriefings are lacking. Intervention research examining the effectiveness of disaster response programs is sorely needed in social work disaster research (Dodds and Nuehring, 1996).

International and Cross-Cultural Issues

The effectiveness of many interventions within social work has not been tested in either cross-cultural or international settings. Little research exists on effective intervention or prevention models for refugees living in camps (Drumm et al., 2003). Most of these models rely on theory instead of empirical validation. Also, disaster preparedness has been neglected in the social work research literature. Preparedness is an important indicator of effective response and recovery. Empirical work in a variety of communities and cultural settings needs to be conducted to directly establish the relationship between preparedness and effectiveness of disaster services organizations (Banerjee and Gillespie, 1994).

Vulnerable Populations

A number of gaps exist in understanding disaster impact and recovery for vulnerable populations and communities. The identification of family and community risk factors in violent victimization is still in its very early stages. It is not clear why some adolescents, for example, are exposed to substantial community violence, while other are not (Rosenthal, 2000). Yet community violence is an insidious source of traumatic stress which is widespread and intensive. It is also not known how often and at what key points children exposed to a natural disaster should be assessed, nor how long they should be monitored for PTSD effects. This is significant because of the considerable costs involved in screening large numbers of children throughout a region impacted by disaster (Kreuger and Stretch, 2000, 2003).

More research is needed to assess the impact of acute and chronic environmental disasters on rural and small communities, which often contain high percentages of low-income residents. This research could be helpful in understanding the relationship between poverty and exposure to toxic waste. Historical research in particular is needed to understand the process of toxic exposure in rural, small, and low-income communities (Soliman, 1996). It is also not known where in the economic/technological production process toxic risk is located (e.g., the production process versus the end-stage of production), and it is unknown if visible versus invisible toxic emissions are more likely to be conceptualized as hazardous (Rogge, 1996).

Examination of the populations residing within the geographic service areas of organizations shows that African American and female-headed households as well as those older than 75 have decreased access to disaster services. However, closer study of the actual clients served by disaster programs would provide a clearer understanding of access to resources. Also, study at the individual or household level would provide a greater understanding of the relative access of different populations to services. This research at the individual level should also examine the interpersonal networks of key volunteers in disaster organizations and how these networks influence access to services (Zakour and Harrell, 2003). Finally, research is needed to assess the effectiveness of joint training programs to connect informal organizations serving vulnerable populations to the larger disaster interorganizational network (Harrell and Zakour, 2000).

SUBSTANTIVE IMPLICATIONS FOR EMERGENCY MANAGEMENT

In this section suggestions from social work research are offered for emergency managers. Research on debriefing suggests that debriefing groups for first responders should be homogeneous, so that each group has, for example, only fire personnel, law-enforcement personnel, or social workers. Anonymity and voluntary participation are also shown to be important predictors of debriefing success (Callahan, 2000). Debriefing concepts and interventions for members of traditional societies, such as Native Americans within the United States, should be combined with traditional world views and healing practices to be maximally effective. To the extent that debriefing groups and group interventions are consistent with traditional religious healing ceremonies, the effectiveness of these interventions may be enhanced (Brave Heart, 2000).

Some suggestions for emergency managers involved with refugee camps are (a) provide assessment and counseling as standard camp services; (b) help refugees access information through casework and electronic data bases; (c) normalize activities partly through provision of education and play for children; (d) strengthen linkages and communication among front-line workers, camp managers, and governmental and relief agencies; and (e) ensure cultural appropriateness of material aid items (Drumm et al., 2003). In addition, emergency managers providing social welfare services should be aware of the multiple dimensions involved in toxic material exposure. Toxic exposure impacts need to be understood at the individual, family, and community levels if interventions are to be effective (Soliman, 1996).

Disaster services organizations should adopt disaster plans and missions that emphasize wider geographic ranges of service delivery, and cooperation with disaster organizations throughout a larger geographic area (Zakour and Gillespie, 1998; Zakour and Harrell, 2003). The stated geographic range of mutual aid is an important predictor of resource flow throughout metropolitan areas during a disaster, and larger service ranges promote links among organizations. Geographically larger service ranges help to provide services to otherwise underserved or isolated disaster victims. Closely related to widening of service range, informal organizations should be included in the officially comprised network of disaster organizations. This could add significant resources in terms of both volunteer personnel providing relief resources, and culturally sensitive knowledge of heavily impacted areas in disasters.

Joint training and disaster planning activities may be the most effective means for inclusion of smaller, less formal organizations (Zakour and Gillespie, 1998; Zakour and Harrell, 2003). In particular, governmental disaster agencies should work closely with community-based non-profits so that the greater service efficiency of governmental agencies can be combined with the community orientation of local non-profits to increase the effectiveness of the disaster services system (Zakour and Gillespie, 1999). Disaster agency managers and directors could use geographic information systems and network analysis software to aid in examining the location of disaster services agencies to insure that vulnerable populations are served by these agencies. Emergency managers should also use the results of network analysis and structural equivalence analysis to identify organizations with no links or few links to the disaster services network. Based on this information, emergency managers could act to bring these organizations into the larger network of disaster organizations, perhaps through joint training exercises or collaborative programs of service delivery (Gillespie and Murty, 1994; Streeter and Gillespie, 1992). These research tools should be part of a comprehensive disaster victim services evaluation strategy which is particularly valuable when used for mitigation and planning purposes (Johnson, Olson-Allen, and Collins, 2002).

Emergency managers should emphasize recruitment, training, and retention of volunteers (Zakour and Gillespie, 1998). High-volunteerism organizations tend to have more cooperative links with other disaster organizations, and these volunteer organizations also are less likely to be limited by geographic distance barriers. High-volunteerism organizations are also able to facilitate network coordination, and rank high on capacity measures. Promoting volunteerism also helps facilitate access to disaster services for vulnerable populations (Zakour, 1996b).

Rural social workers need to build community coalitions to help coordinate human service delivery for timely disaster response. Social workers should educate the community about mutual support, particularly in chronic disasters. Social workers can also develop communication linkages among critical community leaders, both horizontally and vertically. Finally, social workers should help foster resource awareness and disaster preparedness related to resources. Information and referral systems are one example of fostering resource awareness (Sundet and Mermelstein, 1996).

SUMMARY AND CONCLUSIONS

Social work has long been involved in disaster response, and disaster research in social work grew out of sociological and psychological studies of disaster. Both the profession's definition of disaster and the major concerns of disaster research in social work are consistent with the profession's theoretical perspectives and historical mission. Future social work disaster research needs to focus on understanding the effectiveness of disaster interventions including psychological debriefings, particularly for vulnerable populations and communities located in other cultures and nations. Emergency managers should consider suggestions from social work research, which include (a) paying careful attention to the composition of debriefing groups, and to inclusion of traditional religious practices as part of the debriefing process; (b) ensuring that activities are normalized and material items are culturally appropriate in refugee camps; (c) adopting wider geographic ranges of service delivery for disaster response organizations; (d) promoting joint training activities, and sharing geographic information systems and other technologies to improve coordination; (e) emphasizing recruitment, training, and retention of disaster volunteers; and (f) building community coalitions to help coordinate service delivery in disasters. These suggestions are largely related to building social networks, including interorganizational networks, for service delivery and resource sharing during disasters.

Disaster research in social work may offer unique contributions to our knowledge of disasters partly because of the profession's emphasis on disaster impacts on many systems and levels of analysis. Disaster research in social work has examined systems from the family to societal levels, both domestically and internationally. Interventions at the community level have been developed to prevent long-term damage to vulnerable populations such as children, and to prevent disruption in systems including families, groups, and organizations. Individual and organizational behavior, including volunteerism, has been examined for its relationship to coordination and effectiveness of the disaster relevant organizational network. Social work has much to contribute to current and future disaster research and practice.

REFERENCES

Banerjee, M.M. and Gillespie, D.F. (1994). "Linking Disaster Preparedness and Organizational Response Effectiveness." *Journal of Community Practice 1*(3): 129–142.

Barton, A.H. (1969). *Communities in Disaster. A Sociological Analysis of Collective Stress Situations.* Garden City, NY: Doubleday.

Bell, J.L. (1995). "Traumatic Event Debriefing: Service Delivery Designs and the Role of Social Work." *Social Work 40*(1): 36–43.

Bell, J.L., Yoo, S.H., Park, J.Y. and Shin, G.H. (2000). "The Second Injury/Social Wound: The Villagers' Experience of the Crash of South Korea's Asiana Flight 733." *Tulane Studies in Social Welfare 21/22*: 311–321.

Brave H. and Horse, M.H. (2000). "*Wakiksuyapi*: Carrying the Historical Trauma of the Lakota." *Tulane Studies in Social Welfare 21/22*: 245–266.

Callahan, J. (2000). "Debriefing the Oklahoma City Police." *Tulane Studies in Social Welfare 21/22*: 285–294.

Cherry, A.L. and Cherry, M.E. (1996). "Research as Social Action in the Aftermath of Hurricane Andrew." *Journal of Social Service Research 22*(1/2): 71–87.

Cherry, A.L., and Cherry, M.E. (1997). "A Middle Class Response to Disaster: FEMA's Policies and Problems." *Journal of Social Service Research 23*(1): 71–87.

Cosgrove, J.G. (2000). "Social Workers in Disaster Mental Health Services: The American Red Cross." *Tulane Studies in Social Welfare 21/22*: 117–128.

Dodds, S. and Nuehring, E. 1996. "A Primer for Social Work Research on Disaster." *Journal of Social Service Research 22*: 27–56.

Drabek, T.E. 1970. "Methodology of Studying Disaster." *American Behavioral Scientist 13*(3): 331–343.

Drabek, T.E. and McEntire, D.A. (2002). "Emergent Phenomena and Multiorganizational Coordination in Disasters: Lessons from the Research Literature." *International Journal of Mass Emergencies and Disasters 20*(2): 197–224.

Drumm, R.D., Pittman, S.W. and Perry, S. (2003). "Social Work Interventions in Refugee Camps: An Ecosystems Approach." *Journal of Social Service Research 30*(2): 67–92.

Dynes, R.R. (1970). *Organized Behavior in Disaster.* Lexington, MA: D.C. Heath.

Edelstein, M. (1988). *Contaminated Communities – The Social and Psychological Impacts of Residential Toxic Exposure.* Boulder, CO: Westview.

Gillespie, D.F. (1991). "Coordinating Community Resources." In Drabek, T.E. and Hoetmer, G.J. *Emergency Management: Principles and Practice for Local Government,* Washington, DC: International City Management Association. pp. 55–78.

Gillespie, D.F. and Murty, S.A. (1994). "Cracks in a Post-disaster Service Delivery Network." *American Journal of Community Psychology 22*(5): 639–660.

Gillespie, D.F., Colignon, R.A., Banerjee, M.M., Murty, S.A. and Rogge, M. (1993). *Partnerships for Community Preparedness.* Boulder, CO: University of Colorado, Institute of Behavioral Science.

Gillespie, D.F., Sherraden, M.W., Streeter, C.L., and Zakour, M.J. (1986). Mapping Networks of Organized Volunteers for Natural Hazard Preparedness (Report No. PB87-182051/A07). Springfield, VA: National Technical Information Service.

Harrell, E.B. and Zakour, M.J. (2000). "Including Informal Organizations in Disaster Planning: Development of a Range-of-Type Measure." *Tulane Studies in Social Welfare 21/22*: 61–83.

Johnson, K.W., Olson-Allen, S., and Collins, D. (2002). "An Evaluation Strategy for Improving Disaster Victim Services: Blueprint for Change." *International Journal of Mass Emergencies and Disasters 20*(1): 69–100.

Kreuger, L. and Stretch, J. (2000). "Assessing Long-Term Traumatic Stress Disorder and Serving School Children Impacted by Natural Disaster." *Tulane Studies in Social Welfare 21/22*:151–173.

Kreuger, L. and Stretch, J. (2003). "Identifying and Helping Long Term Child and Adolescent Disaster Victims: Model and Method." *Journal of Social Service Research 30*(2): 93–108.

Liu, L.W., Gillespie, D.F. and Murty, S.A. (2000). "Service Coordination as Intergovernmental Strategies in Service Delivery." *Tulane Studies in Social Welfare 21/22*: 85–103.

Mileti, D.S. (1999). *Disasters by Design: A Reassessment of Natural Hazards in the United States.* Washington, DC: John Henry.

Miller, J. (2003). "Critical Incident Debriefing and Social Work: Expanding the Frame." *Journal of Social Service Research 30*(2): 7–25.

Minahan, A. and Pincus. A. (1977). "Conceptual Framework for Social Work Practice." *Social Work 22*(5): 347–352.

Mitchell, J.T. (1983). "When Disaster Strikes . . . The Critical Incident Stress Debriefing Process." *Journal of Emergency Medical Services 8*(1): 36–39.

Murty, S.A. (2000). "When Prophecy Fails: Public Hysteria and Apathy in a Disaster Preparedness Network." *Tulane Studies in Social Welfare 21/22*: 11–24.

Norlin, J.M., and Chess, W.A. (1997). *Human Behavior and the Social Environment: Social Systems Theory.* Boston: Allyn & Bacon.

Padgett, D.K. (2002). "Social Work Research on Disasters in the Aftermath of the September 11 Tragedy: Reflections from New York City." *Social Work Research 26*(3): 185–192.

Peacock, W.G., Morrow, B.H, and Gladwin, H. (Eds.). (1997). *Hurricane Andrew: Ethnicity, Gender, and the Sociology of Disasters.* New York: Routledge.

Peacock, W.G., and Ragsdale, K.A. (1997). "Social Systems, Ecological Networks and Disasters: Toward a Socio-Political Ecology of Disasters." In Peacock, W.G., Morrow, B.H, and Gladwin, H. (Eds.). *Hurricane Andrew. Ethnicity, Gender, and the Sociology of Disasters,* New York: Routledge. pp. 20–35

Phillips, B.D. (1993). "Cultural Diversity in Disasters: Sheltering, Housing, and Long Term Recovery." *International Journal of Mass Emergencies and Disasters 1191*: 99–110.

Phillips, B.D, Garza, L., and Neal, D.M. (1994). "Intergroup Relations in Disasters: Service Delivery Barriers After Hurricane Andrew." *The Journal of Intergroup Relations 21*(3): 18–27.

Pryce, J.K., and Pryce, D.H. (2000). "Healing Psychological Wounds of War Veterans: Vet Centers and the Social Contract." *Tulane Studies in Social Welfare 21/22*: 267–283.

Puig, M.E., and Glynn, J.B. (2003). "Disaster Responders: A Cross-Cultural Approach to Recovery and Relief Work." *Journal of Social Service Research 30*(2): 55–66.

Quarantelli, E. L. (Ed.) (1998). *What is a Disaster? Perspectives on the Question.* New York: Routledge.

Robards, K.J., Gillespie, D.F., and Murty, S.A. (2000). "Clarifying Coordination for Disaster Planning." *Tulane Studies in Social Welfare 21/22*: 41–60.

Rogge, M.E. (1996). "Social Vulnerability to Toxic Risk." *Journal of Social Service Research 22*(1/2): 109–129.

Rogge, M.E. (1998). "Toxic Risk, Community Resilience, and Social Justice in Chattanooga, Tennessee." In Hoff, M.D. (Ed.) *Sustainable Community Development: Studies in Economic, Environmental, and Cultural Revitalization.* Boston: Lewis Publishers. pp. 105–121.

Rogge, M.E. (2000). "Children, Poverty, and Environmental Degradation: Protecting Current and Future Generations." *Social Development Issues 22*(2/3): 46–53.

Rogge, M.E. (2003). "The Future is Now: Social Work, Disaster Management, and Traumatic Stress in the 21st Century." *Journal of Social Service Research 30*(2): 1–6.

Rosenthal, B.S. (2000). "Who is At Risk: Differential Exposure to Recurring Community Violence among Adolescents." *Tulane Studies in Social Welfare 21/22*: 189–207.

Sanders, S., Bowie, S.L., and Bowie, Y.D. (2003). "Lessons Learned on Forced Relocation of Older Adults: The Impact of Hurricane Andrew on Health, Mental Health, and Social Support of Public Housing Residents." *Journal of Gerontological Social Work 40*(4): 23–35.

Sherraden, M.S., and Fox, E. (1997)."The Great Flood of 1993: Response and Recovery in Five Communities." *Journal of Community Practice 4*(3): 23–45.

Soliman, H.H. (1996). "Community Responses to Chronic Technological Disaster: The Case of the Pigeon River." *Journal of Social Service Research 22*(1/2): 89–107.

Soliman, H.H. (In Press). "An Organizational and Culturally Sensitive Approach to Managing Air-Traffic Disasters: The Gulf Air Incident." *International Journal of Mass Emergencies and Disasters.*

Soliman, H.H., Raymond, A., and Lingle, S. (1996). "An Evaluation of Community Mental Health Services Following a Massive Natural Disaster." *Human Services in the Rural Environment 20*(1): 8–13.

Soliman, H.H., and Rogge, M.E. (2002). "Ethical Considerations in Disaster Services: A Social Work Perspective." *Electronic Journal of Social Work 1*(1): 1–23.

Soliman, H.H., and Silver, P.T. (2003). "Preface." *Journal of Social Service Research 30*(2): xiii–xv.

Streeter, C.L. (1991). "Disasters and Development: Disaster Preparedness and Mitigation as an Essential Component of Development Planning." *Social Development Issues 13*(3): 100–110.

Streeter, C.L. (1992). "Redundancy in Organizational Systems." *Social Service Review 66*(1): 97–111.

Streeter, C.L., and Gillespie, D.F. (1992). "Social Network Analysis." *Journal of Social Service Research 16*(1/2): 201–222.

Streeter, C.L., and Murty, S.A. (1996). "Introduction." *Journal of Social Service Research 22*(1/2): 1–6.

Sundet, P. and Mermelstein, J. (1996). "Predictors of Rural Community Survival after Natural Disaster: Implications for Social Work Practice." *Journal of Social Service Research 22*(1/2): 57–70.

Sundet, P.A., and Mermelstein, J. (2000). "Sustainability of Rural Communities: Lessons from Natural Disaster." *Tulane Studies in Social Welfare 21/22*: 25–40.

Thomas, N.D., and Soliman, H.H. (2002). "Preventable Tragedies: Heat Disaster and the Elderly." *Journal of Gerontological Social Work 38*(4): 53–66.

Zakour, M.J. (1994). "Measuring Career-Development Volunteerism: Guttman Scale Analysis Using Red Cross Volunteers." *Journal of Social Service Research 19*(3/4): 103–120.

Zakour, M.J. (1996). "Disaster Research in Social Work." *Journal of Social Service Research 22*(1/2): 7–25.

Zakour, M.J. (1996). "Geographic and Social Distance during Emergencies: A Path Model of Interorganizational Links." *Social Work Research 20*(1): 19–29.

Zakour, M.J., and Gillespie, D.F. (1998). "Effects of Organizational Type and Localism on Volunteerism and Resource Sharing during Disasters." *Nonprofit and Voluntary Sector Quarterly 27*(1): 49–65.

Zakour, M.J. and Gillespie, D.F. (1999). "Collaboration and Competition among Nonprofit and Governmental Organizations during Disasters." pp. 495–506 in *Crossing the Borders: Collaboration and Competition Among Nonprofits, Business and Government.* Alexandria, VA: Independent Sector.

Zakour, M.J., Gillespie, D.F., Sherraden, M.W., and Streeter, C.L. (1991). "Volunteer Organizations in Disasters." *The Journal of Volunteer Administration 9*(2): 18–28.

Zakour, M.J., and Harrell, E.B. (2003). "Access to Disaster Services: Social Work Interventions for Vulnerable Populations." *Journal of Social Service Research 30*(2): 27–54.

Chapter 10

U.S. Disaster Policy and Management in an Era of Homeland Security

Richard T. Sylves

ABSTRACT

Using the tools of policy analysis, presidency studies, and public management research, this study provides an overview of post 9/11 presidential homeland security directives, altered presidential disaster declaration powers, and revisions in the National Response Plan and the National Incident Management System. Since 9/11, the president, aided by a host of federal officials, has largely established and steered homeland security policy. Congress has provided disaster managers new authority, new responsibilities, and regular infusions of funding for purposes set forth in law and policy. Homeland security policymakers have engaged in massive government planning efforts aimed fundamentally at a broad pool of federal, state, and local disaster responders. This study maintains that changes in policy and management have been hugely consequential, both positively and negatively, for the nation's emergency management community.

INTRODUCTION

The terrorist attacks of September 11, 2001 resulted in profound changes in U.S. public policy. In public policy analysis language, the attacks represented the ultimate focusing event (see Birkland, 1997 on focusing events); it impelled the President, Congress, and a host of other government officials to craft and enact laws, policies and procedures affecting domestic preparedness, response, and "consequence management" for terror attacks. The federal government also responded to the events of 9/11 by creating the Department of Homeland Security. The President's issuance of new directives, collective federal efforts that developed new preparedness and consequence management programs, and ongoing federally-directed reorganization and redirection of existing disaster-related programs all owe their impetus to the 9/11 terror attacks (Nicholson, 2005; Tierney, 2005; Bullock et al., 2005; Kettl, 2004). Hurricane Katrina, which struck the U.S. Gulf Coast in late August 2005 triggering the failure of the levy system that protected New Orleans, tested the capacity, adequacy, and limits of disaster policy and management changes made since 9/11.

POLICY ANALYSIS AND PRESIDENCY STUDIES

When political scientists engage in the study of disasters and emergency management, the tools of policy analysis, presidency studies, and the practices of public management help them do their work. "Presidency studies" involves analysis of presidential power and attempts to understand the process of presidential policy-making. Many tools of social science have been applied to the study of the presidency, including psychological theories, decision theory, organizational behavior models, sophisticated econometric techniques, historical analysis, and survey research. Public management refers to an identifiable group of actors in political life who "collectively perform a significant part of the executive function of government (Lynn, 1996, p. 2). The intellectual domain of public management logically encompasses all officials and agencies sharing executive authority and their collective impact on public policy (Lynn, 1996, p. 2).

Policy analysis, presidency studies, and the study of the practices of public management serve as excellent sources of theory, conceptualization, methods, and lesson drawing. This work will use policy analytic techniques to examine presidential authority and decision-making in the realm of disaster. This study is a public policy examination of federal disaster policy and management. A central aim is to describe presidential disaster declaration powers; explain and critique the National Response Plan and the National Incident Management System; and at the same time lay out a path for researchers interested in conducting a policy analysis of this subject.

The 9/11 attacks and the 9/11 Commission Report (2004) both pressed the Bush administration to overhaul disaster management with terrorism and terror-caused disasters as the new priority. The NRP and the NIMS, refashioned largely as a reaction to the nation's sad experience on 9/11, has transformed federal, state, and local disaster planning and response.

U.S. homeland security policymakers have engaged in massive government planning efforts aimed fundamentally at a broad pool of federal, state, and local disaster responders. Since 9/11, the president and federal agency officials have largely established and steered homeland security policy. Congress has provided disaster managers new authority, new responsibilities and regular infusions of funding for purposes set forth in law and policy. Homeland security policy has manifested itself as a colossal, intergovernmental, multiagency, multimission enterprise fueled by widely distributed, but often highly conditional, federal program grants to state and local governments. Planning in homeland security is more than simply reorganization or realignment of existing functions; it is part of the federal government's official policy response to the 9/11 terror attacks.

PUBLIC POLICY STUDIES

Policy may be defined as "a relatively stable, purposive course of action followed by an actor or set of actors in dealing with a problem or matter of concern." (Anderson, 2006, p. 6). Public policies are developed by governmental bodies and officials. "Policies consist of courses or patterns of action taken over time by governmental officials rather than their separate, discrete decisions" (Anderson, 2006, p. 7). Policies emerge from demands posed by private citizens, group representatives, legislators, or other officials. In response to policy demands "public officials make decisions that give content and direction to public policy" (Anderson, 2006, pp. 7–8). Policy also involves what governments do. Policy outputs are in effect actions taken in pursuance of policy decisions and statements.

The Public Policy Process model, or heuristic, maintains that policy involves five stages: policy agenda setting, policy formulation, policy adoption, policy implementation, and policy evaluation (Anderson, 2006, p. 4). These stages overlap and elements of each stage may be found in both the legislative and executive branches of the government. The policy formulation stage involves, "development of pertinent and acceptable proposed courses of action for dealing with a public problem." Policy adoption requires "development of support for a specific proposal so that a policy can be legitimized or authorized." Policy ". . . implementation is the application of policy by the government's administrative machinery" (Anderson, 2006).

What this paper demonstrates, and something that future researchers may want to explore, is that homeland security policy is being simultaneously formulated, adopted, and implemented through the application of homeland security presidential directives, through president-led changes in disaster declaration type and eligibility, and in the NRP and NIMS planning and response activities conducted by federal, state, and local disaster management officials.

Law and policy since 9/11 is that terrorism prevention and consequence management is the primary mission of the Department of Homeland Security. In today's homeland security, terrorism trumps all. Federal emergency management as currently constituted addresses non-terror disasters and emergencies as a sub-category of "domestic incident," in which the definitive incident of note is the terror attack on the United States in September, 2001.

PRESIDENTIAL DISASTER DECLARATIONS

To begin, understanding disaster policy in the U.S., or for that matter in any nation, requires consideration of emergency powers of the nation's chief executive. Before 1950, the federal government had only intermittent involvement in disasters (Birkland, 1997; Platt, 1999; Tierney et al., 2001, p. 201) and presidents were usually peripheral policy actors in disaster management. Occasionally, Congress would pass special relief measures to help people, as well as state and local governments, recover from certain serious disasters. However, since 1950 American presidents have been empowered to declare major disasters and emergencies. In addition, there has been a vast expansion of post-disaster federal government services to individuals, governments and organizations that have experienced disaster (Sylves and Waugh, 1996; Platt, 1999; Waugh, 2000).

Intertwined with presidential powers are matters of national-subnational government relations: often defined in the U.S. as intergovernmental relations. These relations are conducted largely through a web of public management. May's *Recovering from Catastrophes: Federal Disaster Relief Policy and Politics* (1985) was one of the first modern era books done by a political scientist on the subject of disaster and its intergovernmental public management. Gordons's *Comprehensive Emergency Management for Local Governments* (2002) is a more recent contribution to the intergovernmental study of disaster policy and emergency management.

While almost every governor can issue declarations and proclamations of state-level disaster, states often do not provide substantial disaster recovery aid without first securing a special spending measure from the state legislature, drawing down state contingency funds, or raising state taxes (Beauchesne, 1998). Most states have responded to federal inducements like 75/25 federal/state-local matching subsidies of state/local relief and recovery spending when their governors have been able to win presidential declarations of major disaster or emergency (Sylves and Waugh, 1996; NAPA, 1993; Platt, 1999).

When a governor wins a presidential disaster declaration and qualifies for the aid provided by federal programs under a declaration, his or her state and eligible counties often stand to collect very sizable sums of federal money. Some states absorb all, and many others most, of the federally-required state-local matching share; thus, helping their disaster damaged localities avoid most of the burden of the match. Federal money to help pay for post-disaster individual and family assistance and/or disaster unemployment assistance is often substantial (Waugh, 2000; Platt, 1999; Sylves and Waugh, 1996). So too federal funds help these sub-national governments rebuild, repair, or replace damaged state or local infrastructure such as roads, bridges, utilities, public buildings, and so forth. Federal disaster relief under presidential declarations often comes to states and localities as a massive infusion of federal dollars for public works (Platt, 1999).

There have been numerous published studies of presidential disaster declarations before 2001 (Settle, 1990; Sylves, 1996; Sylves, 1998; Dymon and Platt, 1999; Platt, 1999). Relyea crafted an excellent study regarding the history of presidential emergency powers and the legislative and reorganization issues surrounding formation of the Department of Homeland Security in his "Organizing for Homeland Security" (2003). Presidential disaster declarations interlace many features of federal emergency management. A presidential declaration of major disaster or emergency typically activates the NRP and puts various federal, state, and local agencies to work under the NIMS. These will be examined later.

Lawrence E. Lynn, Jr. (1996), asks in his book, *Public Management: Art, Science or Profession*, is public management an extension of policymaking or is policymaking an extension of public management? The answers to these questions are relevant both in presidential declaration decision-making and in homeland security disaster management.

The president's decisions in the realm of disasters and emergencies sit squarely at the intersection of the old politics/administration dichotomy. As Lynn points out, for decades many have disputed whether politics and administration may be kept separate and distinct in the management of government. In addition, Lynn adds that for many scholars there is a division between policymaking and its administration and this distinction continues today, even though the politics/administration dichotomy was discredited 50 years ago. The public policy community is on the politics side and they have an "implementation perspective." Their emphasis is on the politics of management. They concern themselves with social outcomes legitimated by voting constituencies, stakeholders, legislative processes, and so forth. The public administration community has a public interest, public trust focus based on law, regulation, professionalism, value-neutrality, and objectivity. The exercise of presidential authority in matters of disaster declarations and homeland security is an appropriate venue for examining differences, overlaps, and coincidences of politics and administration.

Does each president decide on declaration requests in political terms as an elected representative impelled to maintain office; to ensure and improve re-election chances; and to advance the interests of their political party? Does each president behave in accord with his personal political motives and ambitions, thus encouraging him to rule on governor requests for disaster declarations to reward his political friends and punish his political enemies? Are the president's aims in such decision-making to reinforce his public image and popularity or perhaps to embellish his political and historical legacy? Another question must be posed. Do presidents, as the law suggests, decide on their own whether to approve or turn down governor requests for declarations or is this job delegated to federal disaster agency directors and White House senior staffers? If the president defers to the recommendations of his top disaster managers as he makes these decisions, this may connote his public management, rather than his politically motivated, behavior.

When presidents decide on whether to approve governor requests for declarations of disaster, they have an opportunity to make or shape "public policy." For example, presidents from Dwight D. Eisenhower to George W. Bush have been able to use disaster declaration authority to press forward their respective political ideology on matters of federal-state relations. They have been able to do this on a case-by-case basis. Moreover, President G.W. Bush has responded to the 9/11 terror attack to make homeland security policy through a series of Homeland Security Presidential Directives and by making senior appointments to public management posts integral to homeland security. Following Hurricane Katrina in 2005, and owing to the displacement of thousands of New Orleans flood victims, President G.W. Bush took the unprecedented step of granting presidential disaster declarations to states that absorbed Katrina evacuees, but had no actual disaster damage from Katrina (http://www.fema.gov/news/disasters.fema?year=2005). (See Appendix 1.)

Disaster declarations, as well as presidential directives, often demarcate where the president wants to set the line between public sector (federal) involvement and private market involvement. When a president approves or rejects a governor's request for a declaration, that president is also drawing the line between what is a state and local obligation and what is a federal obligation. When a president approves or rejects a governor's request for a declaration, that president is also drawing the line between what is a public responsibility and what is a private responsibility. Sometimes, when the president denies a governor's request for a declaration, the president is sending a message about each American's "personal responsibility" to prepare for, or insure against, various calamities, especially those that may be foreseeable or expected. Each declaration decision asks the president to determine whether the federal government and national taxpayer should be the "insurer of last resort" for the disaster or emergency. Conversely, each declaration decision implicitly denotes whether the president wants state and local government and/or disaster victims and their insurers, and perhaps private charitable organizations, to collectively carry the burden of disaster relief or whether the federal government is to shoulder the major part of the burden.

In making declaration decisions, presidents work between two extremes. One extreme assumes that the president extends federal disaster relief purely as form of presidential political pork barreling, that is, allocating federal resources in excess of need and for political reasons. Another extreme assumes the president confers or withholds federal disaster relief based on a purely "objective" consideration of public need, this is determined by whether the nature and magnitude of the event conforms to declaration criteria set forth by federal disaster management officials. This second extreme also assumes that the president makes declaration decisions under strict norms of federal budget guardianship.

Each president, constrained by the limits of what law allows, has had some freedom to define "what is" or "what is not" a "disaster" or "emergency." Each president's actions have often expanded the range and definition of what constitutes a disaster or emergency. In this sense, the broad declaration authority presidents have enjoyed has enabled them to shape federal disaster policy, sometimes subtly, and sometimes dramatically.

While disasters and emergencies seem astronomically unlikely in each locality, they are more likely on the state level, and even more probable on the national level. This is because the probability of disaster or emergency occurrence increases as a function of dramatically expanding geographic scale. The point is, on a national scale, there is an almost routine flow of governor requests for federal disaster declarations to the president. The president's serial processing of these requests allows him to steer declaration policy in various directions to achieve his general policy agenda as president. However, the almost weekly flow of declaration requests may encourage a White House routine under which senior staff attempts to relieve the president of the task of intensively pondering each governor request. This routine may mean that presidents and their staffs are likely to defer to the recommendations of their top federal disaster directors.

Because presidents do not make disaster declaration determinations alone, it is fair to ask if federal emergency management officials who advise the president about governor requested declarations behave as political supporters of the president and his administration or as politically neutral or unbiased senior public servants. Public managers collectively perform a significant part of the executive function of government. Public managers who have directed federal disaster agencies from 1953 to the present era of Homeland Security have been politically appointed subject to Senate confirmation, unless they were senior civil servants asked to serve in an "acting" capacity pending selection and confirmation of the presidential appointee.

Unfortunately, the vast majority of top appointed federal disaster agency managers have lacked emergency management experience before they came to the job. An exception is James Lee Witt, an experienced state emergency manager who was appointed by President Clinton to head FEMA (1993–2001). The problem of inexperienced disaster managers was underscored by the example of Michael Brown, a President G.W. Bush appointee (2003–2005). Brown was excoriated and forced to resign for his failures in leadership during FEMA's response to Hurricane Katrina.

If these declaration decisions have been purely administrative, one would expect that the president judges all governor requests for declarations against a scale of need or deservedness. The scale itself must be clear and must set forth objective criteria. However, the U.S. General Accounting Office once criticized the Federal Emergency Management Agency for failing to develop a clearer, more systematically explicit criterion by which to judge the deservedness of governor requests for presidential declarations of major disaster or emergency (GAO, 2001).

When presidents engage in declaration decision-making they have an opportunity to practice distributive politics. "Distributive policies involve the allocation of services or benefits to particular segments of the population." (Anderson, 2006, p.11). In distributive policy costs are widely shared, often through the public treasury, but benefits are targeted. Through disaster declarations presidents create the illusion that there are only winners and no specific losers. Presidents from Eisenhower to G.W. Bush have had considerable discretion and flexibility in judging governor requests in part because declaration criteria have been general or vague. This discretion and flexibility permits presidents to exhibit either public serving executive behavior or politically ambitious distributive behavior.

CHANGES IN PRESIDENTIAL DECLARATION AUTHORITY

Presidential declaration decision-making changed in the months and years after the 9/11 attacks. For example, a new term, "Incidents of National Significance" now encompasses major disasters or emergencies declared by the president (DHS, 2004, 8–9). "Incidents of National Significance" under the NRP are defined as "An actual or potential high-impact event that requires coordination of Federal, State, local, tribal, nongovernmental and/or private sector entities in order to save lives and minimize damage" (DHS, 2004, ix). According to the Final Draft of the NRP, issued June 30, 2004, the Secretary of the Department of Homeland Security (DHS) "can use limited pre-declaration authorities to move initial response resources . . . closer to a potentially affected area" (DHS, 2004, 9).

CATASTROPHIC INCIDENTS

The NRP adds a new category of incident beyond major disaster and emergency. "Catastrophic Incidents" are "Any natural or manmade incident, including terrorism, which results in extraordinary levels of mass casualties, damage, or disruption severely affecting the population, infrastructure, environment, economy, and national morale and/or government functions. A catastrophic event could result in sustained national impacts over a prolonged period of time; almost immediately exceeds resources normally available to State, local, tribal, and private sector authorities; and significantly interrupts governmental operations and emergency services to such an extent that national security could be threatened. All catastrophic incidents are considered Incidents of National Significance" (DHS, 2004, x).

NATIONAL SPECIAL SECURITY EVENTS

The president maintains authority to issue both pre-event and post-event declarations for National Special Security Events (NSSEs). These include "high-profile, large-scale events that present high probability targets" (DHS, 2004, 4) such as the G-8 Summit; the Republican and Democratic national political party conventions; and any other event the president believes may be vulnerable to terror attack. The U.S. Secret Service has been lead agency in managing NSSE's. Presidents have had authority to issue NSSE's since the early 1990s; however, what has changed is that the Secret Service is now part of DHS. Through the president and the Secretary of DHS, NSSEs are now a component of homeland security emergency management.

Annexes to the NRP suggest that the president, under Homeland Security Act of 2002 (P.L. 107-296) authority, is formalizing declaration authority to cover bio-terror, cyber terror, food and agricultural terror attacks, nuclear or radiological incidents, and oil and hazardous materials pollution incidents of national significance, to name a few (DHS 2004, vi–vii). Moreover, under the National Emergencies Act of 2003 (50 U.S.C., 2003, §§ 1601–1651) the president must follow new procedures "for Presidential declaration and termination of national emergencies" (DHS, 2004, 108). The president is obligated to identify the specific provision of law under which he "will act in dealing with a declared national emergency" and under a sunset provision in this law, declarations of national emergency are subject to automatic termination if not renewed by the president (DHS, 2004, 108).

DOMESTIC INCIDENT MANAGEMENT

The Homeland Security Act of 2002, Homeland Security Presidential Directive 5, and the Stafford Act of 1988, justify and provide, according to the NRP, a comprehensive, all-hazards approach to "domestic incident management" (DHS, 2004, 5). Disaster declarations in the post-9/11 era are now conceived as matters of "domestic incident management." All major disasters, emergencies, and catastrophic incidents declared by the president are Incidents of National Significance (DHS, 2004, x). Nonetheless, not every Incident of National Significance "necessarily results in a disaster or emergency declaration under the Stafford Act" (DHS, 2004, 8).

Why are these changes important? The cumulative outcomes of these changes in presidential declaration power are several. The term "Incidents of National Significance" incorporates a range of terror-relevant threats as well as conventional natural and non-terror human-caused disasters and emergencies. The president now possesses much unencumbered authority to mobilize federal, state, and local resources if he concludes that an event of some kind may be a possible target of terrorism; he would start said mobilization by issuing declarations for National Special Security Events. Presidential declaration authority now concedes openly that disasters may have nationally catastrophic consequences, and so presidents may now declare incidents as catastrophes.

All hazards emergency management remains a tenet of federal emergency management, but in the past, all-hazards meant terrorism was one of many possible agents of disaster or emergency. Today, federal emergency management is predicated on terrorism as a paramount threat while other types of disasters or emergencies occupy diminished positions within the all-hazards community. Natural and non-terror human-caused disasters since 1950, with the exception of civil defense against nuclear attack, were rarely considered matters of national security. Owing to the 9/11 terror attacks, homeland security and presidential declaration authority are primarily national security instruments and secondarily instruments of conventional emergency management. Moreover, these changes have significantly increased the range of presidential authority.

HOMELAND SECURITY ORGANIZATIONS AND DIRECTIVES

The September 11, 2001 terror attacks on the U.S. homeland induced Congress and the president to enact new laws, establish a new federal department incorporating some 22 agencies including the U.S. Federal Emergency Management Agency (FEMA), and to create new administrative arrangements. Major events in U.S. history have often triggered major changes in the governmental process (Truman, 1951). The creation of the Department of Homeland Security (DHS) was one of the largest federal reorganizations since President Truman created the Department of Defense in 1947. DHS incorporated approximately 180,000 employees as well as a wide assortment of agencies and offices (Tierney, 2005).

The reorganization merged together agencies (or parts of agencies) with very diverse organizational structures, missions, cultures, and, importantly, diverse ideas about the management of domestic threats and emergencies. Tierney argues that the overall effect of the reorganization has been to expand the role of defense-related and law enforcement oriented agencies, most of which are focused exclusively on terrorism, while correspondingly curtailing the role of agencies working in accord with all-hazards emergency management (2005).

Homeland security presidential directives have transformed the U.S. consequence management policy system since 9/11. Homeland Security Presidential Directive Five (HSPD-5) directed the Secretary of Homeland Security to develop a National Response Plan to integrate all existing federal response plans. In separate works Tierney and Nicholson declare that the two most relevant directives are HSPD-5, "Management of Domestic Incidents," and HSPD-8, "National Preparedness." The stated aim of HSPD-5 was to improve the nation's capacity to respond to domestic disasters by creating a single, comprehensive incident management system. To this end, HSPD-5 mandated the development of a "concept of operations" for disasters that would incorporate all levels of government as well as crisis and consequence management functions within one unifying management framework. The Secretary of Homeland Security was given responsibility for implementing HSPD-5 by developing a National Response Plan and a National Incident Management System (Tierney, 2005; Nicholson, 2005).

Under HSPD-5 directive, all federal agencies were required to adopt NIMS and to make its adoption a requirement for other governmental entities receiving federal assistance. HSPD-8, "National Preparedness," gives the Secretary of Homeland Security broad authority in establishing a "national preparedness goal" and implementing programs to improve "prevention, response, and recovery" operations. Although the directive explicitly calls for actions that address all hazards within a risk-based framework, its major focus is on preparedness for terrorism-related events. Similarly, while HSPD-8 is intended to address issues related to preparedness, a broad term that is generally conceptualized as an integrative and comprehensive process, the directive is mainly concerned with training and equipping emergency response agencies (Tierney, 2005).

HSPD-5 significantly revamped the existing Federal Response Plan (FRP). The FRP had been developed in the late 1980s and adopted in the early 1990s, and had proven effective for coordinating federal resources during and after several major national emergencies, including the 9/11 attacks. At the time the new plan was mandated, the U.S. had an internationally recognized emergency management structure in place (Tierney, 2005) that was compatible with its policy of "shared governance" (May and Williams, 1986). Shared governance meant that federal, state, and local governments had overlapping spheres of authority rather than a hierarchical top-down, command and control, authority relationship. While the NRP did not supplant that framework, it did make several important modifications. Under the NRP, the primary responsibility for managing domestic crises now rests with the Secretary of Homeland Security. The plan also contains language strongly suggesting that the federal government will, in the future, assume more responsibility for directly managing some crises, which significantly modifies "shared governance" policies that assign responsibility for disaster management to local authorities in affected areas (Tierney, 2005).

WHITHER FEMA?

The Federal Emergency Management Agency (FEMA) was formerly an independent agency within the executive branch of government. Owing to an initiative of President Clinton, the FEMA director was accorded *de facto* cabinet status for almost the whole of the Clinton administration. In 2003, FEMA was incorporated into DHS as lead agency for emergency preparedness and response. FEMA, which is the only agency within DHS that is charged specifically with reducing the losses associated with non-terrorism-related disasters, has lost significant visibility and financial and human resources in the reorganization. As a small agency within a massive bureaucracy, its activities are now overshadowed by much larger and better-funded entities within DHS.

According to Haddow and Bullock, here is what FEMA lost when it was absorbed into the Department of Homeland Security:

- Federal Response Plan, in which FEMA had the lead coordinating role, was replaced by a national incident response plan in which FEMA no longer has a lead coordinating role.
- FEMA, within the Emergency Preparedness and Response Directorate (EP&RD) of DHS, will no longer train first responders for terrorist events of any kind, though EP&RD will be responsible for coordinating federal agency response to acts of terrorism. (In July 2005, the EP&RD itself was disbanded and a Preparedness Directorate is to be established that does not contain FEMA. Recently, FEMA will regain the preparedness function.)
- FEMA Director no longer reports directly to the president when governors request presidential declarations of major disaster of emergency. Secretary of DHS has this authority instead.
- DHS Office of State and Local Government Coordination is outside FEMA as is the Office of Domestic Preparedness (ODP) (the one-stop shopping site state and local governments must go to in order to receive DHS grants of all types). ODP is in Border and Transportation Security Directorate of DHS (Haddow and Bullock, 2003). (In 2005, ODP was moved under the umbrella of the Office of State and Local Coordination).

Haddow and Bullock argue that FEMA and emergency management are destined to lose in bureaucratic conflicts over jurisdiction and funding within the DHS because other directorates and divisions of DHS are bigger and have more political influence (2003). For example, much of the responsibility, authority, and budget for preparedness for terrorism events, which might logically have been assigned to FEMA, are now channeled through the Office for Domestic Preparedness (ODP), an entity that was transferred into DHS from the Department of Justice. ODP has taken on many and varied responsibilities, including overseeing preparedness assessments on a city-by-city basis, training, planning, exercises, and the provision of grants to local agencies (Tierney, 2005). ODP manages several important DHS programs, including the Urban Areas Security Initiative, the Homeland Security Grant Program, and the Metropolitan Medical Response System, which was transferred from FEMA (GAO, 2004).

Since ODP had its origins in the Department of Justice, it is not surprising that it defines domestic preparedness primarily in terms of law enforcement functions. For example, ODP's "Preparedness Guidelines for Homeland Security" (GAO, 2005) gives priority to police and other public safety agencies. One effect of ODP involvement has been to institutionalize a system of terrorism prevention and management that is largely separate from the existing emergency management system. Another has been to increase direct "top-down" oversight of local preparedness activities on a scale that had not existed prior to 9/11. For a review of how ODP dispenses federal funds to first responders see, "Emergency Preparedness: Federal Funds for First Responders" (GAO, 2004). ODP, seeking to meet the requirements of HSPD-8, has attempted to advance national preparedness goals for all hazards. ODP officials have tried to use the grants system to encourage states and localities to improve their capability to address a variety of hazards, most of them terrorism related (GAO, 2005).

According to Tierney, the decline in FEMA's prestige and influence in the wake of 9/11 has caused great concern among U.S. emergency management experts. Testifying before the U.S. Congress in March, 2004, former FEMA director James Lee Witt warned that the nation's ability to respond to disasters of all types has been weakened by some post-September 11 agency realignments. In written testimony regarding the loss of cabinet status for the FEMA director and the current position of FEMA within DHS, Witt stated that "I assure you that we could not have been as responsive and effective during disasters as we were during my tenure as FEMA director, had there been layers of federal bureaucracy between myself and the White House" (Witt, 2004).

Tierney offers the following criticism, "As a consequence of the increased flow of resources into law enforcement agencies and counter-terrorism programs from ODP and other sources, preparedness for natural and technological disasters has assumed far less importance on the public policy agenda. Moreover, as agencies based on command-and-control principles assume greater importance in local preparedness efforts, the influence of organizations that focus on hazards other than terrorism and that operate in a broadly inclusive fashion and on the basis of co-ordination, rather than control, has waned" (Tierney, 2005).

WHAT IS THE NATIONAL RESPONSE PLAN?

Development of a National Response Plan was mandated in the Homeland Security Act of 2002 and Homeland Security Presidential Directive #5 (HSPD-5); the NRP was to embody a single comprehensive national approach; advance coordination of structures and administrative mechanisms; provide for direction for incorporation of existing plans with emphasis on concurrent implementation of existing plans; and set forth a consistent approach to reporting incidents, providing assessments and making recommendations to the President, DHS Secretary, and HSC (Yagerman, 2004).

The NRP superseded the Federal Response Plan (FRP), the Domestic Terrorism Concept of Operations Plan (CONPLAN), the Federal Radiological Emergency Response Plan (FRERP), and the Interim National Response Plan (INRP). Many of the familiar concepts and mechanisms associated with these plans were carried over to the NRP. For example, the former FRP Emergency Support Function (ESF) process was retained. New elements were introduced, such as the Homeland Security Operations Center (HSOC), the Interagency Incident Management Group (IIMG), the Principal Federal Official (PFO), and the Joint Field Office (JFO) (Yagerman, 2004).

The Federal Response Plan relied on preparedness. NRP preparedness functions were, and remain today, interdependent and delegated among a pool of more than 26 federal agencies (http://www.dhs.gov/interweb/assetlibrary/NIMS-90-web.pdf p. 122 and Haddow and Bullock, 2003, p. 76–77). Below are 16 current categories of emergency support functions:

1. Transportation
2. Communications
3. Public Works and Engineering
4. Firefighting
5. Information and Planning
6. Law Enforcement and Security
7. Mass Care
8. Resource Support
9. Health and medical services
10. Search and Rescue
11. Hazardous Materials Response
12. Food and Water
13. Energy
14. Public Information
15. Animals and Agricultural Issues
16. Volunteers and Donations

The NRP, as the core plan, is designed to link to an array of national-level hazard-specific contingency plans, such as the National Oil and Hazardous Substances Pollution Contingency Plan (NCP). These plans can be implemented independently during localized incidents or concurrently with the NRP during Incidents of National Significance.

The NRP includes the Base Plan and supporting annexes and appendices. The Base Plan outlines the coordinating structures and the processes applicable in national incident management. The Base Plan includes the concept of operations; roles and responsibilities; specific incident management activities;

and plan management and maintenance (Yagerman, 2004).

Emergency Support Function (ESF) Annexes group capabilities and resources into functions most likely needed during an incident. The ESF Annexes describe the responsibilities of primary and support agencies that are involved in providing support to a state or other federal agencies during Incidents of National Significance. Support Annexes provide the procedures and administrative requirements common to most incidents, such as Public Affairs, Financial Management, and Worker Safety and Health.

Incident Annexes describe the procedures and roles and responsibilities for specific contingencies, such as bio-terrorism, radiological response, catastrophic incidents, and so forth. In many cases, these annexes are typically supported by more detailed supporting plans. The Appendices to the annexes contain other relevant information including terms, definitions, and so forth. Also included is a compendium providing a complete listing and summary of national interagency plans, which serve as support plans to the NRP.

The NRP establishes the national framework for assessing domestic incidents to determine the appropriate level of federal involvement, and for coordinating interagency incident management efforts for events considered "Incidents of National Significance." The definition of Incidents of National Significance is directly based on the criteria established in HSPD-5:

1. When another Federal department or agency has requested DHS assistance;
2. When State/local capabilities are overwhelmed and Federal assistance is requested;
3. When an incident substantially involves more than one Federal department/agency;
4. When the (DHS) Secretary has been directed by the President to assume incident management responsibilities (Yagerman, 2004).

A basic premise of the NRP is that incidents are handled at the lowest governmental level possible. DHS becomes involved through the routine reporting and monitoring of threats and incidents, and/or when notified of an incident or potential incident that is of such severity, magnitude, complexity and/or threat to homeland security that it is considered an Incident of National Significance. When this happens, DHS establishes multiagency structures at the headquarters, regional, and field level to coordinate efforts and provide support to the on-scene incident command structures. Other federal agencies carry out their incident management and emergency response authorities within this overarching framework (Yagerman, 2004).

The NRP has been developed through an interagency process designed to incorporate input from a wide range of stakeholders. The plan was drafted by an interagency writing team based on: (1) guiding principles established by the Homeland Security Council, (2) input from various stakeholder groups, and (3) feedback from multiple rounds of review. Throughout the process, the Homeland Security Council continued to provide guidance and served as the multiagency body for review and approval (Yagerman, 2004).

The NRP work involved many rounds of stakeholder reviews about the base plan and the annexes. Through 3 rounds of review, there were 4,260 total comments on the base plan. More than 70 percent of these were accepted and incorporated into the plan. Stakeholder feedback was the most critical component of the development process and substantially determined the final content of the plan (Yagerman, 2004).

IMPLEMENTATION OF THE NRP

According to Yagerman, if local authorities carry out their duties in accord with the NRP, their capabilities to manage disaster should improve. HSPD-5 mandates federal agency compliance with the NIMS. The NRP is based on mutual support, cooperation, coordinated work effort, and partnerships. The federal government does not compel state and local governments to comply with the NRP. State and local governments are "encouraged" to work within the concept of operations. However, state and local officials understand that their prospects of winning some types of homeland security federal grants is much improved if they agree to join in and comply with the requirements of the NRP (Yagerman, 2004). As the section on the Incident Command System will show, the federal government is in effect buying state and local compliance with the NRP through conditions it sets forth in its grants to these governments.

The NRP and NIMS require that local emergency managers acquire additional training in ICS, exercises, and certifications. Local officials must also put together mutual aid agreements, and modify their standard operating procedures. The NRP is built on existing systems and provides greater clarity on the roles and responsibilities of federal departments and agencies in support of, and in coordination with, State, local, tribal and private sector partners (Yagerman, 2004).

Yagerman declared, "Our goal was too preserve and mirror the existing structure of state emergency operations plans to the degree that we could. There are only three new Emergency Support Functions (ESFs) and some modifications to a few of the existing ESFs. The new structures in the plan – the Homeland Security Operations Center (HSOC), the Joint Field Office, and the Interagency Incident Management Group, primarily involve the Federal partners" (Yagerman, 2004).

The federal government does not dispense grants specifically for compliance with the NRP. Yagerman indicated that "The only grants that I know of are the Emergency Management Planning Grants (EMPG), and I believe that there is enough flexibility in those funds to permit funding NRP related planning activities" (Yagerman, 2004). According to Yagerman, the Office of Domestic Preparedness (ODP) and FEMA's Emergency Management Institute (EMI) are to provide training, educational materials, and exercise opportunities. EMI developed an on-line independent study course to help emergency managers and others gain familiarity with the NRP.

According to Yagerman, the NRP was developed in close coordination with the White House, the interagency team, and state and local stakeholders. The base plan includes annexes regarding the proper roles of private sector and donations management organizations (2004). When events transpire and the NRP is put into effect, officials in the participating agencies prepare after-action reports (AARs) (Yagerman, 2004).

WHAT IS THE NATIONAL INCIDENT MANAGEMENT SYSTEM?

The National Incident Management System was developed by the Department of Homeland Security and deployed March 1, 2004. NIMS was a product of DHS' collaboration with state and local government officials and representatives from a wide range of public safety organizations. NIMS incorporates many existing best practices into a comprehensive national approach to domestic incident management, applicable at all jurisdictional levels and across all functional disciplines. NIMS' aim is to help responders at all jurisdictional levels and across all disciplines to work together more effectively and efficiently. One of the federal fiscal year (FY) 2005 requirements for implementing NIMS is "institutionalizing the use of ICS, across the entire response system." Beginning in federal FY-2006, federal funding for state, local and tribal preparedness grants will be tied to compliance with the NIMS (DHS, 2005).

According to DHS, one of the most important best practices incorporated into the NIMS is the Incident Command System (ICS), a standard, on-scene, all-hazards incident management system already in use by fire fighters, hazardous materials teams, rescuers and emergency medical teams. The ICS has been established by the NIMS as the standardized incident organizational structure for the management of all incidents.

The concept of ICS was developed more than thirty years ago, in the aftermath of a devastating wildfire in California. During 13 days in 1970, 16 lives were lost, 700 structures were destroyed and over one-half million acres burned. The overall cost and loss associated with these fires totaled $18 million per day. Although all of the responding agencies cooperated to the best of their ability, numerous problems with communication and coordination hampered their effectiveness. As a result, the Congress mandated that the U.S. Forest Service design a system that would "make a quantum jump in the capabilities of Southern California wildland fire protection agencies to effectively coordinate interagency action and to allocate suppression resources in dynamic, multiple-fire situations." The FIRESCOPE ICS is primarily a command and control system delineating job responsibilities and organizational structure for the purpose of managing day-to-day operations for all types of emergency incidents.

Although FIRESCOPE ICS was originally developed to assist in the response to wildland fires, it was quickly recognized as a system that could help public safety responders provide effective and coordinated incident management for a wide range of situations, including floods, hazardous materials accidents, earthquakes and aircraft crashes. It was flexible enough to manage catastrophic incidents involving thousands of emergency response and management personnel. By introducing relatively minor terminology, organizational, and procedural modifications to FIRESCOPE ICS, the National Interagency Incident management System (NIIMS) ICS became adaptable to all-hazards-type emergency management.

ICS is, according to DHS, based on proven management tools that contribute to the strength and efficiency of the overall system. The following ICS management characteristics are taught by DHS in its ICS training programs:

- Common Terminology
- Modular Organization
- Management by Objectives
- Reliance on an Incident Action Plan
- Manageable Span of Control
- Pre-Designated Incident Mobilization Center Locations and Facilities
- Comprehensive Resource Management
- Integrated Communications
- Establishment and Transfer of Command
- Chain of Command and Unity of Command
- Unified Command
- Accountability of Resources and Personnel
- Deployment
- Information and Intelligence Management.

In Homeland Security Presidential Directive-5 (HSPD-5), President Bush called on the Secretary of Homeland Security to develop a national incident management system to provide a consistent nationwide approach for federal, state, tribal and local governments to work together to prepare for, prevent, respond to and recover from domestic incidents, regardless of cause, size or complexity.

HSPD-5 states that: "Beginning in Fiscal Year 2005, Federal departments and agencies shall make adoption of the NIMS a requirement, to the extent permitted by law, for providing Federal preparedness assistance through grants, contracts, or other activities. The Secretary shall develop standards and guidelines for determining whether a State or local entity has adopted the NIMS."

The NIMS represents a core set of doctrine, principles, terminology, and organizational processes to enable effective, efficient and collaborative incident management at all levels. In order to provide the framework for interoperability and compatibility, the NIMS is based on a balance between flexibility and standardization. The recommendations of the National Commission on Terrorist Attacks upon the United States (The 9/11 Commission Report 2004) further highlight the importance of ICS. The Commission's report recommends national adoption of the ICS to enhance command, control, and communications capabilities http://www.dhs.gov/interweb/assetlibrary/NIMS-90-web.pdf

The NIMS provides a consistent, flexible and adjustable national framework within which government and private entities at all levels can work together to manage domestic incidents, regardless of their cause, size, location or complexity. This flexibility applies across all phases of incident management: prevention, preparedness, response, recovery and mitigation.

The NIMS provides a set of standardized organizational structures – including the ICS, Multi-Agency Coordination Systems and public information systems – as well as requirements for processes, procedures, and systems to improve interoperability among jurisdictions and disciplines in various areas.

As declared on their department website, Department of Homeland Security officials recognize, "that the overwhelming majority of emergency incidents are handled on a daily basis by a single jurisdiction at the local level. However, the challenges we face as a nation are far greater than the capabilities of any one community or state, but no greater than the sum of all of us working together" (DHS, 2005).

Standing DHS policy is that domestic incident management operations depend on the involvement of emergency responders from multiple jurisdictions, as well as personnel and equipment from other states and the federal government. These instances require effective and efficient coordination across a broad spectrum of organizations and activities. Domestic incident management requires that responders be able to mobilize and effectively use multiple outside resources. They must bring these resources together in an organizational framework that is understood by everyone. They must utilize a common plan, as specified through a process of incident action planning. DHS authorities insist that this will only be possible if all parties unite, plan, exercise, and respond using the common NIMS.

When Homeland Security officially issued NIMS in 2004, then-Secretary Tom Ridge and Under Secretary Michael Brown indicated that they believed state and local agencies would, if they had not already, quickly adopt the ICS (Harrington, 2003). They recognized that in some cities, the fire and police departments have worked together using ICS for years. In other places, only the fire department used ICS. Although law enforcement, public works and public health officials were aware of the concept, many regarded ICS as a fire service system. The NIMS sought to end this discrepancy because HSPD 5 requires state and local adoption of the NIMS conceptualization of ICS as a condition for receiving federal preparedness funding. According to DHS officials, while ICS was first pioneered by the fire service, it was, and still is, at its core, a management system designed to integrate resources to effectively attack a common problem. This system is not exclusive to one discipline or one set of circumstances; its hallmark is its flexibility to accommodate all circumstances (DHS, 2005).

With the exception of the way the intelligence function is handled, the principles and concepts of the NIMS ICS are the same as the FIRESCOPE and National Interagency Incident Management System (NIIMS) ICS. Is there a difference between NIMS ICS and FIRESCOPE/NIIMS ICS? The ICS organization has five major functions, including command; operations; planning; logistics; and finance and administration. In the NIMS ICS, a potential sixth functional area covers an intelligence function. Intelligence supplies responders with analytic work and incident related information, which they share among one another. Intelligence may include national security or classified information as well as operational information such as risk assessments, medical intelligence, weather information, structural designs of buildings, and toxic contaminant levels. In standard ICS, information and intelligence functions are located in the Planning Section. However, in exceptional situations incident commanders (ICs) may need to assign this role to other teams of the ICS organization. Under the NIMS ICS, the intelligence and information function may be assigned in one of the following ways:

Within the Command Staff;
As a unit within the Planning Section;
As a branch within the Operations Section; or
As a separate General Staff Section.

ICS AND HOMELAND SECURITY

State and local government officials are expected to comply with NIMS and incorporate and use ICS across their entire response system. Their ICS training had to be consistent with the concepts, principles and characteristics of the ICS training offered by DHS training entities. Responders who have already been trained in ICS were told by DHS that they do not need retraining if their previous training is consistent with DHS standards.

The Homeland Security NIMS Integration Center (NIC) works with federal and state ICS trainers helping them advance a common understanding and application of the ICS. The Center collaborates with stakeholders at all levels of government and across all response disciplines. The initial staff of the Center came from FEMA, the Office for Domestic Preparedness (ODP), and the Science and Technology (S&T) Directorate. Homeland Security leaders expect the NIMS Integration Center to grow and evolve into a robust, fully integrated center that incorporates additional DHS employees, interagency representatives, and liaisons, as well as state, tribal and local government representatives.

DHS officially indicates that, throughout the transition to the NIMS, it is important to remember why the NIMS and why ICS are critical pieces of the incident management system. Most incidents are local, but when responders are faced with the worst-case scenario, such as 9/11, all responding agencies must be able to interact and work together. According to DHS, the NIMS, and in particular, the ICS component, allow that to happen, but only if the foundation has been laid at the local level. If local jurisdictions adopt a variation of ICS that cannot grow or is not applicable to other disciplines, the critical ties between responding agencies and jurisdictions cannot occur when the response expands (DHS, 2005). As agencies adopt the principles and concepts of ICS as established in the NIMS, the incident command system can expand to meet the needs of the response, regardless of the size or number of responders. The key to both NIMS and ICS is a balance between standardization and flexibility (DHS, 2005).

CRITICISM OF NIMS

Tierney offers an excellent critique of NIMS and several of her arguments are excerpted here. "In mandating NIMS, the plan also institutionalizes the Incident Command System (ICS) as the preferred organizational structure for managing disasters for all levels of government and within all organizations that play (or wish to play) a role in disaster response activities. While numerous U.S. jurisdictions and organizations already use ICS, this directive may nevertheless have problematic consequences. Some critics fault ICS for overly emphasizing command-and-control principles; they also question the wisdom of mandating one particular management framework for the many and diverse organizations that respond to disasters" (Tierney, 2005).

Emergency management policy expert William Waugh observes that ICS "was created utilizing management concepts and theories that are now more than 30 years old" (Waugh, 2003) and that current management theory places much less emphasis on the command-and-control philosophy on which ICS is based. Waugh also notes that ICS is far more compatible, both structurally and culturally, with command-oriented organizations like police and fire departments than with the structures and cultures of the many other civilian types of agencies and groups that play key roles in responding to disasters but that do not operate according to hierarchical and paramilitary principles. In his view and that of other critics, top-down management models like ICS (and now NIMS) are particularly ill-suited to the distinctive challenges disasters present, which call for flexibility, improvisation, collaborative decision-making, and organizational adaptability. The danger is that in mandating a single, standardized management approach that is familiar mainly to command-and-control agencies, the NRP will stifle the capacity to improvise and exclude many entities and groups that make can critical contributions during extreme events (Waugh, 2003; Tierney, 2005).

More broadly, the push toward universal adoption of NIMS and ICS reflects the highly questionable assumption that once a consistent management structure is adopted, preparedness and response effectiveness will automatically improve. Such an assumption ignores many other factors that contribute to effective disaster management, such as ongoing contacts among crisis-relevant agencies during non-disaster times, common understandings of community vulnerability, and the likely consequences of extreme events, realistic training and exercises, and sound public education programs (Tierney, 2005).

According to Tierney, "The growing emphasis on terrorism readiness and ICS principles has led to a concomitant emphasis on first responder agencies and personnel. In current homeland security parlance, the term 'first responder' refers to uniformed personnel (fire, police, and emergency services personnel) that arrive at the scene of a disaster. Missing from this discourse is a recognition that, as numerous studies indicate, ordinary citizens are the true 'first responders' in all disasters. For example, in HSPD-8, a mere two sentences are devoted to the topic of citizen participation in preparedness activities. These new policies and programs may leave vast reserves of talent and capability untapped in future extreme events" (2005).

Many post-9/11 investigations have highlighted problems associated with "stovepiping," or the tendency for organizations and agencies to closely guard information, carry out their own specialized activities in isolation from one another, and resist efforts to encourage cross-agency collaboration (Gilmore, 2005, 9/11 Commission Report 2004). Indeed, DHS itself was created in order overcome stovepipes, better integrate disparate agencies and programs, and improve information-sharing and cooperation. It is ironic, then, that some homeland security initiatives appear to be creating new stovepipes and reinforcing existing organizational and institutional barriers.

For example, while diverse law enforcement agencies at different governmental levels may be making progress in working together on terrorism-related issues, the law enforcement sector itself may have little incentive to take an active role in broader cross-sectoral preparedness efforts. Rather than promoting comprehensive preparedness for all potential threats – including disasters and terrorism – special-purpose initiatives encourage organizations to interact and plan within their own separate spheres and to focus on particular kinds of threats. Large infusions of funds into specialized programs only exacerbate the problem" (Tierney, 2005).

CONCLUSIONS

American homeland security policy is being simultaneously formulated, adopted, and implemented vis-à-vis homeland security presidential directives and through the actions of federal, state, and local homeland security and emergency management officials. This study has examined presidential declaration authority, the NRP, and the NIMS. Remember, approved presidential declarations trigger activation of the NRP and NIMS, which is fundamentally the tactical implementation of the NRP.

Has natural hazards emergency planning been helped or hurt by the federal emphasis on terrorism? At the risk of seeming disingenuous, the answer is both. Natural hazards emergency planning has not been ignored in the massive establishment of homeland security organization and funding. Each new natural disaster serves as a reminder of the value of all-hazards emergency management as an organizing paradigm. Hurricane Katrina devastated the Gulf coast of Louisiana, Alabama, and Mississippi; triggered the failures of levees surrounding New Orleans; and caused arguably one of the nation's most expensive natural disasters. The immensely destructive and widespread hurricane is blamed for over 1,000 deaths and has displaced more than a half a million people for periods ranging from weeks to months. The flaws revealed in after-action reports about the hurricane may highlight aspects of the NRP and NIMS that have to be changed or improved.

Few can deny that state and local emergency management has enjoyed a windfall of federal assistance that has provided training; equipment; funding for large-scale and more realistic exercises and drills; facilitation of planning, and so forth. So far unexplored is the possibility that states, owing to pressures from federal authorities, have passed through to localities more funding and resources than they otherwise would have were it not for federal homeland security funding initiatives. Natural hazards emergency planning is still embedded in the NRP and NIMS. It is difficult to imagine how the nation could have organized for future terror attacks, possibly involving weapons of mass destruction, without revamping its system of emergency management at the same time. Responders to natural disasters and emergencies will most likely be many of the same people expected to respond to the destructive consequences of a terror attack. The expertise and skill sets of people in public health, information technologies, transportation security, national security, and law enforcement, to name a few, now complement the expertise of traditional (if there still is such) emergency managers.

Natural hazards emergency planning has been "hurt" in the sense that FEMA's absorption into a massive new federal bureaucracy with a counter-terrorism mission has fashioned FEMA into a "shadow" of its former self. Every time DHS undergoes a reorganization aimed at integrating its inherited legacy agencies and programs into a more coherent whole, the vestiges of FEMA are parsed or splintered anew. Moreover, a host of new interests now reside in DHS and many of these new homeland security "players" know little or nothing about emergency management. These interests often behave as political pressure groups that work to advance their interest in bureaucratic competitions (Richardson, 1993). Members of the natural hazards community who are supportive of FEMA now face considerable challenges from these new pressure groups.

Because major disasters and emergencies of any type are now subsumed under the rubric "incidents of national significance" emergency management is today very much a concern of national security. Moreover, the president's serial processing of governor requests for declarations allows him to steer declaration policy toward the achievement of his policy agenda. Emergency managers must today work shoulder-to-shoulder with military authorities in a realm of "civil security" (Healy, 2003; Haddow and Bullock, 2003). Law enforcement has also had a very major impact on how homeland security and emergency management is to proceed.

Threats of different forms of terrorism (e.g., bioterror, weapons of mass destruction, sabotage, suicide bombings, etc.) have changed the substance and process of federal disaster response planning. Some local emergency managers have complained that these plans are turning into "encyclopedias and federal requirements for all this additional paper keeps growing" (Yagerman, 2004). The 9/11 disaster, as in many disasters, further centralized authority. The president has developed new forms of emergency management authority for his office. Homeland security laws as implemented by DHS have demanded broad-ranging state and local conformity to federal disaster planning and response. Some emergency managers may conclude officials who have made these changes have neglected to adequately consult the emergency managers before they fashioned these reforms.

Emergency management involves four fundamental phases each of which is related to the others: mitigation, preparedness, response, and recovery. Examining how changes made through presidential directives, disaster declarations, the NRP, and NIMS have affected the nation's ability to address all four phases is beyond the scope of this study. However, a mitigation policy obsessed by a federal effort to prevent terrorism disregards, at great peril, the importance of reducing disaster vulnerability, particularly vulnerability to natural disasters.

Disaster policy is a worthy domain for policy analysis, presidency studies, and public management research. Disaster sociologists have done magnificent work examining social constructions of disaster phenomena in individual, communal, and organizational terms. However, political science has much to offer too; government institutions, laws, public spending, presidential directives, and the political dynamics associated with the operation of a democratic republic are worth considering as well. Disaster policy research encompasses facts and forces deserving of explanation. Focusing events, like 9/11 and Hurricane Katrina produce political consequences that bring change in public policy. Polities unable to adapt to the demands and pressures imposed by major focusing events risk loss of public legitimacy. History has repeatedly demonstrated that erosion of public legitimacy may lead to the ultimate collapse of an entire political system.

National response planning and incident management work continues to remake the nation's system of disaster management work. In accord with fiscal federalism, the DHS uses the conditions of federal grants to state and local governments as both "carrots" and "sticks." Planning, money, and tactical requirements have dramatically re-made and re-shaped the public policy of disaster management in the U.S.

APPENDIX 1

Table 10.1

APPROVALS AND TURNDOWNS OF GOVERNOR REQUESTS FOR PRESIDENTIAL DISASTER DECLARATIONS, 1953–2005

Presidential	*Time-Span*	*Pres. Approvals*			*Pres. Turndowns*			*Turndown Percentage*		
Administration		*Major**	*Emer***	*Total*	*Major<*	*Emerg<<*	*Total*	*Major*	*Emerg*	*Total*
Eisenhower	5/2/53–1/21/61	106	0	106	55	0	55	34%	0%	34%
Kennedy	1/21/61–11/20/63	52	0	52	22	0	22	30%	0%	30%
Johnson	11/21/63–1/21/69	93	0	93	49	0	49	35%	0%	35%
Nixon	1/21/69–8/5/74	195	1	196	102	15	117	34%	94%	37%
Ford	8/5/74–1/21/77	76	23	99	35	7	42	32%	23%	30%
Carter	1/21/77–1/21/81	112	59	171	91	37	128	45%	39%	43%
Reagan	1/21/81–1/21/89	184	9	193	96	16	112	34%	64%	37%
GHW Bush	1/21/89–1/21/93	158	2	160	43	3	46	21%	60%	22%
Clinton	1/21/93–1/21/01	380	68	448	103	13	116	21%	16%	21%
GW Bush	1/21/01–9/22/05	247	98	345	46	12	58	16%	11%	14%
Total		1603	260	1863	642	103	745	29%	28%	28%

*Represents approved Presidential declarations of major disasters, which begin in 1953.
**Represents approved Presidential declarations of emergencies. Emergency declarations begin in 1974.
<Represents President's turndown of a governor's request for a presidential declaration of major disaster.
<<Represents President's turndown of a governor's request for a Presidential declaration of emergency.
Date of declaration checked for each administration to the day.
Source FEMA, DARIS June 1997, FEMIS Dec 2001, DHS EP&R, Justification of Estimates FY04 March 2003.
Source of 9/11/01-9/22/05 turndown data, letter to author from Sen. Carper re: DHS FOI Request 9/22/05.

REFERENCES

Anderson, J.E. (2006). *Public Policymaking*. 6th ed. Boston: Houghton-Mifflin Company.

Beauchesne, A.M. (1998). *A Governor's Guide to Emergency Management.* Natural Resources Policy Studies Division, National Governor's Association Center for Best Practices. Washington, DC: National Governor's Association.

Birkland, T.A. (1997). *After Disaster: Agenda Setting, Public Policy, and Focusing Events.* Washington, D.C.: Georgetown University Press.

Bullock, J.A., et al. (2005). *Introduction to Homeland Security.* Boston: Elsevier Butterworth-Hienemann.

Dymon, U.J. and Platt, R.H. (1999). U.S. Federal Disaster Declarations: A Geographical Analysis." In Platt, R.H. (Ed.), *Disasters and Democracy: The Politics of Extreme Natural Events*, pp. 47–67. Washington, DC: Island Press.

Gilmore, J. S. III (2005). *Management Challenges at the Department of Homeland Security: Hearing.* House Homeland Security Committee, Subcommittee on Management, Integration, and Oversight. U.S. Congress. April 20, 1–47.

Gordon, J.A. (2002). *Comprehensive Emergency Management for Local Governments.* Brookfield, CO: Rothstein Associates.

Haddow, G.D. and Bullock, Jane A. (2003). *Introduction to Emergency Management.* Boston: Butterworth-Heinemann.

Harrington, H. (2003). Transitioning Newsletter. A weekly electronic newsletter with DHS transition information for FEMA employees. A Message from Michael D. Brown. March 05, 2003, 11:03.

Healy, G. (2003). "Deployed in the USA: The Creeping Militarization of the Home Front." *Policy Analysis* No. 303, December 17. Washington, DC: The Cato Institute, 2003.

Kettl, D.F. (2004). *System under Stress: Homeland Security and American Politics.* Washington, DC: CQ Press.

Lynn, L.E., Jr. (1996). *Public Management: Art, Science, or Profession.* Washington, DC: CQ Press.

May, P. (1985). *Recovering from Catastrophe: Federal Disaster Relief Policy and Politics.* Westport, CT: Greenwood Press.

May, P. and Williams, W. (1986). *Disaster Policy Implementation: Managing Disaster under Shared Governance.* New York: Plenum Press.

National Academy of Public Administration (NAPA) (1993). *Coping with Catastrophe: Building an Emergency Management System to Meet People's Needs in Natural and Manmade Disasters.* Washington, DC: NAPA.

National Research Council (NRC) (1999). *The Impacts of Natural Disasters: A Framework for Loss Estimation.* Washington, DC: National Academy Press.

Nicholson, W.C. (Ed.) (2005). *Homeland Security Law and Policy.* Springfield, Il. Charles C Thomas, Publishers.

9/11 Commission Report: Final Report of the National Commission on Terrorist Attacks Upon the United States. (2004). New York: W.W. Norton.

Platt, R.H. (1999). Shouldering the Burden: Federal Assumption of Disaster Costs. In Rutherford H. Platt (Ed.), *Disasters and Democracy: The Politics of Extreme Natural Events*, Washington, DC: Island Press. pp. 11–46.

Platt, R.H. and Rubin, C.B. (1999). Stemming the Losses: The Quest for Hazard Mitigation." In Platt, R.H. (Ed.), *Disasters and democracy: The politics of extreme natural events*, Washington, DC: Island Press, pp. 69–107.

Relyea, H.C. (2003). Organizing for Homeland Security. *Presidential Studies Quarterly*, Vol 33, No. 3 (September), 602-624.

Richardson, J.J. (ed). (1993). *Pressure groups.* New York: Oxford University Press.

Rubin, C.B. and Renda-Tenali, I. (2000). Disaster Timeline: Selected Events and Outcomes 1965–2000. Arlington, VA: Claire B. Rubin and Associates.

Schattschneider, E.E. (1975). *The Semisovereign People.* Hinsdale, IL: Dryden Press.

Settle, A.K. (1990). Disaster Assistance: Securing Presidential Declarations. In Sylves, R.T. and Waugh, Jr., W.L. (Eds.), *Cities and Disaster: North American Studies in Emergency Management*, Springfield, IL: Charles C Thomas, Publishers, pp. 33–57.

Sylves, R.T. and Waugh, Jr., W.L. (Eds.) (1996). *Disaster Management in the U.S. and Canada.* Springfield, IL: Charles C Thomas, Publishers.

Sylves, R.T. (1996). "The Politics and Budgeting of Federal Emergency Management." Sylves, R.T. and Waugh, Jr., W.L. (Eds.) (1996). *Disaster Management in the U.S. and Canada.* Springfield, IL: Charles C Thomas, Publishers, pp. 26–45.

Sylves, R.T. (1998). Disasters and Coastal States: A Policy Analysis of Presidential Declarations of Disaster 1953-97. DEL-SG-17-98. Newark, DE: University of Delaware, Sea Grant College Program.

Tierney, K.J. (2005). Recent Developments in U.S. Homeland Security Policies and Their Implications for the Management of Extreme Events. Boulder, CO: Natural Hazards Research and Applications Center, Institute of Behavioral Science, University of Colorado at Boulder.

Tierney, K.J., Lindell, M.K. and Perry, R.W. (2001). *Facing the Unexpected: Disaster preparedness and Response in the United States.* Washington, DC: Joseph Henry Press.

Truman, D.B. (1951). *The Governmental Process.* New York: Knopf.

U.S. Code of Federal Regulations. (2003). 50 U.S.C., 2003, §§ 1601–1651.

U.S. Department of Homeland Security. The Office for Domestic Preparedness. (2003). *Guidelines for Homeland Security.* Washington, DC: U.S. Department of Homeland Security.

U.S. Department of Homeland Security (DHS) (June 30, 2004). *Final Draft: National Response Plan.* Washington DC: See http://www.dhs.gov/interweb/assetlibrary/NIMS-90-web.pdf

U.S. Department of Homeland Security (DHS) (Dec. 12, 2005). Declared Disasters Archive. http://www.fema.gov/news/disasters.fema?year=2005

U.S. General Accounting Office (GAO) (August 31, 2001). *Disaster Assistance: Improvement Needed in Disaster Declaration Criteria and Eligibility Assurance Procedures.* Washington, DC: GPO. GAO-01-837.

U.S. General Accounting Office (GAO) (May 13, 2004). *Emergency Preparedness: Federal Funds for First Responders.* GAO-04-788T.

U.S. General Accounting Office (GAO) (April 12, 2005). *Homeland Security: Management of First Responder Grant Programs and Efforts to Improve Accountability Continue to Evolve.* Statement of William O. Jenkins, Jr., Director of Homeland Security and Justice (GAO). GAO-05-530T.

U.S. Senate (May 18, 1993). Committee on Governmental Affairs. *Hearing on Rebuilding FEMA: Preparing for the Next Disaster.* 103rd Cong., 1st sess.

Wamsley, G.L., Schroeder, A.D., and Lane, L.M. (1996). To Politicize is NOT to Control: The Pathologies of Control in Federal Emergency Management. *American Review of Public Administration 26*, no. 3, 263–85.

Waugh, Jr. W.L. (2000). *Living With Hazards, Dealing With Disasters: An Introduction to Emergency Management.* Armonk, NY: M. E. Sharpe.

Waugh, W. L., Jr. (2003). Terrorism, Homeland Security and the National Emergency Management Network, *Public Organization Review* no. 3, 373–385.

Witt, J. L. (2004). Testimony before the Subcommittee on National Security, Emerging Threats and International Relations and the Subcommittee on Energy Policy, Natural Resources and Regulatory Affairs, March 24.

Yagerman, B. (2004). EIIP Virtual Forum Presentation – September 15. See http://www.emforum.org

Chapter 11

Public Administration, Emergency Management, and Disaster Policy

William L. Waugh, Jr.

ABSTRACT

Public administrationists have been relatively slow to address the organizational issues in emergency management and the policy dilemmas resulting from the complex intergovernmental context of disaster policy and emergency management. A 1984 workshop helped develop a small community disaster researchers and they have provided much of the core of public administration disaster research over the past two decades. The events of September 11, 2001, however, have drawn many more public administrationists to disaster-related research, but the difficulty gaining access to Homeland Security offices and the lack of transparency in Homeland Security operations has created serious problems for those seeking to conduct rigorous research on organizational structures and processes.

INTRODUCTION

While "[e]mergency management is the quintessential government role" (Waugh, 2000: 3), public administrationists have been slow to address the organizational and policy issues that define the role and the practice of emergency management. While the escalating social and economic costs of disasters since the 1980s, the "war on terrorism" since September 2001, and recent catastrophic disasters have certainly encouraged much greater attention to how we deal with natural and unnatural disasters, the number of public administration researchers involved in disaster policy and emergency management research is still relatively small in comparison to researchers in other social science disciplines. Public administration programs have also been slow to develop courses and curricula, but that gap is also closing as research funding increases and as colleges and universities develop research programs and centers focused on emergency management, Homeland Security, and related policy issues.

Catastrophic natural disasters in the 1960s and 1970s ultimately lead to the creation of the Federal Emergency Management Agency (FEMA) in 1979 and drew the attention of some public administration researchers. To some extent the new agency became a focus of organizational and administrative study, drawing those interested in issues such as intergovernmental relations, federal-state-local fiscal relations, human resource management, executive leadership in emergencies, environmental planning, and organizational coordination. Perhaps to a greater extent, the policy problems engendered by major natural disasters became a focus of those interested in policy design, policy implementation, and policy and program evaluation. The watershed for public administration research in emergency management was a 1984 workshop sponsored by FEMA and the National Association of Schools of Public Affairs and Administration, that expanded the community of public administration researchers and the publication of a 1985 special issue of *Public Administration Review*, the preeminent journal in the discipline, on emergency management. Since that time, administrative imperatives, often driven by catastrophic events, have provided impetus for scholarship and encouraged the development of academic programs in emergency management and related professional fields. The events of September 11, 2001, energized and expanded that community of scholars and educators and the special issue of *Public Administration Review* (2002) published on the first anniversary of the attacks broadened the scope of public administration research to include issues of governance, civil liberties and privacy, and security, as well as organization. The special issue also addressed the policy implications of decisions made in the months following the attack for American democracy and the decisions yet to come, such as the creation of the Department of Homeland Security. Catastrophic events in 2004, principally the Indian Ocean or "Christmas" earthquake and tsunami and the four hurricanes that struck central Florida, have given the profession and the field of emergency management even greater visibility and raised more questions about preparedness, especially alert and warning systems and evacuation, and social and economic recovery.

The need for public administration research in emergency management is clear. Beginning with Hurricane Hugo in 1989, the U.S. has had to deal with a lengthening series of billion dollar disasters and the effectiveness of government programs in managing risks to life and property has become a serious political issue on which the careers of elected and appointed officials have hung. The slow Hurricane Andrew response in 1992 could have cost President George H.W. Bush Florida's electoral votes and the lesson was not lost on President George W. Bush in 2004 when four hurricanes made landfall in Florida during his re-election bid. To avoid electoral loss, the Bush Administration orchestrated an extraordinary response in Florida to demonstrate the effectiveness of federal, as well as state, programs. Thousands were mobilized for the effort, including personnel from FEMA and other federal agencies, temporary employees hired specifically and rapidly for the response, and volunteers from other state emergency management and disaster relief organizations and from within the state of Florida. Indeed, the scale of the response was so great that other states and other communities might be disappointed if similar responses are not launched when they suffer catastrophic disaster. Dealing with such threats to life and property is perhaps the clearest indicator of government effectiveness and failing to address such threats may be construed as failing to govern responsibly.

The attacks on the U.S. in 2001 certainly raised questions about the effectiveness of government officials and programs and the resultant policy and organizational challenges drew the attention of public administrationists. While hurricanes and earthquakes are "acts of God" and officials generally are not blamed for the destruction unless they fail to respond appropriately, officials are blamed when attacks occur "on their watch," when they fail to protect their constituents. Consequently, there is increased political pressure to invest in prevention programs and to measure effectiveness in terms of the number of future attacks. There is also a tendency to cloak offices and programs in secrecy to assure that defenses are not compromised. Less transparency and openness means less access by public administration and other social science researchers, not to mention media representatives and the public at-large. With that lack of access, the critical examination that can improve organizational performance by challenging policy assumptions, developing good performance measures, and evaluating results is extremely difficult. Transparency and openness encourage trust and support, particularly when the research community has come to expect such access from public agencies. Nonetheless, despite the problems studying Homeland Security structures and processes, there are more public administrationists engaged in emergency management and disaster policy research, which may be result of both the growing interest in policy relevant research among funding agencies and the perception in the research community that there are serious organizational and policy problems that need to be addressed.

The attention to emergency management as an area of research interest among public administration researchers also reflects their current interests in organizational networks and the roles of non-governmental organizations in delivering public services. The national emergency management networks are complex webs of public, nonprofit, and private organizations, as well as individual volunteers and ad hoc groups (see, e.g., May and Williams, 1986; Waugh, 2002, 2003). Communities rely upon community and faith-based organizations to help victims of disaster and to support emergency responders and the nation relies upon such groups in major disasters. There is a long tradition of volunteerism in American emergency management and that tradition is still very much alive in the fire service and in many emergency management offices. Most American communities still rely upon volunteer fire departments and many still rely upon volunteer or part-time emergency managers. Many emergency management offices have only a few hundred to a few thousand dollars a year or no public funding at all to spend on planning, communications, and other critical functions. Modern information technologies, even the automated office equipment that most of take for granted, are unavailable unless state or federal agencies provide funding or the emergency managers spend their own money. In short, the networks that deal with the nation's natural and technological hazards and respond when disasters occur are built upon an uneven foundation, some with cutting edge technologies and capabilities and some without even basic technologies and very limited capabilities. It is an organizational underpinning that may fail under the stress of catastrophic events. This is also the foundation upon which the nation's Homeland Security programs rest.

PUBLIC ADMINISTRATION AND EMERGENCY MANAGEMENT EDUCATION AND TRAINING

The number of Master of Public Administration and doctoral programs with specializations in emergency management, Homeland Security, and/or related fields is expanding steadily, likely as a result of increasing research dollars and growing student interest in the field since the 2001 attacks and, now, the 2004 Indian Ocean earthquake and tsunami and Florida hurricanes. There is a natural affinity between public administration and emergency management largely because emergency managers plan, organize, manage human resources, lead, coordinate, review, and deal with budgets. Whether they work in the public, non-profit, or private sector, their organizational responsibilities are critical to their disaster responsibilities. Indeed, most of their time is spent in managing human and financial resources and dealing with other officials and organizations. The discipline of public administration provides a foundation for emergency management educational programs and the discipline is increasingly associated with emergency management research. Of the twenty-nine institutions offering masters-level degrees, concentrations, and/or certificate programs in emergency management, nine are associated with public administration or public affairs programs (see FEMA, 2005). There is much less connection with undergraduate programs in emergency management because there are so few undergraduate public administration programs, but the graduate connection is direct and clear. A 2003 National Science Foundation-funded workshop on the skills and competencies necessary for emergency management identified almost all of the core competencies required in Masters of Public Administration programs, including an understanding of the social and political context, decision making, communication, leadership, analytical skills, budgeting, and human resource management (Thomas and Mileti, 2003: 6). What was not mentioned explicitly but was reflected in the inclusion of qualities like empathy was the *public service ethic*, the desire to respond to public needs or simply *to do good*. The connection is also recognized broadly in the field of emergency management. In order to be a Certified Emergency Manager (CEM©), the leading national credential for professional emergency managers, individuals have to have both general management and emergency management training (IAEM, 2005). The management training may be in business administration or generic administration, but there is some expectation that it be appropriate to the organizational context of emergency management and that most often means the public sector.

PUBLIC ADMINISTRATION AND EMERGENCY MANAGEMENT RESEARCH

The main streams of public administration disaster research come from three sources primarily. Because it is an interdisciplinary field, the boundaries of the public administration disaster literature are very broad and overlap considerably with other disciplines, including, for example, political science, business administration, criminal justice, psychology, history, geography, medicine, civil engineering, and sociology. The literature ranges from the politics of environmental hazard regulation (see, e.g., Mittler, 1989; and Moore and Moore, 1989) to the management of terrorist threats (see, e.g., Waugh, 1984, 1990) to the psychology of evacuation (see, e.g, Riad, Waugh, and Norris, 2001). Research by public administrationists has generally been scattered within the larger social science literature, but that is changing as researchers find more outlets in mainstream journals.

Interest in disaster research among public administrationists was jumpstarted in 1984–1985 with the two-week FEMA/NASPAA workshop at the National Emergency Training Center (NETC) in Emmitsburg, MD, and the special issue of *Public Administration Review* on emergency management. Thirty-three public administration, political science, urban planning, engineering, geography, urban affairs, and criminal justice faculty were invited to the NETC to be educated about emergency management, to discuss the perspectives that their disciplines brought to the study of disasters, and to review a draft of the special issue. Twenty-seven of the participants were from public administration/affairs or political science departments or programs. Some of the participants were contributors to the special issue but most were not. The issue was edited by William J. Petak, a public administration faculty member at the University of Southern California. Two of the articles were written by FEMA officials (i.e., public administrators), including Director Louis O. Giuffrida, and nine of the remaining nineteen articles were written wholly or in part by public administrationists or political scientists. Most of the other contributors were sociologists.

At the conclusion of the Emmitsburg workshop, a few of the participants pursued other interests. The workshop may have been their only foray into disaster research or emergency management. However, more than a dozen of the participants became very active in disaster research and helped found the American Society for Public Administration's Section on Emergency Management, which later merged with the Section on National Security and Defense Policy to become the Section on Emergency and Crisis Management.

Over the past two decades, that initial group of public administration scholars has produced dozens of books, chapters, articles, and research reports. They have also developed courses for FEMA's Higher Education Program. Conceptual frameworks have been developed, theories have been tested, and practice has been described, explained and evaluated. As is common with academic fields oriented toward practice, disaster research in public administration has evolved from single case studies to case analyses informed by theory to multivariate analyses. The methodologies of policy analysis, program evaluation, economic analysis, organizational analysis, network analysis, organizational behavior, etc., have been brought to bear on disaster-related issues and phenomena. Disaster response, hazard mitigation (from building codes to flood control projects), preparedness (from planning to training to simulation), and, increasingly, recovery, as well as everything from the translation of the science of natural and technological hazards into policies to reduce risk to life and property to the management of volunteers and nonprofit disaster relief agencies have become foci of study. How knowledge concerning natural hazards and disasters can inform policymaking concerning the threat of terrorism and how to deal with terrorist incidences is a major theme in current research efforts.

Elsewhere in the discipline, budget scholars are trying to evaluate the funding of the new Department of Homeland Security and the impact of pass-throughs from federal agencies to local first responders through state offices. Human resource management specialists are trying to evaluate the implementation of the Department of Homeland Security's new personnel system and the impact of personnel turnover as older employees retire or seek more hospitable workplaces. Administrative law scholars are trying to examine the new processes adopted by federal, state, and local officials to pursue the "war on terrorism," from procurement practices to due process rights. There are fiscal and financial issues, human resource issues, organizational issues, intergovernmental issues, and every manner of administrative/political issues involved in emergency management and disaster research. Unfortunately, access to officials and information has severely limited research efforts and, as a result, programs and decision processes are not getting the scrutiny necessary to assure accountability and effectiveness. Nonetheless, the research that is being done is increasingly finding its way into the major journals in the discipline. The visibility of the research discipline-wide, as opposed to only being read by specialists, will stimulate even more interest and encourage even more research.

Until recently, within public administration, emergency management and disaster research has tended to be viewed as "ambulance chasing" and, to the extent that the research was largely case studies, it was not a highly visible or respected pursuit among scholars. To some extent, that has been the perception of public administration itself among political scientists and other social scientists (Kettl, 2003) although public management research has become much more quantitative and theoretical in recent years. Public administration research on disasters and emergency management is still largely practice-oriented and too often not theory- or data-driven. But, public administration research on disasters and the practice of emergency management is also becoming more theoretical (see, e.g., Comfort, 1999; Busenberg, 2000; Kapucu, 2004) and more quantitative (see, e.g., Birkland, 1997), but case studies are still far more common than empirical analyses. Comparative analyses are also becoming more common (see, e.g., McEntire and Fuller, 2002; Waugh and Waugh, 2002) and the research community is increasingly international, involving noted scholars such as Dr. Akira Nakamura (Japan), Dr. Neil Britton (Japan), Dr. Alexander Kouzmin (Australia), and Dr. Uriel Rosenthal (The Netherlands), to mention but a few. For disaster policy and emergency management research to have credence in the discipline, it necessarily needs to become more empirical and theoretical. For that research to have credence in the profession, it needs to address practical issues and be presented in forms that professional emergency managers can use. That is, professional education in emergency management needs to include competencies in quantitative analysis and research design so that emergency managers can be good users of the available research.

WHAT IS THE PUBLIC ADMINISTRATION VIEW OF HAZARDS, DISASTERS, AND VULNERABILITY?

The discipline of public administration focuses on policy problems and the management of public and nonprofit organizations. Increasingly, governmental action is through partnerships, networks, and contractual relationships (government services delivered by third parties – nonprofit organizations, private firms, or even other governments – through outsourcing, franchising, and other means). Public administrationists deal with a wide variety of societal risks, from health risks to occupational safety risks to financial risks to national security risks. Environmental hazards are dealt with on several levels – as community issues because most state governments have delegated authority for land-use planning and regulation to county or municipal governments, as state issues because of the potential economic impact of disasters, and national issues because the federal government generally has more resources to deal with environmental problems (although it has been less inclined to be proactive in recent years). Disasters are handled at the local level first and by state and federal agencies when local capacities are overwhelmed. Emergency management is necessary when emergency response involves multiple agencies, jurisdictions, or sectors or the scale of the emergency requires greater coordination. In short, emergency management is a logical focus of public administration research and a logical subject of public administration education because of its importance to society and its sociopolitical and economic context. It is a logical focus because emergency management is a central government role and the effectiveness of programs is critical. It is also a logical focus because emergency management often involves intergovernmental, intersector, multiorganizational responses to situations and events that put lives and property at risk.

Emergency management is also related to policies such as sustainable development (see, e.g., McEntire, Fuller, Johnston, Weber, 2002). FEMA's Project Impact or Disaster Resistant Communities program encouraged the development of hazard mitigation strategies and the adoption of sustainable development practices. Planning for hazards was linked to development issues. Linking emergency management to the larger community environmental concerns can broaden its visibility and its political appeal. The link has been made very clear with the calls for sustainable assistance to encourage development that reduces the risk of future disasters in the coastal communities being rebuilt following the 2004 Indian Ocean earthquake and tsunami. Setbacks from the beach, alert and warning systems, and other measures are being encouraged to increase warning times and reduce structural damage.

WHAT ARE THE DISASTER-RELATED ISSUES IN PUBLIC ADMINISTRATION?

To some extent, disasters also provide a context for the study of more traditional public administration issues. For example, among public personnel specialists, the design and implementation of the personnel system within the Department of Homeland Security (DHS) is a major focus. The DHS system is the extension of the Bush Administration's management/political agenda that predates the September 11th attacks and has driven policy changes in other federal departments and agencies. The Department of Homeland Security provides a new laboratory to see if removing traditional civil service protections from employees, instituting pay-for-performance systems, and loosening controls on hiring result in higher productivity because of greater accountability or greater attrition because of political abuses of managerial discretion or simply because of decreased job security. Setting standards and performance goals are also critical activities in public agencies and the newly formed Department of Homeland Security should be providing lessons for other agencies. There is also growing interest in the funding of Homeland Security programs among public finance and budgeting specialists. Large budget increases and minimal control over expenditures with the Department of Homeland Security are more interesting than the marginal changes in the budgets of other agencies.

The major methodological challenges to doing disaster research, from the perspective of public administrationists, have been several. Studying organizations as they are engaged in dealing with crises has always been a problem. Critical processes need to be examined in the midst of the chaos if they are to be understood and managed more effectively. Organizational dynamics (e.g., adaptation, innovation, learning, etc.), decision making, leadership, communication, interpersonal relations, group dynamics, etc., are much different in the course of disaster than they are in the aftermath. This challenge is much the same as it is for scholars in other disciplines (see, e.g., Stallings, 2004).

For public administrationists, access to officials, offices, and program data is critical. Unfortunately, since the creation of the Department of Homeland Security, access has been a serious and growing problem. The transparency that characterized many public agencies prior to the September 11th attacks is being replaced by a preoccupation with security and distrustful views of the public and its right to information. The openness that facilitated communication between agency and public is affecting agency effectiveness and public trust. The same holds true for researchers. Access that was easily gained prior to September 11th, 2001, and was still largely available during the World Trade Center response and in the early months following the attacks has effectively been cut off in most cases. Public administration researchers, even those within government agencies like the Government Accountability Office and the Congressional Research Service are finding it difficult to follow and evaluate the organizational changes that are accompanying the largest federal reorganization since the creation of the Department of Defense sixty-some years ago. In short, public administration researchers are missing out on the most profound changes in the federal bureaucracy, and the agencies themselves are missing out on the lessons that might be gleaned from current experience and are failing to benefit from historical experience.

WHAT ARE THE CONTRIBUTIONS THAT PUBLIC ADMINISTRATION CAN MAKE TO THE KNOWLEDGE BASE OF EMERGENCY MANAGEMENT?

Public administration research and educational programs could do much to dispel misconceptions about collective behavior in crises (i.e., the panic myth) and increase the effectiveness of emergency responders, emergency managers, and public officials by aiding in the design, implementation, operation, and evaluation of policies and programs to deal with hazards and disasters. The foci of public administration are effective and efficient policies and programs. Public administration research can help agencies become more efficient in the use of public and private resources and more effective in managing within complex networks of public, private, and non-profit organizations and with volunteers.

As well as educating emergency managers, public administration programs can educate and, indeed, are educating other public administrators about emergency management. It is critical that mayors, governors, presidents, and other public and nonprofit officials understand the roles of emergency managers in order for them to support and facilitate those roles.

Lastly, the American Society for Public Administration (ASPA), the leading professional organization for public administrators and public administrationists, has guidelines for professional education that include knowledge of the American political system and its social, legal, and economic contexts. ASPA also has developed a code of ethics to guide public and nonprofit managers and leaders. The practice of emergency management and Homeland Security certainly presents ethical dilemmas that officials need to recognize and address. The profession of emergency management will be much stronger if it is built upon a solid foundation of knowledge, skills, and abilities and ethical standards.

DO THE CONTRIBUTIONS OF PUBLIC ADMINISTRATION OVERLAP WITH THOSE OF OTHER FIELDS OF STUDY?

In some measure, sorting out the contributions of public administration from those of other fields is difficult largely because public administration is an interdisciplinary field. Disciplinary boundaries overlap. However, public administrationists have likely increased attention to the policy and organizational issues in disaster policymaking and emergency management practice. Clearly, sociologists, engineers, psychologists, geographers, and others involved in emergency management and disaster policy research are recognizing the need to provide policy- and program-relevant information as well. But, those issues are the core interests of public administrationists.

WHAT ARE THE GAPS IN KNOWLEDGE IN PUBLIC ADMINISTRATION?

Disaster research in public administration, for the most part, has focused on organizational issues such as how well emergency management agencies have dealt with specific disasters, how decisions to issue presidential disaster declarations are made, how policies and programs are implemented and the effectiveness of policies and programs. Those and related foci reflect the more general interests of public administration researchers and, perhaps, the lack of knowledge of the wider disaster literature. The lack of knowledge may also be reflected in the persistence of some of the common myths that so infuriate disaster researchers, such as the myth of public panic in crises that still seems to exert powerful influence on policymakers. Those who connect public policies with social behavior, such as why people choose to evacuate or not or how people behave in emergencies, have fertile ground for policy studies.

WHAT SUGGESTIONS DOES PUBLIC ADMINISTRATION OFFER TO IMPROVE EMERGENCY MANAGEMENT?

Policy and program effectiveness is a consuming interest in public administration. Legions of scholars are engaged in developing performance measures and benchmarks. Modeling policies and programs and their impacts has evolved from simple univariate and multivariate arrangements to complex, dynamic systems models – even neural models – that capture the interactions among hundreds of variables. The question that remains is "what is effectiveness." In dealing with disasters, there is not always a clear test of programmatic impacts because of the variability in intensity of disasters and the low frequency of major disasters. In policy terms, how much investment in, say, preparedness is enough (Waugh, 1999)? The public administration literature does provide tools for program operation and policy design.

RECOMMENDATIONS FOR PUBLIC ADMINISTRATION AND OTHER DISCIPLINES

Public administration has two primary contributions to make to emergency managers and the disaster research community, as well as to their constituents. First, public administration's current focus on public management practices encourages attention to the efficiency and effectiveness of policies and programs. It is important that emergency management be done well. There should be accountability for policymaking and program operations, guidance for effective allocations of human and financial resources, processes to evaluate programs and to correct problems, benchmarks for program performance, and so on. Public, non-profit, and private emergency management organizations should be organized and operate effectively. All involved in emergency management, individuals and organizations, should also behave ethically. But, it is even more important that the focus of emergency management remain on doing good, not just doing it well (see, e.g., Waugh, 2004).

A positive effect of September 11th has been the focus on the potential impact of terrorism and related issues on American society among scholars, but the focus still tends to be very narrow. American society, including the government, should be a unit of analysis. An exception to the narrow research is Donald Kettl's (2004) work on Homeland Security's impact upon governance in the United States and on administrative processes in general. More attention to the impacts of disaster-related policy, including terrorism-related policy, on societal values, government processes, and economic conditions is very much needed.

Second, greater access to policy and program data would permit examination of the effectiveness and efficiency of actions (and inactions) to deal with hazards and disasters. The Witt-era mantra of "one dollar spent for mitigation saves two dollars in recovery costs" needs to be affirmed (or not) with real data. While there is a risk of losing focus on effects or phenomena that cannot be easily quantified, there is great political advantage to having better cause-effect information. Also, the flexibility of counter-terrorism plans and programs to address natural and other man-made hazards and disasters, including the aftermath of terrorist attack, needs to be examined closely. Has the United States lost capacity to deal with common hazards and disasters in its rush to prevent terrorism? Indeed, are the prevention programs working? Are we safer? Is Homeland Security organized in a way that will facilitate effective action?

Third, disaster and emergency management research needs to be in the mainstream in the discipline of public administration, as well as in other social and behavioral sciences. Although more of the research is finding its way into the major public administration journals, most is still largely scattered in specialized disaster-related journals. Few public administrationists read the disaster journals and the same is likely true of sociologists, psychologists, and other social scientists. Greater exposure to the disaster literature would benefit the discipline of public administration, as well as the community of disaster scholars and educators.

REFERENCES

Birkland, T.A. (1997). *After Disaster: Agenda Setting, Public Policy, and Focusing Events.* Washington, DC: Georgetown University Press.

Busenberg, G.J. (2000). "Innovation, Learning, and Policy Evolution in Hazardous Systems," *American Behavioral Scientist 44* (December): 679–691.

Comfort, L.K. (1999). *Shared Risk: Complex Systems in Seismic Response.* Amsterdam: Pergamon.

International Association of Emergency Managers (2005). "Certification," Website www.iaem.com, accessed December 28, 2004.

Kapucu, N. (2004). "Effective Communication Networks during Emergencies: Boundary Spanners in Multiagency Coordination," Paper presented at the 2004 National Conference of the American Society for Public Administration, Portland, OR, April.

Kettl, D.F. (2002). *The Transformation of Governance: Public Administration for Twenty-First Century America.* Baltimore: Johns Hopkins University Press.

Kettl, D.F. (2004). *System Under Stress: Homeland Security and American Politics.* Washington, DC: CQ Press.

May, P.J., and Williams, W. (1986). *Disaster Policy Implementation: Managing Programs Under Shared Governance.* New York: Plenum Press.

McEntire, D.A., and Fuller, C. (2002). "The Need for a Holistic Theoretical Approach: An Examination from the El Nino Disasters in Peru," *Disaster Prevention and Management 11*(2): 128–140.

McEntire, D.A., Fuller, C., Johnston, C.A., and Weber, R. (2002). "A Comparison of Disaster Paradigms: The Search for a Holistic Policy Guide." *Public Administration Review 62*(May/June): 267–280.

Mittler, E. (1989). *Natural Hazard Policy Setting: Identifying Supporters and Opponents of Nonstructural Hazard Mitigation.* Boulder: University of Colorado, Institute of Behavioral Science, Monograph #48.

Moore, J.W., and Moore, D.P. (1989). *The Army Corps of Engineers and the Evolution of Federal Flood Plain Management Policy.* Boulder: University of Colorado, Institute of Behavioral Science, Special Publication No. 20.

Olson, R.S., Olson, R.A., and Gawronski, V.T. (1999). *Some Buildings Just Can't Dance: Politics, Life Safety, and Disaster.* Stamford, CT: JAI Press.

Riad, J., Waugh, Jr., W.L., and Norris, F. (2001). "Policy Design and the Psychology of Evacuation," *Handbook of Crisis and Emergency Management*, ed. Ali Farazmand. New York: Marcel Dekker.

Schneider, S.K. (1995). *Flirting with Disaster: Public Management in Crisis Situations.* Armonk, NY: M.E. Sharpe.

Stallings, R., (Ed). (2002). *Methods of Disaster Research.* Xlibris/International Research Committee on Disasters.

Sylves, R.T., and Waugh, Jr., W.L. (Eds.). (1990). *Cities and Disaster: North American Studies in Emergency Management.* Springfield, IL: Charles C Thomas, Publishers.

Sylves, R.T., and Waugh, Jr., W.L. (eds). (1996). *Disaster Management in the US and Canada.* Springfield, IL: Charles C Thomas, Publishers.

Thomas, D., and Mileti, D.S. *Designing Educational Opportunities for the Hazards Manager of the 21st Century, Report of the Workshop*, October 22–23, 2003, Denver, Colorado.

Waugh, Jr., W.L. (2004). "The Existential Public Administrator," *International Journal of Organizational Theory and Behavior 6/4*(Fall): 432–451.

Waugh, Jr., W.L. (2003). "Terrorism, Homeland Security and the National Emergency Management Network," *Public Organization Review 3*: 373–385.

Waugh, Jr., W.L. (2002). *Leveraging Networks to Meet National Goals: FEMA and the Safe Construction Networks.* Washington, DC: PricewaterhouseCoopers Foundation for The Business of Government, March 2002.

Waugh, Jr., W.L. (2000). *Living with Hazards, Dealing with Disasters: An Introduction to Emergency Management.* Armonk, NY: M.E. Sharpe Publishers.

Waugh, Jr., W.L. (1999). "Assessing Quality in Emergency Management," In Halachmi, A. (Ed.) *Performance and Quality Measurement in Government: Issues and Experiences.* Burke, VA: Chatelaine Press, 1999. pp. 665–82.

Waugh, Jr., W.L. (1990) *Terrorism and Emergency Management.* New York: Marcel Dekker.

Waugh, Jr., W.L. (1984) *Emergency Management and Mass Destruction Terrorism: A Policy Framework*, Report for the Federal Emergency Management Agency, NASPAA/FEMA Workshop for Public Administration Faculty, Emmitsburg, MD, July 1984.

Waugh, Jr., W.L. and Sylves, R.T. (2002) "Organizing the War on Terrorism." *Public Administration Review*, Special Issue (September 2002): 145–153.

Waugh, Jr., W. L. and Waugh, W.W. (2002) "Emergency Management on the Pacific Rim: From Global Warming to Globalization." *International Journal of Urban Sciences 4*(2): 190–202.

Chapter 12

International Relations and Disasters: Illustrating the Relevance of the Discipline to the Study and Profession of Emergency Management

David A. McEntire

ABSTRACT

The following chapter explores why International Relations is a vital discipline for the study and profession of emergency management. It discusses past, current and potential contributions of this area of academic investigation, and describes its view of disasters and vulnerability. The chapter provides recommendations for future research and mentions how to improve the practice of emergency management from the standpoint of this discipline. Major challenges and opportunities identified in the chapter include the importance of understanding the threat of terrorism and the need for individuals, groups and nations to work together to resolve mutual disaster and development problems at the global level.

INTRODUCTION

International Relations, also known as International Politics or International Studies, is a discipline that investigates the political affairs of nations and the interactions among their multilateral institutions at the global level. The most common issue addressed by scholars in this discipline is interstate conflict. Dougherty and Pfaltzgraff (1990, p. 1) note, for example, that the central question to be addressed is the cause or causes of war. Others, including Wright (1942), Morganthau (1948), Waltz (1959), Bueno de Mesquita (1980) and Vasquez (1990) also agree that hostility and peace are the key subjects of this important field of study. The goal of these and other scholars of International Relations is to seek an understanding of the malady of war in order to find ways to prevent its occurrence.

However, it is worth noting that International Relations also incorporates topics other than interstate conflicts. Viotti and Kauppi (1993, p. 1) remind us:

> despite the adjective international, the field is concerned with much more than relations between or among states. Other actors, such as international organizations, multinational corporations, and terrorist groups, are now all part of what could more correctly be termed world politics. Studies have also focused on factors internal to a state, such as bureaucratic governmental coalitions, interest groups, presidents, and politburos. The discipline ranges from balance of power politics and economic structures at the international level to the ideological and perceptual predispositions of individual leaders.

Related topics of study therefore include international regimes, epistemic communities and foreign policy. For these and additional reasons, International Relations has a close relation to Political Science, History, Comparative Politics and other disciplines in the social sciences.

Keeping the above in mind, some may question the relevance of International Relations to the study of disasters. And, it is certainly true that the theorists and researchers interested in global affairs have not traditionally been involved in Disaster Studies. In fact, with a few exceptions it is difficult to find explicit contributions on the topic from academicians in this field. In spite of this truism, it is worth pointing out that many emergency managers and disaster scholars have been educated in this area. Wayne Blanchard,[1] Frances Edwards,[2] William L. Waugh Jr.,[3] and Richard A. Bissell[4] are only a few of the many examples that can be given. This begs the question: why would it be beneficial to have this type of background if one researches disasters or serves as an emergency manager?

The following chapter explores this question, and attempts to illustrate why International Relations is and must be an integral part of Disaster Studies. It discusses the potential contributions of this field and describes its view of disasters and vulnerability. The chapter concludes with recommendations for future research and mentions ways to improve the practice of emergency management.

THE LINK BETWEEN INTERNATIONAL RELATIONS AND DISASTERS

International Relations has a surprisingly close connection to catastrophic events and the emergency management profession. This proximity is evident on numerous grounds. First, Disaster Studies and emergency management are in many ways the outgrowth of global affairs (Drabek, 1986, p. 2; Quarantelli, 1987). Just as International Relations emerged as a result of the two World Wars at the beginning and mid-point of the twentieth century, disasters likewise became more important to practitioners and scholars during the Cold War era.

After the Nazi and Fascist regimes were pushed back and dismantled in Europe, serious disagreements arose between two former allies: the United States and Russia.

There were mutual concerns about devastation on the European continent, and both parties wanted to avert a similar tragedy in the future. But there were significant differences about how this would be accomplished. The United States desired democratic governments and open economic markets, while Russia felt the need to establish communist political and economic systems. Adding to the fray, leaders in the East were fearful of the continued presence of Western military forces in Europe. These opposing viewpoints and misunderstandings became more pronounced, and the Cold War began. Even though this hostile relationship would not lead to direct and total confrontation, it did lead to small proxy skirmishes and long, drawn-out battles in countries throughout the world (e.g., Korea, Vietnam, Nicaragua, Afghanistan, etc.). Complicating the matter, the United States utilized nuclear weapons to end World War II in the Pacific, and Russia developed similar capabilities a few years later. The threat of "mutually assured destruction" reached a pinnacle in the Cuban Missile Crisis in 1962.

It was during this period that civil defense, the precursor to emergency management, was born (Drabek and Hoetmer, 1991, pp. 9–16). The goal of civil defense was to protect – as far as possible – the government and citizens from the effects of nuclear war. War planners identified ways to evacuate public officials and the American population should missiles be launched from the Soviet Union. Shelter and mass care arrangements were also developed at this time. The proximity of these functions to disasters is readily apparent, and civil defense has had a profound impact on the direction of emergency management in the United States (Quarantelli, 1987; Dynes, 1994). Interestingly, other countries including Russia were influenced in a similar manner during the Cold War (Porfiriev, 1998).

While civil defense was being ingrained in the institutional fiber of the American government, military leaders wondered how the populace would react after a nuclear exchange. Because it would obviously be impossible and unethical to run this test on humans,[5] the government looked to scholars for assistance.

1. Higher Education Project Manager, Emergency Management Institute, Federal Emergency Management Agency, Department of Homeland Security.
2. Emergency Preparedness Director, Office of Emergency Services, City of San Jose, California.
3. Professor, Department of Public Administration and Urban Studies, School of Policy Studies, Georgia State University, Atlanta, Georgia.
4. Associate Professor, Department of Emergency Health Services, University of Maryland, Baltimore County.
5. The government did test the performance and physical impact of nuclear weapons on uninhabited ranges in Nevada and on deserted islands in the Pacific.

Millions of dollars were poured into the social sciences (particularly Sociology) and academic institutions (such as the well-known Disaster Research Center) were created to answer the inquiry. Although people's responses to hazards had been studied years before (Prince, 1920), scholars were now able to utilize these events to illustrate that victims generally exhibited rational behavior in natural disasters (Fritz and Marks, 1954). These findings were shared with military officials, but there has been some reluctance to accept the academic conclusions. For instance, disaster planning and emergency responses have often been based on false assumptions and myths about human behavior (Dynes, 1994). Regardless, emergency management and Disaster Studies owe their existence to international affairs (and the positive and negative impacts have been profound and long-lasting).

A second link between International Relations and disasters lies in the area of international organizations. Scholars in this discipline have tried to understand global institutions (such as the League of Nations and United Nations) as well as non-government organizations (NGOs)[6] involved at the global level. Although the overriding concern of the former institutions has again been the aversion of interstate conflict, these and the later organizations have also focused on issues such as development, trade, education, public health and the environment. There have also been some notable examples of such collectivities operating in the disaster area throughout history. For instance, the International Relief Union was founded in Italy in 1921 and the United Nations Disaster Relief Organization was established in 1971. Both organizations had the goal of coordinating international assistance to disaster-stricken countries. The Red Cross was initially created to deal with the suffering of soldiers on the battlefield, but this organization has expanded its operations to include disaster services. Today, there are literally thousands of voluntary agencies attempting to prevent and respond to disasters (see http://www.interaction.org for examples).

Scholars interested and educated in International Relations have studied the impact of these organizations, and have made recommendations to improve their operations. Green (1977), for instance, noted some major shortcomings of the international disaster relief system and offered suggestions for their resolution. Brown (1979) provided a very good initial examination of what the United Nations was doing to deal with disasters and recommended several measures for increased preparedness. MaCalister-Smith (1985) explored the relation of international law to humanitarian assistance. Individual scholars (Kent, 1987) and various committees (UNA-USA 1977; CIDA, 1979) also described how international actors worked together, and suggested what could be done differently to facilitate more timely and effective responses. More recently, there has been a considerable amount of attention given to the difficulty of dealing with complex emergencies[7] (Minear and Weiss, 1995). International Relations thus provides a general picture of international organizations and describes how they may overcome their weaknesses when confronting disasters around the world.

Research on security indicates a third relationship between International Relations and disasters. In the past, a great deal of emphasis in International Relations was given to arms control, particularly as it relates to nuclear weapons (Sagan and Waltz, 1995). The belief was that international anti-proliferation laws and organizations would do much to avert the negative aspects of nuclear war (e.g., a nuclear winter). In the 1990s, the definition of security expanded, and it was suggested that a nation cannot be totally secure unless it takes into account other issues including natural and other types of disasters (Jacobsen, 1994). Today, the focus of many governments has shifted to an almost exclusive view on terrorism (White, 2002; Kegley, 2003; Pillar, 2001; Simonsen and Spindlove, 2000). Homeland security, as it is now labeled, incorporates intelligence gathering, anti-proliferation campaigns, border control, infrastructure protection, emergency management and other areas (CPAI, 2002). Homeland security is

6. NGOs are also known as voluntary agencies (VOLAGs) and private voluntary organizations (PVOs).

7. A complex emergency is multi-dimensional disaster that includes political (e.g., failed state governments), economic (e.g., poverty), social (e.g., ethnic tensions), physical (e.g., drought/famine) and other (e.g., refugee) dimensions. Examples have been evident in Rwanda, Yugoslavia, Afghanistan and Sudan.

therefore similar in many respects to civil defense (Alexander, 2002). But, homeland security focuses on many different enemies (e.g., individuals, groups and states), and covers a much broader spectrum of threats than nuclear war (e.g., dirty bombs, bio-terrorism, agro-terrorism, cyber-terrorism, etc.).

A fourth affiliation of International Relations to disasters is that catastrophic events affect all countries around the world (McEntire, 1997). All nations are confronted by hazards, although the range of incidents may be dramatically different in various parts of the world (e.g., tornadoes in the United States, major droughts in African nations, and consistent terrorist attacks in the Middle East). Furthermore, a disaster in one nation may have a devastating impact on other countries. The Chernobyl nuclear accident spread radiation around the world. The Kobe earthquake affected the computer market in the United States and in other countries. The complex emergency in Rwanda created severe refugee problems in surrounding states. Disasters are consequently problems of international magnitude.

This brings us to the final connection between disasters and International Relations: these disruptive incidents require a global approach (McEntire, 1997; Mileti, 1999). Scholars have illustrated that the activities in one nation or emanating from the global economic system may increase risk in other countries (Oliver-Smith, 1994). For instance, Union Carbide (a corporation based in the United States) was to blame for the chemical release in Bhopal, India that killed at least 5,000 people. And, since most disasters occur in developing nations and because these countries are least able to deal with their adverse consequences, collaborative efforts must be made toward disaster reduction (McEntire, 1997; McEntire, 2002; McEntire, 2003a). This is especially true in terms of the environmental and development challenges that are jeopardizing the survival of humanity on our planet.

CONTRIBUTIONS OF THE DISCIPLINE

If we accept the premise that International Relations is related to disasters, it follows that we should inquire about the actual or potential contributions of this field in this area. It can be easily argued that International Relations has the potential to provide a number of lessons for Disaster Studies. First, students interested in International Relations have spent a great deal of time investigating decision making in times of crisis. Starting with Allison's thorough exposition (1971) on the Cuban Missile Crisis, there has been a long legacy of research in this area. Other excellent examples of this type of research have been provided by Janis (1972) and Jervis (1976). Obviously, the findings about rational decision making, bureaucratic politics/procedures, groupthink and misperception are equally applicable to natural disasters. Decision-making is a major problem in disasters, but the lessons from International Relations have yet to be fully applied in emergency management.

A second finding in International Relations that could be applied to disasters is from the study of regimes and epistemic communities. In the late 1980s, International Relations theorists began to counter the assumption that cooperation is impossible in our "anarchic" world. It was illustrated that countries could work together to resolve mutual concerns and challenges (Krasner, 1983). The positive impact of international organizations, as well as scholars and policy experts, was uncovered in various subject areas (e.g., arms control, trade, environment). Toward the end of the 1980s, a group of scholars and practitioners met to discuss the disturbing rise in the occurrence and impact of disasters. As a result, the 1990s was dedicated as the United Nations International Decade for Natural Disaster Reduction. Although progress was constrained during the first half of the decade because of its technocratic approach to disasters,[8] there can be little doubt that this recently created regime has brought more attention to a growing problem. Progress has been made in many areas, but there is undoubtedly much more that needs to be done within this circle of policy experts. Ironically, Disaster Studies has not yet grasped the impact of this epistemic community on international disaster management.

8. The Yokohama Strategy and International Strategy for Disaster Reduction have corrected the mistakes and limitations evident in earlier international disaster reduction policies.

A third potential area of contribution is from International Relations' recent discussions about epistemology (see Lapid, 1989; McEntire and Marshall, 2003). International Relations is a relatively young discipline, and Disaster Studies may glean important insight about theory development from this newly established field of investigation. When scholars started to dedicate time and energy to International Relations, there was a heated debate about realist and idealist explanations of global affairs. Many scholars assumed that war was an inevitable feature of the international system, but others disagreed with this set of dismal assumptions. A few decades later, attention shifted to methodological strategies. Some researchers preferred a rigid and quantitative approach while others relied on historical interpretations and in-depth explanations of unique case studies. In recent years, professors of International Relations have critically examined the merit of theoretical values, hypothetical assumptions, and academic conclusions. Disaster Studies is undergoing a similar, but unrecognized and dramatically compressed transition. It has radically shifted its explanation of disaster causes, moving from "acts of God" to the "social construction of disasters" (McEntire, 2001). A recent book has initiated discussion about advantages and disadvantages of various methodologies (Stallings, 2003). And, there are now works that compare the values, assumptions and recommendations of different disaster paradigms (McEntire et al., 2002). One of the best examples of such epistemological work is a book that bears the title "What is a Disaster?" (Quarantelli, 1998). Much more of this type of critical reflection needs to be given to other dilemmas such as:

- What hazards should be studied?
- What phases of disaster should be given priority?
- What actors should be included in academic investigations?
- What variables should be incorporated into current research?
- How can different disciplines integrate findings on disasters?
- To what extent are humans able to eliminate calamities?
- What concepts and policies are most likely to improve understanding and the reduction of catastrophic incidents (McEntire and Marshall 2003).

International Relations should therefore been viewed as a resource as scholars in Disaster Studies tackle these difficult enigmas.

A final contribution of International Relations to Disaster Studies relates to the concept of the "security dilemma." Theorists such as Jervis (1978) have shown that efforts to enhance security might actually produce the opposite result. For instance, as the United States sought produce larger and more accurate warheads, the Soviet Union undertook similar measures. When one side placed nuclear weapons on submarines, the other quickly followed suit. As communists and pro-Western factions took over governments in Asia and Latin America, the Soviet Union and America sent in troops to repel the enemy. The net result of these escalating activities is that security became more fragile during the Cold War. This same lesson could be applied to homeland security. If not approached with forethought, homeland security can increase terrorist threats to the United States from both external and internal sources. Counter-terrorist operations in other countries may generate additional hostility against the United States, and further encroachments on privacy and rights may add to some American's fear of the government. The net result is that certain steps to counter terrorism may actually aggravate the situation. It is surprising that this lesson has also not been sufficiently integrated into recent research in Disaster Studies.

PERSPECTIVES ON DISASTERS AND VULNERABILITY

In light of the events taking place around the world, International Relations would likely be inclined to define disasters in terms of a homeland security focus. Although scholars in the field would concede that natural disasters are more common around the world, the consequences of hijackings, bombings, and weapons of mass destruction (nuclear, biological and chemical) are simply too great to ignore or downplay (Falkenrath, Newman and Thayer, 1998). In keeping with its academic roots, International Relations would therefore point out the potential disruption and devastation of modern terrorism and give priority to civil hazards. This, of course, would create some major conflicts with other disciplines including Geography, Meteorology, and Sociology.

Nevertheless, International Relations would converge with other disciplines in respect to disaster terminology. Many disciplines give preference to the concept of vulnerability (see McEntire 2003b), and International Relations is no exception. Important works such as *America's Achilles Heel* repeatedly mention this term in discussions about homeland security (Falkenrath, Newman and Thayer, 1998). Falkenrath, Newman and Thayer (1998, p. 97) define vulnerability as "a function an adversary's access to a particular weapon type, its ability to use the weapon in an offensive mode, the target's ability to defend itself against this attack, and the consequences of a successful attack." They also go to great length to discuss the need to improve medical and disaster response operations/systems in order to reduce vulnerability. Such findings on vulnerability were reiterated in the National Commission Report on Terrorist Attacks Upon the United States (2004). International Relations would therefore view vulnerability reduction as a vital step for homeland security.

FUTURE AREAS OF INVESTIGATION

In the future, it will be imperative that International Relations scholars engage in additional research for Disaster Studies. The lessons from research on decision making can be easily applied to disaster situations, and there is a great need to improve our thought processes during these types of crisis events. More attention also needs to be given to the impact of epistemic communities on global disaster policies. It is ironic, for example, that Disaster Studies has not fully comprehended the activities of scholars and experts to launch the International Decade for Natural Disaster Reduction. Scholars of International Relations can also help to generate additional knowledge about those international governmental organizations, regional governmental organizations, and non-governmental organizations involved in disasters. We do not have sufficient information about the United Nations Strategy for Disaster Reduction, the Pan American Health Organization's role in disasters, and the International Committee of the Red Cross.

International Relations must also help to generate knowledge about ways to integrate national efforts for disaster reduction, and find methods to improve response operations to calamities around the world. Because complex emergencies are conflict-based events that have far-reaching effects on many nations, scholars in this field seem well-suited to conduct research on these types of disasters. Academicians in International Relations must add to the understanding of terrorism too, as this has been a major focus area in the field and as these intentional disasters are likely to become more frequent and deadly in the future.

Researchers in Disaster Studies should likewise look at International Relations in order to resolve some of the epistemological dilemmas being faced currently. International Relations has already undergone many of the challenges facing disaster scholars, and lessons about explanatory perspectives, methodological strategies, and theoretical values and assumptions can be integrated into research about disasters. It is vitally imperative that disaster scholars give increased attention to "reflexivity" or future theory development will be hindered (McEntire and Marshall, 2003).

A final and crucial topic for future investigation concerns the study of cultures around the world. If we do not understand Islamic terrorism, we will be unable to effectively confront it at the local, national and international levels. Research should therefore be dedicated to uncovering where this ideology comes from, what the movement desires, and how radical Muslims should be dealt with. Unless we increase our understanding of these types of religious beliefs, there will be little hope that terrorism will be prevented and that we will respond to these attacks in a successful manner.

RECOMMENDATIONS FOR EMERGENCY MANAGEMENT

International Relations has three recommendations for the emergency management profession. First, this discipline would assert that the broadening of the emergency management profession to include homeland security (or the expansion of homeland security to incorporate emergency management) is imperative due to the growing threat of international terrorism. If life, well-being and freedoms cannot be guaranteed by governments around the world, other goals such as education, health care, environmental protection and economic development have no meaning.

Second, International Relations would recommend further collaboration to seek a global solution to disaster problems. If natural, technological and civil disasters are caused by and adversely affect all nations, it is vital that an international approach be pursued by all individuals, groups, countries, multinational corporations, and international organizations. The possibility and/or increasingly devastating impact of climate change, bio-hazards (SARS, AIDS, Mad-Cow, West Nile, etc.), technological disasters, population growth, urban development, scarcity of resources, environmental degradation, and other similar problems necessitate a concerted international effort.

The last recommendation for emergency managers is to find ways to properly balance counter-terrorism activities with methods to reduce all types of other disasters. The war on terrorism requires innumerable resources and, if not carefully approached, may preclude actions to foster development and mitigate natural, technological, biological and environmental disasters. On the other hand, if we only concentrate attention and resources on disasters other than terrorism we will be ignoring the very real and present threat posed by Islamic fundamentalism. Those involved in emergency management are faced with extremely difficult choices; there are no simple solutions to the complex problems facing the global community. Accordingly, what is clear is that all types of disaster and development problems must be addressed simultaneously. One-sided and linear solutions are bound to fail in light of the challenges facing us today and in the future.

CONCLUSION

International Relations is a discipline in the social sciences that investigates the causes and consequences of interstate conflict as well as a myriad of other topics. Although scholars in this field have not contributed significantly to Disaster Studies, there are obvious links between this discipline and disaster phenomena. International Relations offers numerous advantages for the study of disaster, it has traditionally given preference to security issues, and now concentrates heavily on the concept of vulnerability. There is still much that we do not know about all types of disasters at the international level. A particular weakness that must be overcome is the lack of understanding about Islamic fundamentalism. Even though International Relations encourages more emphasis on the threat of terrorism, we must not let this overshadow the need to give additional attention to broader, global approaches to promote development and reduce all types of disasters. It is for these reasons that International Relations is closely related to Disaster Studies, and must contribute to and be integrated into the same.

REFERENCES

Alexander, D. (2002). "From Civil Defense to Civil Protection – and Back Again." *Disaster Prevention and Management 11*(3): 209–213.

Allison, G. (1971). *Essence of Decision.* Boston: Little, Brown.

Brown, B. (1979). *Disaster Preparedness and the United Nations: Advanced Planning for Disaster Relief.* New York: Pergamon Press.

Bueno de Mesquita, B. (1980). *The War Trap.* New Haven, Connecticut: Yale University Press.

Campbell Public Affairs Institute. (2002). *Governance & Public Security.* New York: Maxwell School of Citizenship and Public Affairs, Syracuse University.

CIDA. (1979). *Assessing International Disaster Needs.* Washington, D.C.: National Academy of Sciences.

Dougherty, J.E. and Pfaltztgraff, Jr., R.L. (1990). *Contending Theories of International Relations: A Comprehensive Survey.* New York: HarperCollins Publishers.

Drabek, T.E. and Hoetmer, G.J. (Eds.). (1991). *Emergency Management: Principles and Practices for Local Government.* Washington, D.C.: International City Management Association.

Drabek, T.E. (1986). *Human System Responses to Disaster: An Inventory of Sociological Findings.* New York: Springer-Verlag.

Dynes, R.R. (1994). "Community Emergency Planning: False Assumptions and Inappropriate Analogies." *International Journal of Mass Emergencies and Disasters 12*(2): 141–158.

Falkenrath, R.A., Newman, R.D. and Thayer, B.A. (1998). *America's Achilles' Heel: Nuclear, Biological and Chemcial Terrorism and Covert Attack.* Cambridge, Mass.: MIT Press.

Fritz, C.E. and Marks, E.S. (1954). "The NORC Studies of Human Behavior in Disaster." *The Journal of Social Issues*

10(3): 26–41.

Green, S. (1977). *International Disaster Relief: Towards a Responsive System.* New York: McGraw-Hill Company.

Jacobsen, C.G. (1994). *World Security: The New Challenge.* Science 4 Peace.

Janis, I.L. (1972). *Victims of Groupthink.* Boston: Houghton Mifflin.

Jervis, R. (1978). "Cooperation Under the Security Dilemma." *World Politics 30*(2): 186–214.

Jervis, R. (1976). *Perception and Misperception in International Politics.* New Jersey: Princeton University Press.

Kegley, Jr. C.W. (2003). *The New Terrorism: Characteristics, Causes, Controls.* New Jersey: Prentice Hall.

Kent, R. (1987). *Anatomy of Disaster Relief: The International Network in Action.* London: Pinter Publishers.

Krasner, S.D. (1983). *International Regimes.* Ithaca, New York: Cornell University Press.

Lapid, Y. (1989). "The Third Debate: On the Prospects of International Theory in a Post-Positivist Era." *International Studies Quarterly 33*: 235–254.

MaCalister-Smith, P. (1985). *International Humanitarian Assistance: Disaster Relief Actions in International Law and Organization.* Boston: Martinus Nijhoff.

McEntire, D.A. and Marshall, M. (2003). "Epistemological Problems in Emergency Management: Theoretical Dilemmas and Implications." *ASPEP Journal 10*: 119–130.

McEntire, D.A. (2003a). "Causation of Catastrophe: Lessons From Hurricane Georges." *Journal of Emergency Management 1*(2): 22–29.

McEntire, D.A. (2003b). "Searching for a Holistic Paradigm and Policy Guide: A Proposal for the Future of Emergency Management." *International Journal of Emergency Management 1*(3): 298–308.

McEntire, D.A. (2002). "The Need for a Holistic Theoretical Approach: An Examination From the El Niño Disasters in Peru." *Disaster Prevention and Management 11*(2): 128–141.

McEntire, D.A., Fuller, C., Johnston, C.W., and Weber, R. (2002). "A Comparison of Disaster Paradigms: The Search for a Holistic Policy Guide." *Public Administration Review 62*(3): 267–281.

McEntire, D.A. (2001). "Triggering Agents, Vulnerabilities and Disaster Reduction: Towards a Holistic Paradigm." *Disaster Prevention and Management 10*(3): 189–198.

McEntire, D.A. (1997). "Reflecting on the Weaknesses of the International Community During the IDNDR: Some Implications for Research and its Application." *Disaster Prevention and Management 6*(4): 221–233.

Mileti, D. (1999). *Disasters by Design: A Reassessment of Natural Hazards in the United States.* Washington, D.C.: Joseph Henry Press.

Minear, L. and Weiss, T.G. (1995). *Mercy Under Fire: War and the Global Humanitarian Community.* Boulder: Westview Press.

Morganthau, H.J. (1948). *Politics Among Nations: The Struggle for Power and Peace.* New York: McGraw-Hill, Inc.

National Commission on Terrorist Attacks Upon the United States. (2004). *The 9/11 Commission Report.* New York: W.W. Norton & Company.

Oliver-Smith, A. (1994). "Peru's Five Hundred Year Earthquake: Vulnerability in Historical Context." In Varley, A. (Ed.) *Disasters, Development and Environment,* edited by Chichester: John Wiley & Sons. pp. 31–48

Pillar, P.R. (2001). *Terrorism and U.S. Foreign Policy.* Washington, D.C.: Brookings.

Porfiriev, Boris. (1998). *Disaster Policy and Emergency Management in Russia.* New York: Nova Science Publishers.

Prince, S.H. (1920). *Catastrophe and Social Change, Based Upon a Sociological Study of the Halifax Disaster.* Ph.D. Dissertation. New York: Columbia University Department of Political Science.

Quarantelli, E.L. (1998). *What is a Disaster? Perspectives on the Question.* New York: Routlege.

Quarantelli, E.L. (1987). "Disaster Studies: An Analysis of Factors Affecting the Development of Research in the Area." *International Journal of Mass Emergencies and Disasters 5*(3): 285–310.

Sagan, S.C. and Waltz, K.N. (1995). *The Spread of Nuclear Weapons: A Debate.* New York: W.W. Norton.

Simonsen, C.E. and Spindlove, J.R. (2000). *Terrorism Today: The Past, The Players, The Future.* New Jersey: Prentice Hall.

Stallings, R.A. (2003). *Methods of Disaster Research.* Xlibris Corp.

UNA-USA. (1977). *Acts of Nature, Acts of Man: The Global Response to Natural Disasters.* New York: N.C. Scott Printing Company.

Vasquez, J.A. (1990). *Classics of International Relations.* New Jersey: Prentice Hall.

Viotti, P.R. and M.V. Kauppi. (1993). *International Relations Theory: Realism, Pluralism, Globalism.* New York: Macmillan Publishing Company.

Waltz, K.N. (1959). *Man, the State and War.* New York: Columbia University Press.

White, J.R. (2002). *Terrorism.* Belmont, Ca.: Thomson and Wadsworth.

Wright, Q. (1942). *A Study of War.* Chicago: University of Chicago Press.

Chapter 13

Comparative Politics and Disasters: Assessing Substantive and Methodological Contributions

David A. McEntire and Sarah Mathis

ABSTRACT

The following chapter illustrates how the discipline of Comparative Politics may help increase our understanding of disasters in other countries as well as promote more effective emergency management institutions and practices domestically and abroad. In seeking to reach this objective, the nature, goals, history, and background of comparative politics will first be mentioned. The chapter will then discuss the underappreciated method of comparison, and identify a number of subject areas that have been examined or could be addressed by this discipline in the future. The major argument to be made is that the comparative method makes unrecognized contributions to disaster studies and will continue to do so as research advances in the United States and in foreign territories.

> "Nations can only be understood in comparative perspective" (Lipset, 1990, xiii).

> "The significance of disaster . . . is brought sharply into focus when one takes a cross-cultural and international view" (Dynes, 1988, 102).

INTRODUCTION

According to the renowned disaster sociologist, Thomas Drabek, the field of emergency management is currently being professionalized and internationalized (McEntire 2001)..These changes imply that emergency managers are now more knowledgeable than they were in the past, and suggest that there is increased effort to expand this valued area of public service to other countries.

Although a great deal of attention is being directed toward the increasingly recognized profession in terms of new degree programs, additional academic journals and recurring conferences sponsored by emergency management associations, we lack understanding of disasters and emergency management institutions around the world. This not only calls into question the benefit of applying research from the United States to other nations, but it also limits improvements in the field in this country because lessons are not sufficiently drawn from the positive and negative experiences of others. The obvious outcome is that disaster prevention and management is hindered, both here and elsewhere.

With this preface in mind, the goal of the following chapter is to illustrate how the discipline of comparative politics may help increase our understanding of disasters in other countries as well as promote more effective emergency management institutions and practices internationally. In order to reach this objective the nature, goals, and historical background of Comparative Politics will first be discussed. The chapter will then discuss the underappreciated method of comparison, and identify a number of subject areas that have been examined or could be addressed by this discipline in the future. The major argument to be made is that the comparative method makes unrecognized contributions to disaster studies and will continue to do so as research advances across foreign territories.

COMPARATIVE POLITICS AND ITS RELATION TO DISASTERS

The discipline of Comparative Politics is the study of political systems and processes around the world (Hauss, 1997). It is an area of scholarship that is interested in understanding all nations and the political activities that take place within them. This being the case, Comparative Politics is sometimes known as comparative public policy – "the study of how, why, and to what effect different governments pursue particular courses or action or inaction" (Heidenheimer, Heclo and Adams, 1990, 3). Regardless of the actual title of the discipline, Comparative Politics might be the only field of study based on an explicit methodology. Its approach to research includes comparing and contrasting variables to identify why change occurs, what makes for a successful government, and how policy can be made effective. According to Wiarda (1993, 12), Comparative Politics "is particularly interested in exploring patterns, processes, and regularities among political systems." He further adds that students of comparative politics generally undertake the following types of research: studies of one country, studies of two or more countries, regional or area studies, studies across regions, global comparisons, and thematic studies (Wiarda, 1993, 12–15).

As can be seen, Comparative Politics is an offshoot of Political Science, and it initially reflected "significant concern for both historical perspective and the norms of political behavior" (Bill and Hardgrave, 1981, 2). Although this area of scholarship can trace its roots to the late nineteenth and early twentieth centuries, it did not really emerge "as a distinct subfield of Political Science until the two decades between the two world wars" (Rustow and Erickson, 1991, 1). This was a period when scholars became consumed with understanding why conflict broke out in Europe, how new government institutions were fairing, and what could be done differently to ensure political stability and prevent similar events from recurring. After World War II ended and the international community entered the Cold War era, interest in comparative politics grew dramatically. While the United States and the Soviet Union were aligning themselves with their respective allies, scholars began to examine the plethora of countries that made up the Third World. Their goal was to comprehend what these nations looked like and how they might become more like those in the West (or East if you were from the communist block). Comparative Politics thus developed a close relationship with sociology, anthropology, economics and other disciplines in the social sciences.

While comparative politics is related to many fields of study, it has not contributed directly to the study of disasters. Indeed, it would be difficult to find any substantial discussion of disasters by scholars of comparative politics. However, it is interesting to note that Green (1994, 143), Walker (1994, 157), Chirot (1994, 174) and others have traced foment for the revolutions in Iran, Nicaragua and the Soviet Union to natural and technological disasters (e.g., earthquakes, Chernobyl) and the preferential distribution of relief afterwards. Nonetheless, Comparative Politics has remained, for the most part, aloof from disaster studies. But this is not to imply that Comparative Politics could not assist the study of disasters, because there are a number of issues that overlap considerably between the two fields (see Table 13.1 below). Disaster researchers have already

Table 13.1

Subject Area	*Application to Disaster Studies*
Political Culture	What values and attitudes affect disaster policy?
Socioeconomic Status	How do poverty/powerlessness relate to vulnerability?
Interest Groups	Why is apathy towards disasters so common?
Institutions/State	How do governments/agencies operate in disasters?
Public Policy	What makes emergency management effective?
Decision-Making	Why are choices difficult to make in disaster situations?
Development	Does modernization increase/decrease vulnerability?

recognized the value of these issues and have produced some very important findings in these areas (Mileti, 1999; Peacock, Morrow and Gladwin, 1997; Birkland, 1996; Drabek and Hoetmer, 1991; Schneider, 1995; Dror, 1988; Wisner et al., 2004). More research in these subject areas is needed however. For instance, how do cultures around the world view disasters? Why do class relations have such a large impact on disaster vulnerability? What can be done to increase political support for disaster mitigation policies? Are intergovernmental relations problematic in foreign disasters? What steps can be taken to improve emergency management around the world? Do models such as incrementalism, group think, or misperception shed light on decision-making before and after disasters? What is the relationship between development and disasters? These are only a few of the questions that could be addressed by scholars interested in comparative politics.

METHODS

The greatest potential contribution of Comparative Politics to disaster studies is in the area of methods. In fact, Comparative Politics defines itself by "a methodological instead of substantive label" (Lijphart, 1971, 682), and this method may do much to advance the study of disaster. But, what exactly is the comparative method and how does it relate to other research methodologies? What problems are inherent in comparison and how can these be overcome? Finally, what are the benefits of comparative research?

First, the well-known comparativist Arend Lijphart "defines the comparative method as the analysis of a small number of cases, entailing at least two observations, but less than about twenty" (Collier, 1991, 8). Sartori suggests that this analysis of comparing "is both to assimilate and to differentiate" (1991, 246). He then adds:

> If two entities are similar in everything, in all their characteristics, then they are the same entity. If, on the other hand, two entities are different in every respect, then their comparison is nonsensical. . . . The comparisons in which we sensible and actually engage are thus the ones between entities whose attributes are in part shared (similar) and in part non-shared (and thus, we say incomparable). (Sartori, 1991, 246)

Przeworski and Teune (1970) also note, however, that our comparisons may be based on most similar or most different designs.

The comparative method is similar to other methods in the social sciences because much of the subject matter in this area does not lend itself to the scientific rigors of experimentation (Lijphart, 1971). Nevertheless, comparison lies between the case study and statistical methods because of its modest scope. On the one hand, case studies are utilized to describe, generate hypotheses, confirm theory or expose deviant situations. They are relatively easy to conduct, but they do not allow for far-reaching generalizations. On the other hand, the statistical method is employed to control relationships by mathematically manipulating dependent and independent variables. Although statistics approximates experimentation, this type of method can be very time-consuming and expensive (due to the large number of variables involved). The comparative method is thus less difficult to utilize than the statistical method and it also helps to generate stronger conclusions than the case study method.

This is not to say that the comparative method is void of problems. Sartori (1991) has identified five typical problems with this method:

1. **Parochialism** – focusing on one country only and failing to incorporate and build upon prior research.
2. **Misclassification** – placing phenomena into pseudo classes.
3. **Degreeism** – finding it difficult to choose between continuums and categories.
4. **Conceptual stretching** – implying that certain words mean everything (e.g., for ideological purposes).
5. **Incommensurability** – failing to find a common measure for different systems or variables.

But these challenges need not be insurmountable. They can be overcome by increasing the number of cases, reducing the number of variables, and including comparable phenomena in research strategies (Collier, 1991).

In spite of these weaknesses, there are a number of advantages associated with the comparative method. It has been suggested that the "comparative method allows systemic comparison which, if appropriately utilized, can contribute to the assessment of alternative explanations" (Collier, 1991, 10). In other words, comparison helps us to understand, explain, interpret, and verify or falsify generalizations (Sartori, 1991, 244). Furthermore, comparison facilitates "thick description" (Geertz, 1973) and limits "conceptual stretching" (Satori, 1991). Summarizing these points, Collier states:

> Comparison sharpens our powers of description and can be an invaluable stimulus to concept formation. It provides criteria for testing hypotheses and contributes to the inductive discovery of new hypotheses and to theory building (1991, 7).

Is it any wonder, then, that the scientific method is inherently comparative (Lasswell, 1968, 3), or that comparison is regarded to be equivalent to the natural scientist laboratory (Eckstein in Lijphart, 1971)?[1]

Ironically, the discipline of Comparative Politics has been notably slow to fully adopt the comparative method. Macridis asserted in 1955 that the discipline did not live up to its name when it was initially founded. Sartori even declares that not much has changed in the last fifty years:

> Let us squarely face it: normal science is not doing well. A field defined by its method – comparing – cannot prosper without a core method. My critique does not imply, to be sure, that good, even excellent, comparative work is no longer under way. But even the current good comparative work underachieves on account of having lost sight of what comparing is for (1991, 255).

Disaster studies should not make the same mistake.

POTENTIAL AND ACTUAL CONTRIBUTIONS OF COMPARISON

It is evident that comparison enables an understanding of important phenomena. Comparison can help one identify the hazards confront by policymakers, the varying impact of disasters on distinct nations, and the degree of vulnerability in other countries. Comparative work has also been useful to understand emergency management organizations and human behavior around the world. Researchers have likewise produced a number of case and comparative studies which may facilitate understanding of disaster and emergency management internationally. Each of these areas will be discussed in turn.

Hazards Around the World

First, the use of comparison helps us to better understand the disasters that may affect nations around the world. The potential for disaster is growing everywhere, but the types of events experienced are based on each country's geography, their use of technology and many other factors.

For instance, African nations face a vast variety of disasters. In 2003, twenty-eight disasters were declared in Africa by the United Nations. The continent is ravaged by floods, droughts, cyclones, earthquakes, and food security emergencies. Moreover, the AIDS epidemic is running rampant throughout many African nations. Of these, however, eleven were complex emergencies. A complex emergency is often sparked by a natural disaster and/or political, economic, or environmental stress. Complex emergencies are also marked by political or military conflict that impedes response and relief efforts (Minear and Weiss, 1995, p. 17).

While Africa is overwhelmingly afflicted with complex emergencies, Asia declared only two in 2003. Asian nations more commonly face hydrometeorological hazards. Floods have been the cause of disaster situations in Vietnam, Indonesia, China, and Sri Lanka. Typhoons have wreaked havoc in Korea, Fiji, and the Solomon Islands. In addition to floods and typhoons, drought and epidemics are also a common problem for Asian nations.

Europe and the Middle East have had to deal with terrorism as a rising source of disaster. Suicide bombers in England, Spain, and Israel have all forced emergency personnel to reevaluate their methods in mitigating and responding to terrorists. In addition to terrorism, fire, floods, and shipping accidents have been the cause of disasters throughout these areas.

In Latin America, geological disasters are declared

---

1. This is not to suggest that comparison is the best and only method. Peacock is correct to assert that "it would be bordering on methodological arrogance to suggest that certain forms of comparative research, be they characterized as qualitative, quantitative, case study, cross-national, time-series or longitudinal, or cross-sectional surveys, take precedence over others" (1997, 122). In addition, it is necessary to recognize that the research question should logically determine which method is prescribed.

with some frequency. Ecuador, Chile, Costa Rica, and Mexico have had issues with volcanoes. Mexico has also been damaged by earthquakes. Floods, droughts, and hurricanes also pose threats for countries in this area.

In North America, the United States faces hazards such as terrorism, earthquakes, hurricanes, and tornados. Earthquakes are commonplace in California, tornados ravage the Midwest, and hurricanes menace the Eastern and Gulf coasts. Indeed, the variation of climates and geography make it vulnerable to all types of disasters. Terrorism has risen to new heights of awareness since the coordinated attacks of 9/11. Canada also is at risk from similar hazards. In addition, their northern location presents them with severe winter storms.

Impact of Disasters

Disasters have plagued mankind throughout history. Indeed, tales of floods and famines have been passed down for generations. In this modern age, the occurrence of disasters has only become more frequent. The United Nations reports a steady increase of disasters across the globe (UNISDR, 2004). The International Strategy for Disaster Reduction operates under the mandate to "enable all societies to become resilient to the effects of natural hazards and related technological and environmental disasters, in order to reduce human, economic, and social losses" (UNISDR, 2005). As this trend continues it is important to identify how various nations are affected. In comparing disasters in developed versus underdeveloped countries, it becomes clear that the effects of disaster are not uniform.

The UN/ISDR reports that the countries most severely affected by disasters are of low or medium income, and rank low on the scale of human development. Approximately 80 percent of disasters are in predominantly developing areas (Alexander, 1991, p. 212). When disasters strike a developing nation, a high number of human deaths result. The top 25 countries that experienced the highest numbers of people both affected and killed by disasters between 1994–2003 were all developing nations (see appendix A). As an example, the tsunami that hit Asia in December 2004 left close to 200,000 people dead, and 100,000 missing. Mileti and his colleagues say that "losses from natural disasters occur because of development that is unsustainable" (1995, p. 122). This means that land use planning is lacking, that basic needs are not being met, and that the environment is being degraded. Other reports reveal that underdeveloped nations tend to focus their resources on issues apart from disaster preparedness, and only deal with a disaster after it hits (Aleskerov et al., 2005, p. 256).

While disasters strike the developing world with alarming regularity, they also ravage developed nations. However, developed nations are impacted by fifteen percent of disasters, and their death toll accounts for only 1.8 percent of the total deaths (United Nations, 2004). The effects of disasters in developed nations are felt more strongly in the economic sector, although the strength of their economies are better able to absorb such high losses. During the period of 1994–2003, the countries that suffered the highest economic loss were the United States and Japan (see Appendix B). As an example, after the attack on New York City's World Trade Center, the economic impact was felt much beyond the destruction of the buildings. Economic damage and loss estimates range up into the billions of dollars (Cochrane, 2004, p. 293). More recently, the death toll projections from Hurricane Katrina were initially reported in the ten thousands. However, as recovery progressed the toll did not reach the one thousand mark. Instead, the economic factors were more prevalent as major ports in New Orleans were shut down, impacting the shipping and oil companies as well as the tourism industry. Total costs are estimated at $200 to $400 billion. Thus, disasters affect all nations but in very different ways.

The Vulnerability of Nations

The distinct impact of disasters is a result of the nature and degree of vulnerability. Vulnerability is defined as a measure of proneness to disaster along with the ability to effectively withstand or react to their adverse consequences (see Watts and Bohle, 1993; Comfort et al., 1999; Wisner et al., 2004). McEntire (2004) describes this proneness in terms of the liabilities of risk and susceptibility, and he explains that coping ability is determined by the degree of resistance and resilience. This model consequently captures both the positive and negative features associated with the physical and social environments, and includes variables such as land use planning, politics, economics, culture, psychology, engineering, and institutions. Development can also be linked both positively and negatively to vulnerability (McEntire, 2004). Researchers report that countries with middle and low human development have a higher incidence of disasters (see Appendix C), which is particularly evident in 1999. This disparity is a product of social systems being more vulnerable than others.

As indicated previously, developed nations do not reflect casualties as heavily as developing nations. Their vulnerability is lower because of their ability to acquire and employ greater resources. The wealth of developed nations allows them to allocate funds for mitigation and preparedness measures. As an example, studies are often funded in these countries to identify hazard-prone areas and recommend appropriate measures for protection. Elaborate training systems are created to prepare disaster response teams in developed nations. Everyone from first responders to community volunteers can access training to more quickly and efficiently respond to a crisis. Furthermore, education and technology are relied upon in these countries to develop warning systems for the general public.

Australia, Sweden, and the United States are examples of developed nations that have advanced emergency management institutions. Australia's national government has an emergency management program that focuses heavily on using education to reduce vulnerability. The United States is now requiring that communities develop mitigation action plans to address rising disaster losses and it is giving special attention to WMD preparedness. SEMA, the Swedish Emergency Management Agency, takes responsibility to effectively coordinate their society's ability to respond to crises. However, mistakes are still made frequently in developed nations and they have a bearing on vulnerability. For instance, beachfront property is a luxury commodity for the wealthy and such locations are at risk due to hurricanes. People also increase their vulnerability by building their communities on fault lines or near industrial centers. Developed nations do not have perfect emergency management programs.

In comparison to developed nations, developing countries typically lack education, funding, and equipment to reduce their vulnerability. In Botswana, Africa, AIDS spreads quickly because of a lack of education about the transmission of the disease. Developing societies are vulnerable to other hazards because of their impoverished living conditions and weak warning systems. Building codes are rarely established or enforced in developing nations. For instance, squatter towns in Bhopal, India, built near the Union Carbide chemical plant, were partly responsible for the high death rate when poisonous gas leaked from the facility in 1984. Villages on the coast of Thailand, Sri Lanka, and India were washed away during the Tsunami of 2005 because of their dangerous location and primitive construction. Nepal has institutions that focus on landslide management and floods, but they have not established a joint, integrated warning system and vulnerability is not addressed (Paudel, 2003, p. 481).

Both developed and developing nations are affected by technology, industry, and culture. Developed nations are facing increased technological disasters as computers become more integrated into every part of their lives. Developing nations, on the other hand, may lack the familiarity with new forms of technology that could reduce or cause disasters. Each group faces adverse risks associated with hazardous material incidents, even though manufacturing plants are increasingly being moved to the developing world. People and governments in both developed and developing nations continue to make mistakes regarding disasters. They each can be found guilty of downplaying risk, augmenting social susceptibility, relying too heavily on technical remedies, and failing to strengthen emergency management institutions.

Organization of Emergency Management

There are relatively few studies about official disaster organizations and activities around the world, but there are some notable exceptions. Benjamin McLuckie completed one of the earliest comparative studies of official emergency management organizations while he was a Ph.D. candidate at the Disaster Research Center. His study examined the disaster management organizations in Japan, Italy, and the United States (1970). McLuckie's research indicated that, in times of disaster, the United States maintained a more decentralized authority structure when compared with Italy or Japan. He recognized that organizational arrangement can potentially have a dramatic influence on the effectiveness of disaster organizations because it determines the speed of response. Other studies about Russia, the United Kingdom, Australia, and New Zealand, have been conducted by Porfiriev (1999), O'Brien and Read (2005), Gabriel (2002), and Britton and Clarke (2000).

During the Cold War, Russia had a strong emphasis on civil protection because of the threat of nuclear attack from the U.S. Nuclear fallout shelters and evacuation procedures were emphasized because of the immediate crisis and threat of mutual destruction. As Cold War hostilities dissipated, Russia began to produce legislation to revamp emergency management. The Russian government realized that effective emergency management required a structured, developed system. Russia is now integrating additional mitigation and preparedness measures into their programs, thus becoming more pro-active than reactive in their strategies (Porfiriev, 1999b, p. 1).

The United Kingdom labels their emergency management program in terms of a laudable goal: the UK Resilience. In 2004, the United Kingdom passed the Civil Contingencies Act. The Act was responsible for redefining emergency management methods, including roles, responsibilities, training, and powers (O'Brien and Read, 2005). UK Resilience gives local governments the authority to handle issues at their level. However, it is still unclear to whom the local authorities should report and information does not flow smoothly between the national and local authorities. This has a potential to create confusion when disaster strikes. As such, there have been questions about the UK Resilience approach (O'Brien and Read, 2005, p. 356).

Australia, similar to the UK, has developed an emergency management program that focuses on resilience. The Australian Emergency Management Agency (EMA) does not respond directly to emergencies. Rather, as an agency of the commonwealth, it provides resources, finances, training, and research. Australia delegates responsibility for emergency management to individual states and territories. It has only been recently that Australia has begun to focus on prevention and mitigation measures (Gabriel, 2002, p. 296).

In the 1990s, New Zealand conducted a study of its emergency management structure and found significant vulnerabilities that had never been addressed. A task force was created to examine these potential problems. It recommended that the government move "more quickly and farther into areas of professional development" (Britton and Clarke, 2000, p. 146). Decision-makers then redefined the roles and responsibilities of the primary actors involved in emergency management, and created the Ministry for Emergency Management in 1999. The new system was based on principles that included an all-hazards approach, and involvement of volunteer agencies and the community. The local governments were issued the authority to handle emergencies, provided they had the capability (Britton and Clarke, 2000, p. 147).

Britton offers continued insight through his comparative studies of New Zealand, Japan, and the Philippines. In comparison to New Zealand, Japan's overall approach to emergency management is "centralized/directive, fragmented, and reactive" (Britton, forthcoming, p. 10). The Disaster Countermeasures Basic Act of 1959 defines the disaster management policies for each level of government, and this act has been revised in subsequent years (Britton, forthcoming, p. 11). However, after the Kobe earthquake, "rivalry, competition and failure to use designated focal agencies" was evidence that further measures still needed to be taken to strengthen the emergency response plan (Britton, forthcoming, p. 14). As a result, the 1998 Comprehensive National Development Act identifies methods such as improved construction codes and projects, research, and warning systems that will aid in making Japan a safer and more resilient nation (Britton, forthcoming, p. 12).

The Philippines also has a centralized approach to national emergency management. Their system, while more hierarchical than Japan and the U.S., is also fragmented and reactive. President Quezon developed their system after World War II. The National Disaster Coordinating Council does not have its own budget; rather, it operates through various other agencies (Britton, forthcoming, p. 21). While disaster management is typically handled by the central government, the powers to focus on the area of risk reduction are limited. The system lacks a comprehensive, organized, proactive, and participatory structure. The Philippines has been criticized by the

World Bank for its ad hoc approach to disasters (Britton, forthcoming, p. 23).

There are other examples of comparative research. For instance, Newton conducted a study of the United States and Canada. Although located on the same continent, the vastness of both countries affects the natural hazards they each face. Canada, in general, experiences less natural hazards than the U.S. However, their northern location has caused them to be more adaptive to severe weather (Newton, 1997, p. 225). The Canadian emergency management program assigns responsibility to local, provincial, and territorial governments. The emergency management program chooses to focus on preparation initiatives rather than mitigation directly (Newton, 1997, p. 226). The national government sets standards, provides leadership, and attempts to oversee the development of the overall ability to respond to a disaster. In contrast, the Federal Emergency Management Agency of the United States created a Mitigation Directorate in 1993. Consequently, the U.S. has seen a reduction in fatalities and property loses generally attributed to its mitigation programs (Newton, 1997, p. 228). Nevertheless, an argument can be made that the focus on mitigation has waned in recent years due to the threat of terrorism.

These studies reveal that the actors, organization, and activities associated with disaster planning and management will vary according to country. Hazards, culture, history, political objectives, and current events influence emergency management organizations around the world. Unfortunately, comparative studies of emergency management institutions have been limited.

Human Behavior

Cross-national and international studies on emergence provide insight on disaster response in various locations. Studies have been conducted in Mexico, Russia, Japan, and the United States. For instance, after the gasoline explosion in Guadalajara, Mexico, emergent behavior was similar to patterns displayed in the United States. Aguirre et al. (1995) study of that particular disaster underscores the importance of preexisting social organization. Those that responded and began search and rescue efforts were generally immediate relatives and close friends of victims. Scawthorn and Wenger (1990, p. 3) found that the first people to respond in the U.S. and Mexico were part of extending or emergent groups. However, there are also differences that can be noted between the two countries.

In Mexico City's 1985 earthquake, the percentage of people who volunteered their efforts was much smaller than in the U.S. Only 9.8 percent volunteered their services, though in real numbers this accounts for 2 million people (Quarantelli, 1989, p. 2; see also Vigo and Wenger, 1994, p. 239; Wenger and James, 1990, p. 6; Scawthorn and Wenger, 1990, p. 4). The emergent groups in Mexico typically converged from outside of the impacted area. They were generally larger groups and tightly knit socially (Vigo and Wenger, 1994, p. 240; Wenger, 1992, p. 5). Quarantelli (1989, p. 6) likewise found that families played a greater role in sheltering efforts in Mexico than in the U.S.

Behavior in Russia and the United States has also been compared. In Russia, as in the U.S., emergent groups are the first to begin search and rescue efforts (Porfiriev, 1996, p. 223). However, Porfiriev (1996) notes that the organization in Russia is more centralized than those in the U.S. The U.S. government has a fairly decentralized approach, but Porfiriev (1996, p. 96) says his "government provided practically all financial, material and a considerable part of the human resources needed to cope" after an earthquake in the late 1990s. Looting was also reported after this earthquake, a finding that contradicts previous findings in the U.S. (Porfiriev, 1996, p. 223).

Similarities and differences between Japan and the U.S. have also been examined. Both nations exhibit cooperative emergent phenomena after a disaster. The Japanese Mafia (yamaguchigumi) even collaborated with their government after a disaster (Comfort, 1996). It has been noted, however, that emergent behavior is more likely to happen in the United States than in Japan (Drabek, 1987, p. 278).

A final comparative study is between Italy and the U.S. Each nation suffered a major flood: the U.S. in 1993 and Italy in 1994. Both nations were quick to respond to the disaster. Survey results showed that while the Americans affected were knowledgeable and prepared to recover, the Italians were relatively unprepared to cope (Marincioni, 2001, p. 217). In both countries, emergent phenomenon was observed at the community level (Marincioni, 2001, p. 219). Permanent volunteer organizations were particularly useful in both countries. The established social associations were remarkable in identifying and meeting the needs of the affected populations (Marincioni, 2001, p. 219). The examples indicate both similarities and differences in terms of human behavior in disaster, and illustrate the need for further cross-national studies of emergent behavior.

ADDITIONAL CASE STUDIES AND COMPARATIVE RESEARCH

Much, if not most, of the research on disasters uses a case study methodology. Case studies provide valuable information about particular disaster situations. Researchers have examined many cases in the United States. Robert Bolin's (1990) book about the Loma Prieta earthquake is an excellent example. Part one of the book covers the economic and social costs attributed to the quake, as well as information about how it compares to previous earthquakes. Part two looks at organizational behavior, psychological impacts, gender roles, and concludes with a look at shelter and housing issues. The book therefore provides a detailed discussion about emergency management issues after Loma Prieta.

Walter Peacock, Betty Hearn Morrow, and Hugh Gladwin edited the book *Hurricane Andrew: Ethnicity, Gender, and the Sociology of Disaster* (1997). The sociopolitical ecology of Miami before the storm, warning and evacuation procedures, and crisis decision-making are each examined. In addition, sociological response of gendered groups, ethnic and racial inequalities, and a variety of familial responses are also documented. It is a must have for those interested in the topic of social vulnerability.

Beyond September 11th is a third book that can be mentioned. Its contributors examined issues in relation to the engineered environment (as it pertains to buildings and infrastructure), individual and collective behavior, the roles of private sector groups, and finally the public policy and political contexts of the disaster. It is a very comprehensive and multidisciplinary assessment of the 9-11 terrorist attacks.

There are other case studies written about particular functions within a disaster. Benigno Aguirre (1988) discusses the lack of warnings before a tornado in Saragosa, Texas. He relates effectiveness to the existence of a common shared culture, and emphasizes the importance of adapting warning systems to multicultural social contexts. Henry Fischer and his colleagues (1995) focus their research on evacuations. They outline various ways of increasing the likelihood of people heeding an evacuation warning.

Organizations have also been the subject of case studies. John Broullette (1970) discussed how adaptation to disaster demands is best accomplished by organizations that have material and personnel resources maintained and ready to utilize. Will Kennedy (1970) studied the organization and tasks of police departments in times of disaster. E.L. Quarantelli (1970) focused his research on community hospitals and the problems they must address in dealing with disaster scenarios. Martin Smith (1978) examined the response of various religious organizations after a tornado in Ohio.

Researchers have conducted case studies regarding management issues. Brenda Phillips (1993) looks at the impact of gender and diversity in relation to disaster response operations. James Kendra and Trisha Wachtendorf (2003) focus their case study on elements of resilience present in New York City's emergency operations center after the attacks of 9-11 and note that creativity is extremely beneficial. Robert Bolin and Lois Stanford (1998) examine community response to unmet needs after the Northridge earthquake response. Richard Olson, Robert Olson, and Vincent Gawronski (1998) use their case study of the Loma Prieta earthquake to stress that the recovery period offers the greatest opportunity for mitigation. Swaroop Reddy's (2000) article examined Hurricane Hugo and addresses factors that foster mitigation measures during disaster recovery.

Case study research has also focused on disasters outside of the United States. Stuart Batho, Gwyndaf Williams, and Lynne Russell (1999) looked at the terrorist bombing of the Manchester City Centre in Manchester, England. Their study uncovered the emergency responders' ability to deal with the crisis immediately and then transition through the recovery process.

Boris Porfiriev (1996) focused his case study on organizational response to the Sakhalin earthquake. The paper covers social consequences of organizational response based on one of the worst earthquakes in Russia's history. Poor response, or in some cases, the inability to respond on behalf of Russian emergency personnel, resulted in a lack of communication about the earthquakes magnitude and destructive toll. This impeded search and rescue and evacuation efforts, and exacerbated the chaos of the disaster. Francis Terry (2001) looks at the rail disaster in South London, and suggests the importance of exemplary safety standards in such an industry. Pan Suk Kim and Jae Eun Lee (2001) examine emergency management framework in South Korea. Traditionally, emergency management in this country has focused on natural disasters. The authors point out that it is necessary for the Korean Government to also focus on mitigation and response for man-made disasters. During the same year, Habib Zafarullah, Mohammad Habibur Rahman, and Mohammad Mohabbat Khan (2001) studied the disaster management strategies in Bangladesh. These scholars provide an overview of Bangladeshi disaster management as it stands today, and also assess areas of weakness and constraint that should be addressed.

Other case studies indicate the need for a holistic approach to emergency management. McEntire (2003) looks at how social and physical environmental interaction can lead to disasters. He based his work on Hurricane Georges impact on the Dominican Republic. McEntire and Christopher Fuller (2002) further examine the need for a holistic theoretical viewpoint by studying Peru's El Niño disasters. Alpaslan Özerdem and Sultan Barakat (2000) emphasize a similar theme in their investigation of the 1999 Marmara earthquake in Turkey.

Comparative studies have also been utilized to provide valuable information on emergency management practices. Joseph Scanlon (1994) compares the roles emergency operations centers in Canada and the United States. Tricia Wachtendorf (2000) used comparison to study the response of Canada and the United States to the flooding of the Red River in 1997. John Harrold and Hugh Stephens (2001) examined the maritime transportation disaster caused by the explosion of two ships carrying fertilizer in Texas City with the Exxon Valdez oil spill in Prince William Sound, Alaska. Frances Winslow (2001) analyzes the lessons that can be drawn from the accidents at Chernobyl and Three Mile Island. Alice Fothergill, Enrique Maestas, and JoAnne DeRouen Darlington studied vulnerable groups across disasters in the United States. They also compare disaster response and its variation based on factors of ethnicity and race. Dennis Mileti and Eve Passerini (1996) explore three relocation decisions employed after earthquakes. Robert Bolin and Patricia Bolton study the recovery of families in Managua, Nicaragua versus the Rapid City, South Dakota. Richard Olson and A. Cooper Drury (1997) use a cross-national analysis to compare political unrest that occurs throughout societies after various disasters. Olson also teamed up with Vincent Gawronski (2003) to contrast the 1972 earthquake in Nicaragua with the 1985 earthquake in Mexico City. They find that the Nicaraguan earthquake forced the government to shift its focus and goals, while, surprisingly, the Mexico City earthquake had no such effect on its government. Tim Ziaukas (2001) investigates the public relations aspects of disaster response between the Bhopal chemical release with the Exxon Valdez oil spill. Ronald Perry and Hirotada Hirose (1991) look at volcano management in Japan contrasted with the administration activities in the United States. Comparison has therefore been utilized as a methodology in disaster and emergency management research in the U.S. and around the world.

DISCUSSION AND CONCLUSION

It should be readily apparent from this chapter that disaster studies may gain much from the discipline of Comparative Politics. Its subjects (e.g., political culture, socioeconomic status, interest groups, institutions, public policy, decision-making and development) have recently been investigated by disaster researchers and many fruitful avenues of research are being opened as a result. In addition, the method of comparison allows us to comprehend the unique mix of hazards that face countries around the world. It is also useful to understand the impact of disasters on developed and developing nations, as well as alternate explanations for their varying degrees of vulnerability. Comparison likewise elucidates common and divergent behavioral patterns in disasters, and enables a better understanding of emergency management institutions internationally. The comparative method has certainly been used by disaster scholars, with increasing frequency over time. Nevertheless, we do have to be extremely careful to assume that concepts, issues and variables are equivalent across cultures and systems (Peacock 1997, 125).

There are also many opportunities to advance our knowledge about disasters and emergency management activities by fully engaging the comparative method. For instance, we need to learn more about the similarities and differences of complex emergencies, terrorist attacks and other types of disasters. Research is needed on varying levels of vulnerability in the developed and developing worlds, and what we should do about the disturbing trends that are confronting us. Furthermore, there is a great deal of research on disasters in the United States and select other countries (Dynes, 1988). Much less is known about emergency management institutions in Asia, Africa and Latin America. While several important case studies on disasters have been conducted, we lack systematic information about these issues and events in other nations around the world. More research will need to follow the excellent comparative work presented in this chapter.

The major finding of this chapter, therefore, is that researchers must fully recognize the value of comparison and do more to apply this method in their future studies. Effectively utilizing the comparative method will undoubtedly enable us to better comprehend the deadly, destructive and disruptive events we call disasters. Comparison will also improve the practice of emergency management as it permits us to learn from the mistakes and success of others.

APPENDIX A

TOP 25 COUNTRIES IN ABSOLUTE AND RELATIVE VALUES OF PEOPLE KILLED AND AFFECTED 1994–2003

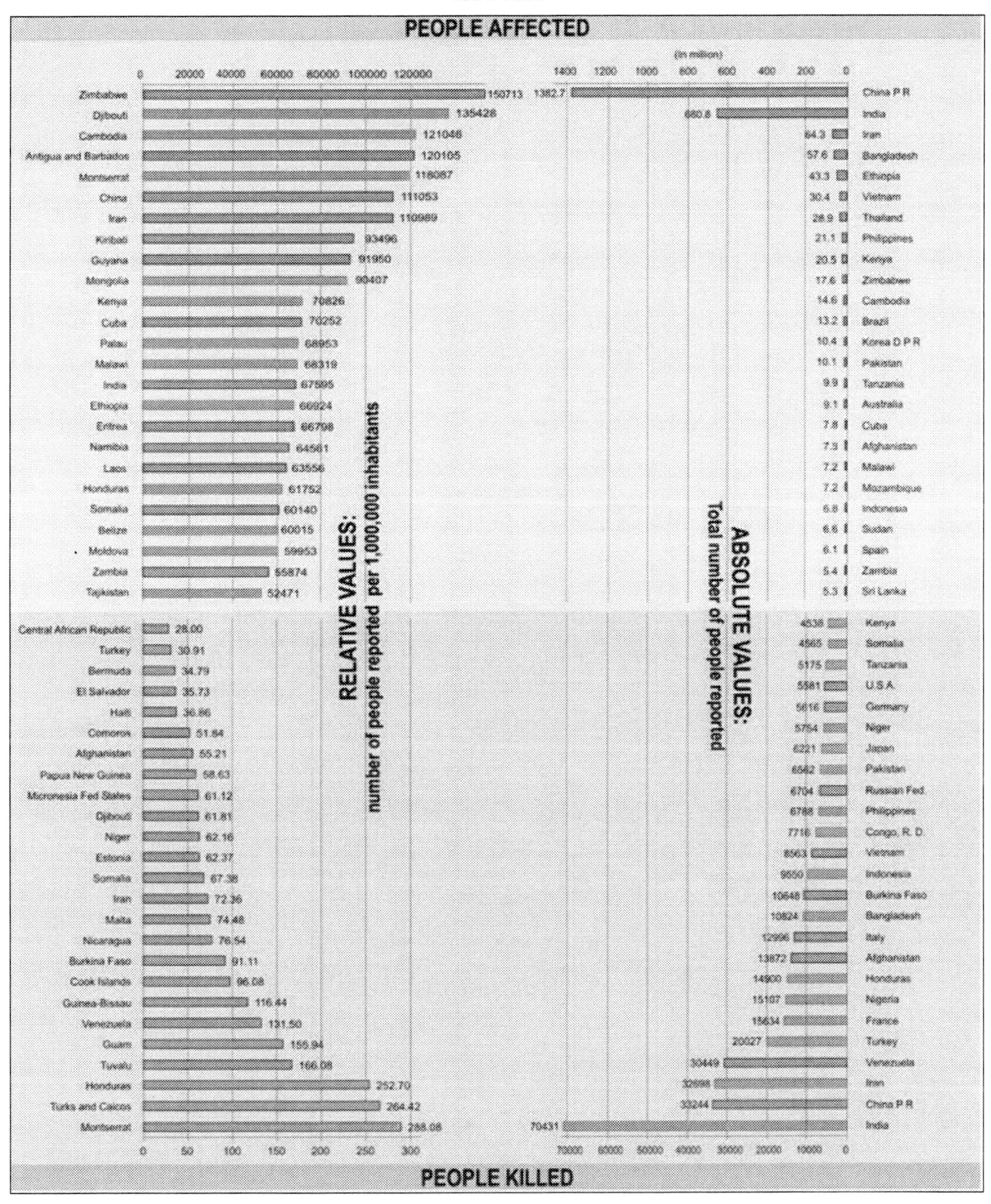

Source of data: EM-DAT: The OFDA/CRED International Disaster Database. http://www.em-dat.net, UCL – Brussels, Belgium.

APPENDIX B

TOTAL AMOUNT OF ECONOMIC DAMAGES REPORTED: ALL DISASTERS 1994–2003 (2003 US $ BILLION)

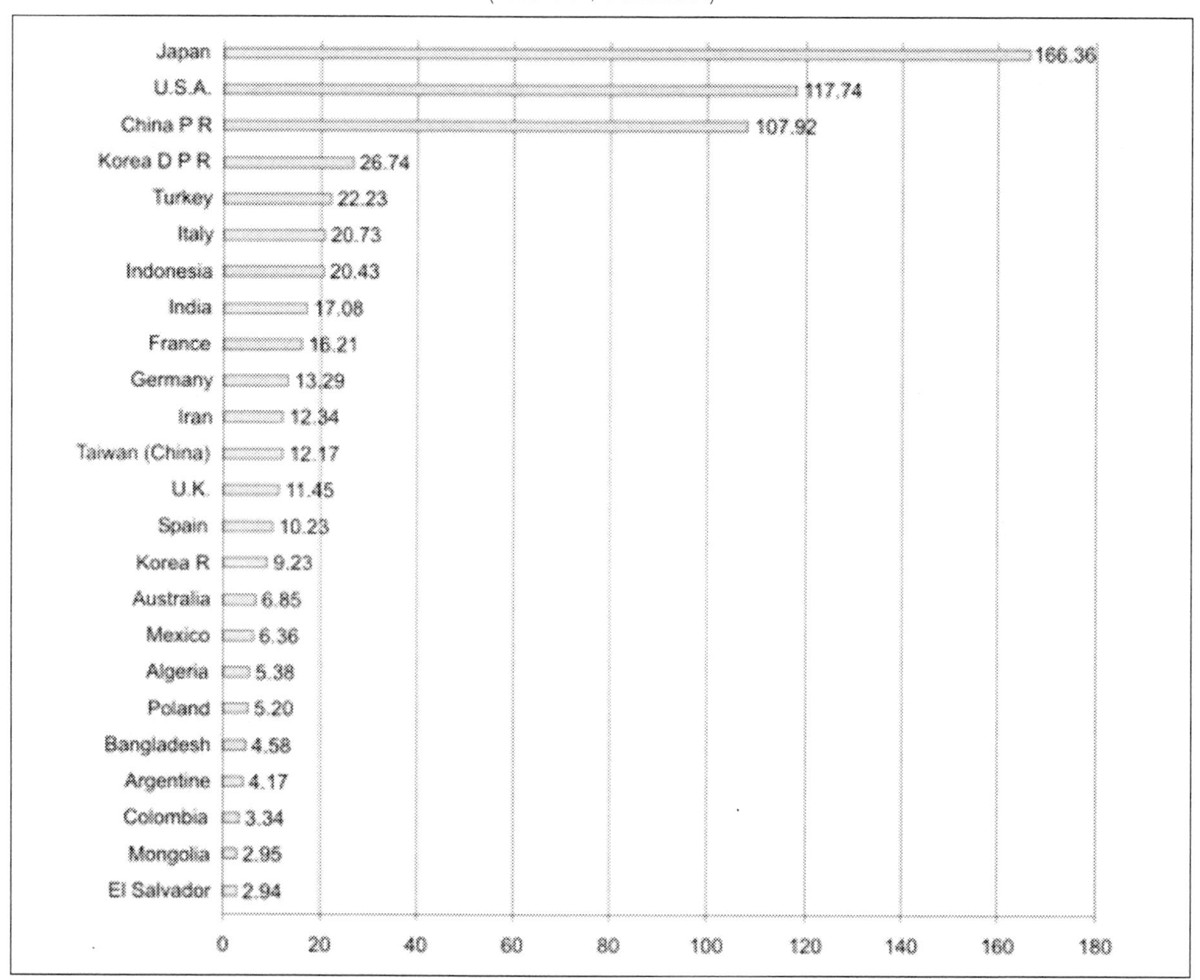

Source of data: EM-DAT: The OFDA/CRED International Disaster Database. http://www.em-dat.net, UCL – Brussels, Belgium.

APPENDIX C

TOTAL NUMBER OF DISASTERS BY YEAR 1994–2003 (ACCORDING TO HUMAN DEVELOPMENT AGGREGATES)

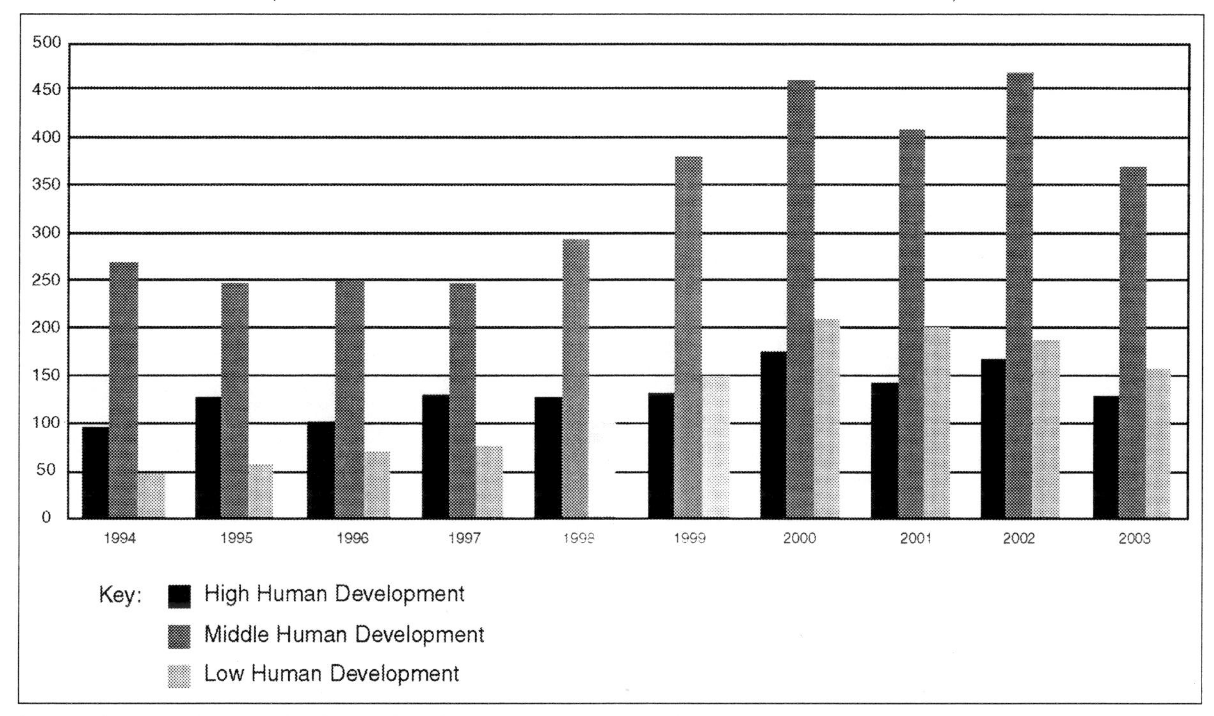

Source of data: EM-DAT: The OFDA/CRED International Disaster Database.
http://www.em-dat.net, UCL – Brussels, Belgium.

REFERENCES

Aguirre, B. (1988). "Feedback from the Field. The Lack of Warning Before the Saragoso Tornado." *International Journal of Mass Emergencies and Disaster 6*(1): 65–74.

Aleskerov, F., Say, A.I., Toker, A., Akin, H., Altay, G. (2005). "A Cluster based decision support System for estimating earthquake damage and casualties." *Journal of Disaster Studies And Policy Management 29*:256.

Alexander, D. (1991). "Natural disasters: a framework for research and teaching." *Disasters 15*: 209–26.

Australian Emergency Management. (2005). "Emergency Management in Australia." Accessed at http://www.ema.gov.au/ on Oct. 2.

Batho, S., Williams, G. and Russell, L. (1999). "Crisis Management to Controlled Recovery: The Emergency Planning Response to the Bombing of Manchester City Centre." *Overseas Development Institute.* 217–231.

Bill, J.A. and Hardgrave, R.L. Jr. (1981). *Comparative Politics: The Quest for Theory.* Lanham, MD: University Press of America.

Birkland, T.A. (1996). "Natural Disasters as Focusing Events: Policy Communities and Political Response." *International Journal of Mass Emergencies and Disasters 14*(2): 221–243.

Bolin, R. (ed.). (1990). *The Loma Prieta Earthquake Studies of Short-Term Impacts.* Boulder, CO: University of Colorado.

Bolin, R. and Stanford, L. 1998. "The Northridge Earthquake: Community-based Approaches." *Disasters 22*(1): 21–38.

Britton, N. and Clarke, G. (2000). "From response to resilience: emergency management reform in New Zealand." *Natural Hazards Review 1*(3): 145–150.

Britton, N. (forthcoming). "National Planning and Response: National Systems" in *Handbook of Disaster Research.* Disaster Research Center: University of Delaware.

Brouillette, J. (1970). "The Department of Public Works' Adaptation to Disaster Demands." *American Behavioral Scientist 13*(3): 369–379.

Bolin, R. and Bolton, P. (1983). "Recovery in Nicaragua and the U.S.A." *International Journal of Mass Emergencies and Disasters 1*:125–152.

Chirot, D. (1994). "The East European Revolutions of 1989." In Goldstone, Jack A. (Ed.) *Revolutions: Theoretical, Comparative and Historical Studies.* Orlando, FL: Harcourt Brace. pp. 165–180

Cochrane, H. (2004). "Economic loss: myth and measurement." *Disaster Prevention and Management 13*: 290–296.

Collier, D. (1991). "The Comparative Method: Two Decades of Change." In Rustow, D.A. and Erickson, K.P. (Eds.) *Comparative Political Dynamics: Global Research Perspectives.* New York: HarperCollins. pp. 7–31.

Comfort, L. 1996. "Self-organization in disaster response: the Great Hanshin, Japan Earthquake of January 17, 1995." *Quick Response Report No. 78*, Natural Hazards Center, University of Colorado, Boulder, CO.

Comfort, L., Wisner, B., Cutter., S. Pulwarty, R., Hewitt, K., Oliver-Smith, A., Wiener, J., Fordham., M., Peacock, W., and Krimgold, F. (1999). "Reframing Disaster Policy: The Global Evolution of Vulnerable Communities." *Environmental Hazards 1*: 39–44.

Drabek, T. (1987). "Emergent Structures," In Dynes, R., De Marchi, B. and Pelanda, C. (Eds), *Sociology of Disasters: Contribution of Sociology to Disaster Research*, Milan: Franco Angeli. pp. 190–259.

Drabek, T.E. and Hoetmer, G.J. (1991). *Emergency Management: Principles and Practices for Local Government.* Washington, D.C.: ICMA.

Dror, Y. (1988). "Decision Making Under Disaster Conditions." In Comfort, L. (Ed.) *Managing Disasters: Administrative and Policy Strategies.* Durham, NC: Duke University Press. pp. 255–273.

Dynes, R.R. (1988). "Cross-cultural International Research: Sociology and Disaster." *International Journal of Mass Emergencies and Disasters 6*(2): 101–129.

EM-DAT: The OFDA/CRD International Disaster Database. (2005). "Top 25 countries in absolute and relative values of people killed and affected 1994–2003." Accessed at http://www.em-dat.net on Sept. 20.

EM-DAT: The OFDA/CRD International Disaster Database. (2005). "Total amount of economic damages reported : all disasters 1994–2003 (2003 US $ billion)." Accessed at http://www.em-dat.net on Sept. 20.

EM-DAT: The OFDA/CRD International Disaster Database. (2005). "Total number of disasters by year 1994–2003." Accessed at http://www.em-dat.net on Oct. 2.

Fischer, H., Stine, G., Stoker, B., Trowbridge, M., and Drain, E. (1995). "Evacuation behaviour: why do some evacuate, while others do not? A case study of the Ephrata, Pennsylvania evacuation." *Disaster Prevention and Management 4*(4): 30–36.

Fothergill, A., Maestas, E.G.M., Derougen Darlington, J. (1999). "Race, Ethnicity, and Disasters in the United States: A Review of Literature." *Disasters 23*(2):156–173.

Gabriel, P. (2002). "The development to municipal emergency management planning in Victoria, Australia." *International Journal of Mass Emergencies and Disasters 20*(3): 293–307.

Geertz, C. (1973). "Thick Description: Toward an Interpretive Theory of Culture." In *The Interpretation of Cultures.* New York: Basic Books.

Green, J.D. (1994). "Countermobilization in the Iranian Revolution." In Goldstone, J.A. (Ed.) *Revolutions: Theoretical, Comparative and Historical Studies.* Orlando, FL: Harcourt Brace. pp. 136–146.

Harrold, J. and Stephens, H. (2001). "From Texas City to Exxon Valdez: What Have We Learned About Managing Marine Disasters?" In Farazmand, A. (Ed.) *Handbook of Crisis and Emergency Management.* New York: Marcel Dekker, Inc. Pp. 231–44.

Hauss, C. (1997). *Comparative Politics: Domestic Responses to Global Challenges.* St. Paul, MI: West Publishing Company.

Heidenheimer, A.J., Heclo, H., and Adams. C.T. (1990). *Comparative Public Policy: The Politics of Social Choice in America, Europe and Japan.* St. Martin's Press: New York.

Homeland Security, National Response Plan. (2005). Accessed at http://www.dhs.gov/dhspublic/interapp/editorial/editorial_0566.xml on Oct. 2.

Kendra, J. and Wachtendorf, T. (2003). "Elements of Resilience after the World Trade Center Disaster: Reconstituting New York City's Emergency Operations Centre." *Disasters 27*(1): 37–53.

Kennedy, W.C. (1970). "Police Departments: Organization and Tasks in Disaster." *American Behavioral Scientist 13*(3): 354–361.

Kim, P.S. and Lee, J.E. (2001). "Emergency Management in Koria: Mourning over Tragic Deaths." In Farazmand, A. *Handbook of Crisis and Emergency Management.* New York: Marcel Dekker, Inc. pp. 501–19.

Lasswell, H.D. (1968). "The Future of the Comparative Method." *Comparative Politics 1*(1): 3–18.

Lijphart, A. (1971). "Comparative Politics and the Comparative Method." *American Political Science Review 65*(3): 682–693.

Lipset, S.M. (1990). *Continental Divide: The Values and Institutions of the United States and Canada.* New York: Routledge.

Macridis, R. (1955). *The Study of Comparative Government.* New York: Doubleday.

Marincioni, F. (2001). "A cross-cultural analysis of natural disaster response: The northwest Italy floods of 1994 compared to the U.S. Midwest floods of 1993." *International Journal of Mass Emergencies and Disasters 19*(2): 209–39.

McEntire, D.A. (2001). "The Internationalization of Emergency Management: Challenges and Opportunities Facing an Expanding Profession." *International Association of Emergency Managers Bulletin* (October): 3–4.

McEntire, D.A. (2004). "Tenets of Vulnerability: An Assessment of a Fundamental Disaster Concept." *Journal of Emergency Management 2*(2): 23–29.

McEntire, D.A. (2004). "Development, Disasters and Vulnerability: A Discussion of Divergent Theories and the need for their integration." *Disaster Prevention and Management 13*(3): 193–198.

McEntire, D.A. (2003). "Causation of catastrophe: Lessons from Hurricane Georges." *Journal of Emergency Management 1*(2): 22–29.

McEntire, D.A. and Fuller, C. (2002). "The need for a holistic theoretical approach: an examination from the El Niño disasters in Peru." *Disaster Prevention and Management 11*(2): 128–140.

Mileti, D.S., Darlington, J.D., Passaini, E., Forest, B.C., and Myers, M.F. (1995). "Toward an integration of natural hazards and sustainability." *Environmental Professional 17*:117–26.

Mileti, D. and Passerini, E. (1996). "A Social Explanation of Urban Relocation after Earthquakes." *International Journal of Mass Emergencies and Disasters 14*(1): 97–110.

Mileti, D. (1999). *Disasters by Design: A Reassessment of Natural Hazards in the United States.* Washington, D.C.: Joseph Henry Press.

Minear, L. and Weiss, T. (1995). *Mercy Under Fire: War and the Global Humanitarian Community.* Boulder, Colorado: Westview Press, Inc.

Monday, J. (ed.). 2003. *Beyond September 11th: An account of post-disaster research.* Boulder,CO: University of Colorado.

Newton, J. (1997). "Federal legislation for disaster mitigation: a comparative assessment between Canada and the United States." *Natural Hazards 16*: 219–241.

O'Brien, G. and Read, P. (2005). "Future UK emergency management: new wine, old skin?" *Disaster Prevention and Management 14*(3): 353–361.

Olson, R.S. and Gawronski, V.T. (1985). "Disasters as Critical Junctures? Managua, Nicaragua 1972 and Mexico City 1985." *International Journal of Mass Emergencies and Disasters 21*(1): 5–35.

Olson, R.S. and Drury, A.C. (1997). "Un-Therapeutic Communities: A Cross-National Analysis of Post-Disaster Political Unrest." *International Journal of Mass Emergencies and Disasters 15*(2): 221–238.

Olson, R., Olson, R. and Gawronski, V. (1998). "Night and Day: Mitigation Policymaking in Oakland, California Before and After the Loma Prieta Disaster." *International Journal of Mass Emergencies and Disasters 16*(2): 145–179.

Özerdem, A. and Barakat, S. (2000). "After the Marmara earthquake: lessons for avoiding short cuts to disasters." *Third World Quarterly 21*(3): 425–439.

Paudel, Prem Prasad, Hiroshi Omura, Tetsuya Kubota, Koichi Morita. (2003). "Landslide damage and disaster management system in Nepal." *Disaster Prevention and Management 12*(5): 413–419.

Peacock, W., Morrow, B.H. and Gladwin, H. (1997). *Hurricane Andrew: Ethnicity, Gender and the Sociology of Disasters.* London: Routledge.

Peacock, W.G. (1997). "Cross-national and Comparative Disaster Research." *International Journal of Mass Emergencies and Disasters 15*(1): 117–133.

Perry, R. and Hirose, H. (1991). *Volcano Management in the United States and Japan.* Greenwich, CT: Jai Press.

Phillips, B.D. (1993). "Cultural Diversity in Disasters: Sheltering, Housing and Long Term Recovery." *International Journal of Mass Emergencies and Disasters 11*(1): 99–110.

Porfiriev, B. (1996). "Social Aftermanth and Organizational Response to a Major Disaster: The Case of the 1995 Sakhalin Earthquake in Russia." *Journal of Contingencies and Crisis Management 4*(4): 218–227.

Porfiriev, B. (1999). "Emergency and disaster legislation in Russia: the key development trends and features." *The Australian Journal of Emergency Management 14*(1): 59–64.

Przeworski, A. and Teune, H. (1970). *The Logic of Comparative Social Inquiry.* New York: John Wiley.

Quarantelli, E. L. (1970). "The Community General Hospital: Its immediate problems in disaster." *American Behavioral Scientist 13*(3): 380–391.

Quarantelli, E.L. (1989). "Human behavior in the Mexico City earthquake: Some implications from basic themes in survey findings," *Preliminary Paper #37,* Disaster Research Center, University of Delaware, Newark, DE.

Rustow, D.A. and Erickson, K.P. (1991). "Introduction." In Rustow, D.A. and Erickson, K.P. (Eds.), *Comparative Political Dynamics: Global Research Perspectives.* New York: HarperCollins. pp. 1–4.

Sartori, G. (1991). "Comparing and Miscomparing." *Journal of Theoretical Politics 3*(3): 243–257.

Sartori, G. (Ed.). (1966). *Social Science Concepts: A Systemic Analysis.* New Haven: Yale University Press.

Scanlon, J. (1994). "The Role of EOCs in Emergency Management: A Comparison of American and Canadian Experience." *International Journal of Mass Emergencies and Disasters 12*(1): 51–75.

Scawthorn, C. and Wenger. D. (1990). "Emergency response, planning and search and rescue," *HHRC Publication 11p,* Hazard Reduction and Recovery Center, Texas A&M University, College Station, TX.

Schneider, S.K. (1995). *Flirting with Disaster: Public Management in Crisis Situations.* Armonk, New York: M.E. Sharpe.

Smith, M. (1978). "American Religious Organizations in Disaster: A Study of Congregational Response to Disaster." *Mass Emergencies 3*: 133–142.

Swaroop, Reddy. (2000). "Factors Influencing the Incorporation of Hazard Mitigation During Recovery from Disaster." *Natural Hazards 22*: 185–201.

Swedish Emergency Management (2005). "SEMA Welcome." Accessed at http://www.krisberedskapsmyndigheten.se/defaultEN____224.aspx on Oct. 2.

Terry, F. (2001). "The 1989 Rail Disaster at Clapham in South London." In Farazmand, A. (Ed.) *Handbook of Crisis and Emergency Management.* New York: Marcel Dekker, Inc. Pp. 491–99.

United Nations, ISDR, 2004. Disaster Occurrence. Accessed at http://www.unisdr.org/disaster-statistics/occurrence-trends-century.htm on Sept. 21.

United Nations. Bureau for Crisis Prevention and Recovery. 2005. *Reducing Disaster Risk: A Challenge for Development.* Accessed at http://www.undp.org/bcpr/disred/rdr.htm on Sept. 21.

United Nations, ISDR, 2005. *Fact Sheet #39.* Accessed at http://www.usaid.gov/locations/asia_near_east/tsunami/pdf/Indian_Ocean_EQ_and_TS_FS39-05.06.05.pdf on Sept. 21.

United Nations, ISDR, 2005. Mission and Objectives. Accessed at http://www.unisdr.org/unisdr/eng/about_isdr/isdr-mission-objectives-eng.htm on Sept. 21.

Vigo, G. and Wenger, D. (1994). "Emergent Behavior in the Immediate Response to the 1985 Mixico City Earthquake and the 1994 Northridge Earthquake in Los Angeles. Proceedings of the NEHRP Conference and Workshop on Research on the Northridge, California, Earthquake of January 17, 1994. *Social Science and Emergency Management, Vol. 4,* 20–22, August, Los Angeles, CA, pp. 237–244.

Wachtendorf, Tricia. 2000. "When Disasters Defy Borders: What We can Learn from the Red River Flood about Transnational Disasters." *Australian Journal of Emergency Management 15*(3): 36–41.

Walker, Thomas W. 1994. "The Economic and Political Background." pp. 147–159 in Goldstone, Jack A. (ed.) *Revolutions: Theoretical, Comparative and Historical Studies.* Harcourt Brace: Orlando, Fl.

Watts, M.J. and Bohle, H.G. 1993. "The Space of Vulnerability: The Causal Structure of Hunger and Famine." *Progress in Human Geography 17*: 43–67.

Wenger, D. 1992. "Emergent and volunteer behavior during disaster: research findings and planning implications," *HHRC Publication 27P,* Hazard Reduction and Recovery Center, Texas A&M University, College Station, TX.

Wenger, D., and T. James. (1990). "Convergence of Volunteers in a Consensus Crisis: The Case of the 1985 Mexico City Earthquake." *HHRC Publication 13P,* Hazard Reduction and Recovery Center, Texas A&M University, College Station.

Wiarda, Howard J. 1993. *Introduction to Comparative Politics: Concepts and Processes.* Wadsworth: Belmont, California.

Winslow, Francis. (2001). "Lessons Learned from Three Mile Island and Chernobyl Reactor Accidents" pp. 481–90 in *Handbook of Crisis and Emergency Management,* edited by Ali Farazmand. New York: Marcel Dekker, Inc.

Wisner, Ben, Piers Blaikie, Terry Cannon and Ian Davis. (2004). *At Risk: Natural Hazards, People's Vulnerability and Disasters.* Routledge: London.

Zafarullah, Habib, Mohammad Babibur Rahman, and Mohammad Mohabbat Khan 2001. "Coping with Calamities: Disaster Management in Bangladesh." pp. 545–58 in *Handbook of Crisis and Emergency Management*, edited by Ali Farazmand. New York: Marcel Dekker, Inc.

Ziakas, Tim. (2001). "Environmental Public Relations and Crisis Management: Two Paradigmatic Cases-Bhopal and Exxon." pp. 245–57 in *Handbook of Crisis and Emergency Management*, edited by Ali Farazmand. New York: Marcel Dekker, Inc.

Chapter 14

The Contributions of Management Theory and Practice to Emergency Management

John C. Pine

ABSTRACT

This chapter takes a look at the impact that management theory and how the basic functions and practice of management as well as the role of the manager and approaches to management have contributed to the practice of emergency management. Current views of management theory stress the changing nature of the external environment and the need to understand and address these external forces for change. The contribution and role of systems theory and contingency theory to the emergency management process is stressed. Although some might view that we do not manage disasters, there is an overlap between the contribution of management theory and emergency management. Management theory stresses the need for effective planning to ensure that organizational goals are obtained. Emergency and crisis management emphasize that effective emergency response and recovery is based on good planning. Building sustainable organizations and communities is a common goal of both management and emergency management. Management and disaster-related issues and concerns along with strategies to improve emergency management practice from the field of management are provided. Finally, recommendations are provided for including emergency and crisis management in management curriculums.

INTRODUCTION

Emergency today is a complex function involving public safety and security, business affairs, public and information affairs, information systems administration, communication technologies, mapping sciences and hazard modeling, legal affairs, and coordination with numerous other organizations. This diverse set of functions and activities requires emergency managers to be effective managers of programs and operational managers of many direct disaster activities. The effective management of both program and operational activities requires an understanding of management principles. This chapter examines the development of management theory and some of the major contributions that management theory has made to the field of emergency management. It discusses some of the major management concepts including the role of the manager, strategic planning, systems theory and contingency theory, which are critical to the practice of emergency management. The overlap between management theory and disasters may be seen in concepts associated with crisis management and the importance of values, diversity, and legal issues to both management theory and emergency management. A solid foundation in concepts of management will form the basis for any emergency management activity.

THE DEVELOPMENT OF MANAGEMENT THEORY AND PRACTICE

The field of management grew in its formalization during the latter part of the nineteenth century and throughout the twentieth century along with the rise of the industrial revolution. The growth of management concepts was needed to guide the growth of industrial manufacturing in the United States and Europe. A similar growth in emergency management theory also evolved in response to the need for theory, concepts and proven practices in response to the devastating impacts of hurricanes, floods, earthquakes, and chemical spills. Our current focus on homeland security is also driving the development of even more concepts in this area.

Management theory provides a sound basis for supporting the emergence of emergency management theory utilizing the management process from planning, organizing, leading and controlling (Fayol, 1916; Mintzberg, 1973; Katz, 1974; Koontz, 1984). Taylor (1911) considered management a process and one that "if approached scientifically" would lead to success. His principles of scientific management initiated a revolution in how we viewed both the process and position of the manager. Many of the early writers in management contended that there was a right way of organizing work and accomplishing tasks (Gilbreth, 1911). Others built on the engineering approaches to acknowledge the impacts of bureaucracies (Weber, 1947). Mintzberg explained the role of the "manager" in directing the organization to achieving goals in a rational manner (1973). The interpersonal, informational, and decisional roles he characterized are mutually applicable to the emergency manager in the public, private and non-profit organizational setting.

The theory of management has grown over the past 100 years evolving from the time and motion studies of engineers to contributions from social scientists, the Hawthorne studies and a behavioral approach to more quantitative approaches that look for the "best" or optimum functioning of an organization or "total quality management (TQM)" (Gabor, 1990). Emergency management has been influenced by the same developments in management theory in utilizing engineering to design the most efficient emergency operations center or emergency response routing for emergency services. The selection of emergency medical and law-enforcement units in response to 911 communication calls and the most recent traffic hurricane evacuation planning suggest that scientific management is applicable to problems today. The ongoing assessment of disaster response programs using quantitative measurement criteria demonstrates that TQM can be used in emergency management.

The behavior scientists have also been involved suggesting the necessity of involving community organizations in planning and mitigation strategies. Finally, emergency management has been influenced by those who stress the need for quality management and the efficient use of resources, even in a disaster.

The development of principles and concepts of management encouraged the formalization of schools of business during the twentieth century. We currently see the establishment of academic programs in emergency management from concentrations, minors, certificates, and even majors from the associate to the advanced doctoral degree programs. The school of hard knocks is quickly evolving into formal academic programs in emergency management and homeland security. One wonders if the future has academic departments or schools of emergency management and homeland security. The key is that the development of professionals in emergency management requires a formal educational process and an intentional exposure to emergency management theory and concepts. Today, over 100 colleges and universities offer some program in emergency management. The standardization of these curriculums will evolve just as similar initiatives grew in response to a need for quality instructional programs.

The contribution of organizational culture theory and the impact of environmental constraints is an important part of the growth of management theory over the past fifty years (Kotter, 1992; Schien, 1985). The impact of changes in organizational culture is so well illustrated in the Federal arena during the tenure of James Lee Witt. He led a charge to change FEMA's culture to one of responsive service delivery and proactive emergency response. The changing environment and the impact of the external environment on organizations is fundamental to business as well as government operations and so important in preparedness and mitigation of hazards/disaster (Tapscott, 1998).

Finally, management has stressed the need to be aware of managing in a global environment (Adler, 1996). Today, we see emergency management emerging from a local approach to one that examines on a regional basis and with the notion of national and international linkages. The need to monitor the external environment not only locally but on an international scale is becoming a more critical element of the emergency management literature.

CONTRIBUTIONS OF MANAGEMENT TO EMERGENCY MANAGEMENT THEORY

Strategic Planning and the Changing Nature of the Organizational Environment

A major contribution of the strategic planning process to management and to emergency management is the need to monitor the nature and changing character of external forces and how they impact the operations of an organization. Environmental scanning clarifies how technology, the law, the press, elected officials, citizens, and the natural environment impact internal operations. Hurricane Andrew provides an excellent illustration of how the external environment changed emergency management theory and practice.

The catastrophic impacts of Andrew in Florida and Louisiana resulted in many changes in FEMA from an increased focus on mitigation and disaster reduction to broader operational planning. Disasters reveal not only the structural strengths and limitations of the physical environment of a community but also how local, state and national response organizations function effectively and ineffectively. Hurricane Andrew also reminded emergency managers that organizational change is often the result of external forces for change. Other external forces for change such as new technologies, laws and regulations as well as community and business needs were major factors pushing for changes in emergency management response and recovery programs, planning tools and approaches to mitigation.

The Role of the Manager

The view of the organization as a system suggests a very special role for managers in the emergency management system. For many years, management theory has suggested a rational or economic technical basis for organizational performance. This is a closed system view and appropriate for the technical level but not for the organizational or institutional level. The view of the open system creates a more difficult role for management. It must deal with uncertainties and ambiguities and must be concerned with adapting the organization to new and changing requirements. Management is a process, which spans and links the various sub-systems.

The basic function of management is to align not only people, but also the institution itself including technology, processes, and structure. It attempts to reduce uncertainty at the same time searching for flexibility.

Management faces situations, which are dynamic, inherently uncertain, and frequently ambiguous. Management is placed in a network of mutually dependent relationships. Management endeavors to introduce regularity in a world that will never allow that to happen. Only managers who can deal with uncertainty, with ambiguity, and with battles that are never won but only fought well can hope to succeed.

Management Systems Theory and Emergency Management

Systems theory evolved from the basic sciences but is utilized in the social sciences including management theory. A system composed of interrelated and interdependent parts arranged in a manner that produces a unified whole is critical in understanding all parts of the emergency management process. Viewing societies as complex open systems which interact with their environment provides such a critical view of the emergency management system (Barnard, 1938).

Systems theory is based on the idea that everything is part of a larger, interdependent arrangement. It is centered on clarifying the whole, its parts, and the relations between them (Von Bertalanffy, 1972). Some critical concepts that are applicable to emergency management include some of the following: open system, sub-systems, synergy, interface, holism, strategic constituencies, boundaries, functionalism, interface, strategic constituencies, feedback and a moving equilibrium. Emergency management is composed of many parts including: local, state and national public, private and non-profit units. These units interact in many independent ways and each has their own constituencies, boundaries, function, and sub-units. The units may interrelate in emergency management activities in an open environment with few organizational barriers or collaborative and cooperative efforts limited by specific organizational policies, rules and procedures. Emergency managers acknowledge that effective emergency response and recovery efforts require the cooperation of the entire community; emergency managers do not operate in isolation but as a part of a large open system.

Effective emergency response and recovery is dependent on cooperation between local public agencies, business enterprises, and community groups. Shelters are often sponsored by public and private schools and operated by the American Red Cross. Evacuation efforts are often supported by community transportation agencies and school systems. Special needs shelters are often staffed by local medical facilities, volunteers, and community organizations. Traffic control and security is a collaborative effort between numerous local law enforcement jurisdictions. Coordination is critical in linking multiple organizational efforts in a seamless response and recovery effort.

An *open system* involves the dynamic interaction of the system with its environment. This theory is fundamental to understanding hazards and emergency management for it maintains that everything is related to everything else. Emergency management has a dynamic relationship with the environment and receives various inputs, transforms these inputs in some way, and exports outputs. These systems are open not only in relation to their environment but also in relation to themselves; the interactions between components affect the system as a whole. The open system adapts to its environment by changing the structure and processes of the internal components.

Systems are composed of *sub-systems.* That is, the parts that form the system may themselves be a system. The emergency management system includes police, fire, and emergency medical agencies; each agency with their own system (sub-system of the emergency services system). The emergence of homeland security makes this concept even more important in understanding how the parts relate and that each part has sub-parts that impact the functioning of the whole.

The combined and coordinated actions of the parts of the system achieve more than all of the parts acting independently. This concept known as "synergy" is critical to the field of management and equally to emergency management. The performance of an enterprise is a product of the interaction rather than sum of its parts, but it is entirely possible for the action of two or more parts to achieve an effect of which either is individually incapable. Synergy is characterized by the whole being greater than the sum of its parts. It explains why the performance of a system as a whole depends more on how its parts relate than on how well each part operates. Indeed, the interdependence of the parts is such that even if each part independently performs as efficiently as possible, the system as a whole may not. Synergy is an important concept for emergency managers in that it emphasizes the need for individuals, as well as departments to work together in a cooperative fashion (Bedeian, 1993). An emergency response is not just a single unit but many different parts that, when effective, understand how they work together to protect public safety and property.

Emergency management, as with the field of management is dependent on conceptual frameworks or models. As an example, management theory suggests that social organizations are *contrived* and constantly evolving and not static mechanical systems. They have structure, but the structure of events rather than physical components, cannot be separated from the processes of the system. The fact that social organizations are composed of humans suggests that they can be established for an infinite variety of objectives and do not follow the same lifecycle pattern of birth, maturity, and death as biological systems. Social systems are made of imperfect systems. The cement which holds them together is essentially psychological rather than biological. They are anchored in the attitudes, perceptions, beliefs, motivations, habits, and expectations of humans.

Management systems theory notes that organizations are not natural as with mechanical or biological systems; they are contrived. They have structure or *boundaries*, but the structure of events rather than physical components. The human and organizational boundaries cannot be separated from the processes of the system. The fact that social organizations are contrived by human beings suggests that they can be established for an infinite variety of objectives and do not follow the same lifecycle pattern of birth, maturity, and death as biological systems. Social systems are made of imperfect systems. The cement which holds them together is essentially psychological rather than biological. They are anchored in the attitudes, perceptions, beliefs, motivations, habits, and expectations of human beings.

A systems approach does not provide a means for solving all problems. It is however, useful for viewing the relationships between interdependent parts in terms of how these relationships affect the performance of the overall system (Kast, 1985; Freemont, 1985). Systems theory provides emergency managers with a critical perspective to view and understand how to prepare for and respond to hazards and mitigate their adverse impacts.

The systems perspective to emergency management integrates the diverse interdependent (or interconnectedness of the system) factors including individuals, groups, formal or informal organizations, attitudes, motives, interactions, goals, status, authority. The job of an emergency manager is to ensure that all parts of the organization are coordinated internally and with external organization that are involved in emergency management activities. The emergency management thus is leading and directing many activities so as to achieve established organizational and community goals. A systems view of management suggests that all parts of the organization are interdependent. For example, if a service unit functions well, but the personnel section does not replace retired staff in a timely manner, the system malfunctions.

The open systems approach recognizes that organizations are not self-contained. They rely on their environment (including the social, political, technological, and economic forces) for life-sustaining inputs and as sources to absorb their outputs. No organization can survive for long if it ignores government regulations, the courts, outside interest groups, private service providers, or elected officials. An organization should be judged on its ability to acquire inputs, process these inputs, channel the outputs, and maintain stability and balance. Outputs are the ends, where acquisition of inputs and processing efficiencies are means. If an organization is to survive over the long-term, it must remain adaptive. System concepts such as subsystems or units within units; synergy or that the group has greater outputs than each single unit, boundaries, holism or viewing the larger context rather than a narrow view, interface, and adaptive organizational mechanisms to change are crucial in marshaling community resources so critical in emergency management. The importance of leadership and adaptive behavior are stressed by many writers (Lewin, 2000; Toffler, 1985; Garvin, 1993; and Sugarman, 2001) who stated that today's leaders including emergency managers must discover ways of creating order in a chaotic world.

Finally, chaos theory suggests that even in general management systems theory, organizations must adapt to complex change and institutionalize institutional learning through feedback systems. Chaos theory states that just a small change in the initial conditions may have significant change in the long-term behavior of the system. The classic example quoted by many to illustrate the concept is known as the butterfly effect.

> The flapping of a single butterfly's wing today produces a tiny change in the state of the atmosphere. Over a period of time, what the atmosphere actually does diverges from what it would have done. So, in a month's time, a tornado that would have devastated the Indonesian coast doesn't happen. Or maybe one that wasn't going to happen, does (Stewart, 1989).

Individual response and rescue efforts in evacuating buildings in 911, illustrate that single acts can have dramatic impacts for a city, a country and the world. On a small local the disaster of 911 required public, private and non-profit organizations to adapt and form new relationships to recover from the terrorist attack. New recovery organizations evolved from local resources as illustrated by the St. Paul Episcopal Church support effort for workers in the World Trade Center pit. The St. Paul response effort evolved from local resources, but was supported by public, private and other non-profit groups throughout the United States. Many studies from the 911 disaster provided lessons learned; each study noted that successful strategies were based on flexible ongoing adaptations to changing events (Kendra, 2003; Rubin, 2001; Sutton, 2002; and Weber, 2002). Chaos theory thus provides the emergency manager with a broad perspective for appreciating how other agencies and external organizations are interdependent with and impact emergency management operations.

Contingency Theory and Approach

Contingency theory suggests that management principles and practices are dependent on situational appropriateness. Luthans (1976) notes that "The traditional approaches to management were not necessarily wrong, but today they are no longer adequate.

The needed breakthrough for management theory and practice can be found in a contingency approach." Different situations are unique and require a managerial response that is based on specific considerations and variables. The appropriate use of a management concept or theory is thus contingent or dependent on a set of variables that allow the user to fit the theory to the situation and particular problems. It also allows for management theory to be applied to an intercultural context where customs and culture must be taken into consideration (Shetty, 1974). Adapting theory to the context is extremely important to a new homeland security international context.

For management and emergency management alike, the successful application of any theory or concept is greatly influenced by the situation. For example, a functional organization structure with many layers of management functions best in stable environmental conditions and routine operations. For emergency management, the operating environment is ever changing and must be flexible to accommodate the many different hazards that a community or business faces. Emergency managers must build an organizational culture and structure that improvises and acknowledges that each disaster is unique. As a result, a more dynamic organizational structure could be structured based on the nature of the problem (hazard) and who needs to be involved and the actions taken (Kreps, 1991). Utilizing an organizational design that is rigidly structured regardless of the situation might not provide the appropriate basis for quick and comprehensive decision making in a crisis or disaster.

THE OVERLAP BETWEEN MANAGEMENT THEORY AND DISASTERS

A major goal of emergency management is to minimize the adverse impact of a disaster on a business, community or large geographic area. The efforts of many organizations to build a more sustainable community, business or country are consistent with emergency management goals of hazard mitigation.

SUSTAINABILITY. In 1987 the United Nations World Commission on Environment and Development coined the term "sustainable development." It defined sustainable development as "meeting the needs of the present generation without compromising the ability of future generations to meet their own needs." This means that while we are harvesting natural resources and developing our land we must do so in a manner that will allow other generations to have at least the same opportunities that we currently have.

Sustainable development is more of a compromise between the traditional standards of conservation and preservation. Conservation suggests that we should use the earth's natural resources while at the same time replacing them for future use. It focuses more on the renewable resources while for the most part ignoring exhaustible resources such as oil and natural gas. At the other extreme is preservation, which suggests that we leave nature alone. These two viewpoints are at opposite ends of the spectrum, which lends itself to small numbers of supporters from the general public. Sustainable development is a kind of middle ground between these two ideologies that is more likely to be accepted by a larger group of people. It is based upon a logical viewpoint that people will not want to diminish their quality of life or standard of living to preserve the environment. It takes into account that the economy will continue to grow and develop but also encourages ways to do this that will have as little negative impact on the environment as possible.

For many years society, the economy, and the environment were all seen as separate entities. The key to understanding sustainability is understanding the way in which these three issues link together. Sustainability deals with quality of life issues as well as achieving balance between the three. In order to be sustainable we must learn to manage economy and society in a way that doesn't harm the environment while at the same time learning to live within our limits and divide resources equitably.

Sustainability also is a fundamental theoretical contribution to our understanding hazards, disasters and their impacts (FEMA, 2000; Livable, 2000; and Burby, 1998). Making rational choices concerning land use, development, and economic development has tremendous implications for dealing effectively with hazards and disasters.

Crisis Management

Management theory has embraced as a part of the

planning process the preparation of contingency plans and crisis management to address threats and hazards (Pearson, 1998). The development of a crisis audit including "What if" questions and contingency plans when things go wrong are critical elements of business planning and analysis (Roberts, 2001). The management literature reflects an appreciation for the need for business to grow more aware of the need to provide some level of protection against an unplanned disaster (Myers, 1999). Management needs to know how to structure strategic planning to include plans to minimize disruptions in operations in times of crisis and disasters. *The Harvard Business Review* published a crisis management series on the best articles relating to disasters and business interruption (2000). Laye's assessment of how to keep business going when catastrophe strikes (2002) is a reflection of the attention that hazards and disasters have had on the literature since 2001.

Values Diversity and the Legal Environment

A critical element in the emergency management is the development of an understanding of potential impacts of a disaster. Vulnerability analysis focuses on physical, political, economic and social vulnerability (Cutter, 2001). Mileti (1999) states that disasters can do more than impose deaths, injuries and economic losses, they can redirect the character of social institutions, alter ecosystems and impact the stability of political structures. Blaikie et al. (1994) note that some groups in society are much more vulnerable to disaster losses and suffer differently; variations of impact from disasters evolve from class, caste, ethnicity, gender (Enarson, 1997), religion (Bolin, 1986), disability, or age (Bolin, 1983). Vulnerability is the susceptibility to hazard, disasters, or risk. And, it can also be a measure of resilience.

In emergency management, there needs to be a balance in examining vulnerability and understand the social, economic and environmental impacts from disasters. Too often we see the damage to structures rather than the immediate and long term impacts of disaster to our environment or social systems. Our organizations must be inclusive and offer balanced perspectives rather than just a single perspective. It is not enough to just examine the economic impacts of flooding or earthquakes on local communities but examine other impacts such as social or environmental. We need to encourage faculty to seek out alternative views in the forms of books, journals, and research reports and expose students to these perspectives.

Management theory shares this view and encourages diversity and non-discrimination in employment and contracting. An appreciation of organizational values and potential conflicts in international operations must be acknowledged and addressed. In the traditional sense, equal opportunity in organizational performance can be applied both internally and externally in business affairs (Thomas, 1990 and Hall, 1993).

Many disciplines have stressed individual privacy in their programs and activities. State and federal privacy provisions are common in health care statues to protect the privacy of individuals. Emergency managers must understand and ensure that staff and volunteers know what personal identification information may be released to the public in disaster response and recovery. Public information cannot be obtained from the Centers for Disease Control (CDC) that provides any indication of the health and well-being of individuals in a community. Much of the data is only released in groups of 100,000 or greater. The aggregation of data is intended to protect the privacy of individuals.

A fundamental element of the practice of emergency management that is also present in the field of management is its evolution from many disciplines from engineering, business, sociology, psychology, political structures, and urban planning to name only a few. Management also grew from many disciplines, especially from engineering (scientific management), psychology, sociology, and quantitative methods. Emergency management draws from many disciplines and suggests that emergency management is an interdisciplinary process. An appreciation of organizational and group dynamics, individual motivation, leadership, program and organizational assessment, and planning are all elements of both the emergency management and the management process.

MANAGEMENT AND DISASTER-RELATED ISSUES AND CONCERNS

The unintended consequences of human action

are described by Chiles (2001). He documents many examples of our failure to adequately manage technology. He shows that chain reaction catastrophes have occurred as the world has grown more technologically complex and our machines have become more difficult to control. He suggests that we may have a false value of technology and do not adequately place limits on its use. Emergency management should also share his suggestion that we acknowledge the potential adverse impacts technology and the need to ensure human assessment of technology.

The terrorist attacks of 2001 have made the business community increasingly sensitive to the impacts of disasters and especially terrorism on domestic and international operations. Risk management is now a part of any large operation and a dependence on insuring risk is no longer the only contingency. Businesses are increasingly looking at avoiding disasters and identifying methods to mitigate disasters.

The insurance industry has adapted to this changing environment by excluding coverage for terrorism in business policies or calculating the potential costs associated with insuring this risk in their plans. Most organizations can no longer afford to insure for this risk. Insurance companies have also reassessed coverage for many natural hazards and taken steps to adequately cover their potential vulnerabilities. The increased costs to public and private organizations for insuring against hazards has increased to the point that it may impact business plans and future strategies.

IMPROVING THE MANAGEMENT IN EMERGENCY MANAGEMENT

The field of management has stressed the need for the development of positive organizational culture and organizational learning. The management environment today and in the future will provide new challenges and organizational responses. The management literature has been sensitive to this need and been quite responsive. Emergency management must also acknowledge the need for organizational learning and the importance of a positive organizational climate to effective operations. Possibly more executive education would support the increasing interdependence between the Department of Homeland Security, the business community, as well as state and local operations.

During the past thirty years, the business community has focused on the importance of quality control and service. Emergency management operations must share this emphasis and adopt methods of organizational assessment and quality control to enhance all elements of the emergency management process.

The management literature has for many years stressed the importance of strategic planning (Drucker, 2002). A greater awareness of the value of environmental scanning and the broader impacts of international affairs on internal operations will be increasingly important to the emergency management community. Business may call on emergency management for help in identifying strategies to cope with a dramatically changing environment.

RECOMMENDATIONS: EMERGENCY MANAGEMENT IN MANAGEMENT CURRICULUMS

Few business schools have embraced the contribution that emergency management theory and practice can make to the success of business operations. As a result, attention to hazards and disaster impacts are limited to crisis management and contingency planning. Few if any schools of business have worked with emergency management curriculums on their campuses and exposed their students to other disciplines that are so much a part of disaster research. Interdisciplinary courses that expose students from throughout the campus to the nature of hazards and disaster impacts are needed. Including students from business programs will expose other hazard oriented coursework to the vulnerability of business operations and impacts well beyond financial considerations. An integrated approach to college and university curriculums will prepare students to understanding the changing nature of hazards and disasters in an increasingly interdependent world.

REFERENCES

Adler, N. (1996). *International Dimensions of Organizational*

Behavior. Cincinnati: Southwestern.

Barnard, Chester I. (1938). *The Functions of the Executive.* Cambridge MA: Harvard University Press.

Bedeian, Arthur G. (1993). *Management.* Chicago: Dryden Press.

Blaikie, P. et. al. (1994). *At Risk.* London: Routledge.

Bolin, R. and Klenow, D. (1983). "Response of the Elderly to Disaster: An Age Stratified Analysis." *International Journal of Aging and Human Development 16*(4): 283–296.

Bolin, R. and Bolton, P. (1986). *Race, Religion, and Ethnicity in Disaster Recovery.* Boulder, CO: University of Colorado.

Burby, R. (Ed.). (1998). *Cooperating with Nature: Confronting Natural Hazards with Land-Use Planning for Sustainable Communities.* Washington, D.C.: Joseph Henry Press.

Chiles, J.R. (2001). *Inviting Disaster: Lessons from the edge of technology.* New York, NY: Harper Business.

Cutter, S.L. (2001). *American Hazardscapes: The Regionalization of Hazards and Disasters.* Washington, D.C.: Joseph Henry Press.

Drucker, P.F. (2002). *The Effective Executive Revised.* New York, NY: Harper Collins.

Enarson, E. and Morrow, B. (1997). "A Gendered Perspective: Voices of Women." In Peacock, W.G., Morrow, B.H. and Gladwin, H. (Eds.). *Hurricane Andrew.* Miami: International Hurricane Center. pp. 116–140.

East, F.E. and Rosenzweig, J.E. (1985). *Management: Systems and Contingency Approach.* New York: McGraw Hill.

Fayol, H. (1916). *Industrial and General Administration.* Paris: Dunod.

Federal Emergency Management Agency (FEMA). (2000). "Planning for a Sustainable Future: The Link Between Hazard Mitigation and Livability". FEMA Report 364. Washington D.C.: FEMA. http://www.fema.gov/mit /planning_toc.htm

Federal Emergency Management Agency (FEMA). (2000). "Rebuilding for a More Sustainable Future: An Operational Framework". FEMA Report 365. Washington D.C.: FEMA. http://www.fema.gov/mit/planning_ toc2.htm

Freemont, E.K. and Rosenzweig, J.R. (1985). *Management: Systems and Contingency Approach.* New York: McGraw Hill.

Gabor, A. (1990). *The Man Who Discovered Quality.* New York: Random House.

Garvin, D.A. (1993). "Building a learning organization," *Harvard Business Review, 71*: 78–91.

Gilbreth, F.B. (1911). *Motion Study.* New York: Van Nostrand.

Hall, D.T. and Parker, V.A. (1993). "Their Role of Workplace Flexibility in Managing Diversity," *Organizational Dynamics, 22* (Summer): 8.

Harvard Business Review (2000). *Harvard Business Review on Crisis Management.* Harvard Business School Press.

Johnson, R.A., Fremont, E.K. and Rosenzweig, J.E. (1963). *The Theory and Management of Systems.* New York: McGraw Hill.

Kast, F.E. and Rosenzweig, J.E. (1985). *Management: Systems and Contingency Approach.* New: McGraw Hill.

Katz, R.L. (1974). "Skills of an effective administrator." *Harvard Business Review.* 90–102.

Kendra, J. and Wachtendorf, T. (2003). "Creativity in emergency response to the world trade center disaster." *Beyond September 11th: An Account of Post-Disaster Research.* Special Publication #39. Public Entity Risk Institute, Fairfax, VA.

Koontz, H. (1984). "Commentary on the management theory jungle – nearly two decades later," in Kooontz, H., O'Donnell, C., and Weihrick, H. (Eds.) *Management: A Book of Readings,* 6th Ed. New York: McGraw-Hill.

Kotter, J. and Heskett, J. (1992). *Corporate Culture and Performance.* New York: The Free Press.

Kreps, G.A. (1991). Organizing for emergency management. In Drabek, T.E. and Hoetmer, G.J. *Emergency Management: Principles and Practice for Local Government.* Washington, D.C.: International City Management Association.

Kunreuther, H. and Roth, R.J. (Ed.) (1998). *Paying the Price: The Status and Role of Insurance Against Natural Disasters.* Washington, DC: Joseph Henry Press.

Laye, J. (2002). *Avoiding Disaster: How to keep your business going when catastrophe strikes.* John Wiley and Sons, Inc.

Lewin, R. and Regine, B. (2000). *The Soul at Work: Embracing complexity science for business success.* Simon & Schuster.

Livable Communities Initiative (2000). "Building Livable Communities: Sustaining Prosperity, Improving Quality of Life, Building a Sense of Community." Washington D.C.: U.S. Government Printing Office. http:// www.smartgrowth.org/library/gore_report2000.html

Luthans, F. (1976). *Introduction to Management: A Contingency Approach.* New York: McGraw Hill p. 28.

Mileti, D. (1999). *Disasters by Design: A Reassessment of Natural Hazards in the United States.* Washington, DC: Joseph Henry Press.

Mintzberg, H. (1973). *The Nature of Managerial Work.* New York: Prentice-Hall.

Myers, Kenneth N. (1999). *Manager's Guide to Contingency Planning for Disasters: Protecting vital facilities and critical operations.* New York: John Wiley & Sons, Inc.

Pine, J.C. (1994). *A Systems View of Emergency Response to Hurricane Andrew.* Boulder, CO: University of Colorado.

Pearson, C.M. and Clair, J.A. (1998). "Reframing Crisis Management," *Academy of Management, 23*: 60.

Roberts, K.H. and Bea, R.G. (2001). When systems fail. *Or-*

ganizational Dynamics, 29: 179–191.

Roberts, K.H. and Bea, R.G. (2001). Must accidents happen? Lessons from high reliability organizations. *Academy of Management Executive, 15*:70–78.

Rubin, C.B., and Renda-Tanali, I. (2001). The terrorist attacks on September 11, 2001: immediate impacts and their ramifications for federal emergency management. *Quick Response Report #140.* University of Colorado.

Schien, E.H. (1985). *Organizational Culture and Leadership.* San Francisco: Jossey Bass.

Shetty, Y. K. (1974). Contingency Management: Current Perspective for Managing Organizations. *Management International Review, 14* (6): 27.

Stewart, I. (1989). *Does God Play Dice? The Mathematics of Chaos.* Harmondsworth, Middlesex: Penguin Books Ltd, pg. 141.

Sugarman, B. (2001). "A learning-based approach to organizational change: Some results and guidelines," *Organizational Dynamics, 30*: 62–75.

Sutton, J. (2002). The response of faith-based organizations in New York City following the world trade center attacks on September 11, 2001. *Quick response Report #147.* Boulder, CO: Natural Hazards Center, University of Colorado at Boulder.

Tapscott, D. (1998). *Growing Up Digital: The Rise of the Net Generation.* New York: McGraw-Hill.

Taylor, F.W. (1911). *Principles of Scientific Management.* New York: Harper.

Thomas, R.R. Jr. (1990). "From Affirmative Action to Affirming Diversity." *Harvard Business Review, 68*: 107–117.

Tierney, K.J. (2001). *Facing the Unexpected: Disaster Preparedness and Response in the United States.* Washington, D.C.: Joseph Henry Press.

Toffler, A.(1985). *The Adaptive Corporation.* New York: McGraw-Hill.

Tosi, H.L., Jr. and Slocum, J.W. Jr. (1984). "Contingency Theory: Some Suggested Directions." *Journal of Management, 10*: 9–26.

Von Bertalanffy, L. (1972). "The History and Status of General Systems Theory." *Academy of Management Journal, 15*: 411.

Weber, M (1947). *The Theory of Social and Economic Organizations.* T. Parsons (Ed.) New York: Free Press.

Weber, R.T. and McEntire, D.A. (2002). Public/private collaboration in disaster: Implications from the world trade center terrorist attacks. *Quick Response Report #155.* Boulder, CO: Natural Hazards Center, University of Colorado at Boulder.

Chapter 15

Gerontology and Emergency Management: Discovering Pertinent Themes and Functional Elements within the Two Disciplines

Kathy Dreyer

ABSTRACT

In this chapter, the confluence of gerontology and emergency management is considered. Both disciplines have facets that are mutually complimentary. The chapter illustrates the psychosocial issues, practical considerations, and the applicability of gerontology within emergency management, and vice versa. Other elements that provide insights, such as future collaborations between the two fields of study, are given consideration, as well as the overlapping features of other disciplines such as public health, psychology, economics, and social work. Potential means of expansion of collaboration between the two disciplines that extends beyond typical natural disasters are also explored.

INTRODUCTION

At first blush, the nexus between gerontology and emergency management may not automatically conjure much of an ostensible affiliation. After all, the study of aging might not be generally associated with the discipline of planning for extreme events such as natural disasters. However, the coalescence of these two domains warrants examination beyond what might automatically engender consideration at first: the vulnerabilities that older adults may possess, in terms of their health status, psychological issues, utilization of services, and financial resources, placed in the context of preparing for and surviving the aftermath of life-threatening, calamitous events. In truth, there have been several connections between gerontology and emergency management; there are also overlapping characteristics between these fields within the domains of public health, psychology, social work and economics. It is advantageous to combine gerontology and emergency management for the mutual benefits each discipline dispenses, as well as to implement those assets that have not been as widely employed, such as the strengths and abilities of older adults to assist in the preparation for and the aftermath of natural disasters.

ADJUVANT BENEFITS INVOLVING GERONTOLOGY AND EMERGENCY MANAGEMENT

The field of gerontology has provided many avenues of collaboration for emergency management. Information about vulnerable populations and their need for additional assistance in the advent and wake of natural disasters is a starting point. Gerontology can offer perspectives on those older adults with special needs (medications, mobility, physical setting, monetary, ability to recover, and predisposition to stress). This information can be extrapolated for use with vulnerable populations such as disabled adults or individuals in frail or poor health. Also, gerontology can provide information on the contingency plans that long-term care facilities use in other emergency situations (coping with power outages, distribution of medications, handling and storing food or disposable products, evacuation procedures for those in wheelchairs or who are unable to mobilize independently). The study of gerontology offers multiple opportunities for providing beneficial insights into assisting vulnerable populations before, during, and after a disaster; however, the field can also serve to demonstrate the benefits of employing the unique perspectives, strengths, and skills of older adults who experienced disasters and survived. Other disciplines can provide additional perspectives within the realms of gerontology and emergency management, in terms of their relevance to vulnerability and disasters; yet, the implementation of how older adults have made accommodations for disaster-engendered losses is a technique that justifies additional research. In other words, viewing and utilizing older adults as survivors, not just victims, is an asset that gerontology can add to the field of emergency management.

PERSPECTIVE ON VULNERABILITY AND DISASTERS

Public Health

A primary consideration of older adults and their experience with natural disasters is their ability to survive with a minimum of property damage or personal injury. From this contemplation a central theme emerges: older adults' ability to exit safely in the event of an emergency. Older adults may be vulnerable in a number of ways. Their physical condition or pathological factors can affect an older person's ability to escape a disaster safely, thus making them more susceptible to the effects of a natural disaster. As Eldar (1992) points out

> elderly persons may have locomotor, sensory or cognitive impairments restricting their activities. Some impairments – such as those of sight or hearing – may limit them in perceiving warnings and emergency instructions; others will reduce their ability to carry out recommended self-protected actions (getting under tables during an earthquake or tornado-shaking) or their speed and agility in leaving a room or building (in a fire). When a disaster occurs, even without collapse or major damage, buildings and their surroundings may become unsafe and previously innocuous elements of the interior environment can become dangerous and result in injuries from broken glass, falling electrical fixtures, moving equipment and furniture. Elderly persons with functional limitations will face greater risks in this changed environment than able bodied and younger individuals. (355)

Obviously, there is a need to assist those older adults with limited functional capacity and chronic impairments; however, not all older adults have physical or health-related difficulties. According to a report from the National Center for Health Statistics (2004), 76 percent of non-Hispanic white, 59 percent of non-Hispanic black, and 63 percent of Hispanic adults, aged 65 and older, reported having health that is good to excellent. The report also demonstrates that among adults aged 65 to 74, 80 percent of non-Hispanic white, 62 percent of non-Hispanic black, and 65 percent of Hispanic origin indicate their health as good to excellent. The same report found that among those aged 65 and older; the majority does not indicate a problem with hearing or seeing. Only 47 percent of men and 30 percent of women in this age group reported trouble hearing; 16 percent of men and 19 percent of women in this age group indicated trouble seeing. These results may suggest that the majority of older adults would be capable of escaping and surviving a disaster.

The surroundings in which older adults live require assessment for potential evacuation routes and exits. Those who live alone require additional attention and support, in regard to evacuating safely, or at least having a plan in case of a disaster. Older adults, who reside in continuing care retirement communities, assisted living facilities, or nursing homes, should be able to rely on the administration of their facility to provide information and direction as to the proper procedures for escaping a disaster. Additionally, the facility should have contingency plans in place to allow for the secure departure of older adults who require additional assistance to exit the facility. In this case, emergency management planners would have an interest in the procedures and preparation plans utilized, for older adults residing alone, and those within a retirement or long-term care facility.

Emergency management personnel responding to disaster situations must be especially careful with those older adults suffering from a variety of ailments who require assistance with keeping track of medications or other items mandatory for maintaining their health. Personnel must also be able to communicate effectively with individuals suffering from visual and auditory impairments, in addition to the emotional response that is bound to occur when a person's home and belongings are damaged or destroyed. The field of emergency management can certainly provide strategies and guidelines for assisting older adults who need help planning for disasters, especially those who have physical impairments or cognitive challenges, as well as those who live alone. There is some information available about emergency procedures relative to older adults living independently at home who may have mobility limitations, physical difficulties, or visual impairments (American Red Cross, 2004; United States Administration on Aging, 1995), to say nothing of the condition of their home environment, and its ability to provide a quick exit in an emergency. These factors can certainly influence a person's ability to respond to a disaster.

Secondly, with advancing age, changes in health status occur. Lung capacity is reduced, which can be problematic in the event of dust and particles in the air after a disaster. A change in hearing is also common as people age. The effect of hearing deficits can impair older adults' ability to hear certain tones or discern announcements given in an area with substantial background noise, which can prevent them from responding to emergency sirens. Another change accompanying aging is that of smell. There may be an inability for older adults to detect spoiled food and, inadvertently consume it (United States Administration on Aging, 1).

Susceptibility to illnesses is another component. Older adults may be more prone to the after-effects of a natural disaster. Tanida (1996) states that after the 1995 earthquake in Hanshin, "many elderly people caught diseases such as 'shelter pneumonia' because of the unhealthy environment and dehydration" (1134). In the event of non-functioning air conditioning or heating units after a disaster, older adults may not be able to perceive the changes in temperature. This can result in hyper/hypothermia. The field of public health could assist both gerontology and emergency management professionals by assisting planning efforts to minimize the effects of disasters, especially those related to environmental hazards.

The effort to alleviate the consequences of natural disasters is the first step in assisting older adults to effectively cope with disasters and remain as physically healthy as possible. Ollenburger and Tobin (1999) indicate that "individuals in poor health and who have difficulty getting around are restricted in the actions they can take to mitigate hazard losses, which can lead to higher stress levels" (66). In addition to preserving physical health, maintenance of older adults' mental health is of crucial importance, given the elements of preparing for and surviving the aftermath of a disaster. There are many components of mental health and vulnerability that are unique to older adults, which compels the use of psychology and its expertise.

Psychology

Older adults can be extraordinarily vulnerable to the effects of a natural disaster, in terms of psychological response. As Langer (2004) illustrates,

> most older people will inevitably suffer multiple losses. Some will need to learn the roles and responsibilities of becoming widows or widowers. Other losses might include loss of mobility or independence. Older adults who develop health problems will necessarily have to adjust from an unrestricted lifestyle to one with greater confinements. As individuals experience a diminution in control and an inability to solve their problems independently, fear – such as fear of the onset of additional health problems, of living alone, or of becoming a crime victim – can become a major concern. How each person copes with these life changes will often depend on individual differences as well as the "modus operandi" that has been the vehicle for overcoming crises in the past (277).

Given the accumulation of other losses, another loss, especially one of this nature, can be more traumatic and devastating. After the Hanshin earthquake in 1995, "suicide by elderly victims occurred at a rate of almost once a month. In addition, some elderly people were found dead in a ditch, presumably because they were lost in a new territory" (Tanida, 1996, 1134). Healthy psychological responses to disasters are important to surviving a disaster, and all of the losses intrinsic to it. Oriol (1999) points out "apathy or helpless stoicism may be among the likely reactions, based largely on the attitude that they will never be able to recover or replace losses ranging from property damage to death of friends or family" (28). Being able to safely evacuate a home environment would certainly improve a person's response to a natural disaster; not knowing how to escape or where the safest place in the home is to survive when a disaster strikes can certainly result in psychological trauma, and can affect the ability to cope with additional losses or in the event of extensive damages to the home environment. The natural state of an older adult's home is also a factor to consider. The home environment should be evaluated for its potential evacuation routes, safety, and ability to withstand impending damage. The safety of the older adult to remain in the home environment is a factor, especially if he/she does not want to leave.

Another significant factor in extenuating losses involves an older adult's financial resources. Langer (2004) notes that "economic status is a powerful determinant of an elderly person's ability to cope; intelligence and education also play a considerable role, as does an individuals' degree of physical and mental health" (278). As such, the realm of economics provides insights into surviving a natural disaster.

Economics

Another potential vulnerability is the financial status of older adults who may not have the means to rebuild or recover from the effects of a disaster that destroys their home and belongings. Those who are already in dire financial straits can suffer more than economic losses. In the devastating 1995 Hanshin earthquake, Tanida (1996) illustrates that "77 percent of victims were crushed to death, and a greater proportion of elderly people died from burns or penetrating injuries. Many of the elderly people lived in inexpensive old wooden houses of their own or in tenements" (1133). To make matters worse, according to Tanida, "elderly people tended to sleep in ground floor rooms, which were especially prone to collapse in this earthquake" (1133). There were over 6,000 deaths caused by the quake, and "about half of the deaths were people aged 60 and older" (Hirayama, 2000, 115). In terms of ability to rebuild and recover from the devastation caused by this earthquake, those older adults who survived faced enormous obstacles. Many of those directly affected could not rebuild for financial reasons, and Hirayama (2000) states "a problem in the disaster area is not the shortage of loans but the lack of credit among the elderly victims" (125).

Typically, retired older adults are living on fixed incomes, and do not have financial means to replace belongings in the event they are destroyed. In particular, older women are especially vulnerable to the effects of financial loss, as they are more likely to be without a pension, especially since they may not have worked for pay during their young adult life. Also, they usually live longer than their male counterparts, suggesting they need more financial resources for survival. Women may suffer more financial devastation if their house is destroyed from the effects of a natural disaster. Ollenburger and Tobin (1999) indicate

> the economics of aging place many women in extremely vulnerable positions which influences their ability to cope with the unexpected consequences of natural disasters. This economic vulnerability of women, who may be the sole support for themselves and/or for their children, had been described as the feminization of poverty. Consequently, natural disasters can perpetuate the poverty trap for women as demonstrated by some of the recent research looking at women and other marginalized groups in hazardous areas (66).

Another part of the problem is that older adults who experience losses of any kind may downplay or their difficulties due to the level of distress caused by having to admit a problem exists. A failure to handle their problem independently makes them reluctant to acknowledge help is needed (Langer, 2004). The devastation caused by a disaster, financially speaking, leads to an examination of what services are available and, more importantly, used, by those in greatest need after a natural disaster strikes.

Social Work

An analysis of the utilization of services designed to assist victims of natural disasters yields unsatisfactory results. For example, in the aftermath of the Hanshin earthquake, Tanida (1996) points out that "elderly people tended not to proclaim their problems unless they were questioned specifically. Thus, the superficial survey by the medical teams failed to notice the problems of elderly people in shelters" (1134). Additionally, as Tanida continues, those who provided care for the older adults were also victimized by the disaster and rendered unable to offer their usual services. Unfortunately, the situation was exacerbated to the extent that those older adults were receiving little, if not inadequate, care for an extended period of time (Ibid, 1135).

The viewpoint that older adults do not seek care after disasters is supported by Langer (2004), who states, "the at-risk elderly do not usually self-refer. If they receive help, it is typically because someone else obtained it for them" (278). There are many reasons for a low utilization rate of services by older adults trying to recover from the effects of a natural disaster. Langer (2004) further explains that

> American culture places a high premium on independence and self-reliance, and, as a result, people feel uncomfortable when they need to ask for assistance. For the current cohort of older adults who survived such hardships as the Depression, this admission might be difficult. Maintaining independence is one of the most frequently occurring life themes. Many older people worry that they will lose their independence and their value in society through reliance on a fixed income or through a loss to maintain their own home (279).

Another difficult aspect of utilization of services involves forced relocation and its relevance to social work. Sanders, Bowie, and Bowie (2003) suggest "geriatric social workers are often a frontline service for older adults who have experienced a forced relocation following a natural disaster. Additionally, the social work profession is involved in the long-term problems that forced relocation can cause for older adults" (25). Forced relocation is especially traumatic for older adults who rely on informal care for support, such as African Americans. A lifetime spent developing a support system within a community, can be abolished in the event a person has to move to another location in order to have a home. For those whose experience with the health care system raises suspicion and distrust, the change in their home is even more problematic.

In light of the health, psychological, economic, and service utilization issues, there are many prospects for the field of emergency management to converge with gerontology to improve not only how older adults experience, survive, and move beyond natural disasters, but how they can and will help others coping with the same event.

RECOMMENDATIONS FOR EMERGENCY MANAGEMENT

First and foremost, the field of emergency management could support gerontology in an effort to apply practical guidelines for older adults in the event of natural disasters. A guide or directions for preparing for the advent or aftermath of an emergency could be developed collaboratively, and could involve public health professionals, especially in the case of long-term care facilities. Saliba, Buchanan and Kington (2004, 1436) note that "nursing facilities often are overlooked as a health resource and generally are not incorporated into disaster-relief plans." To exacerbate this problem, nursing homes may experience an overload in capacity of residents in the event that hospitals are damaged by a disaster and hospital personnel attempt to transfer their patients to the safety of a facility (Ibid, 1438). Again, social workers frequently encounter this scenario as they attempt to assist older adults in their quest for assistance. Given the fact that public health, social work, emergency management, and gerontology professionals have a stake in this situation, it is only natural that these disciplines work together to address and resolve this critical situation.

The discipline of emergency management could provide the impetus for the gerontological field to focus more energy on assisting older adults before, during, and after events such as these. Most of the gerontological literature and research appears to be directed at the psychosocial effects of natural disasters on older adults, but there is little, if any, information provided on how to prepare for an event. There is no doubt that natural disasters can be traumatic, devastating, and tragic, especially for older adults. Concentrating more information as to how best to prepare beforehand would go a great way in helping to mitigate the effects of such an episode.

Also, the field of gerontology would certainly help the emergency management discipline address the unique effects a natural disaster can impose on older adults. The unique health, psychological, economic, and service-related issues facing older adults before, during, and after a natural disaster, could be mitigated through the collaboration of gerontology and emergency management professionals to address those issues.

Finally, older adults who have coped with natural disasters and the losses inherent to them can certainly provide insights into how to help other elders survive these events. By working with these individuals, emergency management professionals can determine the most effective and beneficial means to assist those older adults in need of assistance when disasters occur.

FUTURE AREAS OF INVESTIGATION

A realm in which gerontology would benefit from the field of emergency management is the handling of disasters that affect older adults more frequently. Extreme temperatures, both hot and cold, have a significant impact on older adults, as the ability to perceive changes in temperature decreases with age. The heat wave that occurred in Chicago (1995) provides a stark reminder of how vulnerable older adults that live alone can be (Klinenberg, 2002).

Also, the prevention of fires in the home, particularly in older adults' homes would be useful. According to the United States Fire Administration (2001), fires and burns are the sixth leading cause of death among older adults, who are also "far more likely than the rest of the population to die or be injured in a fire" (4). The health hazards posed by this type of emergency, to say nothing of the financial and emotional costs, are worthy of consideration. In any case, the classification of what constitutes a disaster can be expanded to include droughts, freezing temperatures, and fires. These scenarios can have a psychological and financial effect on older adults, as well as affecting their health, and usually generates a need for assistance, similar to other natural catastrophes.

Although these calamities typically do not generate as much attention or publicity as a hurricane, tornado, or earthquake, older adults are still at risk. The efforts of gerontology and emergency management experts could certainly provide much needed solutions to these and other dangerous situations. Despite the importance of assisting older adults after calamitous events, it is critical to also understand the unique strengths, talents, abilities, and resources that older adults possess, and not simply believe they are only victims of natural disasters. Maintaining a balance between identifying the singular needs of older adults after a natural disaster has occurred and meeting those needs, while also realizing the strengths and abilities of these older adults for their potential to assist in the aftermath of a natural disaster, is imperative.

Most importantly, older adults can also be recruited for their expertise from experiencing disasters in the past. Additionally, there are many older adults who volunteer for the American Red Cross; in fact, "sixty-five percent of American Red Cross volunteers are age fifty-five or above" (Oriol, 1999, 31). Based on the willingness of older adults to volunteer, not to mention their experience with disasters, it seems quite natural to deploy them into service after a disaster. Older adults offer strength, familiarity, and assurance to other older adults, especially if they have previously experienced and survived a natural disaster. By their nature, older adults are survivors. As the Project COPE (1992) "Voices of Wisdom: Seniors Cope with Disaster" Videotape so eloquently stated:

> Senior citizens today are a sturdy, reliable generation. We've proven time and again our ability to survive everything from the Great Depression to world wars and the threat of nuclear holocaust. We've lived through droughts, floods, and all sorts of other natural disasters. We've given birth, supported our families, and stood by our loved ones through personal and financial losses. We are proud, tough, and resilient.

The discipline of emergency management can provide the field of gerontology with an opportunity to promote one of the country's best natural resources – older adults – to assist others after a natural disaster.

REFERENCES

American Red Cross. (2004). Disaster Preparedness for Seniors by Seniors. Retrieved on November 8 from (http://www.redcross.org/services/disaster/0,1082,0_9_,00.html)

Childers, C. (2003). Age and Disaster Vulnerability. In Federal Emergency Management Agency, Higher Education Project's Social *Vulnerability Approach to Disasters* Retrieved January 26, 2005, from: http://training.fema.gov/emiweb/edu/completeCourses.asp

Drabek, T.E. (1999). Understanding disaster warning responses. *Social Science Journal, 36*(3): 515–523.

Eldar, R. (1992). The needs of elderly persons in natural disasters: Observations and recommendations. *Disasters, 16*(4): 355–358.

Hirayama, Y. (2000). Collapse and reconstruction: Housing recovery policy in Kobe after the Hanshin great earthquake. *Housing Studies, 15*(1): 111–128.

Kelly, J. I. (2003). How to develop a disaster action plan for older, distant relatives. *AARP Bulletin.* Retrieved on November 8, 2004 from (http://www.aarp.org/bulletin/yourlife/Articles/0505_sidebar_11.html)

Kilijanek, T. S. and Drabek, T.E. (1979). Assessing long-term impacts of a natural disaster: A focus on the elderly. *The Gerontologist, 19*(6): 555–566.

Klinenberg, E. (2002). *Heat wave: A social autopsy of disaster in Chicago.* Chicago: University of Chicago Press.

Langer, N. (2004). Natural disasters that reveal cracks in our social foundation. *Educational Gerontology, 30*: 275–285.

National Center for Health Statistics. (2004). *Health, United States, 2004.* Hyattsville, MD.

Ollenburger, J.C. and Tobin, G.A. (1999). Women, aging, and post-disaster stress: risk factors. *International Journal of Mass Emergencies and Disasters, 17*(1): 65–78.

Oriol, W. (1999). Psychosocial issues for older adults in disasters. Retrieved September 30, 2005 from http://media.shs.net/ken/pdf/SMA99-3323/99-821.pdf

Poulshock, S. W. and Cohen, E.S. (1975). The elderly in the aftermath of a disaster. *The Gerontologist, 15*(4): 357–361.

Project COPE. (1992). "Voices of Wisdom Videotape and Brochure." Ventura County, California.

Saliba, D., Buchanan, J., & Kington, R.S. (2004). Function and response of nursing facilities during community disaster. *Research and Practice, 94*(8): 1436–1441.

Sanders, S., Bowie, S.L., and Bowie, Y.D. (2003). Lessons learned on forced relocation of older adults: The impact of Hurricane Andrew on health, mental health, and social support of public housing residents. *Journal of Gerontological Social Work, 40*(4): 23–35.

Shoaf, K. L., Sareen, H.R., Nguyen, L.H., and Bourque, L.B. (1998). Injuries as a result of California earthquakes in the past decade. *Disasters, 22*(3): 218–215.

Tanida, N. (1996). What happened to elderly people in the great Hanshin earthquake. *British Medical Journal, 313* (7065): 1133–1135.

Ticehurst, S., Webster, R.A., Carr, V.J., and Lewin, T.J. (1996). The psychosocial impact of an earthquake on the elderly. *International Journal of Geriatric Psychiatry, 11*: 943–951.

Tyler, K. A., & Hoyt, D.R. (2000). The effects of an acute stressor on depressive symptoms among older adults. *Research on Aging, 22*(2): 143–164.

United States Administration on Aging, & the Kansas Department on Aging. (1995). Emergency preparedness manual for the aging network. March 1995.

U. S. Fire Administration. (2001). Older Adults and Fires. Volume 1, Issue 5; December 2001.

Vogt, B.M. (1991). Issues in nursing home evacuations. *International Journal of Mass Emergencies and Disasters, 9*(2): 247–265.

Chapter 16

Public Health and Medicine in Emergency Management

Richard A. Bissell

ABSTRACT

Public health and medicine have many parallels with emergency management in terms of overall goals, basic conceptual models, and many operational modalities. The health sector and emergency management (EM) both strive to protect the public from maladies using organizational and science-based tools. Both health and EM are multidisciplinary in their scientific underpinnings, although health has gone much further than EM in developing its own scientific disciplines. Despite many common values and foci, as well as having many similar operational characteristics, health and emergency management have mostly failed to share their tools and personnel and have not collaborated smoothly in preparing for, and responding to mass emergencies. This chapter addresses the medical/public health approach to addressing threats to human well-being, and outlines specific tools that health and emergency management need to share in order to enhance their future collaboration toward better population outcome in emergencies and disasters.

INTRODUCTION

If we ever thought it was "acceptable" that emergency managers did not know much about the health sector and how it responds to threats and real events, that time abruptly and permanently disappeared with the recognition of bioterrorism as a serious hazard. This chapter provides an overview of the basic concepts upon which medicine and public health practitioners base their work, and their use of science to establish appropriate interventions when health is threatened. We establish similarities between EM and health, and describe some of the differences in operational modalities, hierarchical structure, and approaches to thinking about emergencies. We conclude the chapter with an overview of specific lessons EM may take from the health sector, and ways in which EM can help the health sector become more efficient at emergency preparedness and response. The goal of the chapter is to help the emergency manager understand how to productively interact with the health sector as a full and meaningful partner.

OVERVIEW OF PUBLIC HEALTH AND MEDICINE

Problem Orientation

Medicine and public health are complementary disciplines of science and application within the larger field of health care. In the Western world, allopathic medicine typifies what we think of as representing the discipline of medicine: the application of science-based techniques and technologies to the art of healing individuals who have become ill or injured, and use of similar technologies to help keep individuals from becoming ill. The science disciplines incorporated include many aspects of biology, chemistry, physics and psychology. Public health differs from medicine in that it has as its focus a population of people, not individuals. It is also science-based and multidisciplinary, including applications of medicine, epidemiology, biostatistics, sociology, anthropology and psychology. The field of public health bases much of its work on strategies of prevention, and when prevention cannot be achieved, mitigation of illness and injury is pursued through the efficient and effective application of medical and other societal resources to limit or reverse pathological processes. The boundaries between medicine and public health are often blurred, even to those who work in health care. Good medical practice can contribute significantly to the public's health, and good public health practice can contribute significantly to the effectiveness of medical resources. Though the stereotype is limited in many specific applications, the general concept is that public health is prevention- and population-oriented while medicine concentrates on the process of curing individuals.

Both medicine and public health are problem-oriented. Medicine, at least in some primitive form, has been around ever since Homo Sapiens recognized that they could intervene with ill or injured individuals to comfort them or help them heal. Public health, as a specific endeavor, first came about as a result of large numbers of people becoming seriously ill at the same time during the great plagues of renaissance and early industrial Europe. Early tools included quarantine and isolation. The development of biostatistical methods (John Snow, 1855) and germ theory (Green and Kreuter, 1990) led to major increases in the power of public health to make a significant impact on disease mitigation. While the tools and techniques of public health have widespread application in acute, chronic, and slow-onset conditions, the very basis for developing public health came from the need to respond to health disasters.

Conceptual Models

The practitioners of all disciplines conduct their work based on conceptual models that define the discipline's understanding of causal relationships between the phenomena that are of concern to the discipline. These conceptual models help determine where it is that participants in the discipline believe they can make interventions that may help improve life, or productivity, or income. For example, one of the core elements of modern medicine is the role that microbes ("germs") play in initiating and sustaining what we call infectious diseases . . . that is, that certain microbes are capable of invading the body and causing reactions that can make us quite ill, or even kill us. This germ theory component of the conceptual model of medicine and public health allows us to develop interventions against the microbes as a way of both preventing and curing certain illnesses. "Germ theory" is but one of several conceptual models that help form the thinking of health care practitioners. Others worth mentioning here are:

HUMAN-ENVIRONMENT RELATIONSHIPS. From the times of Socrates, and perhaps even earlier, health care practitioners believed that the relationship between humans and their environment affected human health status. Although we do not today focus primarily on the elements of earth, air, fire and water as determining the health of populations, they are among the many variables that contribute to our health. Since we derive our most basic life-supporting substances from the physical environment, such as air, food, and water, it is clear that the quality of the environment has a primary effect on human health. Environments that do not provide sufficient quantity or quality of these life-supporting substances will result in poor human health, or even the inability to support human life. On a more subtle level, deviations from normal balances in the environment can lead to deviations in human health (McMichael and Powles, 1999; Hubert, 2001).

The relationship between humans and their environment is two-way. Humans can, and do, change and affect the physical environment we live in. One way we do this is to create artificial environments (the human-built environment of structures). Another way is by physically changing topography, vegetation patterns, water flow, or by depositing in the soil, air or water chemicals and other substances that would not naturally exist there. Whether in the natural or human-built environments, the way we treat the environment affects its ability to support our health.

MULTIPLE CAUSALITY. Good health (or bad health) is not the result of a single cause (or independent variable, to use the research terminology). In order to remain healthy, we need an adequate quantity of a broad variety of nutrients, clean air and water, physical exercise, social connectedness, and good choices (our own behavior) (Wrosch and Scheier, 2003). Any one variable can bring a person's health rapidly down, such as a bullet to the brain or ingestion of a quantity of cyanide, but these single variables are most often related to other variables, often behavioral in character. For example, dealing illegal drugs increases the probability that one will end up with a bullet in the brain. Good health is clearly the result of multiple things going right; bad health is also usually the result of multiple things going wrong.

EXPOSURE AND VULNERABILITY. In order to contract an infectious disease, you need to be exposed to the microbe that causes the disease. However, some people are exposed and never become ill, while others may die from the same exposure. If we call the person who is exposed a "host," the host may have certain vulnerabilities or strengths that alter the outcome of the exposure. The host may have inherited genetic traits that limit his or her vulnerability to a certain class of microbes, or may have previous experience with the specific microbe, and thus have an immune-response system that is poised and ready to fight off the microbial invader.

Exposure to a given substance may have completely different effects on health, depending on the quantity of the exposure. For example, we cannot live without water, but we also know that ingestion or retention of too much water can make us seriously, even fatally, ill. Many of the medications we use have a dose at which they are effective, and a level at which they are toxic. Exactly what levels are effective differs according to host-intrinsic factors such as metabolic rates or the health of the individual's liver and kidneys (Oskvig, 1999). Thus, throughout medicine and public health, we recognize that both external and internal factors affect how healthy we are, even in apparently identical environments. Our behavior affects both our exposures (e.g., smoking, alcohol, overeating) and our internal capacities (e.g., exercise, sleep and nutritional patterns affect the effectiveness of the immune-response system) (Zacny, 1990; White, Johnson and Buyske, 2000).

PRIMARY, SECONDARY, AND TERTIARY PREVENTION. Much like emergency management, public health conceptualizes various levels of activity designed to prevent or decrease potential harm to humans. In public health terms, *primary prevention* is the act of making sure something will not happen. It is equivalent of *prevention* in emergency management. Vaccination is a public health example of primary prevention . . . it prevents disease from occurring even if exposure happens. *Secondary prevention* means minimizing the harm that occurs once a disease or injury affects an individual or population. In medicine this is usually seen as curative care; in public health it may mean using epidemiologic and disease control tools to minimize the spread of an epidemic. Note that this concept includes components of the emergency management concepts of *mitigation* and *response.* The public health concept of *tertiary prevention* refers to actions taken to help individuals who have been injured or ill to regain full capacity to live normal lives. This is similar to components of the emergency management concepts of both *recovery* and *rehabilitation.* As is the case with emergency management, many health care practitioners will incorporate into their recovery and rehabilitation programs aspects of primary prevention, such as the program at Maryland's Shock Trauma Center designed to prevent trauma victims from becoming repeat patients, by teaching them life skills that will enable them to support themselves without having to deal drugs on the street.

Medicine and public health both incorporate the concepts just described, and, like emergency management, both strongly subscribe to the concept that things happen for a reason . . . that is, that there are causal relationships that lead to maintenance of, or deviation from good health. Health care personnel, like emergency managers, believe that well-placed interventions into a known causal relationship can change the course of events and prevent or minimize human harm. Research is needed to clarify the causal relationships and test the effectiveness of proposed interventions.

Operational Modalities

Whereas emergency management may be seen as being primarily organized and operated by government entities, with some action taken by private commercial organizations and individuals at the neighborhood and family level, medicine and public health in the United States are much more fractured and complicated in their structure and organization.

In the United States, medicine is usually a private service provided by physicians, nurses and other health care personnel within a privately-owned and operated entity. It may be as small as a solo-practice family physician's office, or as large as the Kaiser-Permanente health maintenance organization. Because government funds a substantial portion of medical care through programs such as Medicare and Medicaid, it has an important regulatory role in structuring and supervising medical services, but only directly supplies a small proportion of the medical care Americans receive (Centers for Medicare and Medicate Services, 2005). Within the private organization of medicine, there is an extreme variation of management styles and hierarchical structure. Hospitals tend to be structured with clear authority lines, while clinics and private practices may be much more horizontal in their structure. In terms of one-on-one patient care, the physician leads the decision making; however, government is increasingly limiting the options from which the physician can choose through control of reimbursement and approved treatment regimens. Nevertheless, at its core, medical care decision making tends to be autocratic, although subject to review and suggestions from others.

Organizationally, public health more closely resembles emergency management. Most public health leadership and major decisions take place within a government agency, although individual actions may take place in private agencies, schools, neighborhoods and families. While medicine and public health are both science-based, the breadth of public health is generally thought to be too great for unilateral autocratic decision making, resulting in a decision-making model that is much more based on research and scientific consensus. Scientific consensus is often based on research that is statistically reported as significant at the 95 percent confidence interval. This lengthy decision-making process clearly has implications for EM interactions with the public health establishment during emergencies.

Public health authorities have ultimate responsibility for virtually the entire health sector, and are the primary policymakers and regulators in health care. Hospitals and outpatient care practitioners fall under public health regulations. Health departments are responsible for a great variety of health activities, often organized into groupings centered around maternal and child health, disease control and disease prevention, food and water safety, epidemiologic services and investigations, environmental health services, occupational safety, licensure of health care practitioners, laboratory services, vital records, and health education. Additionally, the federal Centers for Disease Control and Prevention, and the U.S. Public Health Service now require health departments to construct and operate a jurisdictional health sector disaster plan, to help improve the interaction within the various parts of the health sector during emergencies, as well as interaction between the health sector and emergency management.

HISTORY OF THE HEALTH SECTOR AND DISASTERS

The relationship between the health sector and disasters is both central to the development of key components of the health sector, and is a core component to organized human response to disasters. Long before there was organized emergency management there were attempts to organize the health sector to both respond to disaster events, and prevent epidemics.

Many of the most deadly disasters in human history have not been sudden-onset cataclysmic events like earthquakes or tsunamis, but rather epidemics. The Black Plagues of the Middle Ages and Renaissance variously took between a third and half of all Europeans living at that time to an early death (The Black Death, 2005; The Black Death, 1348; Ziegler, 1991). Millions died, including entire populations in some areas of Europe. Even within the last century, a pandemic (widespread epidemic) of influenza took more than 20 million lives (Kolata, 2001). As Europeans in the late Middle Ages and Renaissance came to discard notions that all their tragedies came as a result of displeasing God, they looked for understanding of the causal relationships that might exist on the physical plane, relationships in which they could intervene. Long before "germ theory" became accepted in the 1800s, health practitioners recognized that diseases could be transmitted from one individual to others. This recognition led to some of the earliest public health tools, isolation and quarantine, which still remain powerful tools that we may have to employ in future mass outbreaks.

During the extended time of the great plagues, government came to change its ethic regarding its responsibility to citizens (Defoe, 1722). It was no longer sufficient to just protect citizens from foreign invasion or domestic crime. Governments now recognized a responsibility to do what they could to organize response to epidemics that took more lives than war and crime combined (McNeill, 1976). Early attempts at data collection and analysis had the aim of keeping track of where deaths were occurring and at what volume, often based on parish-level statistics. Tracking these data helped demonstrate the transmissibility of contagion, and frequently showed densely populated cities to be at higher risk than sparsely populated rural areas. Based on this information, governments were in a better position to suggest steps citizens might take to protect themselves. Moving on into the 1800s, northern German states created "Sanitätspolizei" (sanitation police), whose responsibility it was to ensure that rules regarding cleanliness and the handling of food and waste were followed, to ensure the public's health (Haase, 1989). In the 1850s, British physician John Snow used data collected on victims of a serious cholera outbreak in London to detect the location of cholera transmission (two contaminated water taps) and, thus, create an intervention that worked (closed the water taps) (Crosier and Snow, 2005). The number of new cases in that area of London virtually disappeared after the taps were closed, thus demonstrating the strength of combining statistical analysis and medical knowledge to control epidemics. With his book, Dr. Snow (1855) created the new field of epidemiology, and greatly strengthened our ability to intervene against major threats to population health, even when we sometimes do not know the exact causal relationships.

War and Public Health

It was not only in the realm of contagion that the health sector made major contributions to disaster response. Warfare tops the list of human-caused disasters and has, since the dawn of civilization, provided ample opportunity for humans to devise ways of responding in an effective and organized fashion to large numbers of injured victims requiring care at the same time. While Napoleon's chief army surgeon Dominique Larrey is often credited with the development of "triage" and other tools for bringing order to the previously chaotic activity surrounding the response to the injured on a battlefield in the early 1800s (Brewer, 1986; Rignault and Wherry, 1999), there are historical reports of Roman field physicians developing and practicing similar techniques almost 2,000 years earlier. Larrey's principles and techniques were utilized in the American Civil War (Brewer, 1986; Barnes, 1900) for the benefit of wounded combatants, and by numerous militaries during the First and Second World Wars. It was not until the end of World War II that military and civilian health care providers began applying what had been learned on the battlefield to civilian casualties of the war, and later to victims of natural disasters.

By the end of the Korean War, some of the improved practices of wartime medicine filtered back to U.S. civilian application in both routine emergencies and disasters. These new techniques included both organization of response forces, as well as numerous new techniques for medically managing hemorrhaging, airway insufficiency and shock. It is important to note that warfare has also driven significant developments in public health, including the development of vector control methods (e.g., mosquito control), mass food distribution, and emergency water supply management.

During the post-War period, particularly beginning in the 1970s, civilian response to medical needs of disaster victims began to take a two-stage approach, based on the timeliness of the victims' needs. The first stage is characterized by mass application of trauma care personnel and resources, followed typically within a day or two by the introduction of public health personnel who begin addressing the needs for potable water, sanitation, shelter, reliable food, disease control, and mental health care. Furthermore, in the 1970s researchers began to use the tools of epidemiology to describe and evaluate the true health consequences of various types of disasters (Lechat, 1976; De Ville de Goyet and Lechat, 1976; Western, 1972) and later used this same information to predict what kinds of health care resources would be needed for response to particular disasters, as well as evaluate the adequacy of completed responses (Bissell, 1984). This process is still underway, as the health sector tries to remain true to its scientific foundations while striving to improve its preparedness for, and response to disasters of many kinds.

The terrorist attacks of 9/11/2001 and the subsequent anthrax episodes brought a new level of awareness to government as a whole, and specifically to the health sector, regarding the primary roles of public health and medicine in our nation's preparedness and response to disasters. The federal government became aware of the lack of coordination between the health sector and emergency management, and recognized the paucity of funds available to public health sector emergency preparedness. Progress is being made in this regard, but it will take years of funded effort before we have the levels of proficiency and equipment needed to be truly "ready."

One of the barriers to readiness is the traditional lack of communication, common vocabulary, or even trust between the public health and emergency management communities. Public health personnel typically have no training in the vocabulary and concepts of emergency management, and tend to make their decisions in a time-consuming consensus process. Recent research demonstrated that emergency management personnel feel distanced from health sector personnel who typically have higher social and academic ranking (Beissell, Pinet, Azur and Paluck, 2004). Emergency management personnel also lack understanding of the concepts and operations of the health sector (hence the need for this chapter you are now reading!). While there is increasing cooperation between the health and emergency management sectors in terms of preparedness and response to rapid onset disasters, there has be no input from emergency management to date in some of the world's major slow-onset disasters, such as the HIV/AIDS epidemic in some parts of Africa.

HEALTH AND EM CONVERGENCES AND DEPARTURES

While the organized health sector is much older than organized emergency management, EM and health have developed along similar lines. Both hold as their core value the protection of humans and promotion of their well-being in the face of hazardous events. Both fields of work base their core conceptual model on an understanding that human well-being is a function of a multidisciplinary causal string of events. That is to say, our well-being is the result of access to a healthy combination of food, water, clean air, and shelter, as well as social support, education, and so forth. Both fields recognize that access to those resources is affected by the application of politics, economics, the physical sciences, engineering, and the social and behavioral sciences. Both fields also recognize that departure from well-being, whether in an individual or a population, is also the result of a causal string of events, and that understanding the causal relationships can lead to strategies for targeted interventions that can protect people, or help them recover.

Both EM and public health see their work as taking place in stages that overlap and reinforce each other. While the basic concepts of the stages are virtually identical, the vocabulary is different . . . with the public health vocabulary of primary through tertiary prevention being a bit difficult to crack open for outsiders. Both sectors recognize that human vulnerability is a combination of external factors, the hazards to which we are exposed, and internal factors, such as the status of our immune response systems or, at a community level, the design of our structures or status of our public safety services.

Emergency management and the health sector both base their planning and preparedness activities on accruing and using information regarding natural and human-caused phenomena that can harm people . . . with increasing attempts to share this information with others outside of EM or health. While emergency management and health both use science to acquire and analyze the information needed to plan and execute effective interventions, EM could learn a lot from the health sector's use of science to study the effectiveness of it's own actions. In medicine, so-called "evidence-based" practice has started to make significant inroads in the effort to abolish treatment modalities that are ineffective (Rivara, Thompson and Levy, 2004).

While both fields use scientific findings to inform their planning and preparedness activities, public health has recently moved toward full utilization of scientific research on the exact effects of different kinds of disasters on human beings in order to tailor the response as precisely as possible (Bissell et al., 2004). Emergency management has lagged behind in incorporating these findings that come from health-based researchers, although it has at least a 30-year history of incorporating findings from geography, geology, meteorology, architecture, hydrology, and sociology. The annual Natural Hazards meeting at the University of Colorado perhaps best typifies the ongoing effort to make emergency management a field that is multidisciplinary and science-based. If EM were to include in its understanding of the human consequences of disaster more of the findings coming from public health researchers, it would enhance collaboration between the public health and emergency management sectors when dealing with events as broad in scope as hurricanes and bioterrorism. Emergency managers would be better able to gear up for public health requests, and would better understand why health officials set certain priorities.

A key task for response managers is to know how much of what kinds of human and material resources are needed, and to where they should be sent. Emergency managers typically request a damage assessment and then calculate needs from that information. Health response managers have moved to incorporating the tools of epidemiology to conduct a *needs assessment* rather than just a damage assessment, based on a combination of information sources, including data purposely collected using rapid survey techniques, and data from on-scene clinicians and public safety personnel. These data are calculated, based on epidemiologic data reported from other disasters of the same type, to predict the kinds of health care that will be needed and in what volume. Recent rapid needs assessment data collection tools also include mechanisms for assessing gaps in available resources, and allows for estimates of the validity of the information upon which the assessments are made (Wetter and Bissell, 2004; Malilay, Flanders and Brogan, 1996). EM can benefit from adopting some of the techniques used by public health in conducting rapid needs assessments, and the overall response will benefit if EM and public health do a better job of real-time sharing of their event-specific information.

In emergency field operations, it is typical that emergency managers and health sector personnel will each assume capabilities and willingness on the part of the other, often without confirming either. Joint planning is intended to help erase this difficulty, but the effort to overcome this problem will be greatly enhanced if personnel from each sector take the opportunity to come to understand the different authority structures, organizational cultures, and operational modalities and capabilities of the other sector. Neither public health nor EM has allocated adequate research money or course preparation time to effectively analyze differences and impediments to smooth collaboration between the sectors.

Two barriers have been identified by this author and are currently under investigation with some minor funding from the Centers for Disease Control and Prevention and the Maryland Department of Health and Mental Hygiene (Bissell et al., 2004). The first barrier is a social distance that was reported by emergency managers in a multiregional survey, in which respondents indicated that they had a difficult time working with health sector personnel who come to the table with advanced degrees and an air of superiority. The second barrier was identified during the 2001 anthrax scares, in which it became obvious that public health personnel and emergency managers have *very* different decision-making styles. Public health's slow, deliberate consensus-building approach to making decisions is the antithesis of what is often needed in a rapidly developing emergency. Barbera, Macintyre, Bissell and others are currently researching and developing courses to help train public health personnel in the art and science of emergency decision making, using some of the concepts developed and taught at the Emergency Management Institute for emergency managers.

SPECIFIC PUBLIC HEALTH TOOLS POTENTIALLY HELPFUL TO EM

In this section we look at some of the tools that have been developed in the health sector that may prove helpful for emergency management. Because this chapter is designed for EM personnel, we are not including here the many EM tools that would benefit the health sector.

RNA. We have already been mentioned above the need for the development and utilization of mechanisms for conducting *rapid needs assessments*, instead of relying on the much less illustrative damage assessment currently commonly in use by emergency management personnel. While pioneered by public health personnel, the move from damage assessment to a widely useful needs assessment tool for emergency managers would require input from several sciences in order to enhance predictive power.

HORIZONTAL INFORMATION SHARING. Public health has always required a certain amount of information sharing between health agencies, particularly related to the so-called "reportable diseases." However, this process has tended to be vertical, with only the top-level agencies receiving all the information. As it became clear that bioterrorism is a challenge not served well by vertical communications, numerous federal and state public health agencies have joined together to form real-time horizontal information exchange networks so that agencies can gain an overview of what is going on in other units and jurisdictions. This allows agencies to be better able to spot trends, and to prepare for health problems that may be coming their way. CDC's network is open only to health department personnel who have a high level of clearance. However, Maryland's FRED system incorporates all of the state's health departments, EMS agencies, hospitals and emergency managers, to allow cross-agency and cross-discipline horizontal communication. This kind of real-time horizontal networking could prove extremely valuable to EM practitioners at all levels.

COMMITMENT TO RESEARCH. While commitment to research may not seem, at face level, to be a tool, the information produced by research clearly is. Research in medicine and public health has enabled us to understand causal relationships ranging from the subatomic level to the population level, and although there is still much to be learned, we successfully use this understanding to effectively intervene to decrease harm and promote health in the most trying of circumstances. Practice is driven by research, although the process of integrating research findings into the way things are done in the field is usually accompanied by lag time and some trial and error. Practice in emergency management is usually informed by previous experience and information provided by various disciplines that have their own research base, i.e., geology, meteorology, sociology. Emergency management lacks commitment to a strong research base of its own . . . which may be both the cause of and result of a lack of funding to both train emergency management researchers and fund their work. Our allied scientific disciplines (geology, sociology, meteorology, etc.) provide us with significant information upon which to plan and act, but do little to help us sort out what is effective in field application, and what is not, and under what conditions.

RECOMMENDATIONS

Emergency management and the health sector are natural allies that have, seemingly, only recently begun to recognize each other. They share the same basic goal of protecting the public, and share many of the same basic concepts. They are often both called upon to intervene on behalf of the public during the same emergencies. EM has developed a system for emergency decision-making and response management that public health is only beginning to learn. The health sector has developed science-based understandings that greatly increase the precision with which we intervene at the biological level, and is increasingly using research methods to assess and improve the strategies we employ for emergency responses. In order to enhance the power of health-EM collaboration in the future, we offer the following recommendations:

1. The two fields need stronger integration at various levels: planning, mitigation, preparedness and field operations. In order to be successful, this must take place within jurisdictions. The two fields must also improve integration by sharing research methods and findings. One of the first coordinated research topics should be to describe commonalities and then to identify barriers to collaboration and evaluate methods of developing real time synergies between the two disciplines.
2. Integration will be substantially improved if students of EM and medicine/public health learn about the concepts and operations of the other. EM students need to learn about the health effects of disasters of various types, as well as learning about how the health care system responds to disasters and other threats.
3. Short courses, perhaps online, should be developed, targeted at current practitioners of EM and health, to accomplish the aims of item b. above for those who are already mid-career.

4. EM students and practitioners should take health's example and dedicate more energy to studying the causal relationships that lead to human harm, and identifying efficient intervention points and strategies. We know much of the physics and mechanics of how people get hurt, but still lack information on why people will put themselves in harm's way, or fail to take protective action when faced with danger. We also lack information on the effectiveness of certain strategies when applied to a public who must implement them. For example, DHS Secretary Ridge's attempt to describe to the American public strategies for sheltering in place was met with disbelief and derision. We apparently did not have the research available that would allow us to predict the public's reaction. What aspects of Ridge's message were ineffective and why? Where would be the appropriate point to intervene again with a similar message, and how must it be presented in order to be effective? When public health faced similar issues related to obtaining the public's cooperation on public health issues (ranging from boiling water after disasters to using birth control correctly), the field took direction from research conducted by Everett Rogers and others on the diffusion of innovations and other interventions (Rogers, 2003).

5. Specific EM interventions need to be researched for their effectiveness, both in a general sense, and in terms of what specific conditions may make a particular intervention effective or not. For example, we draw up plans to evacuate nursing home patients and other "special needs" populations prior to the arrival of hurricanes. Such action can carry considerable risk for individuals who are already compromised. In terms of actually saving or risking lives, under what circumstances is an evacuation called for? How powerful must the storm be and what characteristics must it have? How do the intrinsic characteristics of the nursing homes affect the decision? How must evacuation transportation be designed so that compromised patients can be adequately supported while in transit? What characteristics of the targeted temporary shelter enhance the probability of good outcome for the special needs visitors? Under what circumstances is it safer for the special needs patients to stay in place? What is the impact on the rest of the surrounding population if resources are utilized evacuating nursing home and other special needs patients? These are not idle questions of no meaning to emergency managers and without consequence for the affected populations. These questions are researchable and such research may provide significant assistance to EM personnel who are required to work with the health sector in devising and implementing a workable policy.

6. EM and public health personnel need to collaborate on researching and assessing the true risks and vulnerabilities of local populations to identified hazards. One of the big stumbling blocks to truly good integrated local EOP development is that EM personnel do not have the tools to accurately assess the health impacts (and hence response needs) of specific local hazards. Local and regional epidemiologists and emergency medical personnel can assist with this process. The resultant information can be integrated into both preparedness and mitigation activities.

7. Good *rapid needs assessment* requires information from both the EM and health sectors. Work needs be done to develop efficient means for both sectors to collect and share the information needed, to avoid duplication and enhance the quality of information available to emergency decision-makers in both sectors.

8. Public health needs to work on developing and teaching good emergency decision-making skills. EM models of decision-making could form a solid base for public health personnel; effective collaboration between experienced EM specialists on decision-making could significantly enhance the process of bringing important decision-making tools and strategies to public health personnel and students.

9. The currently perceived social stature distance between EM and health sector personnel will be decreased if the field of EM makes a concerted effort to encourage its personnel to acquire baccalaureate, or even graduate degrees.

10. Both fields will benefit from taking a longer-run view of hazards and vulnerability, given our rapidly changing world. If we are able to lengthen the time frame of our planning to, say, the next 50 years, the need for cross-sector collaborative work will become immediately obvious. For example, it is clear that there will be far less petroleum available 50 years from now (Appenzeller, 2004; Odum and Odum, 2001). With a larger population demanding energy and petroleum-based chemical products from

a smaller base resource, what will be the effects on the public's health? Food production and distribution? Ability to respond to emergencies? Appropriate research now may be key to developing workable mitigation and response strategies for significant challenges that we will face, if we are able to now take a longer-term view.

SUMMARY

EM and the health sector have similar goals and conceptual models. The health sector has a long history of basing its tools, techniques, and strategies on solid research, and is now bringing that history to issues related to emergency public health. Neither EM nor health personnel have sufficient understanding of the methods and techniques of the other, and each has tools that would contribute to the other's success in emergencies. More importantly, given the compatible skills and overlapping foci of EM and health care, it is imperative that the two sectors actively collaborate with each other throughout the full emergency management cycle.

Acknowledgment

I would like to thank Drew Bumbak, MS, Brian Maguire, DrPH, Stephen Dean, Ph.D., and Sarah Edebe, MS for their invaluable content and editing assistance in the development of this chapter.

REFERENCES

Appenzeller, T. (2004). "The End of Cheap Oil." *National Geographic Magazine*, June.

Bissell, R.A. (1984). *Health and Hurricanes in the Developing World: A Case Study in the Dominican Republic.* Unpublished dissertation, University of Denver Graduate School of International Studies.

Bissell, R.A , Pinet, L. Azur, M. and Paluck, J. (2004). "Barriers to Collaboration Between Emergency Management and the Health Sector." Presented at the National Disaster Medical System annual meeting, Dallas TX, May.

Bissell, R.A, Pinet, L., Nelson, M., and Levy, M. (2004). "Evidence of the Effectiveness of Health Sector Preparedness in Disaster Response." *Family and Community Health 27*(3), 193–204.

Brewer, LA 3rd. (1986). "Baron Dominique Jean Larrey (1766-1842). Father of Modern Military Surgery, Innovator, Humanist." *Journal of Thoracic Cardiavascular Surgery 92*(6): 1096–98.

Carsten ,W. and Scheier, M.F. (2003). "Personality and Quality of Life: The Importance of Optimism and Goal Adjustment." *Quality of Life Research:V12* No. 0, 59–72.

Centers for Medicare and Medicaid Services. (2005). http://www.cms.hhs.gov/publications/overview-medicare-medicaid/default3.asp. Accessed 2 June.

Crosier S. and Snow, S. (2005). *The London Cholera Epidemic of 1854.* CSISS Classics. http://csiss.org/classics/content/8 Accessed 30 May.

Defoe, Daniel. (1722). *A Journal of the Plague Year.* London.

De Ville de Goyet, C. and Lechat, M.F. (1976). "Health Aspects in Natural Disasters". *Tropical Medicine*, (6) 152–157.

Green, L.W. and Marshal W. Kreuter (1990). "Health Promotion as a Public Health Strategy for the 1990s." *Annual Review of Public Health, 11*:319–34.

Haase, H.H. (1989). "Public Health and Medicine: A Leading Article by Friedrich Oesterlen in 1860." *Zeitschrift Gesamte Hygiene. Dec; 35*(12): 726–7.

Kolata, G. (2001). *High Drama About a Lethal Virus: "Flu." The Story of the Great Influenza Pandemic of 1918 and the Search for the Virus that Caused It.* New York: Simon and Schuster.

Lechat, M.F. (1976). "*The Epidemiology of Disasters.*" 1976. Proceedings of the Royal Society of Medicine.

Malilay, J., Flanders, W.D., and Brogan, D. (1996). A modified cluster-sampling method for post-disaster rapid assessment of needs. *Bulletin World Health Organization, 74*(4): 399–405.

Markel, H.S. (2001). "Man's Place in Nature - Past and Future." 2001. In Eckert, E. and Krafft, T. (Eds.) *Understanding the Earth System.* Berlin: Springer Verlag.

McMichael, A.J. and Powles, J.W. (1999). "Human Numbers, Environment, Sustainability and Health." *BMJ* Oct. 9; 319 (7215): 977–80.

McNeill, W.H. (1976). *Plagues and Peoples.* Anchor Press.

Odum, H. and Odum, E.C. (2001). *A Prosperous Way Down: Principles and Policies.* Boulder, Colorado: The University Press of Colorado.

Oskvig, R.M. (1999) "Special Problems in the Elderly." *Chest.* May; *115*(5 suppl): 1585–1644.

Rignault, D. and Wherry, D. (1999). "Lessons From the Past Worth Remembering: Larrey and Triage." *Trauma, 1*(1), 86–89.

Rivara, F.P, and Thompson, D.C. (2000). "Systematic Reviews of Injury-Prevention Strategies for Occupational Injuries: An Overview." *Am J Prev Med. 18*(4 Suppl): 1–3. For a broader discussion of evidence-based medicine, see also http://ebem.org/index.php.

Rogers, E. (2003). *Diffusion of Innovations.* 5th Edition. Free Press.

Snow, J. (1855). *Dr. Snow's Report. Parish of St. James, Report of the Cholera Outbreak in the Parish of St. James, Westminster, During the Autumn of 1854.* Churchill, London, pp. 97–120.

Surgeon General Joseph Barnes. (1900). *Medical and Surgical History of the War of Rebellion (1861–1865).* US Acts of Congress Publications

The Black Death: Bubonic Plague (2005). http://www.themiddleages.net/plague/html. Accessed 2 June.

The Black Deathm,1348. (2005). http://www.eyewitnesstohistory.com/plague.htm

Western, K.A. (1972). *The Epidemiology of Natural and Manmade Disasters: The Present State of the Art.* Unpublished dissertation, London School of Hygiene and Public Health.

Wetter, D. and Bissell, R.A. (2004). "Rapid Needs Assessment for Health Sector Use in Disasters.

White, H.R, Johnson, V. and Buyske, S. (2000). "Parental Modeling and Parenting Behavior Effects in Offspring Alcohol and Cigarette Use: A Growth Curve Analysis." *Journal of Substance Abuse, 12*(3): 287–310.

Zacny, J.P. (1990). "Behavioral Aspects of Alcohol-Tobacco Interactions." *Recent Dev Alcohol, 8*:205–19.

Ziegler, P. (1991). *The Black Death.* Stroud, Gloucestershire; Wolfsboro Falls, NH: Sutton Press.

Chapter 17

Who's in Charge Here? Some Observations on the Relationship between Disasters and the American Criminal Justice System

Robert J. Louden

ABSTRACT

Since the beginning of time the world has experienced a wide range of disasters. Responsibility for organizing and directing responses to disasters has varied over time and from place to place. The core functions of the American criminal justice system were established between 1776 and the adoption of the U.S. Constitution, particularly the Bill of Rights in 1789. However, it was not until 1967 that our federal government produced a schematic that graphically presented both the process and the major decision points of the criminal justice system. Although disaster-related activity has been present and accepted as a central function of many criminal justice agencies, it did not appear in this significant document. A brief overview of the American criminal justice system is offered. The aftermath of the terrorist attacks of 09/11/01 and the hurricanes of 2005 have illuminated many problems and concerns confronting the criminal justice system as a major component of government response to disasters. Practical experiences in NYC and New Orleans are highlighted. Broad-based recommendations for research are suggested.

INTRODUCTION

As with any discussion involving links between disasters and a given discipline, the relationship between this subject and the criminal justice system are extensive and complex. In our society criminal justice is perhaps the ultimate multidisciplinary discipline. At a minimum, aspects of the law, political science, public health, public management, psychology, and sociology influence the practical, tactical and legal activities of criminal justice system agencies on a daily basis. All of these sometimes complementary and sometimes contradictory interactions converge in disaster planning and response. Disaster in this paper is regarded to be a non-specific event and includes natural and man-made incidents.

There are numerous overlapping practical associations that one must be concerned with when considering this topic. One set of issues is relevant to the mandate of an individual criminal justice organization; a second is the interconnected bureaucratic concerns of police, the courts and corrections. Another is the involvement of the machinery of criminal justice with the broader community that it is part of. Within this context one must also remember that given the nature of governmental and political subdivisions in the U.S.: federal, state, county, and local, linkages are far-reaching and potentially confusing if not conflicting. The concerns of issues related to "states-rights" and "home-rule" have influenced the development of criminal justice agencies, and their organizational mandates, throughout the country. Although our federal government does not directly control most of the criminal justice agencies in the U.S., there are several ways in which local policy and practice may be influenced by Washington through various court decisions, rule settings, investigative bodies, oversight mechanisms and funding. All of this has had an impact upon the role of law enforcement in emergency management in the United States.

This paper reflects on the topic of disaster and the discipline of criminal justice from a number of perspectives: inter and intraoperations of criminal justice agencies, and the collaboration or lack there of, between criminal justice agencies and other aspects of governance. Portions of this paper are anecdotal in nature, based on the author's participant-observer status during an active 21-year career in policing and a subsequent 18-year career in higher education, primarily involving criminal justice and protection management programs.

The three components of the criminal justice system – police, courts, and corrections – have their own responsibilities and concerns when faced with a disaster. This paper presents many issues from a police perspective since they are generally considered the gatekeepers for the rest of the system and they are also the boots-on-the-ground among the first responders to disasters which usually impact on the operations of the courts and correctional institutions. Occasionally, conflict will arise between law enforcement agencies and others responsible for disaster-related response activities. In other cases, there will be lawlessness in disasters. Both situations present a challenge to be overcome by future emergency managers.

THE CRIMINAL JUSTICE SYSTEM

First, let's consider an abbreviated and selective criminal justice history. It was not until The President's Commission on Law Enforcement and Administration of Justice published their *The Challenge Of Crime In A Free Society* in 1967 that we could view ". . . in simplified form the process of criminal administration and the many decision points along its course" (pp. 7–9). Obviously, our nation had components of a criminal justice system in place since the founding of the Republic, but President Johnson indicated in forming his Commission in 1965 that there was a "depth of ignorance about [criminal justice];" the country needed a comprehensive examination of crime and justice to identify where we were and the direction that we should be headed in. The 340-page document "embodies all the major findings we have drawn from our examination of every facet of crime and law enforcement in America" (forward). It does not include discussions of disasters, with the exception of a brief consideration of some social, economic and criminal aspects of riots (pp. 37–38). Riots, not unlike other disasters, are mostly spontaneous outbursts, precipitated by a spark-event and every bit as complicated as the social problems and conditions present in a community.

The President's Commission presented some 200 Recommendations (pp. 292–301), but none of them dealt with disasters. The Commission also issued nine task force reports. The one most appropriate to a consideration of disasters is *Task Force Report: The Police* (1967). In that 239-page document, the treatment of disaster is limited to issues related to "Handling Crowds, Demonstrations, and Riots." The topics were mostly concerned with police-community relationships and Constitutional questions (pp. 192–193). Police and subsequently court and correction agency reaction to looting and violence are the focus of the very brief discussion. Riots and related activities are thus distinctive types of disaster requiring both specific and generalized responses. Twenty-first century police agencies, as they have been for more than 150 years, are often the first responders to a variety of actual and potential disaster scenarios. However, the *Task Force Report* admonition that "the police have little control over the social, economic, or other factors which create riots" (p.193) could well have been written in 2006 in a discussion about the response of the criminal justice system to disasters.

In *Disasters and Democracy* (1999), Rutherford Platt cited a 1987 California Court of Appeals case, *First English Lutheran Evangelical Church* v. *County of Los Angeles* (p. 143): "If there is a hierarchy of interest the police power serves – and both logic and prior cases suggest there is – then the preservation of life must rank near the top" (originally reported as 258 Cal. Rptr. at 904). Although this is obviously a reference to the overall police power of our government in a Constitutional context, the concept could be interpreted as applicable to certain roles of government in dealing with disasters, often operationalized by criminal justice agencies.

Cops

Sir Robert Peel, the British Home Secretary in the early nineteenth century wanted a particular attention to order in society and was instrumental in forming the first professional police force in 1829. His police force was organized according to nine principles which demonstrated a commitment to the public. It is generally accepted that his organization for the London police, originally tested in Ireland, became the model for the first police agencies formed in the northeastern United States. The British police role became ours (Pfiffner, 1967). Disaster-related activity was a common police function from the earliest days in the UK and in the U.S. Commentary on characteristics of present-day policing would not be meaningful without reflecting on the formation of organized policing in the U.S.

For instance, Lee (1901) noted that the police are the only members of the public who are paid to give full-time attention to duties which are incumbent on every citizen in the interest of community welfare. Although this is obviously an out-dated characterization, such police duties often involved response to disasters.

The fire service, of course, has a longer history of response activity, but then as now the majority of fire fighters are volunteers who are not always as readily available as full-time paid police officers. Interestingly, Boston hired the first paid fire captain in the country (IMCA, 1979) almost one hundred years before the first police forces in the U.S. were formally organized in Boston and New York.

This nation's police departments are therefore charged by law, administrative code and ready availability with providing certain services to the population of a given political or geographical division. Depending on geographic location, correctional institutions and courts are located within a given law enforcement agency's area of responsibility.

According to federal government statistics there are more than 17,000 local and state police departments in the country; more than 50 federal agencies have law enforcement authority. Even though the NYPD employs almost 40,000 sworn officers, the next largest city has only about one-third that complement of officers. Not even 100 local agencies have at least one thousand officers and less than 1000 departments employ a minimum of 100 officers (BJS, 2003). We are largely a nation of small town, not national policing. In contrast, most democratic nations have one, or at most a few, principal law enforcement organizations which have nationwide jurisdiction.

A forty-year old publication of the University of Connecticut noted that "police duties and responsibilities are predicated upon the customs, traditions, and demands of the community served." This has resulted in a "wide diversity of tasks" (Goldstein, 1966, p. 1). Among those tasks are various response protocols to disasters. As the 1967 *Task Force Report: The Police* noted "The police are the part of the criminal justice system that is in direct daily contact . . . with the public. The entire system – courts and corrections as well as the police – is charged with enforcing the law and maintaining order" (p. 1). The American Bar Association (ABA, 1974) reported that included among the "Major current responsibilities of police" were (excluding four items):

1. to aid individuals who are in danger of physical harm;
2. to facilitate the movement of people and vehicles;
3. to assist those who cannot care for themselves;
4. to identify problems that are potentially serious governmental problems;
5. to create and maintain a feeling of security in the community;
6. to promote and preserve civil order; and
7. to provide other services on an emergency basis. (pp. 3–4)

Each of these responsibilities may be considered as applicable to the role of law enforcement in preparing for and responding to disasters.

Egon Bittner (1975) discussed two interrelated aspects of policing that could also be equated to their role in dealing with disasters: ". . . the police are nothing else than a mechanism for the distribution of situationally justified force in society." And "The American city dwellers repertoire of methods for handling problems includes one known as 'calling the cops'" (p. 39). Bittner advances the concept that police are called upon to respond to a situation and do something about which something must be done. It has been stated many times in the literature of policing that they are viewed as an immediately available resource in times of emergency and crisis.

Post 9-11 law enforcement activities in America span a range from the traditional, often characterized as community policing programs, to enhanced intelligence gathering and analysis to military-style tactical units to search, rescue and recovery. Obviously, not all U.S. police agencies perform all of these functions and the capacity of a local law enforcement agency to do more without commensurate increases in resources is problematic. One law enforcement concern is that intelligence collection or investigative follow-up may be compromised if a criminal justice agency is not 'lead-agency' in disasters resulting from criminal or terrorist actions (see Louden, 2005).

This practical and potentially problematic concern has framed part of the discussion of the role of criminal justice in disaster response at least since the 1993 bombing of the World Trade Center, and continued with the 1995 Oklahoma City bombing, the explosion of TWA 800 and certainly 9/11/01. The final chapter of *The 9/11 Commission Report*, "How to Do It? A Different Way of Organizing the Government" (pp. 399–428) may serve as a good starting point to explore not only intelligence and defense issues but also the broader question of 'Who is in charge here?"(see also Police Executive Research Forum, October 2, 2001.)

Corrections

Correctional institutions present special problems and concerns when exposed to disasters. A disaster may endanger the lives and well-being of inmates and staff, yet under most circumstances the inmates must continue to be controlled. Faced with the high number of individuals incarcerated in jails and prisons throughout the country, the potential for disasters impacting on the correctional population is highly probable. Protection of life is a standard of care imposed on all agencies charged with the custody of others. In addition to protecting inmates, government officials must keep in mind that a perception that prisoners may not be under proper control could generate fear in the surrounding community. Persistent attention to issues of operational security and rumor control are crucial to a competent response to correction-related disaster. A potential positive consequence of a disaster in or near a correctional institution is that non-violent inmates may be available to assist in rescue, recovery and rebuilding efforts.

There is a lack of published research dealing exclusively with corrections and disasters situations. For example, see The New York State Special Commission on Attica (1972); Useem and Kimball (1989); Wills-Raftery (1994); Useem, et al. (1995); and Flin and Arbuthnot (2002). Additionally, the special population nature of inmates found in disaster related literature is appropriate to assess in the context of the criminal justice system.

Prisoners are sentenced to jails or prisons as the result of a judicial process. Much of what subsequently happens to prisoners in the U.S. is determined or at least reviewed by a court. This role is only one part of the potential effects of disasters on our judicial branch of government; local, state or federal.

Courts

As with correction-related disasters, there had not been a great deal of material published on the relationship of disasters and the courts prior to the bombing in Oklahoma City in 1995. The federal General Services Administration, the landlord for the federal judiciary, became more actively involved in prevention, mitigation and response activities for court buildings as a result of the bombing of the Alfred P. Murrah federal building. However, in 1998 the American Bar Association devoted an entire issue of the *Judges' Journal* (Vol. 37, No. 4) to disaster-related court experiences and concerns. This 40-page special volume is a mix of articles and case studies, mostly natural disaster related. It deals with local, state and federal jurisdictions and includes the impact of disasters on the mechanics and logistics of trial calendars, "time-sensitive-progression" aspects of case flow; processing of new arrests and other criminal and civil actions; managing the jury process, production of transcripts and other case files; and the potential distress of court personnel.

AN EXAMPLE FROM NEW YORK CITY

New York City experiences suggest a range of problem areas that deserve consideration and resolution. Less than one year after 9/11/01, the *New York Daily News* quoted NYC Police Commissioner Ray Kelly as saying that between $500 and $700 million was needed for training, resources and equipment to fight terrorism. He further noted that "There was $3.5 billion languishing in Congress, and we are not going to get anything before the coming [2002] 9/11." Likewise, *Newsday* attributed similar sentiments to the director of the NYS Police Chief's Association, who indicated, "Not one additional resource has been provided even though all first-response comes locally." Both of these views were against the backdrop of comments by President Bush, as reported in *The New York Times*, ". . . he would not spend $5.1 billion approved by Congress last month for domestic security. . ." (see Louden, 2002).

Cordes (1971) noted that when rules and procedures are being devised it becomes necessary to clear up obvious misconceptions, re-examine relationships, and review the responsibility of each toward the other. This process should foster a redefinition of the collective obligation to approach and in some cases attack society's problems. This caution was especially cogent in 2005 New York City – the battle of the badges. This is a perennial problem: Which agency, fire or police or emergency management, should perform a particular task, and/or be in charge of what?

Prior to the first administration of Mayor Rudolph Giuliani in 1994, the city-wide function generally referred to as emergency management was actually an organizational entity within the headquarters of the New York Police Department (NYPD). It had been that way for many years, including 25 years ago when the NYPD Office of Civil Preparedness was designated as the Director of the NYC Mayor's Emergency Control Board (Urban Academy, 1979). The Office of Civil Preparedness was a precursor of city-wide emergency management. The NYPD coordinated activities of all city agencies, including FDNY, utilities and supplementary services during a declared emergency.

Giuliani created a separate and distinct Office of Emergency Management (OEM), in part to settle problems and disputes between NYPD and FDNY. He believed that "the city had to reorganize its response to emergencies" (Giuliani, p. 315). A decade later the separate OEM still exists, but serves more as a research, resource and facilitating entity than the "overarching organizational structure that was equipped to coordinate many different departments" that the Mayor had envisioned (p. 315).

The NYPD also maintained a highly specialized tactical element, the Emergency Service Unit (ESU). Originally formed in 1930 as the "riot-squad," ESU evolved into a diverse group of EMT certified police officers which also involved Special Weapons and Tactics (SWAT) and a variety of rescue and recovery-related activities; building collapse, subway derailment, hazmat. A commonly heard expression among NYPD personnel is "when the public needs help they call the police; when the cops need help they call ESU." Companion elements within the NYPD Special Operations Division (SOD) were Aviation, Harbor, and Scuba. The rescue and recovery functions of ESU and several of the other SOD tasks are often assigned to Fire and Rescue Units in many jurisdictions across the country; not generally a police function. The New York Fire Department (FDNY) also fielded rescue and recovery units, marine and scuba personnel. One argument as to why the NYPD was so heavily involved in such specialized non-enforcement duties, besides long-standing traditions, was the relationship between the police SOD and the pre-Giuliani police-based city-wide emergency management apparatus.

Historically, both agencies responded to the same emergency situations. Sometimes the agencies worked the problem together and other times they worked in competition with each other. In extreme cases, fists flew and arrests were occasionally a result! Periodically, politicians or the press, often prompted by one side or the other, call for a modification in arrangements; substantive change seldom occurred.[1]

The attacks of 09/11/01, predictably, brought OEM, FDNY and NYPD, among others, to the horrific scene; we all know the toll in death and destruction. Although often working side-by-side, some aspects of the smoldering rivalry continued and intensified

1. An advisory comment! This author worked very closely with ESU personnel over a span of many years, particularly in their tactical, SWAT, role during hostage, barricade and siege-type situations. In my view, their training and actual deployment to non-enforcement rescue and recovery related activities helps to prepare them to be more competent tactical officers. They gain practical experiences, improved patience and a greater reverence for life.

over the following four years. As Clyde Haberman (2005) noted in the *New York Times*,

> Take the Fire and Police Departments. Having both suffered terribly, they seemed headed after Sept.11 for a new era of comity. Their ancient antagonisms? The occasional testosterone-fueled fistfights at disaster scenes? The failed communication that may have contributed to the 9/11 death toll? They belonged to the unlamented past, or so some thought (April 26, 2005, p. B1).

Haberman's column was one of many reactions to a change in ". . . one part of a (sic) emergency response plan known formally as the Citywide Incident Management System" (Confessore, 2005).

In testimony before the NYC Council, Glenn P. Corbett, an engineer, volunteer fire fighter and professor of fire science at John Jay College of Criminal Justice in NYC noted that, "Instead of taking a giant leap forward through correcting long-standing major flaws in New York City's emergency response protocols, the plan "takes several steps backward" (Corbett, 2005). Corbett has stated that absent exigent circumstances, a police agency should not be in command of most disaster scenes. He also indicated that he does not believe that there is another jurisdiction in the country that is organized like NYC for response to disasters.

Regardless of the genesis of rivalry between FDNY and NYPD, Mayor Bloomberg has spoken. OEM still exists as a separate governmental entity but does not have a lead-agency role in responding to disasters. The new Citywide Incident Management System (CIMS) delineates primacy to one agency or another depending on events and expertise (see www.NYC.gov). The NYC plan is designed to meet the mandate of the NIMS. A major point of demarcation is the nature of the disaster. Terrorism automatically begets law enforcement command and control while natural and non-terror man-made disaster adheres to the mayor's newest plan. Neither NYPD nor FDNY are totally pleased with the new system, but both agencies have pledged to support the mayor and adhere to the guidelines. In fact, there have been, and continue to be, numerous examples of cooperation and coordination among and between the agencies. For example, a task force of FDNY and NYPD personnel, coordinated by NYC OEM, recently participated in WMD exercise at the Center for National Response in Standard, W.Va. (McGeehan, 2005). Likewise, two helicopter crashes during 2005 in the East River in NYC resulted in a well-coordinated multiagency response.

COMPATABILITY AND CONTROVERSY

A different, longer-standing, example of positive police and fire agency collaboration may be found in the Report of the Joint Fire/Police Task Force on Civil Unrest (FEMA, 1994). At the time it was noted that ". . . for the first time in recent memory, police and fire executives were able to sit and work together, to voice their concerns and wishes, and to come to a basic agreement on how personnel can best cooperate in the field" (p. iii). It was also noted in their report that: "A successful collaboration among agencies will depend on:

1. Compatibility of the agencies
2. Adaptation of a common technical terminology
3. A strong joint command structure
4. Regularly scheduled joint training exercises
5. Effective mutual aid agreements." (p. 4).

When considering the appropriateness of these five points one may consider whether there is an ideal organizational structure for disaster response, or might dynamic leadership be more fundamental? And, under what circumstances should criminal justice personnel be in charge at the scene of a disaster?

In these times of increased threat and response, natural and man-made, some factors about preparing for potential role change become apparent. This nation has a variety of government, volunteer and private sector mechanisms that respond to disaster. Some are the result of legal mandate, others are the result of long-time practical experiences and some will continue to be ad hoc. Accordingly, the most important issue to be considered in any response is "Who is in charge?" In order to be able to properly address that simple-sounding question a number of items must be considered. First, there is a need to specify the new reality and determine what is to be done about it. Next, a review of legal and administrative directives is appropriate. These two items should identify (1) the nature of the problem(s), (2) the legislated parties responsible for response; (3) the actual parties involved in attempts at resolution; and

(4) suggested changes in mandate or practice which would provide an improved conclusion. Another basic factor for any role change is to specify appropriate organizational change, if any, and delineate the nature of resources needed and of training required. Adequate and appropriate funding is a must.

Some jurisdictions rely on an operational version of OEM, incorporating Incident Command principals. Others have adopted a Public Safety model (Nickerson, 2005). Whatever the configuration, the controversy will continue.

NEW ORLEANS AND THE FUTURE

There has not been sufficient time since hurricanes Katrina and Rita to assess the long-term implications of the relationship of disaster and criminal justice response. However, preliminary comments can be made.

The police department and its personnel were heroic and selfless in many ways, but also experienced serious bouts of disorder in many forms including desertion, pilfering and brutality. Communications systems failed dismally forcing 120 - 911 operators to evacuate, resulting in thousands of emergency calls to go unanswered (Connolly, 2005). Evidence stored at police headquarters, including DNA in rape kits and bags of cocaine was destroyed by flood water, severely limiting the chance of proceeding with trials for hundreds of defendants; innocent and guilty alike (Peristein, 2006). Additionally, emergency plans were not up to the task, and many routine duties have not been resumed some six months later.

Many court operations have been impacted by the distress of the police department and further negatively influenced by damage to physical structures, computer systems and lack of personnel. In October 2005, the Orleans Parish District Attorney reported that he was close to shutting down operations, impacting on 3,000 criminal cases, due to a lack of resources caused by the hurricane (Randolph, 2005).

Corrections, like the police and the courts, can boast of many positive contributions to the people of New Orleans. They, too, have many problems. Standards of care were not always adequate, prisoners absconded, and facilities are in need of major rebuilding. There were reports of inmates dying due to flood water in confined spaces (see Asbury Park Press, 2005); and, escaped and prematurely released inmates were transported to other jurisdictions at government expense (see Tucker, 2005). Some inmates were evacuated and released early while others were detained beyond their sentence due to the destruction of records and lack of personnel (Millhollon, 2005).

The entire criminal justice system appears to be in dire need of a major influx of personnel, equipment and other resources and plans and training that may simultaneously provide for a return to some normalcy and prepare for the next worse case scenario.

We can state, based on media accounts and government hearings, that: organized policing was not prepared for the vastness of the tasks; the court systems are in a state of turmoil with long-term implications for the delivery of justice, for defendants and victims alike; correctional inmates were in many cases positive contributors to response strategies and in other instances the cause of additional concern as special populations in need of special care and consideration. Criminal, organizational and procedural investigations and reviews are in progress. The findings of these deliberative bodies will make meaningful contributions to the questions of compatibility and controversy that are present in the criminal justice disaster response discussion.

CONCLUSION AND RECOMMENDATIONS

This paper provides a selective overview of the discipline and practice of criminal justice in America. It also notes that there is a direct relationship between criminal justice and disaster response. Sometimes they are well ordered and complementary, and at other times rife with confusion and conflict.

This document was framed, in part, by considering a series of questions: What recommendations does criminal justice have for the study of disasters? What recommendations does criminal justice have for emergency management; what are the gaps in knowledge? Hopefully, answers to those questions are contained in the preceding material. Working backwards through the questions, the gaps are still vast, but the right questions are beginning to be asked, prompted in large part by the New York City experience with 9/11 and the New Orleans experience with Katrina.

Criminal justice, as a discipline and as a system, has had a great deal of experience with formulating and implementing emergency management plans. It is my opinion that an ongoing impediment to smoother future operations is territoriality and turf issues, at all levels of government coupled with sometimes conflicting mandates concerning the primacy of response, rescue, recovery in the context of intelligence and investigative evidence gathering and collection. Criminal justice as an academic discipline must also assume an active leadership role in collaboratively conducting empirically sound research projects, across scholarly disciplines and across geographic, political and bureaucratic boundaries; the problems are not going away.

New York City, of course, is unique in many ways; no other jurisdiction has comparable resource availability or parallel life and death experiences. New Orleans, on the other hand, is still reeling from the aftermath of Katrina. Many of the ongoing problems in both cities are criminal justice related. The New York City example, historically and especially post-9/11 should be viewed as a living laboratory which affords bureaucrats and researchers an opportunity to review and challenge both traditional and unique organizational arrangements and resource allocation for disaster response. Likewise, New Orleans should be subjected to examination, criminal, civil and academic to contribute answers to the question: What went wrong?

Disaster research must focus on the practical, tactical and legal aspects of the appropriate role of criminal justice. Among other facets, the research agenda, privately and publicly funded, must reflect on organizational configurations, human factors, technological applications, governmental relationships.

REFERENCES

ABA. (1974). *Standards Relating To The Urban Police Function.* Washington, DC: American Bar Association.

American Bar Association. (1998). *Judges' Journal. 37* No.4 Judges' J.

Asbury Park Press. (2005). "Sheriff denies allegations of death, abandonment at New Orleans Jail." (p. A12).

Bittner, E. (1975). *The Functions of the Police in Modern Society.* New York: Jason Aronson.

Bureau of Justice Statistics. (2003). "Local Police Departments 2000." Washington, DC: US Department of Justice.

Confessore, N. (2005). "Mayor Says It's Best To Let Police Control Terror Scenes." *New York Times* (http://query.nytimes.com/mem/tnt.html?tntget=2005/04/23).

Connolly, C. (2005). "Thousands of Katrina Calls Went Astray." *Washington Post.* (www.washingtonpost.com/11/07/2005)

Corbett, G. (2005). Testimony, NYC Council, Committee on Fire and Criminal Justice Services.

Cordes, J.G. (1971). "Why the Police Exist," unpublished paper, New York: John Jay College of Criminal Justice.

FEMA. (1994). "Report of the Joint Fire/Police Task Force on Civil Unrest: Recommendations for Organization and Operations During Civil Disturbance.

Flin, R. and Arbuthnot, K. (Eds.). (2002). *Incident Command: Tales from the Hot Seat.* Cornwall, UK: Ashgate Publishing.

Goldstein, B. (1966)."Non-Police Duties of Today's Policemen," *Connecticut Government,* Institute of Public Service, University of Connecticut, June.

Giuliani, Rudolph W. 2002. *Leadership.* New York: Miramax Books.

Haberman, C. (2005). "What's Normal About Security and Harmony?" *New York Times.*

ICMA. (1979). *Managing Fire Services.* Washington, DC: International City Management Association.

Lee, M. (1901). *A History of Police in England.* London: Metheun,

Louden, R.J. (2002). "Bringing the Boys in Blue Together: A Year After 9/11 Special Report, Observations on Attorney General John Ashcroft's Record on the Anniversary of 9/11" in *Legal Times,* Vol.XXV, Number 35, p. 51, September 9, 2002.

Louden, R.J. "Policing Post-9/11" in *Fordham Urban Law Journal,* Vol. XXXII, Number 4, pp. 757–765, July 2005.

McGeehan, P. (2005). "In an Old West Virginia Tunnel, a Team of City Rescuers Rehearses for the Worst." *New York Times.*

Millhollon, M. (2005). "La. Frees, assists Orleans inmates." *The Advocate.* (http://2theadvocate.com/cgi-bin/printme.pl 9/19/05).

New York State Special Commission on Attica. (1972). *Attica.* New York: Praeger.

Nickerson, B. (2005). "White Plains' Lesson for NYC Cops, Firefighters." *The Journal News.* (http://www.thejournalnews.com/apps/pbcs.dll/frontpage.)

Perstein, M. (2006). "When the flood ruined DNA samples at NOPD headquarters, it washed away hope for inmates trying to prove their innocence." *The Times-Picayune.*

Pfiffner, J.M. (1967). "The Function of the Police in a Democratic Society," Unpublished paper, University of Southern California, 1967.

Platt, R.H. (1999). *Disasters and Democracy: The politics of extreme natural events.* Washington, DC: Island Press.

Police Executive Research Forum. (2001). "Local Law Enforcement's Role in Preventing and Responding to Terrorism" Discussion Draft, Washington, DC.

President's Commission on Law Enforcement and Administration of Justice. (1967). *The Challenge of Crime In A Free Society.* Washington, DC: US Government Printing Office.

President's Commission on Law Enforcement and Administration of Justice. (1967). *Task Force Report: The Police.* Washington, DC: US Government Printing Office.

Randolph, N. (2005). "Orleans DA says he may have to shut down." *The Advocate.* (www.2the advocate.com/cgi-bin/printme.pl 10/24/2005)

The Constitution of the United States.

Tucker, E. (2005). "Criminals Among Katrina Refugees Sought." *Washington Post* (www.washingtonpost.com 9/22/05).

The 9-11 Commission. (2004). *Report: Final Report of the National Commission on Terrorist Attacks upon the United States, Official Government Edition.* Washington, D.C.

Urban Academy. (1979). "New York City Mayor's Emergency Control Board: Emergency Management Plan."

Useem, B. and Kimball, P. (1989). *States of Siege.* New York: Oxford University Press.

Useem, Bert, et al. (1995). *Resolution of Prison Riots.* Washington, DC: NIJ.

Wills-Raftery, D. (1994). *Four Long Days: Return to Attica.* New York: American Life Associates.

Chapter 18

Economic Applications in Disaster Research, Mitigation, and Planning

Terry L. Clower

ABSTRACT

This chapter examines the contributions of the economics discipline to disaster research, mitigation, and planning. Economics offers modeling techniques for assessing the impacts of disasters, theories of development for understanding the choices that individuals and firms make in selecting residential and business locations, approaches for risk and vulnerability assessment in insurance and disaster planning, and policy insights in each of these areas that are affected by the political economy. The chapter gives particular attention to common and emerging techniques for assessing the indirect economic impacts of disaster events offering an assessment of the strengths and weaknesses of each analytic approach.

INTRODUCTION

Economics as a specific discipline, its many sub- and closely related disciplines, and research techniques pervade the systematic study of disasters and their human, social, and monetary impacts. The goal of this chapter is to provide the non-economist a look into how this discipline shapes scholarly and public understanding of disaster impacts, the roles that economic information can play in the *realpolitik* of disaster management and response, and reviews methods of analysis in assessing the impact of disasters.

To accomplish this goal, specific data analysis techniques used to estimate the economic impacts of disasters are presented along with a description of how this information is used to address issues of resource allocation and disaster avoidance. Discussion is presented on issues relating to disaster insurance and the contribution of economics to risk analysis. Finally, the chapter offers suggestions for expanded or new research approaches that economists should undertake to further contribute to the discipline of disaster management. But first, a historical perspective of the role of economics in disaster research.

A BRIEF TIME IN HISTORY

Assessing the economic impacts of disasters is a very recent systematic field of study. Disasters have been, and continue to be, human tragedies. History books tell us that more than 2,000 died in the Johnstown, Pennsylvania flood in 1889, the eruption of Krakatoa in 1883 – described as the first catastrophe of the communications age (USGS, 2005) – and the resulting tsunami killed more than 30,000, the Galveston hurricane of 1900 killed more than 6,000 of the island's residents, and, of course, the 1,503 lives lost in icy North Atlantic waters on that "night to remember" in 1912. These numbers represent horrific human tolls and each also represents economic losses in the tens or hundreds of millions of dollars. But, it is the loss of life that catches our attention.[1]

1. Of course, these human losses pale in comparison to many of the great disasters such as the 1976 Northeastern China earthquake that killed 240,000, the 40,000 killed in Northwestern Iran in 1990, the 2004 Indonesian tsunami with a death toll exceeding 300,000 and the 1918–1919 flu pandemic that claimed an estimated 30 million lives (Becker, 2005; World Book, 2005).

In addition, the provision of public monies to help those affected by disasters has been a comparatively recent occurrence. Historically, government policy and/or public sentiment simply did not support monetary aid to disaster victims. Barnett (1999) cites an example from 1887 where President Grover Cleveland in response to an emergency request for $10,000 in aid to Texas drought victims noted that there is no constitutional basis for public funds to be used to offset individual suffering as a result of a disaster. Barnett also observes that even though public policy had changed by 1915 with the advent of federal disaster relief grants and loans, it was many years before the public at large found the receipt of these grants and loans socially acceptable.

Concentrating on the loss of life to describe the magnitude of disasters, combined with public attitudes about the costs of disasters being borne by individuals, there was little demand for comprehensive economic assessments of disasters. One of the few early assessments of the economic impacts of a disaster was published in 1920 estimating the impacts of the Halifax ship explosion of December 1917 (Scanlon, 1988). Little else appears in the academic literature for more than 40 years, but during those years, public policy and public attitudes about disaster relief changed. With these changes came demand for information about the size (impact) of disasters from an economic perspective. If there were to be programs to provide aid to victims of disasters, then the impacts must be quantified.

The development of warning systems broadcast over radio networks and later television gave vital information that has saved innumerable lives. In addition, investments in infrastructure, enhanced construction techniques required by modern building codes, and other physical capital have with one notable exception resulted in fewer deaths due to disasters. For example, following the 1900 hurricane, Galveston Island and almost every structure on the island were raised several feet. Hurricanes have hit Galveston since 1900 but never with anything near the human losses of the 1900 event. As this chapter is being completed, recovery is underway for hurricane Katrina. In the largest disaster to hit a U.S. city since the San Francisco fire, New Orleans, a city of 450,000, was inundated by flood waters after sections of the Mississippi River levee system failed due to storm-related flooding (the City of New Orleans sits several feet below sea-level and has been a high-risk area for flooding since its founding in the late 18th century). Inefficient and ineffective government response is being blamed for many of the city's low-income population not being evacuated. Whether through inability to evacuate, or unwillingness by individuals to evacuate, over 200,000 people were still in the city when the levees broke. Still, less than one-half of one percent of the population perished.

Even with record numbers of people moving into relatively hazardous areas, such as the Florida coast or mudslide prone hills in central and southern California, until Katrina we have seldom seen more than a few deaths in the U.S. related to natural disasters since the early parts of the twentieth century. To justify ongoing public aid to victims and expenditures for disaster preparedness and management, efforts turned to estimating the economic impacts of disasters.

POLITICAL ECONOMY OF DISASTERS

Prior to the twenthieth century political economy was the proper name for the discipline of economics. In today's context it means the convergence of politics and economics. In the previous section changing public policy in the U.S. is illustrated by comparing President Cleveland's strict interpretation of the constitution with the later advent of federally funded grants and loans to aid victims of disasters. The economic considerations were, in many respects, the same, but our policy (political) approach had changed. Economic analysis is at the heart, but is far from the whole, of the *realpolitik*[2] of disasters. From the time of the nation's founding through 1950, the U.S. government enacted 128 pieces of legislation providing relief, mostly in the form of in-kind donations, for

2. *Realpolitik* (literally the "politics of reality" in German) typically refers to a pragmatic, non-idealistic approach to international politics. In my usage it refers to the pragmatic application of politics, influenced by economic considerations, to disaster policy implementation.

victims of disasters (Barnett, 1999). By the 1950s, the U.S. had gone through a fundamental shift in the expected role of government. From the New Deal policies of the 1930s through the G.I. Bill providing for a college education to veterans of World War II, liberal ideas of government responsibility to the nation's citizens was in its ascendancy. The Disaster Relief Act of 1950 and the Small Business Administration Act of 1953 both offered standing programs for disaster relief (Barnett) requiring economic analyses to support budget projections.

Programs of the Great Society of the 1960s and afterwards also included elements of disaster relief and mitigation in housing and introduced formal civil rights considerations in disaster management and planning. In 1953, the federal government provided just 1 percent of total disaster relief spending. By the mid-1970s that percentage had risen to more than 70 percent (Barnett, 1999).

Political considerations also influenced the distribution of private relief money. Prior to Hurricane Camille in 1969, the American Red Cross distributed disaster assistance based on economic need. After being heavily criticized in the press and in some political circles, the Red Cross standardized their rules for funds eligibility and removed economic need as a criterion. This can be seen as a reflection of the growing size and political influence of the middle class in the U.S. after World War II. As observed in surveys conducted by Leitko et al. (1980), middle-class victims of disasters view relief as "a corrective to a naturally induced injustice" (page 735) and tend to demand larger amounts of relief regardless of their own resources. This liberal approach to the distribution of disaster relief has survived the increasingly conservative nature of other public assistance in the U.S. since the early 1980s. Leitko, et al. observed that the public does not see disaster relief as welfare. How else can one politically account for general acceptance at the national level for potentially providing grants and low interest loans to wealthy families whose homes in gated Florida communities are damaged by hurricane events?

The good news about the surprisingly liberal attitudes of the U.S. electorate towards disaster relief and mitigation is that we have had a steady, if not sufficient, stream of economic resources for disaster mitigation, preparation, and management. Of course, this has also been influenced by the *realpolitik* of 9-11 and there will certainly be a shift in federal government spending policy due to failures and perceived failures that led to the New Orleans/Katrina disaster – at least in the short run. The bad news is that federal intervention is increasingly distorting economic decisions at the local level. Kunreuther (1998) notes local governments are not seeking own-source solutions to disaster response needs, such as private sector insurance, because of perceived certainty of federal resource availability. These distortions are also apparent in residential real estate markets.

One of the clear reasons the costs of disasters have escalated rapidly in recent years is a function of the level of development in high risk areas. For example, in 2003, 153 million people, 53 percent of the U.S. total population, resided in coastal counties – an area that comprises just 17 percent of the mainland U.S. land mass (Crossett, et al. 2004). The coastal population has increased by 33 million in 23 years representing a rapid increase in population density and significant development intrusions into barrier islands and marsh lands that offer natural protection from storm events. Moreover, this population growth does not reflect the growth in the number of second and vacation homes, hotels, and resorts that increasingly fill the coastal landscape. The political reality is that local and state governments are willing to trade the potential for more expensive disaster events, which is offset by federal assistance, for tax base growth.

There are three other ways that political economy approaches can help explain the level and distribution of disaster mitigation and planning funding. The first is the political dimension of who qualifies for post-disaster assistance. The release of federal grants and loans for disaster relief is based on the declaration by the President that a specified region, most often a county, is a "disaster area." The general public largely thinks this designation is about damage to buildings, homes, and infrastructure along the lines of the Fujita Scale of tornadic damage. However, it is the impact the disaster event has on local government that forms the basis for a disaster declaration. In theory, local government or state government are supposed to provide disaster assistance. If demand for assistance and services exceeds local capacities or if local government revenues are substantially threatened, then Federal resources are engaged. If a hotel is damaged by a tornado, as in the case of Fort Worth, Texas, in 2000 (see McEntire,

2002 for a description), local government experiences losses in revenue from sales taxes, hotel occupancy taxes, and property taxes and thus local government's ability to provide services and recovery aid is diminished.[3] This interesting quirk of U.S. disaster policy is keenly felt by victims of certain types of disasters. Tornadoes can cause widespread damage qualifying the area for Federal disaster assistance. However, if the tornado destroys only houses located along one block, it is doubtful that the revenue of local government would be severely impacted and those victims will not qualify for federal assistance – even though their individual loss is as great as any individual in a much larger disaster.

In practice, presidential disaster declarations can be overt acts of political largess or electioneering. Sylves (1996) noted that in the winter of 1995, President Clinton waived qualification rules repeatedly in making federal funds available to residents and businesses in California as a result of two flood events. The fact that California had a Democratic governor at the time and holds the largest number of electoral votes in presidential races is assumed to have played a role in Mr. Clinton's decision. In the spring of 1996, widespread flooding causing substantial damage occurred across Pennsylvania, yet only six counties were declared eligible for federal disaster assistance. Governor Tom Ridge publicly threatened consequences in the fall elections for federal officials "playing games with Pennsylvania." Very quickly, 58 of 67 Pennsylvania counties received federal disaster area status (Platt, 1999). Platt describes the political influence in federal assistance as "disaster gerrymandering."

Public policy also affects private insurance approaches to economic mitigation of disaster impacts. The insurance industry remains one of the most heavily regulated industries in the U.S. with many states having oversight bodies approving rates based on allowable underwriting profitability.[4] The problem is that the event horizons for disasters are often long, meaning that the insurer's premiums should account for building risk event reserves over several years. However, accumulating risk event reserves can appear as profits in the short run and are thus the targets of regulators looking to deliver politically popular rate decisions. Moreover, accounting rules and taxing policies on retained earnings hurt insurers' ability to build risk reserves (Andersen, 2004). Together, these policy factors result in wide fluctuations in disaster insurance availability[5] and premiums with the lowest availability/highest rates following disaster events. These higher rates discourage private sector adoption of own-source risk mitigation increasing the dependence on federal level solutions (Klein, 1998). In addition, closely timed disaster events, such as 2004 hurricane season in Florida just 11 months after the last major storm event, place further strains on insurance provider resources that are reflected in subsequent premiums.

The final political economy dimension to disaster research covered here is the potential for overt political considerations in the distribution of disaster relief. Though they specifically studied an Australian case, Butler and Doessel (1980) claim that politics can influence disaster relief in a federalist system of governance. Sverny and Marcal (2002), Scanlon (1988), and McEntire and Dawson (forthcoming) also discuss the politics of disaster relief and preparedness. Some disaster mitigation projects could be considered little more than pork-barrel politics. Moreover, as suggested earlier, there is more than a little of the political economy of wealth redistribution in some disaster policies in the U.S.

Several techniques and approaches will be presented in the remainder of this chapter for estimating the economic impacts of disasters and disaster planning and management. However, the application of the findings of these analyses remains an exercise in political economy.

MEASURING DISASTER LOSSES

Economists are rarely called on to estimate the direct physical damage caused by disasters. This is a

3. If the building is destroyed or substantially damaged, then taxable property values for improvements and business personal property decrease.
4. See Klein, R. (1998) for an excellent introduction into insurance industry regulation in the U.S. and regulatory impacts on disaster insurance.
5. Payouts for claims associated with Hurricane Andrew put several insurance and reinsurance providers out of business.

job for engineers, architects, construction specialists, and others. These damages include property damage to buildings and infrastructure, debris removal, and the cost of emergency protective services (McEntire and Cope, 2004). It is the losses associated with employment income and indirect losses that occupy the efforts of economists in the field of disaster research. Though there is some disagreement among scholars as to exactly what counts as indirect costs, they include the loss of business activity due to reduced activities at damaged firms, loss of income in secondary and tertiary employment, and business disruptions not directly attributable to damage. For example, if a manufacturing firm is damaged sufficiently to disrupt production, then they will not require trucking services to deliver raw materials or pick up finished goods, which may impact the employment of drivers. Rose (2004) illustrates indirect effects with the example of a utility plant being damaged resulting in utility customers (businesses) not being able to operate. Cochrane (2004) uses the comparatively simple definitions that direct damage is property damage plus lost income, and indirect damage is anything else. Rose, along with other researchers cited in his study, find that direct and indirect business interruption losses can be as large as physical losses. Of course, the degree of impact of a disaster depends in large part on the scale of the analysis.

Macroeconomic Analyses

Macroeconomic analysis considers economic events and activities at a national or at least state scale. Dacy and Kunreuther (1969) held that the total national cost of a disaster is the replacement value of the property damaged, regardless of the presence of a relief program. Even when other costs are included, it is a matter of simple division to see that disaster impacts rarely have a meaningful impact on a national economy. Whatever the damage, the divisor is very large. As noted by Mileti (1999), capital markets are simply too large to be disturbed beyond a short period of time by natural disasters.[6] The notable exception would be sustained droughts in countries with an agrarian-based economy (Albala-Bertrand, cited in Horwich, 2000). Nobel Laureate Gary Becker (2005) has noted that even the pandemic flu of 1918–1919 had no major effect on the world economy. To illustrate how this can be, Horwich (2000) offers an example based on the Kobe earthquake.

The Great Hanshin earthquake struck Kobe, Japan on January 17, 1995. In the earthquake and subsequent fires, more than 100,000 businesses were destroyed, 300,000 individuals became homeless, and 6,500 people were killed with total damages estimated at $114 billion (Horwich, 2000). The damage estimate represented about 2.5 percent of Japanese gross domestic product (GDP) in 1995. Yet within 15 months manufacturing was operating at 98 percent of the pre-earthquake trend, all department stores and 78 percent of small shops had reopened within 18 months, and trade at the port was operating close to pre-earthquake levels within one year (Horwich, 2000; Landers, 2001). That is a remarkable recovery based on GDP impacts. Similarly, Hurricane Katrina destroyed a sizable proportion of the economic capacity of Louisiana and Mississippi, but these states combine to represent less than 2 percent of U.S. GDP. The most recent data from the U.S. Department of Labor estimates that Katrina took 230,000 jobs from directly affected areas, but that total national employment for the month of September declined by only 35,000 – little more than a statistical blip on the economic map (Balls and Swann, 2005). Horwich suggests it is more telling to consider the impact of a disaster on economic potential as opposed to economic activity.

Economic potential can be measured by the level of capital stock including unused capacity in the economy. For example, other Japanese ports took on much of the trade activity while Kobe was under repair. In addition, Horwich (2000) suggests that human capital is the dominant economic resource and that, horrible as the losses were, 99.8 percent of the population in the earthquake impact zone survived. Horwich includes the economic value of life at $2 million per person, plus the $114 billion damage to capital stock, to estimate the capitalized value of the

6. Worthington and Valakhani (2004) using an autoregressive moving average model found temporary shocks to the Australian All Ordinaries Index from brushfires, cyclones, and earthquakes, though the direction of the impacts (positive or negative) varies.

Hanshin earthquake on Japan at $127 billion ($114 billion +(6500*$2 million)). Horwich calculates Japan's total resource value by capitalizing GDP (about $5 trillion in 1995) at a real interest rate of 3 percent for a total of $167 trillion ($5 trillion/0.03), which includes the value of a highly skilled workforce. Using this approach, the Great Hanshin earthquake had a total impact of 0.08 percent of the economic potential of the Japanese economy. Much of the economic activity lost due to the physical damage was regained in the form of rebuilding and repair. While Horwich makes some heroic assumptions in these calculations, they offer a clear indication of the resilience of the economy of large industrialized nations. However, even smaller nations, in terms of economic output, appear to possess economic resilience to disaster events.

One week after the Sumatra tsunami of 2004, the Indonesian and Malaysian stock markets had gained value from the pre-disaster level, the Thai stock market declined only slightly, and the Sri Lankan markets were off a few percent (Becker, 2005). Tavares (2004) using an ordinary least squares regression analysis calculates that natural disasters lower U.S. GDP by 0.052 percent per year.[7] Of course, the same may not hold true for smaller nations with more specialized economies.

In addition to the previously mentioned agrarian-based economies, Auffret (2003) finds that natural disasters are an important determinant of economic volatility in Caribbean economies, which is attributed, in part, to consumption shocks due to underdeveloped or ineffective risk management mechanisms. Of course, tourism-based economies are subject to market responses to disaster events – or predictions of disaster events – over which they have little control.

The other factor that minimizes the impacts of most disasters is their short duration. Waters recede, storms pass, and eventually droughts break. But for some types of disaster, the threat of an event can have a long-term effect on macroeconomic performance – specifically the threat of terrorism. Tavares (2004) estimates that the continuous threat of terrorist attacks reduces gross domestic product in Israel by 4 percent. The Basque region of Spain, which has seen decades of separatist terrorist activities, loses about 10 percent of its potential economic activity due to the threat of terrorism. Terrorism impacts national economies in 3 ways: (1) increased risk decreases business insurability meaning that risk is not spread across a greater number of economic actors, (2) trade costs are increased leading to lower levels of international transactions, and (3) increased public and private spending for security and defense decreases capital available for investment (Tavares). Hobijn (2002) estimates that increased security costs incurred after 9-11 has reduced U.S. economic activity by 0.66 percent.

One area of national level impacts that has received press coverage, but little academic analysis to date, is the impact of disasters on the U.S. energy industry. In 2004, hurricanes in the Gulf of Mexico substantially damaged that region's oil and gas production and transmission capacity. Winds and high waves toppled or dislodged the moorings for offshore rigs, and hurricane-spawned underwater mudslides destroyed sections of transmission pipelines. This damage resulted in lower domestic energy supplies that increased the market price for oil and gas and was reflected in the cost of gasoline, diesel, and fuel oil that rippled throughout the U.S. economy.[8] Damage sustained by refineries located in the New Orleans region along with off-shore oil production losses as a result of Katrina and subsequent flooding is currently blamed for adding as much as 40 cents to the price of a gallon of gasoline at the pump. These impacts, though temporary, should be formally assessed.

The resilience a given economy has to disaster events is, of course, largely dependent on national resources committed to mitigation, planning, and response. Horwich (2000) reports comments by noted disaster researcher Fred Cuny stating that if the earthquake that hit San Salvador had instead hit San Francisco, it would have rattled the china, not killed 1,500 people. As national income rises, disaster costs tend to rise, but relative costs as well as the number

7. This does not include the impacts of Hurricane Katrina. In comparison, Tavares (2002) found that currency crises decreased average economic output by 1.9 percent in the nations included in his model.

8. Access to fuel became a problem for FEMA in getting supplies to victims of the Florida hurricanes of 2004.

of lives lost decrease (Dacy and Kunreuther, 1969; Freeman et al., 2003).

However, aggregated analyses at the macroeconomic level miss the intensity of regional and local impacts that create comparative winners and losers when disaster strikes. In addition, macro level analysis often fails to identify and address disaster impacts and vulnerability across populations at differing income levels. As an overall economy gains wealth, it is often the case that low income populations are forced to reside in lower-cost/higher-risk areas compounded by their inability to afford insurance (Barnett, 1999; Scanlon, 1988; Vatsa, 2004). The stark, often horrific, images of the low-income victims of Hurricane Katrina and their disproportionate death rate and loss of most all worldly goods has brought into focus how disaster events can disproportionately affect the poorest segments of our population. Even when the national or regional economy recovers in terms of production and employment, specific localities, groups, and individuals may still be paying the price of disasters.

It is said that all politics are local. Given the earlier assertion that politics intertwines disaster economics and policies, it is reasonable to assume that the politics of disaster are often local. This is one reason why the preponderance of studies examining the economic impacts of disasters are conducted at the local or regional level.

Microeconomic Analysis of Disaster Impacts

It is well documented that the cost of disasters are rising, though care must be taken when making comparisons across time and when translating impacts across different currencies. Mileti (1999) reports the following disaster cost estimates based on a review of several studies:

• Loma Prieta Earthquake	1981	$10 Billion
• Hurricane Hugo	1989	$6 Billion
• Hurricane Andrew	1992	$20 Billion
• Northridge Earthquake	1994	$25 Billion

Mileti also cites an analysis that looked back at the 1923 Tokyo earthquake and estimated total damages at $1 Trillion in 1995 U.S. Dollars (USD). That estimate stretches credulity considering that just a few years after the earthquake Japan had the excess economic capacity to begin a massive military buildup. As noted, the damage estimate of the Kobe earthquake in 1995 was $114 billion USD based on simple currency exchange rates. However, if purchasing parity adjustments are made to the damage estimate, Horwich (2000) reports the cost estimate is $64 billion – a 44 percent reduction. Nonetheless, there is pervasive evidence that disasters are becoming more economically costly. Current assessments of private and public liabilities for rebuilding New Orleans in the wake of Hurricane Katrina exceed $200 billion. Yet, there are mitigating factors and evidence to suggest that the impacts are not always as large as advertised.

Tomsho (1999) reports on one of the most common factors that complicates the assessment of damage costs of disaster events – the Jacuzzi effect. The Jacuzzi effect occurs specifically when homeowners add new or improved features to their dwellings during disaster repairs. Of course, this ability to rebuild and restructure is one of the primary reasons that post disaster regional economies often improve their performance in the long term. As noted by Horwich (2000): "Restored economies will not be a replica of the pre-disaster economy. Destruction of physical assets is a form of accelerated depreciation that hastens adoption of new technologies and varieties of investment" (page 530). In addition, federal grants and low-interest loans act as economic stimuli with effects similar to transfer payments. The Charleston, South Carolina economy received $370 million in unexpected income after Hurricane Hugo in 1989 that helped the local economy to perform better than expected in 7 of the 10 quarters following the disaster event (Tomsho). But, as suggested earlier, the overall effect masked a great deal of disruption and volatility. Some businesses were permanently destroyed while new businesses opened. New Orleans, a city that had been suffering economic decline for decades prior to Katrina, may never recover its economic base beyond tourism and petro-chemicals according to some forecasters.

The main contribution that micro-economic analysis can bring is an understanding of the dynamics of the economic churn that is sparked by a disaster event. Which industries are most heavily impacted? Which are most likely to gain? A few years ago while riding in a taxi in Derry, Northern Ireland

(Londonderry if you are of loyalist persuasion), the driver observed to me that the first people on the scene of a terrorist bombing in his city are often the construction contractors preparing their repair bids. Even if this is a bit of an Irish yarn, it clearly points out that some industries and businesses will see potentially huge increases in their business activities resulting from disasters. By understanding the dynamics of the total economic impacts of disasters, we can more efficiently allocate disaster response resources so that those in need are the ones that are served. In addition, through predictive models using this information, we can make better decisions regarding disaster preparedness and pre-event mitigation strategies (Mileti, 1999; Gordon et al., 2005).

There are several data analysis techniques used to assess the indirect and income effects of disasters. These techniques include surveys, econometric models, Box-Jenkins time series analyses, input-output models, general equilibrium models, and economic accounting models (Cochrane, 2004; Chang, 2003; Zimmerman et al., 2005).

Surveys provide direct information from those impacted or in close association with those directly affected by disasters. They can be flexible in design to accomplish simple data gathering (How much will it/did it cost to rebuild your facility?) to more in-depth approaches (How did you finance your rebuilding? Have you lost customers because of down time? Are you looking to relocate your business?). Tierney (cited in Rose and Liao, 2005) uses surveys to assess impacts on businesses of the 1993 midwest floods and the Northridge earthquake. The largest problem with survey approaches is non-response bias. The researcher cannot know if the respondents are truly representative of the broader population of disaster victims. Given the psychological trauma associated with disasters, the researcher would have to be diligent in assessing response reliability – respondents' answers may change if questioned immediately after the event versus 6 months later. There could be issues of strategic behavior in the responses such as exaggerating losses in the hope of attracting additional aid. There are also potential logistics problems with surveys. Researchers may not have access to the disaster area immediately and may be unable to locate victims later. Moreover, the most appropriate survey medium would likely be in-person interviews, which are expensive and time-consuming. Still, surveys offer the best opportunity for obtaining direct, relevant data.

Econometric modeling approaches can be used when there is substantial data readily available for the affected region. Using a variety of regression techniques, the fully-partialed[9] effects of a disaster event can be modeled as an intrusion on a series of data. However, data availability can be a problem. Much of the economic data that would be used are gathered and published with substantial lags, this approach may not be practical until 2 or 3 years after the event. Of course, predictive models can help us understand post-disaster dynamics, but most econometric approaches do not easily account for product substitution, immediate changes in the imports of goods, or the non-linear nature of production functions inevitable when an economy receives a significant shock. Still, several researchers have offered credible analyses using regression techniques including Ellison, et al., (1984), Cochrane (1974), and Guimaraes, et al., (1993), among many others.

One econometric modeling approach is to use variations of hedonic pricing models. Hedonic models (derived from the term hedonism) account for preferences in purchasing decisions. These models are most commonly used in real estate research to described why some homes are more desirable (higher priced) even when other factors such as size and features are the same. MacDonald et al., (1987) use a hedonic model to assess housing value impacts of being located in a flood-risk area. Brookshire et al., (1985) examined hedonic price gradients based on earthquake safety attributes for housing. This modeling approach could add valuable insights into consumer behavior, especially if standard housing price models are adapted for longitudinal studies to examine changing hedonic factors in cases of recurring disaster events, such as housing prices in Florida after multiple major hurricanes.

---

9. Fully-partialed means that other factors affecting the economy are controlled for statistically so that the estimates relate to only those costs associated with the disaster.

A variation on the intrusion model method is an Auto-Regressive Integrated Moving Average (ARIMA) model. This analytic technique takes a Box-Jenkins approach to time series analysis. A Box-Jenkins analysis uses previous values of the study variable to predict the next value. Data analysis software packages use complex algorithms to account for secular trends in the data (are overall prices rising or falling?), the correlation between current and previous observations, seasonal variations, and other factors. For example, in examining the impacts of a tornado event on local retail sales, the analyst considers trends and patterns in a series of relevant data. The ARIMA model would control for a trend that total retail sales have generally risen over several years, the seasonal variations for Christmas, back-to-school, and other especially busy times, and the fact that if a retailer is successful one month, they will likely be successful the following month. The ARIMA model provides a prediction for what retail sales should be, which can then be compared to what actually happened after the disaster. The difference is an estimate of the disaster's impact on retail sales. The biggest weakness of this approach is being able to account for confounding concomitant events – such as a large retailer closing about the same time as a disaster for unrelated reasons. Because ARIMA models do not require the gathering of data for large numbers of relevant variables, the approach is very cost-effective. Enders et al., (1992) uses an ARIMA model to assess losses in the tourism industry due to terrorist events while Worthington and Valakhani (2004) use this modeling technique to estimate the impacts of disasters on the Australian All Ordinaries averages. Due to its relative simplicity but powerful analytical strengths, this data analysis methodology should be more widely used in disaster research.

Input/output (I/O) models are based originally on the work of Wassily Leontief in the 1930s in which the flow of goods across industries are captured using transaction matrices. For any given commodity there are raw materials, goods, and services purchased as inputs in the production process. Based on economic surveys, we know, on average, which industries produce which commodities and services. These models then provide a description of how demand-satisfying production creates upstream and downstream economic activities. For example, a writing pad is made of backing, paper, ink for the lines, and glue to bind the pages. There are firms that produce each of these inputs. In addition, the paper converter (manufacturer of goods converted from raw paper) hires accountants, computer services firms, and trucking companies, buys advertising space in trade publications, and purchases a host of other goods and services to support its business operations. The I/O models then use data from government organizations such as the Bureau of Labor Statistics to reflect relationships between labor demand for production activities and prevailing salaries, wages, and benefits to estimate not only the value of economic activity associated with a given level of production for a commodity, but how many jobs are supported and how much is paid in labor earnings.

National-level I/O models can be adjusted for regional economies by allowing for some activities to "leak" out of the economy. If the ink used to print lines on a tablet is not produced locally, then spending for that good does not impact the local economy and the related jobs and income are created elsewhere. However, being more precise, in a large regional economy there is likely to be at least one company that makes the ink, but that does not mean that company gets 100 percent of local market ink sales. Therefore, the regional I/O models estimate the proportion of total spending for intermediate goods that stay in the regional economy (expressed as regional purchasing coefficients). An I/O model may or may not include the economic activities (purchases) of households, though most do. The models produce three types of impact assessments: direct, indirect, and induced. Direct effects can be thought of as direct purchases by the industry being described. Indirect effects include purchases by related companies in the supply chain, such as the ink manufacturer buying office supplies from a local retailer. The induced effects capture the economic activity created by employees spending a portion of their earnings in the local economy for goods and services. When you add the direct, indirect, and induced impacts, expressed as coefficients, you can get a total effect greater than 1.0, which is the economic multiplier. For example, demand for $100 worth of writing pads in the Houston economy could create a total of $160 worth of local economic activity when all three types of impacts are summed.

Unfortunately, the multiplier effect works when production is added and when production is lost. If the paper converter's plant is damaged or destroyed, the related indirect and induced impacts spread across the regional economy.

The popularity of I/O modeling approaches has grown with the use and affordability of personal computers. There are two major off-the-shelf I/O models available on the market. One is produced by the Bureau of Economic Analysis of the U.S. Department of Commerce, and the other is called an IMPLAN model developed by the Minnesota IMPLAN Group. Both models are cost-effective and offer modeling capability at the county level. The IMPLAN model allows the user more flexibility in adjusting regional purchasing coefficients and offers estimates of economic activity at a highly disaggregated level – as many as 528 different industry categorizations. In addition, at the basic level, I/O models are relatively easy to use and can be used to quickly obtain an initial impact estimate.

The greatest weaknesses of I/O models are that they are static (measuring economic relationships at a particular point in time), the highly disaggregated impacts sometimes require heroic assumptions, and they are linear. If a new firm has come to town, or an existing firm has departed since the data base year, the regional purchasing coefficients may be wrong. Because detailed data for individual firms is masked in economic surveys, calculating very detailed industry estimates requires using national level data that may not accurately reflect local economic relationships. Finally, I/O models do not easily account for product substitutions, and the coefficients are fixed, which likely will not reflect reality in the aftermath of a disaster. Nonetheless, if used appropriately I/O models can provide reasonable estimates, not exact calculations, and are a valuable addition to the disaster researcher's toolkit. For an example of I/O modeling in disaster research see Rose, et al. (1997) in which the indirect regional economic effects are simulated for an earthquake event that damages electricity generating infrastructure.

I/O models can also include social accounting matrices (SAM) that expand the I/O model calculations to include transfer payments, value-added accounting, and the ability to examine distributional impacts across households at various income levels. Cole (2004) uses a SAM I/O model to project potential impacts of damage to the electric industry in upstate New York to aid regional disaster planning.

Another adaptation of I/O modeling uses econometric techniques to address some of the weaknesses noted above. The improvements include better coefficients that more accurately reflect local economic conditions and the ability to alter those coefficients to adjust for the structural economic changes that would attend a major disaster. This approach iteratively feeds back and forth from the I/O to the econometric portions of the model. Of course, the increased complexity and accuracy come with a price. The base models are more sophisticated than typical I/O models and thus are substantially more expensive. In addition, operating and adjusting the parameters is not typically accomplished by the end-user without extensive training and experience. Greater input data requirements and sophisticated user input mean that this model requires more time to complete an impact analysis. Therefore, these hybrid models usually do not offer details for as many industries as covered by I/O models.

The most widely used commercially available econometric-I/O hybrid model is REMI. However, a review of the disaster literature did not find any published articles using this model. Nonetheless, many state economic planning bodies have contracted access to the REMI model that could be used for disaster planning and impact analysis. For example, a REMI model could assess the regional and state level economic impacts of a tornado where repair services are being performed by a combination of firms previously located in the local economy, firms that open a permanent office in the region, and firms that send in "guest workers" for as long as there is sufficient demand.

Another recent adaptation of an I/O model was developed by the Center for Risk and Economic Analysis of Terrorism Events at the University of Southern California. This model begins with an IMPLAN model of the Los Angeles area (multiple counties), then applies a regional disaggregation model to allocate induced impacts across the region at the municipal level. The disaggregation model uses journey-to-work and journey-to-non-work (shop) transportation matrices that also account for intraregional freight flows (Gordon et al., 2005). However, because the base data of IMPLAN does not reach the sub-county level, this model aggregates the 509 IMPLAN industry sectors into 17 sectors. Still, this modeling approach could improve our ability to forecast or estimate how the economic impacts of a disaster event affect individual municipalities in a large metropolitan area. For example, Gordon et al. use the model to assess where the greatest economic disruptions would occur within the Greater Los Angeles area if there were terrorist attacks on the ports of Long Beach and Los Angeles.

The methodology being increasingly used in disaster research over the past few years has been computable general equilibrium models (CGE). Advocates of this modeling approach assert that CGE models are much more accurate than I/O models because they can incorporate a range of input substitutions and different elasticities of supply and demand can be applied across different tiers of economic activity (Rose and Liao, 2005). If a given input in a production process is no longer available in a post disaster environment, but can be easily imported from another region, then the CGE model more accurately estimates the direct, indirect, and induced effects of this change. However, this level of flexibility is very data intensive. Therefore, CGE models rarely cover more than a few industrial sectors. In addition, CGE models emphasize equilibrium states – a situation not likely to be the case in the aftermath of a significant disaster.[10] Among recent disaster-related research, Wittner et al. (2005) use a dynamic regional CGE model in a simulation modeling exercise on the effects of a disease or pest outbreak, while Rose and Liao (2005) demonstrate how CGE models can be used to value pre-event mitigation. Rose (2004) reviews at least three other studies that use CGE models for analyzing disaster impacts and policy responses. Because of its intensive data requirements and practical limitation on the number of industries that can be effectively analyzed at one time, CGE approaches to disaster impact modeling are better suited to *a-priori* assessments of potential impacts for planning purposes.

FEMA offers an impact assessment software that uses a combination of I/O, hybrid-I/O, and CGE modeling approaches to estimate direct and indirect economic impacts of disasters. The HAZUS-MH model is available for download from the FEMA website, but does require a geographic information system (GIS) model for input and output operations (FEMA, 2005). The HAZUS model is highly flexible allowing users to do a relatively quick and simple analysis using preprogrammed assumptions about the local economy (not recommended), to having to engage in detailed data gathering that would likely require the services of subject matter experts. The portion of the model that estimates indirect economic disaster impacts starts with IMPLAN data matrices and then employs adjustment algorithms similar to those described for hybrid-IO and CGE models. While the HAZUS model does offer many solutions to the problems of I/O impact analysis, it does not offer much in the way of industry detail aggregating the total regional economy into 10 basic industrial sectors that correspond to 1-digit Standard Industrial Code classifications. The HAZUS technical manual, available by request from FEMA, offers a case study based on the Northridge earthquake as well as simulation studies showing applications of the HAZUS model.

Finally, the economic accounting approach to estimating the impacts of disaster events differ from other approaches covered in this section in that it explicitly includes the valuation of human life and injuries. The economic accounting approach also draws from other methodologies to estimate business losses using case-based analysis (surveys), GDP estimates (econometric), or I/O models. These two elements are then added to estimates of physical losses to estimate the total economic impacts of a disaster (Zimmerman et al., 2005). The greatest challenge for the economic accounting method is valuing human life. The U.S. National Safety Council uses a loss of life value of $20,000 compared to the Environmental Protection Agency that calculates the value of lost lives at $5.8 million each. The Special Master for the Department of Justice overseeing claims related to the terrorist attacks on the World Trade Centers has used life values ranging from $250,000 to $7 million (Zimmerman et al.).

---

10. Rose (2004) offers a partial solution to this weakness in CGE modeling.

There are a number of weaknesses in the study of the economic impacts of disasters pointed to by many of the researchers cited above. Mileti (1997) and Cochrane (2004) both lament that most disaster impact studies only include losses that can be measured in transactions. The loss of historic monuments, memorabilia, cultural assets, and the hidden cost of trauma are rarely quantified (Mileti, 1997). In addition, Cochrane cautions against confusion over causality of a post-event loss, using too limited a time frame, and double counting losses among others. McEntire and Dawson (forthcoming) have called for formalizing an approach to document volunteer disaster responders' efforts. These researchers note that volunteer time can be used in federal grant matching requirements. Standardized methods of valuing volunteer time should be used in calculating the total economic impacts of a disaster event. While volunteers do not draw compensation, the time they spend in disaster response does have an opportunity cost.

Even with some weaknesses, there have been great strides in the analytic approaches to estimating the economic impacts of disaster events at the macro- and micro-economic levels. The challenge is to continue to improve the accuracy of our impact models, while keeping the methods computationally reasonable and having the ability to provide timely information to disaster management planners, political leaders, and responders.

INSURANCE

While insurance has its own academic and professional research literature, economics provides data, modeling techniques, and research methodologies to the study of disaster-related insurance markets. Most obviously are the techniques described above for estimating damage, especially indirect damage, following a disaster event. For example, of the $32.5 billion in insurance payouts as a result of the 9-11 terrorist attacks, $11 billion was for business interruption claims (Kunreuther and Michel-Kerjan, 2005). On the cutting edge of research techniques, Chen et al. (2004) employ neural network modeling to help predict house survival in Australian bushfires. In addition, simulation modeling for risk and economic losses is being used to establish premium levels, the degree to which risk spread is required, and the viability of insurance related derivative instruments (Andersen, 2004). One of these derivative instruments is an interesting market-based approach for addressing insurer exposures to the rising costs of disasters.

Catastrophe bonds (cat-bonds) are investments meant to spread the risk of insurance loss due to disaster events. As explained by Andersen (2004), these bonds are issued (sold) to investors. The proceeds of the bond sales are placed in high-grade investments that are relatively liquid (can be sold quickly) and have low interest rate sensitivity to serve as collateral for debt service payments. The holding entity issues insurance contracts and receives income from the policy premiums. Insurance claims are paid from policy proceeds as well as the investment portfolio resources. At maturity, the investors receive the full principal of the bond only if insurance payouts have not been made. For example, cat-bonds have been issued to spread the risk of insuring against FIFA's potential losses if the 2006 World Cup (soccer) tournament in Germany had been cancelled due to terrorism. Unfortunately, Kunreuther and Michel-Kerjan (2005) note that cat-bonds have not been broadly accepted by the market. Similarly, the Chicago Board of Trade and the Bermuda Commodity Exchange both tried issuing disaster-related financial derivatives through options and futures contracts but saw little market interest and have subsequently stopped trade in these financial instruments.

Political economy elements can also be seen in the disaster insurance market. Given huge losses and uncertainty about further attacks, the terrorism reinsurance market effectively stopped functioning in the months immediately after 9-11. Recognizing the connection between business growth and availability of insurance, Congress passed the Terrorism Risk Insurance Act (TRIA) of 2002 that provides up to $100 billion of reinsurance coverage for international terrorism events in the U.S. (Kunreuther and Michel-Kerjan, 2005).[11] However, once it became clear that no further attacks were imminent, the insurance market

11. Unless reauthorized, TRIA expires in 2005.

re-established itself. Brown et al. (2004) judge that TRIA has been, at best, value neutral for insurers and is seen as an impediment to market-based solutions by companies in the banking, construction, transportation, and other industries. Policymakers' concern about insurance market responses in the immediate aftermath of disasters has been a subject for discussion since the early 1990s, which saw huge industry losses in consecutive years as a result of hurricane Andrew (1992), the midwest floods (1993), and the Northridge earthquake (1994) (Barnett, 1999). But, aside from TRIA, there has been no meaningful congressional action on these concerns.[12] At the state level, California and Florida have created risk pools to promote insurance availability in their disaster-prone areas (Barnett, 1999).

REGIONAL DEVELOPMENT THEORY

There is a small but growing literature drawing connections between regional development and disaster planning, though the efforts are far from concerted. McEntire (2004) calls on disaster researchers to integrate development theory into their own research. As an example, he draws on the works of Max Weber and Karl Marx to show potential insights into disaster studies. Perhaps one of the greatest opportunities is to use current regional development thought to help explain consumer behavior in the face of disaster risk.

For many of us who do not live in the great state of Florida, we wonder why the state continues to have a fast growing real estate market in light of repeated disaster events over the past several years. A preliminary attempt at providing an explanation for this phenomenon requires multiple research disciplines. First is the acknowledgement from the social-psychology field that researchers do not understand peoples' responses to low probability events (Ganderton et al., 2000). However, it can be reasonably hypothesized that individuals expect either government or insurance resources to make them effectively whole in the case of disaster. This is supported by Kleindorfer and Kunreuther's (2000) finding that even with low costs and reasonable time periods for investment recovery, most consumers will not spend money for risk mitigation measures. Moreover, the probability of sustaining life-threatening injuries is likely perceived as virtually nil – at least when considering loss of life incident rates resulting directly from hurricanes. So that may explain why individuals are willing to risk hurricane damage to gain the environmental and recreational amenities of the Florida peninsula.[13] But that begs the question of why do businesses locate where there is a greater risk of physical damage and activity loss to go along with higher costs for insurance coverage?

Richard Florida offers a potential explanation in his writing about the "creative class." Florida (2002) asserts that business site location decisions are increasingly driven by the presence of cultural, recreational, and environmental amenities. In other words, site locations used to be based on proximity to raw materials and/or markets, now it is more about being in a location where potential employees want to live. Therefore, businesses locate where they have the greatest advantage in attracting the most talented workers, even if it is in an area with a higher probability of a disaster event. While Florida's theories are not universally accepted by regional economists, there is supporting evidence in the behavior of some firms. This suggests that the level of economic exposure to disasters will continue to rise until individuals perceive greater disincentives for moving to disaster prone areas. Berz (1994) and other researchers have called for greater use of building restrictions in coastal and riparian zones and other market interventions to slow growth in disaster prone areas; however, there is little political support for these suggestions.

The danger of wildfire losses from increasing development encroachment into forested lands are another notable disaster risk. The social, political, and economic conditions that can lead to greater exposure to catastrophe in areas that attract residential development because of environmental amenities are

12. Interestingly, Kunreuther and Michel-Kerjan (2005) report that more companies are purchasing terrorism insurance because executives fear they could be sued under provisions of the Sarbanes-Oxley Act if their firm suffers an uninsured attack.
13. The obvious question deals with the impact of Katrina on individuals' location decisions.

illustrated by Diamond (2005) for the Bitterroot River Valley of Montana.

It has been suggested that the threat of terrorism will impact urban land forms. Glaeser and Shapiro (2002) note that there are three types of effects that the threat of terror can have on urban design: promoting density, promoting dispersion, and increasing costs of transportation. Density in urban design is promoted through the psychology of safety in numbers. A highly dense population center offers a safe harbor where individuals enjoy mutual protection. Conversely, the same high population density makes cities a more efficient target for terrorists, which suggests that dispersing urban centers is appropriate. The third factor considers average transportation costs that favor high density urban designs. Glaeser and Shapiro conclude that, with a few exceptions, these factors balance out, and the threat of terrorism does not materially affect urban form. Rossi-Hansberg (2004) suggests that in theory, bidrents in areas with a higher probability of physical destruction would decrease to account for increased risk and thus impact property investment decisions and change the physical structure of a city. However, this theoretical approach does not appear to fully account for insurance and government assistance – suggesting again that current government disaster policy may be supporting increasingly inefficient real estate markets in disaster prone areas.

DISASTER PLANNING

In addition to the applications of economic theory and research techniques described previously, there are a few other ways that the discipline of economics contributes to disaster planning. For example, Rose (2004) has offered measures of economic resilience – the capacity of an economy to absorb or diminish the effects of shocks – that can enhance the ability of planners and disaster responders to enable individuals and communities to avoid some potential losses. The distribution of mitigation funds could be made based on measures of economic vulnerability and event risks (Adrianto and Matsuda, 2004; Cole, 2004).

Of course, the threat of terrorism occupies much of the efforts of disaster planners in our post-9/11 political environment. Data analysis techniques from the economic discipline are being employed to assess the risk and responses to the threat of terrorism such as spectral analysis to examine cycles of events, vector autoregressive techniques for quantifying patterns of attack, and game theory approaches for predicting the likelihood of attacks and the effect of deterrence strategies (Lapan and Sandler, 1988; Sandler et al., 1991; Arce and Sandler, 2005; Averett, 2005).

Finally, with increases in funding for disaster planning in the past few years, there is need for disaster planners to have access to the knowledge of regional economists and economic development theory and practice. For example, Dekle et al. (2005) have developed a site location tool to assess potential locations for disaster recovery centers. While the physical location of disaster recovery centers and centers for disaster research will continue to be influenced by the political economy, we can hope that sound, practical reasons, such as promoting the effectiveness of the delivery of disaster response services, will remain the primary site location factor.

CONCLUSIONS

Offering the reader a reasonably brief overview of the use of economic research methods and techniques for the study of disasters and disaster management inevitably results in omissions, incomplete descriptions, and failure to recognize the contributions of many talented and insightful scholars. However, this chapter has presented an overview of the contributions of the economic discipline to understanding the costs of disasters, the analysis of private insurance markets, and theories and research techniques used in various phases of disaster management planning. In addition, it has illustrated how political considerations affect disaster policy and the distribution of relief funds. Even so, there is a great deal left in the field of disaster management that could be aided through the application of economy theory and research techniques.

Mileti (1999) specifically calls for the creation of a national database of losses and vulnerability that would serve as a communications feedback loop for communities, researchers, emergency managers, and government. There have been some that have suggested standardizing the approach used to estimate economic losses from disasters. However, from a practical standpoint, it is better to allow for flexibility in research technique for two reasons. First, the choice of cost estimation technique should consider the information need – how fast are the estimates needed, on what scale, and to what depth? Second, standardization will certainly serve to stifle innovation in new, probably better, ways to assess the economic impacts of disasters.

We should continue to employ economic theory and modeling to address issues of efficiency, equity, and consistency in disaster mitigation and response. Under current policies the overall scope of a disaster has too great of an influence in deciding the funds made available to individuals in need. In addition, disaster costs are rising due to rapidly growing populations in coastal and other high risk areas, local zoning and building codes that do not adequately address disaster risks, increasingly inefficient real estate markets that are distorted by spreading the risk of locating in disaster prone areas to all taxpayers, and spin-offs of a growing economy such as increases in the shipment and use of hazardous materials.

Addressing critical information needs to disaster planners, policy makers, and responders will continue to challenge economists. Working in concert with researchers from disciplines such as sociology, geography, anthropology, engineering and others, economists can address information needs and offer guidance on maximizing our ability to mitigate disaster impacts.

REFERENCES

Adrianto, L. Matsuda, Y. (2004). Study on assessing economic vulnerability of small island regions. *Environment, Development and Sustainability, 6*(3), 317.

Andersen, T. (October, 2004). International risk transfer and financing solutions for catastrophic exposures. *Financial Market Trends, 87*, 91–120. Accessed May 6, 2004 at http://proquest.um.com/pqdweb?did=739604541&sid=1&Fmt=4&cliemtld=87&RQT=309&VName=PDQ

Arce, D. and Sandler, T. (2005). Game-theoretical analysis. *The Journal of Conflict Resolution, 49*(2), 183–200.

Auffret, P. (2003). *High consumption volatility: The impact of natural disasters.* World Bank Policy Research Working Paper 2962. The World Bank: Washington, DC.

Averett, S. (2005). Building a better bulwark. *Engineer, 37*(2), 24–29.

Balls, A. and Swann, C. (October 8, 2005). Little sign of Katrina damage as employment figures hold up. *Financial Times*, 3.

Barnett, B. (1999). US government natural disaster assistance: Historical analysis and a proposal for the future. *Disasters, 23*(2), 135–155.

Becker, G. (January 4, 2005). And the economics of disaster. *Wall Street Journal*, A-12.

Berz, G. (1994). The insurance industry and IDNDR: Common interests and tasks. *Natural Hazards, 9*, 323–332.

Brookshire, D., Thayer, M., Tschirhart, J., and Schulze, W. (2001). A test of the expected utility model: Evidence from earthquake risks. *Journal of Political Economy, 93*(1), 369–389.

Brown, J., Cummins, J., Lewis, C., Wei, R. (2004). An empirical analysis of the economic impact of federal terrorism re-insurance. *Journal of Monetary Economics, 51*(5), 861.

Butler, J. and Doessel, D. (1980). Who bears the costs of natural disasters? An Australian case study. *Disasters, 4*(2), 187–204.

Chang, S. (1984). Do disaster areas benefit from disasters? *Growth and Change, 15*(4), 24–31.

Chang, S. (2003). Evaluating disaster mitigations: A methodology for urban infrastructure systems. *Natural Hazards Review, 4*, 186–196.

Chen, K., Jacobson, C., & Blong, R. (2004). Artificial neural networks for risk decision support in natural hazards: A case study of assessing the probability of house survival from bushfires. *Environmental Modeling and Assessment, 9*(3), 189.

Cochrane, H. (1974). *Social science Perspectives on the Coming San Francisco Earthquake: Economic Impact, Prediction, and Reconstruction.* Natural Hazard Research Working Paper No. 25. Institute of Behavioral Science, University of Colorado: Boulder.

Cochrane, H. (2004). Economic loss: Myth and measurement. *Disaster Prevention and Management, 13*(4), 290–296.

Cole, S. (2004). Performance and protection in an adaptive transaction model. *Disaster Prevention and Management, 13*(4), 280–289.

Crossett, K., Culliton, T., Wiley, P. Goodseep, T. (2004). *Population trends along the coastal United States: 1980–2008.* National Ocean Service: Washington, DC. Accessed January 23, 2006 at http://www.oceanservice.noaa.gov

Dacy, D. and Kunreuther, H. (1969). *The Economics of Natural Disasters: Implications for Federal Policy.* The Free Press: New York.

Dekle, J., Lavieri, M., Martin, E., Emir-Farinas, H., and Francis, R. (2005). A Florida county locates disaster recovery centers. *Interfaces, 35*(2), 135–139.

Diamond, J. (2005). *Collapse: How Societies Choose to Fail or Succeed.* Viking: New York.

Ellison, R., Milliman, J., and Roberts, R. (1984). Measuring the regional economic effects of earthquakes and earthquake predictions. *Journal of Regional Science, 24,* 559–579.

Enders, W., Sandler, T., and Parise, G. (1992). An econometric analysis of the impact of terrorism on tourism. *Kyklos, 45,* 531–554.

FEMA (2005). Multi Hazards Loss Estimation Software. See http://www.fema.gov/hazus/hz_index.shtm.

Florida, R. (2002). *The Rise of the Creative Class and How It's Transforming Work, Leisure, Community and Everyday Life.* Basic Books: New York.

Freeman, P., Keen, M., and Mani, M. (2003). *Dealing with Increased Risk of Natural Disasters: Challenges and Options.* IMF Working Paper No. 03/197. International Monetary Fund, Washington, DC.

Ganderton, P., Brookshire, D., McKee, M., Stewart, S., and Thurston, H. (2000). Buying insurance for disaster-type risks: Experimental evidence. *Journal of Risk and Uncertainty, 20*(3), 271–289.

Glaeser, E. and Shapiro, J. (2002). Cities and warfare: The impact of terrorism on urban form. *Journal of Urban Economics, 51,* 205–224.

Gordon, P., Moore, J., Richardson, H. and Pan, Q. (2005). *The Economic Impact of a Terrorist Attack on the Twin Ports of Los Angeles-Long Beach.* A report by the Center for Risk and Economic Analysis of Terrorism Events. University of Southern California, Los Angeles.

Guimaraes, P., Hefner, F., and Woodward, D. (1993). Wealth and income effects of natural disasters: An econometric analysis. *Review of Regional Studies, 23,* 97–114.

Hobijn, B. (2002). What will homeland security cost? *Federal Reserve Bank of New York Economic Policy Review.* New York: Federal Reserve Bank of New York.

Horwich, G. (2000). Economic lessons of the Kobe earthquake. *Economic Development and Cultural Change, 48*(3), 521–522.

Klein, R. (1998). Regulation and catastrophe insurance. In Kunreuther, H. & Roth, R. (Eds.) *Paying the Price: The Status and Role of Insurance Against Natural Disasters in the United States.* pp. 171–208. Joseph Henry Press: Washington, DC.

Kleindorfer, P. and Kunreuther, H. (2000). Managing catastrophe risk. *Regulation, 23*(4), 26–31.

Kunreuther, H. (1998). Introduction. In Kunreuther, H. & Roth, R. (Eds.) *Paying the Price: The Status and Role of Insurance Against Natural Disasters in the United States.* pp. 1–16. Joseph Henry Press, Washington, DC.

Kunreuther, H. and Michal-Kerjan, E. (2005). Terrorism insurance 2005. *Regulation, 28*(1), 44–51.

Kunreuther, H. and Miller, L. (1985). Insurance versus disaster relief: An analysis of interactive modeling for disaster policy planning. *Public Administration Review, special issue,* 147–154.

Landers, P. (October 9, 2001). Kobe disaster offers clues on rebuilding. *Wall Street Journal.* A19.

Lapan, H. and Sandler, T. (1988). The political economy of terrorism. *The American Economic Review, 78*(2), 16–21.

Leitko, T, Rudy, D. & Peterson, S. (1980). Loss not need: The ethics of relief giving in natural disasters. *Journal of Sociology and Social Welfare, 7*(5), 730–741.

Lotterman, E. (1997). Receding flood waters reveal huge damage assessment. *Fedgazzette, 9*(2), 1–2.

MacDonald, D., Murdoch, J. and White, H. (1987). Hazards and insurance in housing. *Land Economics, 63,* 361–371.

McEntire, D. (2002). Coordinating multi-organizational responses to disaster: Lessons from the Marh 28, 2000 Fort Worth tornado. *Disaster Prevention and Management, 11*(5), 369–379.

McEntire, D. (2004). Development, disasters, and vulnerability: A discussion of divergent theories and the need for their integration. *Disaster Prevention and Management, 13*(3), 193–198.

McEntire, D. and Cope, J. (2004). *Damage assessment after the Paso Robles (San Simeon, CA) earthquake: Lessons for emergency management: Quick Response Research Report #166.* Natural Hazards Center at the University of Colorado, Boulder.

McEntire, D. and Dawson, G (forthcoming). "Operating in an Intergovernmental Context." Chapter, International City/County Management Association.

Mileti, D. (1999). *Disasters by Design: A Reassessment of Natural Hazards in the United States.* Joseph Henry Press, Washington, DC.

Mills, E. (2002). Terrorism and US real estate. *Journal of Urban Economics, 51,* 198–204.

Okuyama, Y. (2004). Modeling spatial economic impacts of an earthquake: Input-output approaches. *Disaster Prevention and Management, 13*(4), 297–306.

Platt, R. (1999). *Disasters and Democracy: The Politics of Extreme Natural Events.* Island Press, Washington, DC.

Rose, A., Benevides, J. Chang, S. Szczesniak, P., and Lim, D. (1997). The regional economic impact of an earthquake: Direct and indirect effects of electricity lifeline disruptions. *Journal of Regional Science, 37*(3), 437–458.

Rose, A. (2004). Defining and measuring economic resilience to disasters. *Disaster Prevention and Management, 13*(4), 307–314.

Rose, A. and Liao, S. (2005). Modeling regional economic resilience to disasters: A computable general equilibrium analysis of water service disruptions. *Journal of Regional Science, 45*(1), 75–112.

Rossi-Hansberg, E. (2004). Cities under stress. *Journal of Monetary Economics, 51*, 903–927.

Sandler, T., Enders, W. & Lapan, H. (1991). Economic analysis can help fight international terrorism. *Challenge, 34*(1), 10–17.

Scanlon, J. (1988). Winners and losers: Some thoughts about the political economy of disaster. *International Journal of Mass Emergencies and Disasters, 6*(1), 47–63.

Sverny, S. and Marcal, L. (2002). The allocation of federal funds to promote bureaucratic objectives: An empirical test. *Contemporary Economic Policy, 20*(3), 209–220.

Swanson, S. (June 26, 2005). Funding long-term restoration in next challenge in tsunami relief. *Baltimore Sun.* Accessed at www.baltimoresun.com July 3, 2005.

Sylves, R. (1996). The Politics and Administration of Presidential Disaster Declarations: The California Floods of Winter 1995. Quick Response Report #86. Natural Hazards Center at the University of Colorado, Boulder. Assessed at: http://www.colorado.edu/hazards/qr/qr86.html

Tavares, J. (2004). The open society assesses its enemies: Shocks, disaster, and terrorist attacks. *Journal of Monetary Economics, 51*(5), 1039–1070.

Thissen, M. (2004). The indirect economic effects of a terrorist attack on transport infrastructure: A proposal for a SAGE. *Disaster Prevention and Management, 13*(4), 315–322.

Tomsho, R. (October 5, 1999). "Anthill" economics: How natural disasters can change the course of a region's growth. *Wall Street Journal*, A1.

USGS (2005). *Earthquake Hazards Program.* United States Geologic Service, Washington, DC, available at http://quake.wr.usgs.gov

Vatsa, K. (2004). Risk, vulnerability, and asset-based approaches to disaster risk management. *International Journal of Sociology and Social Policy, 24*(10/11), 1.

Wittner, G., McKirdy, S., and Wilson, R. (2005). Regional economic impacts of a plant disease incursion using a general equilibrium approach. *Australian Journal of Agricultural and Resource Economics, 49*(1), 75.

World Book (2005). Most deadly earthquakes in history. Downloaded May 31, 2005 from www2.worldbook.com/wc/features/earthquakes/html/mostdeadly.htm

Worthington, A. and Valadkhani, A. (2004). Measuring the impact of natural disasters on capital markets: An empirical application using intervention analysis. *Applied Economics, 36*, 2177–2186.

Yezer, A. (2002). The economics of natural disasters. In Stallings, R. (Ed), *Methods of Disaster Research.* Xlibris, Philadelphia.

Zimmerman, R., et al (2005). *Electricity Case: Economic Cost Estimation Factors for the Economic Assessment of Terrorist Attacks.* A report by the Center for Risk and Economic Analysis of Terrorism Events. University of Southern California, Los Angeles.

Chapter 19

Emergency Management and Law

William Charles Nicholson

ABSTRACT

The following chapter relates the history of law and emergency management, discusses vulnerability and steps to be taken for its reduction, defines various concepts from a legal perspective, and examines gaps in knowledge between the two fields. The chapter also notes how law may improve emergency management and identifies considerations that are paramount to the future. The major argument to be presented is that law and emergency management are inherently intertwined and that legal norms in the disaster field are changing and having a significant impact on the profession.

INTRODUCTION

In many ways, emergency management could not exist without the law. In the United States, legal enactments provide the authorities and funding for emergency management. Definitions of critical emergency management terms have been established in legal enactments. Although their interaction may be difficult at times, lawyers and emergency managers need one another. A major obstacle is the mutual ignorance that all too often characterizes their relationship. When attorneys, emergency managers, and leaders of units of government take the time to build a relationship that encompasses all phases of emergency management, the result can be shelter from liability as well as greater life safety and improved property protection.

HISTORY OF LAW AND EMERGENCY MANAGEMENT

The history of disasters in the United States is intertwined with the law (FEMA, 2005a). On the federal level, as early as 1803, Congress enacted legislation to provide relief from a severe fire in a New Hampshire town. The Congressional Act of 1803 is generally thought of as the first piece of disaster legislation. During the next century, specific legal enactments authorized funding for the response to disaster events one incident at a time. The 1930s brought about an organized federal approach to disaster law. The Reconstruction Finance Corporation was authorized to generate disaster loans for the repair and reconstruction of some public facilities after an earthquake. This authority was extended later to other varieties of disaster. The Bureau of Public Roads, under a 1934 law, was empowered to provide funding for highways and bridges damaged by natural disasters. Another important piece of legislation, the Flood Control Act, expanded the authority of the U.S. Army Corps of Engineers to put into effect flood control projects. This approach to disaster assistance improved on the prior "one at a time" practice of creating legal authority. Yet problems remained. The ever-increasing size of the national government meant that sometimes federal agencies with different pieces of disaster authority found themselves working at cross purposes. As a result, Congress enacted legislation requiring better greater cooperation between federal agencies and authorizing the President to coordinate these activities.

The subsequent history of disasters reveals that they steadily grew in both number and magnitude. The federal government was faced with enormous disasters in the 1960s and early 1970s. The Federal Disaster Assistance Administration, which was located in the Department of Housing and Urban Development, coordinated these efforts. Hurricanes Carla (1962), Betsy (1965), Camille (1969), and Agnes (1972), as well as large earthquakes in Alaska (1964) and San Fernando in California (1971) put natural disasters in the forefront of national attention, and resulted in legislation. The 1968 National Flood Insurance Act gave homeowners new assistance, while the 1974 Disaster Relief Act regularized the procedure for issuance of Presidential disaster declarations.

Despite this legal progress, there was still not a unified framework for emergency and disaster practices. By the 1970s, disasters, hazards and emergencies were the business of over 100 federal agencies. On the state and local level, similar structures were in place. The result was a confusing welter of groups and efforts that often competed with or duplicated one another. At the request of the National Governor's Association, President Jimmy Carter moved to consolidate federal emergency functions.

In 1979, President Carter's issued an executive order unifying federal disaster activities under the newly created Federal Emergency Management Agency (FEMA). FEMA incorporated many bodies, including the Federal Insurance Administration, the National Fire Prevention and Control Administration, the National Weather Service Community Preparedness Program, the Federal Preparedness Agency of the General Services Administration and the Federal Disaster Assistance Administration from HUD. Civil defense moved to FEMA from the Defense Civil Preparedness Agency in the Department of Defense.

In the aftermath of the first attack on the World Trade Center (1993) and the Oklahoma City bombing (1995), FEMA's "all-hazards" approach to disaster management was overshadowed by a concentration on homeland security matters. The Homeland Security Act of 2002 (HS Act) united 22 federal agencies, programs and offices, including FEMA, to create the Department of Homeland Security (DHS). Creating DHS was another legal step in unifying disaster preparedness and response. DHS' mission focuses on terrorism, including prevention, vulnerability reduction, minimizing damage, and assisting in recovery from terrorism attacks (107th Congress, 2002, § 1(a-c)). Also included in the Department's responsibilities is carrying out all functions of entities transferred to the Department, including acting as a focal point regarding natural and manmade crises and emergency planning (107th Congress, 2002, § 1(d).

Some experienced emergency management observers believe that the focus at DHS is too terrorism-oriented (Nicholson, 2003a), with troubling impact on the all-hazards preparedness mission that FEMA has traditionally espoused (Waugh, 2002). This is an issue that has two sides, but whatever perspective one endorses, to a great extent the argument revolves around the nature of legal enactments and their interaction with policy. From the view of statutory construction, however, the fact that the Department's terrorism responsibilities are listed as the first three parts of its mission while other hazards are lumped together in fourth place means that Congress intended DHS' terrorism responsibilities to be more important than those dealing with other hazards.

DEFINING AND REDUCING VULNERABILITY

The National Response Plan (NRP) (National Response Plan, 2005a), and the National Incident Management System (NIMS) (National Incident Management System, 2005a) do not define "vulnerability." In the FEMA publication *Building Design for Homeland Security*, "vulnerability" is defined as "any weakness that can be exploited by an aggressor or, in a non-terrorist threat environment, make an asset susceptible to hazard damage" (FEMA, 2005b). The publication discusses vulnerability assessment as well as what steps to take once vulnerabilities have been identified in order to mitigate against the identified threat.

The persuasiveness of authority for the term "vulnerability" is somewhat less than if it were defined directly in the NRP or NIMS. Its promulgation by FEMA and general use in the profession, however, indicate that an American Court under the commonly accepted business practice doctrine (discussed at greater length below) would find them influential.

Recently, a pair of Australians made an interesting suggestion for an increased role for legal enactments in vulnerability reduction (Handmer and Monson, 2004). Their approach features a definition of vulnerability as "a multi-faceted concept incorporating issues of livelihood, housing, security, and gender, among many others" (Handmer and Monson, 2004). The piece suggests that a link between vulnerability and law exists when laws set out rights to adequate housing and livelihood, for example. In addition to the familiar constraints of public and private law, social norms, custom, and international law are posited as having the potential to regulate vulnerability (Handmer and Monson, 2004). The article focuses on human rights as found in national public law, since such laws have been enforceable by the citizenry against their government. Enforcing other types of law is a much less certain endeavor.

Vulnerability is a "function of susceptibility to loss and the capacity to recover" (Handmer and Monson, 2004). Due to their more positive connotations, some prefer the terms resilience or capacity to vulnerability. The most vulnerable people are those whose basic human needs, like adequate food, shelter, health care, and education, are unmet. These needs are defined by the piece as "fundamental human rights."[1] The rights-based approach works from the bottom – originating with the affected groups – as opposed to from the top – through government, Courts, and experts. The approach identifies the sources of vulnerability (failure to meet certain rights) and contains a way to reduce them (through legal enforcement of rights).

The article posits that international law may provide a method for expansion of enforceable human rights, through more inclusive interpretation. For example, it suggests that the right to life, liberty and security of every person under the *Universal Declaration of Human Rights* might expand to include protecting the "security of the person" from other harm, like natural disasters. Such an approach overemphasizes the force of international law, whose power extends only to those matters by which individual nations agree to be bound. Nations unilaterally may change their adherence to such agreements, other than in matters of torture and genocide. The article acknowledges an "implementation gap" on human rights, even in wealthy nations as well as the virtual impossibility of enforcing naked (that is, without incorporation into domestic law) international law. While some may espouse universal human rights, their practice is far from uniform around the world.

Three South African cases are interesting illustrations of the authors' premises. The first establishes a constitutional obligation to provide disaster relief, but states that a hearing is not required for all who object to the way relief is given (Handmer and Monson, 2004). The exceptional circumstances in a disaster allow the government to forego more onerous procedures than would normally apply to decision making. While the United States has never held disaster relief to be a constitutionally protected right (and the possibility of that ever happening in the U.S. is highly unlikely, to say the least), the ability of the government to avoid procedural inconveniences is well established here.[2] The second decision revolves around access to housing and health care. Like many other nations with constitutions established or heavily revised in the second half of the twentieth century,[3] South Africa's constitution lists a range of rights to be provided within its available resources, including housing and health care. The residents in this case were squatting on private land, from which they were cruelly ejected, after which they were relocated into intolerable conditions. They appealed to the Constitutional Court. That tribunal held that, despite the challenges in enforcing them, "these are rights, and the Constitution obliges the State to give effect to them" (Public Law, No Publication Year).

The third case discusses the right to treatment for HIV patients. Some scholars view AIDS as a type of disaster (Varley, 1994). Clearly, the illness's effect on public health budgets has been disastrous. This case provided that South Africa had the constitutional obligation to provide HIV treatment to pregnant women to help prevent transmitting HIV to their unborn children. The South African Constitution recognizes a right to access to public health care services and requires the state to take reasonable steps, within its available resources, to achieve the progressive realization of this right (Public Law, No Publication Year). The Court found that the government was not going far enough in making appropriate medication available.

The South African cases illustrate how far a country may go in guaranteeing and enforcing human rights that go well beyond those afforded in the United States. Other nations with similar constitutions might pursue the same approach. In Europe, human rights established by the European Union cannot be enforced in the European Court of Human Rights, which enforces the European Convention on Human Rights. That convention does not recognize, for example, a right to adequate housing or health care. The best approach in Europe, as well as in Australia, is posited to be through legislation rather than Constitutional change (Public Law, No Publication Year). This is because, as in the United States, it is very difficult to amend the Constitution.

Parenthetically, it must be observed that the desire to resist the faddish causes of the moment and preserve existing property and other legal relationships is an important reason that Constitutions are difficult to amend.

Also, as a practical matter, establishing the redistribution of wealth in the manner envisioned by the expansion of fundamental human rights to include disaster relief, housing, and medical care is most likely to result in national bankruptcy for those countries that decide to put it into action. The limit placed on such services by the "progressive realization" language cited by the South African Court decisions may mean that the process of bankruptcy will be prolonged rather than immediate, but that does not make it less probable.

THE LEGAL PERSPECTIVE

As might be expected, hazards, disasters, and emergency management have definitions established by law. Definitions are found in various locations, most importantly including glossaries in the National Response Plan (NRP) (National Response Plan, 2005) and the National Incident Management System (NIMS) (National Incident Management System, 2005). States also define some of these terms. Federal and state law also determines responsibilities for preparedness.

When finalized, the NRP and NIMS will be the end product of a process that began with the passage of the HS Act of 2002. On February 28, 2003 President Bush issued Homeland Security Presidential Directive 5 (HSPD 5) (The White House, 2003). HSPD 5 directs all Federal agencies to take specific steps for planning and incident management. HSPD 5's major goal is to establish a single, comprehensive approach to domestic incident management. The effect of this unified approach will be efficient and effective operation of all levels of government as regards disasters. The Directive specifies the lead agencies for terrorism events and other major disasters. HSPD 5 directs all Federal agencies to work together with DHS to institute the NRP and NIMS. NIMS is the operational portion of the NRP (Homeland Security Presidential Directive 5, 2003). In this manner, legal authority for creating the NRP and NIMS flows from the HS Act of 2002 through HSPD 5 to DHS (Nicholson, 2003b). Failure to comply with the mandates of the NRP and NIMS subjects emergency response and emergency management groups to sanctions, in the form of losing federal grant funds (Homeland Security Presidential Directive 5, 2003).

Given that the NRP and NIMS establish enforceable standards, their definitions have the effect of law for those entities that do not wish to lose their federal funding. For the few entities that do not elect to preserve their federal funding, the NRP and NIMS definitions will also have legal effect as industry standards. The "commonly accepted business practice" doctrine operates to establish elevated standards of care when a large number of similarly situated concerns take supplemental actions. Here, adoption of NRP and NIMS by an overwhelming majority of emergency management groups would be strong evidence to a Court that it should hold all emergency management organizations to these norms.

The NRP and NIMS define "hazard" as something that is potentially dangerous or harmful, often the root cause of an unwanted outcome (National Response Plan, 2005; National Incident Management System, 2005). To define "disaster," the NRP and NIMS refer to the Stafford Act's definition of a "major disaster" as:

> Any natural catastrophe (including any hurricane, tornado, storm, high water, wind-driven water, tidal wave, tsunami, earthquake, volcanic eruption, landslide, mudslide, snowstorm, or drought) or, regardless of cause, any fire, flood, or explosion, in any part of the United States, which in the determination of the President causes damage of sufficient severity and magnitude to warrant major disaster assistance under this act to supplement the efforts and available resources of States, local governments, and disaster relief organizations in alleviating the damage, loss, hardship, or suffering caused thereby. (National Response Plan, 2005; National Incident Management System, 2005)

States typically have their own definitions of disaster.[4]

The NRP and NIMS do not define "emergency management." Two online courses offer definitional assistance that is consistent. The FEMA online course Introduction to Emergency Management does not offer a simple designation. Rather, it discusses the nature of comprehensive emergency management, building from the simple image of a homeowner responding to a broken water pipe and a flooded basement. The course sums up the modern emergency management's focus as follows: "Today the emphasis is on the protection of the civilian population and property from the destructive forces of natural and man-made disasters through a comprehensive program of mitigation, preparedness, response, and recovery" (FEMA, 2005c). FEMA's on line Principles of Emergency Management course defines the term rather straightforwardly as "Organized *analysis, planning, decision-making, and assignment of available resources* to mitigate, prepare for, respond to, and recover from the effects of all hazards" (FEMA, 2005d). States also have their definitions for emergency management, which correspond to the federal approach described above.[5]

The persuasiveness of authority for the term "emergency management" is somewhat less than if it were defined directly in the NRP or NIMS. Its promulgation by FEMA and general use in the profession, however, indicate that a Court under the above discussed commonly accepted business practice doctrine would find it influential.

The HS Act of 2002 is somewhat contradictory in setting out responsibilities for preparedness. The role of FEMA is defined to include its Stafford Act functions as well as reducing the loss of life and property and protecting the nation from "all hazards" by leading and supporting the nation in a comprehensive, risk-based emergency management program.[6] The law tasks the Office of Domestic Preparedness (ODP) with terrorism preparedness, in contrast to FEMA, which is specifically entrusted with preparing for and mitigating the effects of non-terrorist-related disasters in the United States.[7] The statute's reader wonders whether the "all hazards" language is mere window dressing, given the division of roles in the description of ODP's tasking.

The end result must be confusion to emergency management, similar to that engendered by Congress' decision to break off planning for release of extremely hazardous substances (EHS) from "all hazards" emergency management in 1986. The federal Emergency Planning and Right to Know Act (EPCRA) requires state and local units of government to split off an important part of emergency management to another entity. EPCRA is contained in the Superfund Amendment and Reauthorization Act of 1986 (SARA Title III) (USC, 2005, §§ 11001–11050). EPCRA mandates that a State Emergency Response Agency (SERC) must ensure planning for EHS releases (USC, 2005, § 1001(a)). The SERC creates emergency planning districts and superintends Local Emergency Planning Committees (LEPCs), (USC, 2005, § 1001(b) (c)) which do the planning for EHS releases (USC, 2005, § 1001(a)).

Emergency management has been able to incorporate LEPC plans as annexes to emergency operations plans. LEPCs and emergency management generally work well together. It remains to be seen whether the same approach will be applied as successfully to terrorism planning at all levels of government.

THE INTERACTION BETWEEN LAW AND EMERGENCY MANAGEMENT

The relationship between law and emergency management may be characterized as one mutual need. Mitigation in particular is an area where the two disciplines have the potential to interact very well. Regrettably, in spite of the fact that the law creates emergency management, in general the understanding of emergency managers and lawyers may be described as mutual ignorance. Some are not even aware that their activities are governed by both federal and state law (Pine, 1991).

Many business and government leaders are uninformed regarding the laws that control their behavior. Sometimes, emergency managers may pay no attention to the law. They may vociferously declare themselves to be "too busy saving lives and protecting property to bother with all that legal mumbo jumbo." Such an attitude is peculiar, given emergency management's "all hazards" character. Analysis of the instructive resources accessible to most emergency managers, however, renders their stance more comprehensible. Despite the fact that emergency management is a legal product on the federal, state and local levels, FEMA's educational materials have historically been deficient regarding coverage of legal issues (Nicholson, 2003c). As a result, matters of liability constitute the greatest unanticipated and unexamined vulnerability that emergency management confronts.

One characteristic shared by top attorneys is their knowledge and understanding of their customer's industry in general, as well as the specifics that set the client's business apart from the rest. Sadly, emergency management lawyers possess minimal assets outside of statutes and interpretations thereof as source material to utilize when counseling their clients on even simple legal matters (Nicholson, 2003d). The wise attorney will recognize his or her ignorance and pursue knowledge of the client's organization. It will be necessary to hunt up resources that go beyond the Continuing Legal Education that lawyers usually see as their main font of information. One way for the emergency management lawyer to go farther is to enroll in the Emergency Management Professional Development Series offered by the Federal Emergency Management Institute (FEMA, 2005e).

HOW LEGAL ADVICE CAN IMPROVE EMERGENCY MANAGEMENT

An integrated emergency management system is composed of a conceptual framework that increases emergency management capability through networking. To achieve increased capability, there must be prior networking, coordination, linkages, and partnerships. There must also be creative thinking about resource shortfalls. All hazards threatening a community must be identified so that needs may be compared with resources (FEMA, 2005e). This process requires emergency managers to be pro-active risk managers as opposed to reactive risk ignorers.

Given the all-encompassing nature of law's relationship with emergency management as discussed above, potential liability is a hazard that confronts all emergency management organizations and the units of government that they serve. Potential claims in the aftermath of disasters include wrongful death, negligent planning or actions during the disaster, civil rights violations resulting from improper use of authority, exceeding the scope of proper practice for emergency management, failure to properly distribute aid, monetary damages resulting from loss of business during an evacuation, and many more.

The advance networking required to address the legal hazard entails joining together with legal counsel to avert prospective liability. Properly trained legal counsel may offer beneficial input prior to the emergent event that gives rise to possible liability. Networking with legal counsel in such a manner defines "litigation mitigation" (Nicholson, 2003e). Litigation mitigation has three complimentary objectives:

1. reduced exposure to legal claims;
2. improved life safety; and
3. enhanced property protection.

Typically, legal counsel looks at the first element as his or her main concern. To an emergency manager, all three components are of critical significance. Actually, life safety and property preservation are natural byproducts of legal protection.

If litigation mitigation makes so much sense, one may inquire as to why it is not more prevalent. Several impediments prevent litigation mitigation. A few are intentionally inflicted, while others are the product of the natural evolution of groups with different traditions.

One important obstacle to pro-active connections with attorneys results from groups' usual approach to the use of legal counsel. By tradition, governmental employees look on the attorney like a "legal fire fighter." The lawyer gets the call following the legal conflagration's eruption. All too frequently, the client only contacts the attorney after the arrival of legal documents indicating commencement of a lawsuit. The other side of fire fighting is fire prevention, just as the other side of emergency response is mitigation. For emergency management, the attorney could prove to be the equivalent of a fire inspector. Like the inspector, the attorney often may recognize the tinder for a legal inferno and highlight economical approaches that might reduce the hazard.

To the person in the street, lawyers may have an almost priestly appearance – they employ their incomprehensible terminology and execute their esoteric rites. Unfortunately, some attorneys revel in the feeling of exclusivity. In the same way, certain emergency managers use acronyms with meanings shrouded in obscurity to those not initiated in the fraternity. Clearly, unusual language and rituals distinguish both groups from the laity as well as each another. This method is the antithesis to the networking, coordination, linkages, and partnerships essential to creating an integrated emergency management system. It is essential for legal counsel and emergency manager to rely on and comprehend each other as equal partners for litigation mitigation to be successful.

Attorneys also find themselves unable to locate resources that explain the process for engaging in litigation mitigation. Their professional training at law schools does not include this topic, in general. The majority of such institutions rely on a "case study" approach first created hundreds of years ago. Another barrier is the emphasis placed by law schools on if a case is ready to take to Court. Undertaking a proactive pursuit such as mitigation litigation is contrary to the training and long-standing traditions of the law.

Another factor in lessening the likelihood of networking partnerships between emergency managers and lawyers is the developing nature of the legal market. On one hand, corporate clients insist on a firm that can "do it all," resulting in ongoing pressure to make firms ever larger. Simultaneously, business clients often look at firms as sources of skilled operators who can manipulate the legal system rather than as places to go for trustworthy counselors. Often, companies pay no attention to lawyers' guidance unless it matches with what they wish to hear. The networking and trust needed for a quality litigation mitigation association do not exist in such circumstances. Rather, the lawyer is viewed as a remote, unfamiliar "legal information engineer" who is a tool instead of a partner (Caplan, 2003).

Another major obstacle between attorneys and emergency managers is cost. Although an emergency manager will typically have access to the advice of a city or county attorney, he or she may need more specialized assistance, which may cost a significant amount. A specialist attorney's professional advice may run as much as several hundred dollars an hour, or more. Clearly, lawyers run more per hour than some other types of mitigation. Still, considering the downside of a lack of litigation mitigation, one is hard-pressed to see a valid argument against this enhanced safety. An option that may provide for cost controls is negotiating an arrangement with a firm to provide discounted hourly rates in return for an assured number of hours yearly.

From the government side, several significant obstacles exist. Smaller units of government may have attorney advisors who are local practitioners. These lawyers often collect decreased hourly wages for their government work compared with what they collect for normal hourly fees. The result of this arrangement may be that government work gets done only when ordinary business is lacking. This schedule often results in a conflict with the unit's desire for the attorney's help. Another difficulty is that rural units hardly ever make available funding for Continuing Legal Education (CLE) for their lawyer employees or contractors. The untrained country attorney may prove to be unable to give the high-level guidance the unit needs. The lack of CLE training necessary to bring the lawyer up to speed merely emphasizes the difficulty. The lawyer in such circumstances might be in danger of committing malpractice by advising beyond his or her ability. In some cases, the attorney advising the local unit receives the contract for legal services thanks to political activism instead of any actual knowledge of the practice area. A unit's attorney is often appointed on a political basis, even when the City Manager is a merit position. The effect of this situation may not be ineffective counsel, but it may result in regular changes. The potential exists here for recurring legal fees for getting on top of things as well as conflicting legal guidance.

FUTURE DEVELOPMENTS

In late 2004, the 9-11 Commission officially endorsed adoption of the National Fire Protection Association 1600 "Standard on Disaster/Emergency Management and Business Continuity Programs, 2004 Edition" (NFPA, 1600) (NFPA, 2004) as the national preparedness benchmark. On December 17, 2004, (The White House, 2004) President Bush signed the Intelligence Reform and Terrorism Prevention Act of 2004 (IRTP Act), (Intelligence and Terrorism Prevention Act, 2004) which recognizes NFPA 1600 as a "voluntary" national preparedness norm. NFPA 1600 and other documents such as the Capability Assessment for Readiness (CAR), constitute the core of the Emergency Management Accreditation Program (EMAP) (EMAP, 2005). The accreditation procedure includes application, self-assessment, on-site assessment by an outside review team, committee and commission review of conformity with the EMAP Standard, and re-certification every five years. Adoption of NFPA 1600 is not yet mandatory. It might be said that EMAP is well on its way to becoming the United States' *de facto* emergency management standard. As other emergency management programs are accredited under the standard, it becomes more likely that a Court might hold all emergency management groups to the NFPA 1600 criterion. One nationally known expert believes that "synergy is already building between NFPA 1600, EMAP and the NIMS Integration Center. It's just a matter of time before they are incorporated into NIC's requirements."[7] In fact, mandatory adoption of NFPA 1600 into NIMS will be an important part of standards to be set by the NIMS Integration Center.[8] Emergency management professionals will need to understand and comply with the full dimension of their legal obligations under NIMS, including NFPA 1600 and EMAP.

Other legal needs that will doubtless receive more attention in the future include liability issues in the aftermath of terrorism events. In the wake of the 9-11 attacks, insurance policies are being re-written to exclude terrorism coverage or to make premiums for adding it prohibitively expensive. The result is that businesses and units of government find themselves to be self-insured for this huge potential liability. Every state should examine its immunity statutes to see if the exclusions for third party acts are broad enough to protect from liability from terrorism events. Going hand in hand with that examination, of course, will be the need for units of government to ensure that their steps to provide for the safety of visitors and employees in their offices are appropriate for the dangers involved. As with the private sector, this will involve examining what similar units of government are doing, and being at least as safe. For example, state governments that do not provide for screening of people and parcels entering their premises might be exposed to liability in the event of a suicide terrorist bomber entering their premises unchallenged and causing death or injuries.

Interstate and intrastate mutual aid are currently the focus of examination by emergency response and emergency management across the nation, with NEMA making significant efforts to ensure that common language is available for both. Mutual aid agreements will be required under NIMS in the near future.[8]

Notwithstanding the obstructions discussed previously, litigation mitigation offers benefits that make its adoption highly advisable. As a mitigation step, it is a natural part of comprehensive emergency management. This approach is challenging. Both the attorney and the emergency management client must be willing to commit to a partnership that is based on mutual trust and respect. Both groups need to obtain knowledge of pertinent legal standards so that they may support one another. The attorney must understand the client's business in order to provide the best legal advice. Litigation mitigation must be actively practiced to fully address vulnerability to the hazard of liability.

Whether a rights-based approach to vulnerability is a trend that will grow in the future or a fad confined to nations with significant socialist leanings is a matter well worth following. Most likely, an impetus to create such rights worldwide will result in significant resistance from those nations whose assets are likely to be redistributed by such an approach. As those nations (like the United States) are powerful, their opposition may make difficult reaching a common agreement on universal enactment of fundamental human rights.

CONCLUSION

Law and emergency management are inextricably bound together. In the United States, emergency management law has a history of over 200 years. That history reveals emergency management law as ever more all embracing. Some nations have taken emergency management much farther, incorporating disaster relief, housing, and health care as "fundamental human rights" protected by Constitutional guarantee. This approach changes emergency management law into a social engineering standard that acts to redistribute wealth.

Emergency management legal standards in the United States are in the process of evolution as well. The NRP, NIMS, NFPA 1600, and EMAP are all working together to bring national uniformity to the practice of emergency management.[9] In the U.S., legal norms for emergency management focus on greater professionalization and better execution of traditional functions. Concentrating on the nuts and bolts of emergency management, rather than creating new rights that incidentally affect the discipline, appears to be the direction of future legal development in the United States.

NOTES

1. Id. at 46. One feels constrained to point out that such needs are typically "rights" only in socialist countries.
2. See, e.g., Indiana Code 10-14-3-12 (d) (1) (2005).
   (d)In addition to the governor's other powers, the governor may do the following while the state of emergency exists:
   (1) Suspend the provisions of any regulatory statute prescribing the procedures for conduct of state business, or the orders, rules, or regulations of any state agency if strict compliance with any of these provisions would in any way prevent, hinder, or delay necessary action in coping with the emergency.
3. E.g. Zimbabwe, Zambia, Algeria, Angola, Armenia, and India. "Public Law in Vulnerability Reduction" 54.
4. See, e.g. Indiana Code 10-14-3-1(a) (2005).
   "Disaster"
   Sec. 1. (a) As used in this chapter, "disaster" means an occurrence or imminent threat of widespread or severe damage, injury, or loss of life or property resulting from any natural or manmade cause.
   (b) The term includes the following:
   (1) Fire.
   (2) Flood.
   (3) Earthquake.
   (4) Wind.
   (5) Storm.
   (6) Wave action.
   (7) Oil spill.
   (8) Other water contamination requiring emergency action to avert danger or damage.
   (9) Air contamination.
   (10) Drought.
   (11) Explosion.
   (12) Riot.
   (13) Hostile military or paramilitary action.
   As added by P.L.2-2003, SEC.5.
5. See, e.g., Indiana Code 10-14-3-2 (2005).
   "Emergency management"
   Sec. 2. As used in this chapter, "emergency management" means the preparation for and the coordination of all emergency functions, other than functions for which military forces or other federal agencies are primarily responsible, to prevent, minimize, and repair injury and damage resulting from disasters. The functions include the following:
   (1) Firefighting services.
   (2) Police services.
   (3) Medical and health services.
   (4) Rescue.
   (5) Engineering.
   (6) Warning services.
   (7) Communications.
   (8) Radiological, chemical, and other special weapons defense.
   (9) Evacuation of persons from stricken areas.
   (10) Emergency welfare services.
   (11) Emergency transportation.
   (12) Plant protection.
   (13) Temporary restoration of public utility services.
   (14) Other functions related to civilian protection.
   (15) All other activities necessary or incidental to the preparation for and coordination of the functions described in subdivisions (1) through (14).
6. HS Act § 507 ROLE OF FEDERAL EMERGENCY MANAGEMENT AGENCY.
   (a) IN GENERAL.–The functions of the Federal Emergency Management Agency include the following:
   (1) All functions and authorities prescribed by the Robert T. Stafford Disaster Relief and Emergency Assistance Act (42 U.S.C. 5121 et seq.).
   (2) Carrying out its mission to reduce the loss of life and property and protect the Nation from all hazards by leading and supporting the Nation in a comprehensive, risk-based emergency management program.
7. HS Act § 430 (c) RESPONSIBILITIES.–The Office for Domestic Preparedness shall have the primary responsibility within the executive branch of Government for the preparedness of the United States for acts of terrorism, including–
   (6) as the lead executive branch agency for preparedness of the United States for acts of terrorism, cooperating closely with the Federal Emergency Management Agency, which shall have the primary responsibility within the executive branch to prepare for and mitigate the effects of nonterrorist-related disasters in the United States;
8. Telephone interview (May 9, 2005) with Kay C. Goss, CEM, Electronic Data Systems Corporation, US Government Solutions, Senior Advisor for Homeland Security, Business Continuity Planning, and Emergency Management Services.
9. Interview with Acting Director Gil Jamieson, NIMS Integration Center, Washington, DC (June 3, 2005). "I see the prospects of their being part of NIMS for emergency management. I have met with both the NFPA 1600 committee and the EMAP people, and I endorse the process, but it needs to evolve and be more inclusive of NIMS."

REFERENCES

Caplan. (2003). "Law Firms Become Big Business as Well." *Wilmington News Journal.* April 18, at A 17.

EMAP. (2005). Recent and Upcoming Activities, Accessed at http://www.emaponline.org/ on October 20, 2005

FEMA. (2005a). Retrieved from http://www.fema.gov/about/history.shtm. on October 20, 2005.

FEMA. (2005b). "Vulnerability Assessment." Accessed at http://www.fema.gov/pdf/fima/155/e155_unit_iv.pdf on October 10, 2005

FEMA. (2005c). "Emergency Management: Setting The Scene." Accessed at http://www.training.fema.gov/emiweb/downloads/is1_Unit1.pdf on October 15, 2005.

FEMA. (2005d). "Principles of Emergency Management. FEMA's Independent Study." Accessed at http://www.training.fema.gov/emiweb/downloads/IS230.doc on October 17, 2005.

FEMA. (2005e). "Professional Development Series." Accessed at http://www.training.fema.gov/emiweb/PDS/ on October 7, 2005.

Handmer, J. and Monson, R. (2004). "Does a Rights Based Approach Make a Difference? The Role of Public Law in Vulnerability Reduction," *International Journal of Mass Emergencies and Disasters 22*(3): 43. (November 2004).

Homeland Security Presidential Directive 5. (2003). "Paragraph 16." February 28.

Intelligence Reform and Terrorism Prevention Act. (2004). S.2845 ENR. Accessed at http://thomas.loc.gov/cgi-bin/query/D?c108:4:./temp/~c108150E9Q:: [Henceforth IRTP Act].

National Incident Management System 127. (2005). Accessed at http://www.fema.gov/nims/ on October 23, 2005.

NFPA. (2004). "NFPA 1600 Standards on Disaster/ Emergency Management and Business Continuity Programs." Accessed at http://www.nasttpo.org/NFPA 1600.htm October 14, 2005.

Nicholson, W. C. (2003a). *Emergency Response and Emergency Management Law.* 236–238, Clarles C Thomas Publisher, Ltd, Springfield, IL.

Nicholson, W. C. (2003b). "Integrating Local, State and Federal Responders and Emergency Management: New Packaging and New Controls," *Journal of Emergency Management 1*(15), 15.

Nicholson, W. C., (2003c). "Legal Issues in Emergency Response to Terrorism Incidents Involving Hazardous Materials: The Hazardous Waste Operations and Emergency Response ("HAZWOPER") Standard, Standard Operating Procedures, Mutual Aid and the Incident Command System." *Widener Symposium Law Journal 9* (2): 295, 298–300.

Nicholson, W. C., (2003d). *Emergency Response and Emergency Management Law: Cases and Materials.* Charles C Thomas Publisher, Ltd, Springfield, IL.

Nicholson, W. C. (2003e). "Litigation Mitigation: Proactive Risk Management in the Wake of the West Warwick Club Fire." *Journal of Emergency Management, 1*(2).

Pine, J. (1991). Liability Issues, Chapter 11 of Emergency Management, Principles and Practice for Local Government. International City Management Association, Washington, D.C.

The National Response Plan 63. (2005). Accessed at http://www.dhs.gov/dhspublic/interapp/editorial/editorial_0566.xml on October 17, 2005

The White House. (2003). "Homeland Security Presidential Directive 5, Subject: Management of Domestic Incidents." February 28.

The White House. (2004). "President Signs Intelligence Reform and Terrorism Prevention Act." December 17.

United States Code. (2005). §§ 11001–11050; § 11001 (a, b, c); § 11003 (a).

Varley, A. (1994). Disasters, Development and the Environment.

Waugh, W. L., Jr. (2002). "The "All-Hazards" Approach Must be Continued 2" *Journal of Emergency Management, 1*(2): 11.

107th Cong. (2002). Homeland Security Act of 2002, H.R. 5005, (enacted) [hereinafter HS Act] § 1(a-d).

Chapter 20

Environmental Management and Disasters: Contributions of the Discipline to the Profession and Practice of Emergency Management

John R. Labadie

ABSTRACT

This chapter explores the contributions that environmental management can make to the theory and practice of emergency management. It first examines environmental management as a distinct field of practice and draws parallels in the diversity of academic backgrounds and routes of entry common to both fields. A brief history of the environmental movement in the U.S. is followed by a discussion of the concept of "disaster" in the context of environmental management and emergency management, and an acknowledgement of the significance of environmental degradation as a contributing factor in disaster effects. The chapter notes the domestic and international regulatory imperative that embeds emergency management solidly in the practice of environmental management, and it concludes by identifying areas where environmental management and emergency management can and should interact more positively for mutual benefit and support.

INTRODUCTION

The disciplines of environmental management and emergency management share many of the same concepts, issues, processes, and concerns. Yet they come into contact only rarely, and then usually it is only a glancing blow. Parts of environmental management include risk assessment, hazard identification, spill response, and emergency/contingency planning – all activities that are central to the practice of emergency management. Other parts of the field address such issues as water quality, protection of flora and fauna, and general health of the ecosystem – all of which may be affected by decisions and actions taken in the pursuit of emergency management.

The editor's original assignment for this chapter centered on Environmental Science but, like any good student, I have re-written the exam question just a bit. I found that focusing on Environmental Science is a bit too restrictive and not sufficiently informative. Accordingly, I have modified the scope of the chapter to focus on Environmental Management, which includes Environmental Science, Environmental Engineering, Ecology, and related disciplines. This focus gives a more well-rounded view of the environmental field and its potential contributions to emergency management.

I have long been a practitioner in both fields. I have, therefore, approached the information, concepts, and arguments discovered in researching this chapter from the perspective of those in both fields who are confronted, on a daily basis, with the need to act and make decisions that have immediate practical effects. I have tried to focus on practical applications as opposed to policy formulation. Considering the spectrum of "environment" – small-scale waste management at one end, global warming and climate change at the other – this article focuses on the part from the mid-line (wherever that is) on down.

This chapter does not pretend to an exhaustive discussion of environmental issues, nor does it explore all of the current thinking and research in emergency management. Rather, it focuses on those areas where the two disciplines overlap and can interact to mutual benefit. It also does not trespass on other discrete fields such as safety, meteorology, public health, or law even though there are explicit interactions and interpenetrations between these fields and both environmental management and emergency management.

UNDERSTANDING THE TERMS

Environmental management is somewhat of a portmanteau term that comprises many of the more academically accepted disciplines. It brings together elements of science, engineering, policy, assessment, and auditing, as well as basic down-in-the dirt/air/water analysis and action. At one end of the spectrum lies the realm of environmental policy and regulation; at the other end lies what has been described as "blue-collar science." Here is a quick definitional tour:

- Ecology – a consensus definition on the web is "The study of the relationships between living organisms and their environment."
- Environmental Science "comprises those disciplines, or parts of them, that consider the physical, chemical and biological aspects of the environment. . . . it transcends disciplinary boundaries and is concerned with the interactions among processes each of which is best described by a particular discipline. It is the study of natural cycles and systems and their components" (Allaby, 1996).
- Environmental Engineering is the application of science and engineering principles to improve the environment, to provide healthful water, air and land for human habitation and other organisms, and to enhance the remediation of polluted sites.
- Environmental Management is the planning and implementation of actions geared to improve the quality of the human environment. It is public and private organizations actively dealing with environmental issues on a daily basis.

The boundaries among these fields are neither straight nor rigid, and they permit migration to and from a number of other disciplines – consider "environmental health" or "environmental toxicology."

One can come to a career in environmental management from many directions. In preparing this chapter, I took an informal poll of 38 co-workers and colleagues (in both public and private sectors) who are active in some aspect of the environmental management field. I asked for their academic experience (degree and major; advanced degrees). I was expecting a broad range of backgrounds, and I was not disappointed. Though I make no claim to statistical rigor, the following table (see next page) shows the considerable variation in academic backgrounds among the respondents.

The BS degree predominates but not overwhelmingly so. Environmental Science/Studies represent but a fraction of the academic majors represented. The "other" category is all over the map. Graduate study is equally varied: Engineering, Forest Science, Biopsychology, Environmental Management/Natural Resources, Psychology, History, and Law – to name a few.

The point here is that there are many roads into the practice of environmental management, just as the field itself covers a wide variety of disciplines, activities, and sub-specialties. I imagine that a similar survey of practitioners in the emergency management field would show a similar variability in backgrounds.

BRIEF HISTORY OF THE ENVIRONMENTAL MOVEMENT

The Environmental Movement (or "Environmentalism," another popular term) is a relative youngster, of uncertain parentage. Some of its roots lie in the ethics of conservation and preservation that arose in the late nineteenth and early twentieth centuries. With the closing of the American frontier came the desire to protect the "wild lands" on the one hand and, on the other, to produce as much as possible without spoiling the land. Preservation and "right use" were the prevailing ideas regarding the land and water. Wilderness protection in the '20s and '30s was aimed at "setting nature apart" as a national repository of aesthetic, ecological, recreational, and regenerative riches for the benefit of urban citizens (Gottlieb, 1993).

Table 20.1
ENVIRONMENTAL MANAGEMENT ACADEMIC BACKGROUND

Degree	*Undergraduate Major*	
N = 38	Engineering	History/Humanities/Liberal Arts –3
	• Civil – 1	
BS – 23 (1 person has 2)	• Chemical – 1	Other (1 (each)
	• Mechanical – 1	• Business
BA – 14	• Environmental – 1	• Hotel Management
		• Nat. Resources Mgmt.
B Phil –1	Environmental	• Forestry
	• Science – 3	• Antropology
AA/Cert. – 1	• Studies – 2	• Education
		• Env. Administration
MS – 8	Biology – 4	• Mathematics
		• Planning
MA/MPA – 5	Fisheries – 2	• Agricultural Technology and Management
JD/PhD – 4	Geology – 3	• Marine Science
		• Environmental Health
	Chemistry – 4	• Physics
		• Political Science

On the other side of the family lie efforts through the 1920s to address the human health and environmental hazards of the increasingly urban/industrial life of many Americans. Alice Hamilton, Professor of Industrial Medicine at Harvard, published *Industrial Poisons in the United States* (1920), detailing the hazards affecting industrial workers and urban residents. The public health profession after WWI began to attack such problems in the cities as contaminated water supplies, poor waste collection/disposal systems, and air pollution. Labor unions also promoted reform through environmental advocacy for and by workers.

The unrestrained urbanization and industrialization after WWII brought with it more discretionary income, more leisure (due to automation, labor saving devices, etc.), increased automobile ownership and use – leading in part to more travel and a greater appreciation of environmental surroundings and amenities. At the same time, the growth of suburbia and the proliferation of affordable housing led to more intensive use of land and resources. Construction of wastewater treatment plants did not keep up with population growth and density, with the result that raw sewage flowed into streams and lakes, causing pollution and eutrophication and the death of aquatic wildlife. More consumption created more municipal waste that was disposed of in landfills, open dumps, or incinerators with no pollution control equipment.[1]

A number of disparate ideas, trends, and discontents came together in the late 60s and early 70s to create (among other things) what we know as the modern Environmental Movement in the United States. Rachel Carson's *Silent Spring* had sensitized people to corporate and governmental lack of concern over the effects of pesticides on humans, birds, and animals. Pollution of land, air, and water were becoming increasingly obvious – including such spectacles as the Cuyahoga River catching fire, the Santa Barbara oil spill, to name a few (Speth, 2004, pp. 82–83). At the same time, New Left critiques of political ideas and concepts, philosophies and the dubious fruits of technological progress (not to mention anti-war, anti-government protests) provided a matrix for

1. Flippen, J. B., "Richard Nixon and the Triumph of Environmentalism" in Warren 2003, pp. 272–289.

environmental consciousness and action: ". . . for many in and around the New Left, environmentalism came to be associated with the search for alternative institutions and a new way of life" (Gottlieb, 1993, p. 97).

The passage of the National Environmental Policy Act in 1969 and the celebration of Earth Day around the nation in 1970 (surpassing the organizers' expectations in the extent and enthusiasm of participation[2]) engendered both a significant body of legislative and regulatory action on environmental matters and an explosion of popular enthusiasm, interest, and activity for environmental protection and quality.

> In just four years, between 1970 and 1974, an extraordinary range of legislative initiatives, regulatory activities, and court action came to the fore. These established a broad and expansive environmental policy system centered around efforts to control the environmental by-products of the urban and industrial order. Through this system, a vast pollution control, or environmental protection industry was created, including engineering companies, law firms, waste management operations, and consulting firms specializing in environmental review, standard setting, or other new environmental procedures (Gottlieb, 1993, p. 125).

This environmental protection industry, along with the growth of technical expertise within regulatory agencies, placed the environmental quality debate on a more scientific, technically-based foundation from which to confront the full range of environmental problems facing society. The growth of new environmental expertise and technical competence, emphasizing technical solutions, has largely focused on "end-of-pipe" solutions (i.e., dealing with the problems after they have already been created). Some laws do have a pollution prevention focus, but most efforts, until fairly recently, have concentrated on dealing with the waste already created.

Efforts by the Reagan Administration to roll back environmental regulation were met with a revitalized effort by environmental organizations, citizens' groups, victims of pollution effects, and academics to ensure that the gains of the 70s would not disappear. Though progress had been made in some areas (e.g., urban air quality), enhanced attention to and more rigorous investigation of environmental contamination had discovered even more – and more threatening – problems (Dunlap, 1992, p. 5). The proliferation of grassroots environmental movements in 1990s brought together a broad cross-section of class, occupational, and income groups pursuing local action against toxic industrial and disposal sites, landfills, and treatment plants. This has grown into a concerted focus on environmental justice/equity, environmental quality as an issue of civil rights, gender, ethnicity, and empowerment.

Global environmentalism has proceeded along much the same route, although perhaps not as soon or as quickly. Significant events during the 1980s (Bhopal, Sandoz, Chernobyl, Exxon Valdez) got public attention and influenced governments to increase their rhetoric, if not immediate action. Popular protest eventually created an agenda item for international affairs, and the status of the environment became an object of political action and legal prescription. More recently, governments and NGOs have focused on sustainable development and the deleterious effects of environmental degradation (Dunlap, 1992).

An important, though still quite controversial, development within the environmental community is the "Precautionary Principle," which assigns the burden of proof to those who want to introduce a new technology, particularly in cases where there is little or no established need or benefit and where the hazards are serious and irreversible. Growing out of European environmental policies in the late 1970s, the Precautionary Principle notes a potential environmental or human health hazard, emphasizes the scientific uncertainty that exists regarding the possible result, and therefore asserts the need for preventive, precautionary action as opposed to immediate implementation.[3] Though popular in Europe, the Principle is criticized by many as limiting technological

---

2. A number of accounts have estimated that 20 million participated across the country, in addition to 2,000 colleges and universities. See Flippen.

3. "When an activity raises threats of harm to human health or the environment, precautionary measures should be taken even if some cause-and-effect relationships are not fully established scientifically." From the *January 1998 Wingspread Statement on the Precautionary Principle.*

progress, hindering the introduction of new products (e.g., genetically-modified crops), and generally contributing to an anti-business bias (Foster, Vecchia, and Repacholi 2000).

To return to the focus of this chapter, one can draw parallels between the growth and development of the environmental management profession and that of emergency management. Civil defense became disaster preparedness, which became all-hazards preparedness, which grew into emergency management with an increasing focus on mitigation (prevention/reduction) as opposed to response and recovery ("end-of-pipe"). Considerable research into human response to disasters, effective planning concepts, and information management and communications has provided a technical foundation for the practice of emergency management. Practitioners have shed the "helmet and armband," "retired military" image and have taken advantage of increasing opportunities for formal education, with a discrete body of knowledge, and certification in the field.

"ENVIRONMENTAL" IN THE DISASTER CONTEXT

The environment is often seen as the agent/cause of a disaster or perhaps as the carrier. In an earthquake or a flood, for example, the "environment" behaves in ways that bring harm to the communities affected by them – one suddenly finds the environment sitting in one's living room. However, people make choices – farming practices, use and procurement of fuels, selection of building materials and sites, etc. – that significantly affect their vulnerability to environmental disasters (Aptekar, 1994; May, et al., 1996). This view mirrors the idea that disaster is a social construct formed by the interaction of human development with natural processes. An earthquake is a disaster only when it impacts the human infrastructure (Mileti, 1999; Cutter, 2001; Burton, 1993; Varley, 1994).

But the environment also interacts with human society in complex ways. Floods may damage natural habitats and ecosystems; forest fires may harm forest ecosystems and damage the biotic stock in an area. Yet, floods are necessary to renew and enrich riparian corridors and wetlands and to recharge aquifers; forest fires thin out undergrowth that could fuel larger fires, and they can re-vitalize biodiversity (Sauri, 2004). Floods can clog wastewater treatment plants, causing the release of untreated sewage into water bodies; floods can also mobilize contaminants and industrial chemicals that then flow downstream and possibly into those same aquifers.

Thus, an "environmental" hazard may be difficult to define, and there can be a fine distinction between an environmental hazard (i.e., water out of control – a flood) and an environmental resource (i.e., water in control – a reservoir). It can often be a matter of perception regarding deviations about the norm – too much rain is a flood; too little is a drought (Smith, 1996, p. 11). Some definitions of environmental hazard emphasize the acute and short-term at the expense of the chronic and long-term (droughts desertification, erosion), for example:

> . . . extreme geophysical events, biological processes and major technological accidents, characterized by concentrated releases of energy or materials, which pose a largely unexpected threat to human life and can cause significant damage to goods and the environment (Smith 1996, p. 16).

There is a growing understanding of environmental degradation as a contributing factor in disaster effects – i.e., an exacerbating factor in damage, it worsens impact on victims and makes recovery more difficult. One example:

> Although the largest danger facing Turkish urban areas is earthquake, numerous other hazards exist. Improper handling of solid wastes causes explosive methane build-up, endangers the physical environment, reduces property values and destroys the scenic and tourist values of highly visited areas. . . . Near the larger cities, many bodies of water are so polluted that they are no longer suitable for recreational use. High levels of heavy metals are found in harbor catches, and massive fish kills are common. Marine accidents release massive, toxic discharges, sometimes causing explosions that destroy buildings and facilities. Dangerous chemicals enter the urban food chain . . . urban rivers are polluted . . . agricultural chemicals and waste water have contaminated precious aquifers. . . (Parker, Kreimer, and Munasinghe 1995, p. 13).

The most recent example occurred in the South Asian tsunami – long-term damage to coral reefs and degradation of mangrove swamps in some areas reduced the capacity of natural systems to absorb or cushion the kinetic energy of the tsunami surge.

Deleterious effects of degraded environmental conditions are felt most keenly (though not exclusively) by the poor, residents of shantytowns, "favelas," and other marginal or hazardous areas. They are clustered on steep slopes subject to flash floods and erosion, in dwellings built of substandard materials, with poor water and waste disposal systems. Natural disaster effects can be greatly magnified by the poor environment in which these people live.

According to Pelling (2003b), there is a tendency to focus on technical and engineering issues in addressing environmental problems or issues and to discount the influence of social characteristics on susceptibility to environmental risk. This bias toward technological and physical solutions (e.g., flood walls, or leachate mitigation systems) can encourage development in hazard areas when, in fact, hazards can surpass the margin of safety provided by technological solutions.

"DISASTER" IN THE ENVIRONMENTAL CONTEXT

The field of emergency management tends to focus more on harm to the human environment and the built environment and to pay less attention to the larger environment in which humans and structures exist. Also, the emphasis is on the more acute disasters (like earthquakes or chemical spills) and less on the slow-developing problems with chronic effects (e.g., Minamata or acid rain) or on acute events with long-lasting consequences (e.g., Bhopal, or the Tisza River). This no doubt reflects the understandable orientation of emergency management professionals to the needs of planning for and response to the immediate effects of a disaster and the desire for speedy restoration to something approaching the *status quo ante.*

Environmental professionals take a somewhat more comprehensive view, considering not only the human and built environments but also the matrix in which they exist. Environmental concerns include not only humans but also plants and animals, water and air quality, the fate and transport of environmental contaminants, the toxicology of human and animal effects, and the exposure and vulnerability (both acute and chronic) of the affected biota. All of these concerns can – and should – contribute in some way to the practice of emergency management before, during, and after disasters.

Consider the scope of environmental hazards/disasters in the table below.

Environmental management confronts all of these hazards, in one manner or another, and brings the full range of scientific, technical, and managerial skills and techniques to bear on preventing mitigating, or responding to their effects. Of course, the definitions of "emergency" and "disaster" are a bit different in the environmental field: "An environmental emergency

Table 20.2
ENVIRONMENTAL HAZARDS/DISASTERS

Acute	*Chronic*
✓ Floods	✓ Sedimentation and siltation
✓ Oil spills on water bodies	✓ Air pollution
✓ Hazardous materials spills (to soil)	✓ Genetic mutation in indicator species
✓ Hazardous materials spills (to water bodies)	✓ Drought[4]
✓ Landslides	✓ Global warming
✓ Containment failure (mine spoils, industrial wastes)	✓ Deforestation and flooding
✓ Fish kills	✓ Loss of wetlands

4. Meteorological drought – a departure from anticipated mean rainfall. Hydrological drought – a reduction in available water relative to local demands. Agricultural drought – changes in timing, frequency or intensity of the rainfall that have specific implications for crop yield. (Dolcemascolo 2004, p. 16)

is a tanker truck full of acid overturned and spilling in the middle of town. An environmental disaster is that same tanker spilling into a wetland or a river."[5]

Environmental hazards are not independent of other types of hazards, and one may lead to the other or make the other worse. For example, floods can degrade water quality, release chemicals and other contaminants from impoundments or containers (or even float off the containers themselves to lodge in someone else's backyard). Earthquakes can cause transportation spills, industrial chemical releases through infrastructure damage, or damage to containment. Destruction of the World Trade Center released asbestos, respiratory irritants, polycyclic aromatic hydrocarbons (possible carcinogens), pulverized metals, and god-knows-what-else into the atmosphere, affecting rescue and recovery workers and undoubtedly contaminating the surrounding area (Mattei). As we have seen in the example from Turkey, above, environmental hazards may only be waiting for a triggering event to make a natural disaster even worse.

THE REGULATORY IMPERATIVE

Starting with the National Environmental Policy Act, 28 major environmental protection laws were enacted between 1969 and 1986. Environmental legislation since 1986 has generally focused on expanding or extending (and, in some cases, clarifying) existing laws. A number of these laws – and their implementing regulations – specifically address emergency planning and/or response in some way. These include:

- Resource Conservation and Recovery Act (RCRA, 1976) – Requires hazardous waste facilities to prepare and maintain emergency plans to prevent or respond to releases of hazardous wastes
- Comprehensive Environmental Response Compensation and Liability Act (CERCLA or "Superfund") – includes requirements for emergency plans during cleanup actions on uncontrolled waste sites
- Clean Water Act & Oil Pollution Act (1990) – requires Spill Prevention, Control, and Countermeasures plans be prepared by certain facilities storing petroleum fuels
- Emergency Planning & Community Right-to-Know Act (EPCRA or SARA Title III) – directed states and local governments to establish planning and coordination bodies to carry out emergency planning for chemical emergencies in their jurisdictions
- Clean Air Act, section 112r, Risk Management Program – requires certain industrial facilities to prepare an "off-site consequence analysis" for releases of certain chemicals and to prepare emergency plans in coordination with local response agencies
- Executive Order 12856 (1993) – directs Federal facilities to comply with EPCRA regarding public notification of chemical use and emergency planning

States have enacted their own set of environmental laws and regulations that parallel, enhance, and extend Federal regulations. In many cases (California and New Jersey are good examples), state regulations are more strict than Federal requirements.

An example of European regulatory action, affecting more than one country, is the Seveso Directive. A 1976 explosion at a chemical plant in a small town near Milan, Italy released a large cloud of dioxin that affected a large portion of Lombardy region. The explosion and aftermath, including the botched response, led to creation of the European Community's Seveso Directive in 1982.

> A central part of the Directive is a requirement for public information about major industrial hazards and appropriate safety measures in the event of an accident. It is based on recognition that industrial workers and the general public need to know about hazards that threaten them and about safety procedures. This is the first time that the principle of "need to know" has been enshrined in European Community legislation (Mitchell, 1996, part 4).

Much of the Seveso Directive is analogous to EPCRA with additional elements addressing what

5. Or another perspective: "An environmental disaster is the county deciding to build a wastewater treatment plant half a mile upwind from my house."

are called, in the U.S., "worker Right-To-Know" laws. All of this happened several years prior to the Bhopal disaster, which was part of the impetus for passage of EPCRA in the United States.

Though not regulatory in nature, there are a number of international standards regarding environmental management that specify a requirement for emergency plans. The most widely recognized is the ISO 14001 standard for Environmental Management Systems, one element of which states, "The organization shall establish and maintain procedures to identify potential for and respond to accidents and emergency situations, and for preventing and mitigating the environmental impacts that may be associated with them" (ANSI/ISO 14004-1996, Section 4.4.7).

The objective of the *Guiding Principles for Chemical Accident Prevention, Preparedness and Response* (2003), published by the European Organisation for Economic Co-Operation and Development (OECD) is to "provide guidance, applicable worldwide . . . to prevent accidents involving hazardous substances and to mitigate the adverse effects of accidents that do nevertheless occur." This set of principles covers much the same ground as the EPCRA, the Risk Management program regulations, hazardous materials transportation regulations, and various environmental and safety standards in the U.S. Chapter 5 mandates emergency planning for protection of environmental media as well as for the protection of population.

THE NEXUS OF ENVIRONMENTAL MANAGEMENT, DEVELOPMENT, AND DISASTER RISK

Considerable research and analysis has been done by the European Union and the United Nations to illuminate the connections among environmental hazards, sustainable development strategies (especially in the poorer countries), and disaster response and management. *Living with Risk* (2004), produced by the UN International Strategy for Disaster Reduction, puts it most succinctly:

> The environment and disasters are inherently linked. Environmental degradation affects natural processes, alters humanity's resource base and increases vulnerability. It exacerbates the impact of natural hazards, lessens overall resilience and challenges traditional coping strategies. Furthermore, effective and economical solutions to reduce risk can be overlooked. . . . Although the links between disaster reduction and environmental management are recognized, little research and policy work has been undertaken on the subject. The concept of using environmental tools for disaster reduction has not yet been widely applied by practitioners (p. 298).

The UN International Strategy for Disaster Reduction also focuses on the transboundary nature of disasters and the importance of a "harmonized approach" to the management of pollution of river basins, seismic hazard areas, and volcanoes (*Disaster Reduction and Sustainable Development* 2003, p.13). This issue is perhaps less salient in the United States, due to the extent of Federal disaster management and response.

Researchers in the Swedish Embassy in Bangkok have sought to link environmental programs with disaster risk in the context of sustainable development. They ask:

- How can investments in environmental management and sustainable development also reduce disaster risk?
- Is there a *prevention dividend*[6] that accrues from wise land use planning and development programs?
- Can *prevention dividends* be measured; and, how might the ability to estimate these added values enhance policy and program planning? (Dolcemascolo, 2004, p.1)

Although they find evidence for positive answers to these questions, they acknowledge that more research and analysis is necessary in order to capture the rather elusive cost/benefit parameters of disaster reduction and sustainable development.

ZONES OF CONVERGENCE

Living with Risk (2004, p. 303) outlines ways to integrate environmental and disaster reduction strategies:

6. *". . . the values of foregone disaster losses that accrue from well designed and implemented disaster risk reduction measures, including environmental management and sustainable development initiatives."*

- assessment of environmental causes of hazards occurrence and vulnerability
- assessment of environmental actions that can reduce vulnerability
- assessment of the environmental consequences of disaster reduction actions
- consideration of environmental services in decision-making processes
- partnerships and regional approaches to land use and nature conservation
- reasonable alternatives to conflicts concerning alternative uses of resources
- advice and information to involve actors in enhancing the quality of the environment.

Within this context, there are a number of areas where environmental management and emergency management can and should interact more positively for mutual benefit and support.

Both fields would benefit from continuing and supporting the current movement in the disaster community from "reactive" disaster response to active risk management and from iterative recovery to pro-active mitigation and prevention. Parallel efforts would transition the environmental field from contaminant clean-up to risk reduction and pollution prevention, from discrete issues management to environmental management systems, and from flood control to floodplain management (see Philippi, 1996).

Integration of sustainability considerations into disaster mitigation and recovery can exploit the considerable overlap between environmental management and disaster management. Planners and practitioners in both fields must recognize that the overall objectives of these fields implicitly promote sustainable communities. Sustainability should be considered both prospectively (in sustainable development planning and mitigation) and retrospectively (in response and recovery). This integration would incorporate and enhance current trends toward "holistic disaster recovery" (also "sustainable recovery") that emphasize betterment of the entire community, including environmental improvement and enhancement, through the recovery process (*Holistic Disaster Recovery*, 2001). *Living with Risk* (2004) is even more direct: "Disaster reduction specialists should be encouraged to anticipate environmental requirements under applicable laws and to design projects that address these requirements, coordinating closely with environmental institutions" (p. 302).

Environmental management professionals can make considerable contributions during the mitigation and recovery phases of emergency management. They can identify possible improvements and enhancements as well as things to avoid. More importantly, after enhancements or improvements are in place, they can monitor and assess environmental performance indicators to ensure that goals are met. Environmental assessments should be integrated into emergency planning processes, following the Environmental Impact Statement model mandated by the National Environmental Protection Act. Environmental Impact Statements should (but currently do not) specifically include disaster-hazard considerations. Rapid environmental assessments should be conducted as part of disaster damage assessment and should be an integral part of response/recovery considerations (Kelly, 2001).

Both environmental managers and emergency managers must be cognizant of the importance of environmental justice/equity issues in the context of hazard and vulnerability. Hazards of any type have a disproportionate impact on the poor and disadvantaged. A number of thorny equity issues are coming to a head in the environmental management world, among them: industrial plant and landfill siting; development in industrial or depressed areas; residential settlement on slopes or in other marginal areas; higher population density; immigrants and language differences; differential access to social services and information sources. Most of these issues have not yet been adequately addressed in emergency management planning or community dialogue.

The United States (and, to a certain extent, other nations) has become sensitized to the possibility that terrorists might attack with Weapons of Mass Destruction (nerve agents, bioweapons, "dirty bombs"). The unpleasant reality is that terrorists don't have to try that hard to create death and destruction. The ubiquitous gasoline tanker would make a handy (and easily procured) bomb; there are over 20,000 chemical plants in the U.S. that contain enough extremely hazardous materials to require reporting under EPCRA. The existence and availability of these and other so-called "weapons of convenience" will require a much closer and more explicit cooperation between environmental professionals and emergency managers to: assess the immediate and long-term threats; to identify both mitigation and response strategies; and to manage long-term recovery and clean-up operations.

ENVIRONMENTAL MANAGEMENT AND THE FOUR PHASES OF EMERGENCY MANAGEMENT

At the most practical and operational levels, environmental professionals can contribute directly to the practice of emergency management throughout all four phases.

Mitigation

- Inventory environmental assets
- Identify environmental projects and environmental enhancement opportunities as part of hazard mitigation planning
- Integrate local land use and growth management ordinances into hazard mitigation planning
- Assist in developing Hazard Mitigation Plans
- Establish and monitor Environmental Performance Indicators for assessing progress & operation of environmental projects conducted as part of mitigation
- Identify and monitor environmental regulatory requirements
- Identify and monitor government funding programs to support environmental mitigation actions

Preparedness

- Identify/assess environmental vulnerabilities and threats (analysis of Environmental Impact Statements)
- Review emergency management policies, plans, and procedures for potential environmental impacts
- Assist in emergency preparedness planning (especially for hazardous materials incidents)
- Assist in the activities of the Local Emergency Planning Committee (under EPCRA)
- Assist in developing response procedures to ensure that environmental factors/hazards are addressed

Response

- Conduct Environmental Impact Assessments
- Identify environmental threats/damage related to disaster effects
- Assist in response to environmental hazards or emergencies (e.g., orphan drums; releases; etc.)

Recovery

- Identify environmental damage
- Identify possible recovery options for environmentally sensitive areas
- Identify enhancements to environmental assets/resource for recovery
- Monitor Environmental Performance Indicators (short & long-term)
- Assist in and monitor debris removal and clean-up (e.g., for hazardous materials/wastes problems)
- Identify and monitor environmental regulatory requirements
- Identify and monitor government funding programs to support environmental mitigation actions

CONCLUSION

Considering the extent to which the concepts and practices of environmental management and emergency management overlap and interpenetrate, it is surprising to see the number of publications that examine how the two fields should (and by implication do not) interact in mutually-supportive ways. The most common connection between the two fields usually occurs in the setting of Local Emergency Planning Committee activities to plan for response to chemical spills and releases. Emergency managers largely ignore the full range of environmental issues. When issues and practices do come into conflict (e.g., thinning the forests; use of dispersants in chemical spills), it is vital that environmental management have a seat at the table. Both disciplines must cooperatively seek solutions that will maximize environmental quality as well as meet the needs of disaster preparedness and recovery.

Researchers and practitioners in other countries – particularly in the European Union and Asia – have paid considerably more attention to the effects of environmental conditions on disaster vulnerability and to the necessity of adding an environmental consciousness to the planning for and implementation of disaster response. We have much to learn from our colleagues overseas regarding the most fruitful cooperation between environmental management and emergency management.

REFERENCES

Allaby, M. (1996). *Basics of Environmental Science.* New York: Routledge.

Aptekar, L. (1994). *Environmental Disasters in Global Perspective.* New York: G.K. Hall; Toronto: Maxwell Macmillan Canada; New York: Maxwell Macmillan International.

Burton, I., Kates, R.W. and White, G.F. (1993). *The Environment as Hazard.* New York: The Guildford Press.

"Community cooperation in the field of civil protection." (1999). European Commission, Directorate-General XI, Environment, Nuclear Safety and Civil Protection. Luxembourg: Office for Official Publications of the European Communities; Lanham, Nd.: Bernan Associates [distributor].

Cutter, S.L. (ed.). (2001). *American Hazardscapes: The Regionalization of Hazards and Disasters.* Washington, DC: Joseph Henry Press.

Dayton-Johnson, J. (2004). "Natural Disasters and Adaptive Capacity." OECD Development Centre Working Paper No. 237. http://www.oecd.org/dataoecd/30/63/33845215.pdf

United Nations International Strategy for Disaster Reduction. (2004). *Disaster Reduction and Sustainable Development: Understanding the Links between Vulnerability and Risk to Disasters Related to Development and Environment.* http://www.unisdr.org/eng/risk-reduction/wssd/DR-and-SD-English.pdf

Dolcemascolo, G. (2004). *Environmental Degradation and Disaster Risk.* Prepared for the Embassy of Sweden/Sida Bangkok. Asian Disaster Preparedness Center. http://www.sida.se/content/1/c6/03/03/92/Environmental%20Degradation%20and%20Disaster%20Risk.pdf

Dunlap, R.E. and Mertig, A.G. (1992). *American Environmentalism: The U.S. Environmental Movement, 1970–1990.* Philadelphia: Taylor & Francis.

Ellis, D.V. (1989). *Environments at Risk: Case Histories of Impact Assessment.* Berlin; New York: Springer-Verlag.

Environmental Management Systems – General guidance on principles, systems and supporting techniques. American Society for Quality. ANSI/ISO 14004-1996.

European Commission, Environment Directorate-General. (2002). "EU focus on civil protection: coping with catastrophes – coordinating civil protection in the European Union." Luxembourg: Office for Official Publications of the European Communities.

Environmental Protection Agency. (1993). "Executive Order 12856: federal compliance with right-to-know laws and pollution prevention requirements." August 3. Washington, DC: U.S. Environmental Protection Agency.

Foster, K.R., Vecchia, P. and Repacholi, M.H. (2000). "Science and the Precautionary Principle." *Science,* May 12, 2000, 979–981.

Gee, D. (2001). *Late Lessons from Early Warnings – the Precautionary Principle 1896–2000.* Environmental Issue Report #22. European Environment Agency. Luxembourg: Office for Official Publications of the European Communities.

Gottlieb, R. (1993). *Forcing the Spring: The Transformation of the American Environmental Movement.* Washington, DC: Island Press.

Holistic Disaster Recovery: Ideas for Building Local Sustainability after a Natural Disaster. (2001). Natural Hazards Research and Applications Information Center, University of Colorado, Boulder, CO.

"Implementation guidelines for Canadian Environmental Protection Act, 1999." Section 199, authorities for requiring environmental emergency plans. Environmental Emergencies Program (Canada).

Jackson, S.L. (1997). *The ISO 14001 Implementation Guide.* New York: Wiley & Sons.

Kelly, C. (2001). "Rapid Environmental Impact Assessment: A Framework for Best Practice in Emergency Response." Benfield Greig Hazard Research Centre, University College, London. http://www.benfieldhrc.org/SiteRoot/disaster_studies/working_papers/workingpaper3.pdf

Lachman, B.E. (1997). *Linking Sustainable Community Activities to Pollution Prevention: A Sourcebook.* Washington, DC: Rand Critical Technologies Institute. http://www.rand.org/publications/MR/MR855/

United Nations. (2004). *Living with Risk: A global review of disaster reduction initiatives.* Inter-Agency Secretariat of the International Strategy for Disaster Reduction (UN/ISDR). http://www.unisdr.org/eng/about_isdr/bd-lwr-2004-eng.htm

Lundgren, R.E. and McMakin, A.H. (2004). *Risk Communication: A Handbook for Communicating Environmental, Safety, And Health Risks.* Columbus, OH: Battelle Press.

Mattei, S. (no date). *Pollution and Deception at Ground Zero: How the Bush Administration's Reckless Disregard of 9/11 Toxic Hazards Poses Long-Term Threats for New York City and the Nation.* Sierra Club. http://www.sierraclub.org/groundzero/report.pdf. http://www.sierraclub.org/groundzero/summary.asp

May, P.J., Burby, R.J., Ericksen, N.J., Handmer, J.W., Dixon, J.E., Michaels, S., and Ingle Smith, D. (1996). *Environmental Management and Governance: Intergovernmental Approaches To Hazards and Sustainability.* London; New York: Routledge.

Mileti, Dennis S. 1999. *Disasters by Design: A Reassessment of Natural Hazards in the United States.* Washington, DC: Joseph Henry Press.

Mitchell, James K. (ed.). 1996. *The Long Road to Recovery: Community Responses to Industrial Disaster.* Tokyo - New York - Paris: United Nations University Press. http://www.unu.edu/unupress/unupbooks/uu21le/uu21le00.htm

OECD Guiding Principles for Chemical Accident Prevention, Preparedness and Response. 2003. OECD Environment, Health and Safety Publications. Series on Chemical Accidents, No. 10. http://www.oecd.org/dataoecd/10/37/2789820.pdf

Parker, Ronald, Alcira Kreimer, and Mohan Munasinghe (eds.). 1995. *Informal Settlements, Environmental Degradation, and Disaster Vulnerability: The Turkey Case Study.* Geneva, Switzerland: International Decade for Natural Disaster Reduction (IDNDR); Washington, DC: World Bank.

Pelling, Mark (ed.). 2003a. *Natural Disasters and Development in A Globalizing World.* London; New York: Routledge.

______. 2003b. *The Vulnerability of Cities: Natural Disasters and Social Resilience.* London; Sterling, VA: Earthscan Publications.

Philippi, Nancy S. 1996. *Floodplain Management: Ecologic and Economic Perspectives.* San Diego, CA: Academic Press; Austin, T.: R.G. Landes.

Piasecki, Bruce. 1995. *Corporate Environmental Strategy: The Avalanche of Change since Bhopal.* New York: J. Wiley & Sons.

______, Kevin A. Fletcher, and Frank J. Mendelson. 1999. *Environmental Management and Business Strategy.* New York: J. Wiley & Sons.

Reducing Disaster Risk: A Challenge for Development. 2004. United Nations Development Programme, Bureau for Crisis Prevention and Recovery. New York. www.undp.org/bcpr

Saurĩ Pujol, David. 2004. *Mapping the impacts of recent natural disasters and technological accidents in Europe.* Luxembourg: Office for Official Publications of the European Communities. http://reports.eea.eu.int/environmental_issue_report_2004_35/en/accidents_032004.pdf

Shrivastava, Paul. 1987. *Bhopal: Anatomy of a Crisis.* Cambridge, MA: Ballinger Pub. Co.

Smith, Keith. 1996. *Environmental Hazards: Assessing Risk and Reducing Disaster.* London; New York: Routledge.

Speth, James. 2004. *Red Sky at Morning: America and the Crisis of the Global Environment.* New Haven; London: Yale University Press.

Tickner, Joel A. (ed.). 2003. *Precaution, Environmental Science, and Preventive Public Policy.* Washington, DC: Island Press.

van Aalst, Maarten, and Ian Burton. 2002. "The Last Straw; Integrating Natural Disaster Mitigation with Environmental Management." Disaster Risk Management Working Paper Series No. 5. The World Bank. http://www.worldbank.org/hazards/files/last_straw_final.pdf

Varley, Anne. 1994. *Disasters, Development and Environment.* Chichester; New York: J. Wiley.

Wall, Derek. 1994. *Green History: A Reader in Environmental Literature, Philosophy, and Politics.* London; New York: Routledge.

Warren, Louis S. (ed.). 2003. *American Environmental History.* Malden, MA: Blackwell Pub.

Weir, David. 1987. *The Bhopal Syndrome: Pesticides, Environment, and Health.* San Francisco: Sierra Club Books.

Wisner, B. & J. Adams (eds.). 2002. *Environmental Health in Emergencies and Disasters: A Practical Guide.* Geneva: World Health Organization. http://www.who.int/water_sanitation_health/hygiene/emergencies/emergencies2002/en/

Chapter 21

Communication Studies and Emergency Management: Common Ground, Contributions, and Future Research Opportunities for Two Emerging Disciplines

Brian K. Richardson and Lori Byers

ABSTRACT

This chapter highlights common ground between the Communication Studies and Emergency Management disciplines. It briefly describes the Communication Studies discipline, and its historical roots. Then, areas of common ground between the two disciplines are explored. The chapter also offers areas of Communication Studies inquiry that have informed the study of disaster and emergency processes. Areas of future research between the two disciplines are discussed before the chapter concludes with recommendations of how to improve the practice and study of emergency management from the perspective of the Communication Studies discipline. This chapter identifies major opportunities for future research including sensemaking and the importance of narratives in disasters, compliance-gaining and persuasion strategies, and the effects of stressful communication and events on people.

> "Any attempt to establish a unified command on 9/11 would have been further frustrated by the lack of communication and coordination among responding agencies."
>
> – The 9/11 Commission Report

INTRODUCTION

Emergency management scholars have long recognized the roles of communication within disaster. Scholars in Communication Studies have long viewed disasters as particularly rich contexts within which to study communication processes. Despite this mutual interest, the two disciplines have had little do with each other in an integrated fashion. The metaphor that comes to mind is two ships passing in the night. Perhaps, close enough to acknowledge one another with a flash of light but not yet traveling in the same direction. It is our hope that this chapter will illuminate common ground between the disciplines so that integrated research can more readily occur. To that end, this chapter will overview the role the Communication Studies discipline can play in understanding disaster. First, the Communication Studies discipline will be introduced and briefly described. Next, we will discuss definitions of "disaster" within our discipline. The next two areas of the paper will describe common ground between the Communication Studies and Emergency Management disciplines, and areas of communication research that could inform disaster scholarship, respectively. The paper will conclude with directions for future research and some recommendations for emergency management from the Communication Studies' perspective.

Communication Studies

The discipline of Communication Studies examines the symbolic transmission of meaning in a variety of contexts. Departments in communication originated as offshoots of English departments; instead of analyzing literary texts, the focus at that time centered on the rhetorical study of speeches, primarily those political in nature. During the 1950s and 60s, inspired by studies of persuasion, some members of the discipline took a decidedly social scientific turn, investigating the role of communication in the creation of identity and social roles, in the process of persuasion, and in the context of decision-making. Today, most departments in Communication Studies continue to divide themselves along those two lines: rhetoric and social science. Studies in rhetoric now extend beyond the analysis of speeches to include a broader variety of texts, such as media, social movements, and war memorials, to name a few. Researchers taking a more social scientific focus can be categorized broadly into the areas of interpersonal, health, intercultural, and organizational research, all of which examine communication in a variety of social contexts including, but not limited to, organizations, education, healthcare, and families. Relevant to the current discussion, research in Communication Studies examines issues such as the transmission and processing of information (Sutcliffe, 2001), communication networks (Monge and Contractor, 2001), individual and organizational identity (Cheney and Christenson, 2001), sociopolitical environments (Finet, 2001), leadership (Fairhurst, 2001), decision making (Seibold and Shea, 2001), social support (Burleson and MacGeorge, 2002), persuasion (Dillard, Anderson, and Knobloch, 2002), power (Mumby, 2001), conflict (Oetzel and Ting-toomey, 2001; Roloff and Soule, 2002), and technology (Fulk and Collins-Jarvis, 2001; Walther and Parks, 2001).

Defining Disaster

While studies of disasters appear to be increasing within Communication Studies, an agreed-upon definition of the term "disaster" remains elusive, particularly one grounded in communication theory. What is clear is that disasters are generally considered a subset of the crisis communication literature and are often approached as they relate to organizations. In their review of the crisis communication literature, Seeger, Sellnow, and Ulmer (1998) draw on Quarantelli in distinguishing crises from disasters, suggesting that crises are organization-based, while disasters are "non-organizationally based events generated by natural or mass technological forces" (p. 233). Additional review of the literature reveals a fragmented definition of "disaster." Indeed, a number of Communication Studies scholars employ the term in reference to organizational crises. For example, Tyler (1992) characterizes the Exxon Valdez accident as a disaster, just as Rowland (1986) does in reference to the Space Shuttle Challenger explosion. However, neither offers a conceptualization of the term to distinguish it from an organizational crisis.

Benetiz (2004) offers a glimpse of how "disaster" may ultimately be defined within Communication Studies when he suggests that while present conceptualizations of natural disasters tend to emphasize only physical characteristics and negative impacts of disasters, socio-economic processes such as human decisions, governmental policies, and economic development models should also be considered. As Rogers et al. (1995) argue, "when a major event threatens the stability of a system, it forces the members of the system to construct new and changing meanings of their community" (p. 676). Considering the Communication Studies discipline's long-standing interest in the social construction of reality (Berger and Luckmann, 1966; Rogers et al., 1995), it is likely that a definition or conceptualization of what constitutes a "disaster" will include the way society, and especially subsets of it, co-create the meaning of disaster through processes including language, discourse, power, and politics.

COMMUNICATION STUDIES AND DISASTERS: COMMON GROUND

Communication and Disaster

Communication Studies and Emergency Management scholarship share considerable common ground though little of it is integrated. First, both the communication and emergency management literature routinely highlight the role of human and technological communication within disaster contexts. For instance, in their detailed review of four disaster case studies, Dynes and Quarantelli (1977) generate no fewer than 294 propositions on disaster communication. Auf der Heide (1989) observes that "one of the most consistent observations about disasters is that communication is inadequate" (Ch. 5). Recent disasters have further highlighted the role of communication within disasters. As Liebenau (2003) points out, "The destruction of the World Trade Center . . . brought to public attention the many different critical roles that communications play when disaster strikes" (p. 45). Indeed, communication plays a part in each of the hazard cycle's four phases.

First, mitigation efforts will be enhanced when disaster personnel and agencies develop communication networks, increase the flow of relevant information, and share ideas. Fischer (1998) recognizes that new communication technologies, such as electronic mail and chat rooms, facilitate dialogue about disaster mitigation among experts and community leaders throughout the world. Furthermore, Benetiz (2004) recommends utilizing communication in designing and evaluating effective channels and mechanisms of interaction between local and national governments and other organizations to facilitate natural disaster mitigation. Regarding the second phase, preparedness, Auf der Heide (1989) identifies a number of communication-related reasons for public and governmental apathy, including a lack of awareness, underestimation of risk, social pressures, opposing special interest groups, difficulty substantiating the benefits of preparedness, and ambiguity of responsibility. He contends that effective public communication programs could overcome apathy. For example, he describes how the San Francisco Fire Department aligned with the mayor's office and developed a successful information campaign that secured the community's financial backing, leading to the creation of an emergency response plan that included improvements in the city water system, fire fighting capability, and an emergency operations center.

Regarding the response phase, Ronsenthal, Hart, and Charles (1989) suggest that "in crisis situations, there is a considerable increase in the volume and speed of upward and downward communication" (p. 19) and "controlling the information flow" (p. 20) proves problematic. Some information must be released to the public, generating additional communication concerns. For example, Sorenson (2000) contends that message construction constitutes a critical component influencing public response to hazard warning systems. Warning message characteristics such as consistency and repetition, specificity, and legitimacy of the warning's source have all been recognized as critical factors in the effort to alert citizens (Auf der Heide, 1989; Drabek, 1986; Drabek, 1985; Quarantelli, 1982). Also, effective and timely communication between organizations during response has been acknowledged as vital by a number of scholars (Auf der Heide, 1989; Drabek and McEntire, 2002; Toulmin, Givans, and Steel, 1989). To this end, Granatt (2004) describes and advocates public information and warning partnerships (PIWPs), which are voluntary partnerships in the United Kingdom between governmental, nonprofit, and media organizations. Such partnerships integrate the resources and support of their members and recognize that "public confidence and safety depend on timely, clear, coordinated information" (p. 360). He describes how one PIWP was successfully used to avert public panic when an extortionist threatened to poison English water supplies in 1999.

Finally, the recovery phase includes those processes that restore individual and community lifelines after the immediate emergency has run its course. Disaster management and communication studies both find interest in examining how social and cultural issues influence responses to disaster. Due to gearing services to the dominant majority, disaster aid has been denied (Yelvington, 1997), and people who were homeless before disasters have been denied shelter in favor of the newly homeless (Phillips, 1996). Gender roles are even more stringently enforced during times of disaster (Anderson, 1994; Scanlon, 1997). For example, Scanlon (1997) points out that after a mudslide forced the evacuation of a small town in British Columbia, women, much more than men, were instructed to leave town, regardless of their occupation, their skill level, or whether or not they had families.

Additionally, Tierney, Lindell, and Perry (2001) highlight the importance of intercultural communication within recovery. Specifically, they observe that minority groups may have special dietary requirements, may include non-native English speakers who have difficulty understanding forms written in English, and may be cautious about seeking support from agencies for fear of being deported. Each of these issues could be described as communication problems that routinely exist when members of different cultures attempt to interact when discussing risks or during an emergency (Quigley, Handy, Goble, Sanchez, and George, 2000).

Crisis Communication

A second area of common ground between communication and disaster scholars concerns an interest in sudden, cataclysmic events that generate great potential for harm. Much of the Communication Studies research into such phenomenon addresses crisis communication, particularly from an organizational perspective. While some scholars distinguish between crises and disasters (Seeger et al., 1998), the two share much in common. Disasters invoke a sense of urgency in response, close observation by the news media, and an interruption in normal working or living conditions, all conditions noted by Williams and Treadaway (1992) as characteristic of crises. Rosenthal, Charles, and Hart (1989) and Coombs' (1999) include natural disasters as a specific type of crisis. Communication scholars have conducted "crisis communication" research, albeit usually from an organizational perspective, on such disasters as the 1994 Los Angeles earthquake (Fearn-Banks, 2002), the Mann Gulch and Storm King Mountain Fires (Alder, 1997), the 1994 South Canyon fire in Colorado (Larson, 2003), the 1997 Red River Valley flood in Minnesota and North Dakota (Sellnow, Seeger, and Ulmer, 2002), and the 9/11 attack on the World Trade Center (Noll, 2003).

The crisis communication literature consists of two distinct branches: practitioner-oriented application and theory-grounded understanding (Seeger et al., 1998). The practitioner-oriented approach addresses issues such as (in) effective media response strategies, group decision-making, interorganizational relationships and communication, and long-term recovery activities. For example, Huxman and Bruce (1995) investigate Dow Chemical's response to charges that they produced Napalm during the Vietnam War, pointing out that Dow failed in its attempts to subvert negative attention. Similarly, Williams and Treadaway (1992) scrutinize Exxon's failure to develop a proactive public relations campaign immediately after the spill, instead providing brief and inaccurate messages about the extent of the spill. In addition, Exxon did not take responsibility for the spill, and instead chose to scapegoat two of its employees. By limiting initial information and not taking responsibility for the spill, Exxon damaged its corporate image. Also investigating the Valdez oil spill, Sellnow (1993), by conducting rhetorical analyses of three speeches delivered by Exxon's president, analyzes the ethical issues involved in Exxon's use of scientific evidence and arguments to justify their actions after the Valdez oil spill. Furthermore, Tyler (1992) points out Exxon's 11 major mistakes, including antagonizing local fishermen and other area residents, understating the seriousness of the disaster, and not keeping public promises made to the public. Taken together, these articles provide a list of crisis communication's "do's" and "don'ts" from which other organizations could learn.

Theoretical examinations of crises address them as internal and environmental forces, which affect organizing and communicating processes and systems (Seeger et al., 1998). In his examination of the Mann Gulch and Storm King Mountain Fires, Alder (1997) refers to the "paradox of obedience" to explain how "crew members obeyed orders that should have been questioned and ignored directives with which (they) should have complied" (p. 111) resulting in disorganization and ultimately human death. Other scholars view crises from sensemaking perspectives (Larson, 2003), and use chaos theory to explain how emergency personnel assume that traditional methods of prediction are adequate in contexts which belie predictability (Sellnow et al., 2002). Two other areas which disaster and communication scholars share common ground include interest in technology and decision-making processes.

TECHNOLOGY AND DISASTER

Technology affects essentially every aspect of society; likewise, technology now plays an integral role in disaster processes. Both Communication Studies and Emergency Management scholars are aware of and demonstrate interest in these effects. Technologies promise to facilitate discussion about mitigation efforts, identification of potential hazards, connectivity between response organizations of all types, warning messages to the public, and diffusion of important disaster information (Carey, 2003; Drabek, 1991; Fischer, 1998; Tierney et al., 2001).

Communication Studies scholars, for example, were particularly interested in the role of new communication technologies during and immediately after the September 11 attack on the World Trade

Center. Cell phones played an important role in every aspect of the September 11 attacks: the plane hijackers used cell phones to stay in contact with each other, people on board the planes used cell phones to communicate with family and emergency operators, people inside the towers after the attacks used cell phones to contact friends, family, and emergency networks, and emergency personnel used cell phones to help coordinate rescues (Dutton and Nainoa, 2003).

Most people learned of the September 11 attacks via television (Cohen, Ball-Rokeach, Jung, and Kim, 2003). Carey (2003) found that the World Wide Web and e-mail were important sources of information but took a back seat to television and radio sources immediately following the attacks. In one survey, just six percent of respondents reported using the web as their primary news source. One problem recognized during the immediate aftermath of the attacks was an inability to access many websites because of heavy net traffic. However, in subsequent weeks, visitors utilized news websites in numbers two to three times their normal usage. Of course, the Internet continues to improve in its capabilities, so it would probably experience less difficulty with heavy traffic today, and certainly less so in the future.

Technology also plays an important role in maintaining contact with family and friends after disasters. Carey (2003) notes that instant messages and e-mail served the important purpose of checking on the safety or friends and relatives on September 11. Similarly, Sellnow, Seeger, and Ulmer (2002) point out that residents evacuated after the 1997 Red River Valley flood in Minnesota and North Dakota used radio station broadcasts to reconnect with family members. With improvements in wireless technology, mediated communication during and after disasters will inevitably increase.

Organizations also utilize technology during disasters. In their study of internet-based crisis communication, Perry, Taylor, and Doerfel (2003) observe that a majority of the 50 organizations they studied used the Internet to communicate to the public and the media during a crisis. This trend was consistent regardless of what type of organization the authors examined, or what type of crisis had occurred. Just as technology can aid in crisis and disaster response, its role as a source for disaster must also be determined. Thus, recognizing the risks associated with society's dependence upon various types of technology is essential in understanding the potential for technological disasters. Waugh (2000) argues that society becomes more fragile as our dependency upon increasingly complex technological systems increases. In 2003, blackouts in major U.S. cities evidenced this concern when energy and cellular phone systems became inoperable leaving large numbers of people incommunicado and without basic transportation services. Furthermore, the use of technologies during disasters may result in ironic outcomes. For example, the Internet may be a source for, and foster quick dissemination of, rumors and false information which confound disaster response efforts (Quarantelli, 1997). And while many consider technology a tool for education, it may also be used for educating those who would be responsible for disaster. The Internet's ability to disseminate information widely may expose potentially sensitive information, such as natural gas pipeline routes for example, to those most likely to utilize it for nefarious reasons, particularly terrorists (Carey, 2003). Ravalut (2003) argues that the "September 11 terrorist attacks demonstrate that global communications saturated with images of the United States can dramatically backfire when used by global terrorists for their own ends" (p. 209). At the same time that technologies are valuable tools for disaster mitigation and response, their potential as a source of disaster deserves examination.

DECISION-MAKING PROCESSES AND DISASTER

A final area of common ground between Communication Studies and Emergency Management scholars concerns decision-making processes including those occurring prior to, during, and after disasters. Disaster scholars have long recognized the unique qualities and pressures of making decisions within emergency contexts. Rosenthal et al. (1989) assert that in crisis situations decision-making becomes increasingly centralized and informal. Additional issues include dealing with uncertainty (Sorenson and Mileti, 1987), concerns that a decision to warn the public will cause panic (Auf der Heide, 1989), and the potential for information overload (Quarantelli, 1997). Within the crisis communication literature, researchers examine decision-making for its role in causing crises, how it is affected during a crisis, and its function in post-crisis investigations (Seeger et al., 1998). Niles (2004) argues that faulty decision-making processes at NASA contributed to both the Challenger and Columbia disasters. She goes on to say that efforts to change NASA's culture after the Challenger explosion to a culture more open to disagreement and dissent ultimately failed. This failed attempt created conditions that made another space shuttle tragedy increasingly likely.

Organizational emergencies and crises have been recognized as creating unique conditions for decision-makers. Such high-pressure situations may create what Janis and Mann (1977) term hypervigilance, which may cause decision makers into two faulty states: a state of temporary paralysis in which no decision is made, or a hasty decision based upon the first available option. Of course, either option may serve to heighten the crisis. One fundamental aspect of decision-making involves power and influence. After investigating the Challenger explosion, Gross and Walzer (1997) assert that much of the blame rested on the lack of ethical persuasion tactics; people were coerced to launch the Challenger even though it was against their better judgment. In his analysis of mountain fires at Mann Gulch and Storm King, Alder (1997) argues that firefighter deaths could be attributed to a breakdown in authority; people refused to follow orders during crucial moments. Alder (1997) suggests that while organizations should endeavor to create participative cultures that encourage discussion, in times of crisis leaders should be given the power to make unquestioned decisions.

INFORMING EMERGENCY MANAGEMENT THROUGH COMMUNICATION STUDIES

In addition to sharing much in common, it is likely that the Communication Studies and Emergency Management disciplines could inform one another. This next section of the paper will detail three primary particular areas that Communication Studies could contribute to Emergency Management scholarship. These areas include research into sensemaking and narrative, studies of persuasion and compliance gaining, and the effects of stressful events and communication on people.

Sensemaking and Narrative

Much research in Communication Studies investigates the role of communication and sensemaking in various contexts. Sellnow, Seeger, and Ulmer (2002), in their study of the 1997 Red River Valley flood in Minnesota and North Dakota, observe that people found it difficult to make sense of what happened, due to "facing an experience that was completely unanticipated and beyond their previous experiences" (p. 281). Examining how people make sense of seemingly nonsensical situations, for which few might possess a frame of reference, proves an interesting and worthy area of study.

A growing area of research in Communication Studies analyzes how collective memory and public commemoration of disaster and tragedy serve as rhetorical strategies to aid in sensemaking and coping during the grief and healing process. For example, Jorgenson-Earp and Lanzilotti (1998) studied shrines created after the Oklahoma City bombing in 1995 and in Dublane, Scotland in 1996 after a man walked in to a school, shooting and killing sixteen children and one teacher, and injuring 14 others. In both locations, people created shrines to memorialize the dead. Jorgenson-Earp and Lanzilotti argue that bringing objects to the shrines mitigated the voyeuristic and touristic aspects of visiting the sites, instead inviting "the viewer to add to the text, to help write the final story of the tragedy, and thereby regain control over the scene" (p. 160). The objects left at the shrines symbolized continuing relationships with the children who died, many objects included the names of family members left behind, often including gifts and messages from family members to the deceased child.

Another research area within Communication Studies that could contribute to emergency management involves the use of narrative. Jerome Bruner (1990) argues that narrative constitutes one of the most fundamental modes of thought and understanding; people understand the world, and their experiences in the world, through stories. Telling stories is an essential element of human experience (Fisher, 1987); we story our lives, and what we choose to tell stories about reflects what we value. Narratives create meaning (Feldman, 1990) and facilitate sensemaking (Coopman et al., 1998); narratives help us to make sense of the past and present, while projecting the future (Polletta, 1998). Narratives serve both a communication and sensemaking function by enabling individuals to interpret the world and convey that interpretation to others.

The use of narrative proves especially beneficial for understanding how individuals "make sense" of, recall, and participate in the social construction of how people experience and conceptualize disaster. Weick and Browning (1986) contend that narratives aid individuals in the comprehension of complex environments. Similarly, Maines (1993) posits that narratives serve as tools that people use to understand troubling events when meanings are not clear. Bridger and Maines (1998) recognize that narratives "are an important interpretive and rhetorical resource that is drawn upon in times of crisis and rapid change" (p. 320).

From a methodological standpoint, narrative analysis offers a unique way of accessing the way disaster witnesses, survivors, and victims make sense of and describe the impact on their lives. Additionally, narrative analysis serves to facilitate understanding of the manner in which an afflicted community participates in the social construction of meaning. Krug (1993) illustrates the power of storytelling after interviewed the residents of four small towns in Arkansas, after a climatologist predicted that an earthquake would occur along a fault zone in the vicinity of the towns. Krug argues that stories contributed to the sensemaking of the people in the towns; many people spent most of their days gathered together telling stories of the potential for earthquake, and in the end relied more on the stories of the community than the spectacular stories delivered through the media. Krug's study points out the potential role that stories play in persuasion, leading to the next suggestion for emergency management research.

PERSUASION AND COMPLIANCE-GAINING

Communication Studies scholars have long been interested in how individuals or groups get other individuals or groups to change their attitudes or behaviors. Fifty years of research produces a plethora of information on persuasion strategies and goals. Research in compliance-gaining strategies indicates that effective, long-term persuasion is relatively slow (Miller and Burgoon, 1978), and that both verbal and nonverbal communication are influential in compliance-gaining strategies (Segrin, 1993). Persuasion plays an important role in effective leadership, and effective leaders recognize that both supervisors and employees participate in persuasive tactics (Deluga, 1988). Prosocial rather than antisocial approaches to persuasion tend to prove more effective and influence perceptions of competence (Johnson, 1992). In general, successful compliance gaining proves somewhat complex, and research indicates that utilizing a variety of approaches brings the most success.

Key conclusions from Communication Studies' persuasion and compliance gaining research has not examined the context of emergency management, but this area of research would prove interesting. For example, persuading citizens to heed warning messages and securing their compliance with evacuation or stay-at-home messages function as critical components of disasters. As Alder (1997) points out, encouraging organizational cultures that foster participation, recognizing the reciprocal role of persuasion in organizations is effective during day-to-day operations, yet organizations should also investigate ways leaders can persuade employees to comply with directives during times of crisis. When considering terrorism, researchers should examine the compliance-gaining tactics of terrorist leaders, studies that inevitably would also entail investigation of ideology.

EFFECTS OF STRESSFUL EVENTS AND COMMUNICATION ON PEOPLE

Communication scholars are also interested in how interpersonal relationships, or perceived social support, influence our personal and professional lives. Disasters are particularly stressful times, fostering states of uncertainty, fear and fright. Research in a variety of contexts indicates that people experience more negative symptoms, both psychologically and physiologically, without the strength of interpersonal support (Albrecht, Burleson, and Goldsmith, 1994). Much research in emergency management focuses on responses during disasters and on utilitarian responses after disaster. How communication and relationships aid in coping and survival during and after disasters also deserves attention.

In the context of disasters, for example, research has indicated that social support is an important coping strategy for emergency dispatchers (Jenkins, 1997). However, relationships are not always positive. For example, research has found that domestic violence increases during disasters (Clemens, Hietala, Rytter, Schmidt, and Reese, 1999). In their investigation of the 1997 Red River Valley Flood, Davis and Ender (1999) found that if a marital relationship was strong before the flood, the couples perceived their marriage to be stronger after the flood; if the relationships were perceived as weak before the flood, they were perceived as even more weak after the flood.

People cope with disaster in a variety of ways. For example, one research study explored the phenomenon of graffiti that occurred after the 1997 Red River Valley Flood. Hagen, Ender, Tiemann, and Hagan (1999) termed graffiti after a disaster as "catastroffiti," and functioned initially to mark one's property, but also functioned to express emotions such as frustration and hope. Jorgenson-Earp and Lanzilotti's (1998) study of shrines created after the Oklahoma City bombing and in Dublane, Scotland discussed earlier in this chapter point to the different ways people cope with tragedy. More studies of how people respond to disaster would provide more information on aiding in the coping and healing process.

DIRECTIONS OF FUTURE INQUIRY

Future research should focus on developing a better understanding of "terrorist culture," particularly in an age of increasing globalization. Globalization, with its promises of greater economic, technological, and democratic rewards, deserves examination of how people throughout the world perceive globalization, particularly Islamic fundamentalists, other potential terrorists, and marginalized groups. Communication Studies' scholars are well positioned to examine how messages of globalization are being interpreted across the world. It would be of great value to determine how these messages are altered in order to contribute to a particular ideology. Likewise, the rhetoric of terrorist leaders and its effects on potential followers should be investigated. Scholars who contributed to Noll's (2003) examination of the Crisis Communications lessons of September 11 have already begun some work into this area which should serve as a springboard for future study.

Another area of future inquiry should address interorganizational relationships and communication. As globalization proceeds, an increasing number of organizations in diverse cultures will work together. If disaster strikes at or near these organizations, unique communication challenges may need to be overcome, such as what happened in the Union Carbide leak in Bhopal, India, discussed earlier. Additionally, organizations and individuals are increasingly linked via communication technology. These linkages necessarily require interdependence of people from a variety of backgrounds, communication styles, and ideologies. As the number of connected organizations are forced to cope with disaster issues, it is important to understand the challenges of their communication with each other so communication can be facilitated.

Communication should be examined not just for its role in preventing, responding to, and recovering from disaster. Researchers should also study communication for its role in creating or being a source of disaster. Communication Studies and Emergency Management scholars should explore communication's contribution to issues such as mob mentality, riots, and panicked evacuations. As mentioned above, communication also plays a significant role in terrorists' ideology formation, recruitment, and propaganda. Each of these areas could contribute to potential disaster situations and are worthy of future exploration. Finally, we must continue to examine society's increasing dependence upon new communication technologies. Does our dependence upon technological systems make us more or less vulnerable to disasters? How will response efforts be affected when communication technologies break down? Can communication technologies cause disasters through the rapid transmission of false information? Each of these questions and others must be explored.

RECOMMENDATIONS FOR EMERGENCY MANAGEMENT

Communication Studies would offer several recommendations to the emergency management profession. First, this discipline would suggest that "communication" not be seen solely from a functional, or utilitarian, perspective. Rather, consider communication's symbolic functions particularly within disasters contexts. For example, when a government agency releases a hazard warning or evacuation message, the message has both content and symbolic functions. While the content element may be precise, the symbolic elements may be more powerful. One could imagine members of a marginalized community, already mistrusting of government agencies, treating a warning message much differently than a community that is fully confident in such agencies. Such phenomenon must continue to be included in emergency management research and practice. Communication Studies would also propose a partnership between our discipline and Emergency Management scholars. Considering the wide variety of research areas, within our discipline, the possibilities of integrated scholarship are limitless. As mentioned at the outset of the chapter, Communication Studies scholars research an incredibly diverse subject area, primarily because communication is substantive to so many areas of individual, group, organizational, and societal life. For example, as globalization moves forward, we must understand how it is communicated and interpreted throughout the world. Such efforts may help us comprehend terrorism, particularly its recruiting, propaganda, and diffusion processes.

Finally, we would recommend that more disaster-related studies privilege multiple voices, beyond those of researchers and disaster managers. Qualitative techniques, such as long-term observation, or ethnography, interviewing, focus groups, and document and artifact analysis provide valuable means of data collection and capture subjects' own words, perceptions, and experiences. Disasters especially effect individuals and communities in unique ways and it is critical for research subjects to express their experience from their perspective. Communication Studies has benefited greatly in recent years by including qualitative research in its repertoire of methodological approaches and would recommend that Emergency Management scholars consider doing likewise.

CONCLUSION

The Communication Studies discipline examines the symbolic transmission of meaning in a variety of contexts, including disasters and crises. Specifically, Communication Studies has examined a multitude of ways communication contributes to our understanding of disaster processes, before, during and after disasters. While both Communication Studies and Emergency Management scholars seem to realize the value of one another's disciplines, little integrated research between the two has been conducted. Yet, there are a number of areas where such integration could occur. These areas include, but are not limited to, decision-making processes, technology and disaster, sensemaking and narrative, stress and social support, and persuasion and compliance gaining, all within the context of disaster. Joint studies on these areas and others can spotlight the common ground of both disciplines, and open new directions for future research.

REFERENCES

Abrams, J., O'Connor J., and Giles, H. (2004). "Identity and Intergroup Communication." Pp. in *Handbook of International and Intrcultural Communication*, edited by William B. Gudykunst and Bella Mody. Sage Publications, Thousand Oaks, CA.

Albrecht, T L., Burleson, B.R. and Goldsmith. (1994). "Supportive Communication." Pp. 419–449 in *Handbook of Interpersonal Communication*, edited by M. L. Knapp and G. R. Miller. Sage Publications, London.

Alder, G. S. (1997). "Managing Environmental Uncertainty with Legitimate Authority: A Comparative Analysis of the Mann Gulch and Storm King Mountain Fires." *Journal of Applied Communication Research 25*: 98–114.

Anderson, M. B. (1994). "Understanding the Disaster-Development Continuum: Gender Analysis is the Essential Tool." *Focus on Gender 2*(1): 7–10.

Auf der Heide, E. (1989). *Disaster Response. Principles of Preparation and Coordination.* The C. V. Mosby Company, St. Louis.

Benitez, J. L. (2004, November). Natural disasters and Communication: Consideration for communication strategies and research in Central America. Paper presented at the meeting of the National Communication Association, Chicago, IL.

Berger, P. L. and Luckmann, T. (1966). *The Social Construction of Reality: A Treatise in the Sociology of Knowledge.* Doubleday, New York.

Bridger, J. C. and Maines, D.R. (1998). "Narrative Structures and the Catholic Church Closisngs in Detroit." *Qualitative Sociology 21*(3): 319–340.

Bruner, J. (1990). *Acts of Meaning*. Harvard University Press, Cambridge, MA.

Burleson, B. R. and McGeorge, E. L.. (2002). "Supportive Communication." Pp. 374–422 in *Handbook of Interpersonal Communication*, edited by Mark L. Knapp and John A. Daly. Sage Publications, Thousand Oaks, CA.

Carey, J. (2003). "The Functions and Uses of Media during the September 11 Crisis and its Aftermath." Pp. 1–16 in *Crisis communications. Lessons from September 11*, edited by A. M. Noll. Rowman & Littlefield Publishers, New York.

Carey, J. W. (2003). "Globalization isn't New, and Antiglobalization isn't Either: September 11 and the History of Nations." Pp. 199–204 in *Crisis communications. Lessons from September 11*, edited by A. M. Noll. Rowman & Littlefield Publishers, New York.

Cheney, G. and Christensen, L.T. (2001). "Organizational Identity: Linkages Between Internal and External Communication." Pp. 231–269 in *The New Handbook of Organizational Communication. Advances in Theory, Research, and Methods*, edited by Jablin, F. M and Putman, L.L.. Sage Publications, Thousand Oaks, CA.

Clemens, P., Hietala, J.R., Rytter, M.J., Schmidt, R.A. and Reese, D.J. (1999). "Risk of Domestic Violence after Flood Impact: Effects of Social Support, Age, and History of Domestic Violence." *Applied Behavioral Science Review* 7(2): 199–206.

Cohen, E. L., Ball-Rokeach, S.J., Jung, J.Y. and Kim, Y.C. (2003). "Civic Actions after September 11: A Cmmunication Infrastructure Perspective." Pp. 30–44 in *Crisis communications. Lessons from September 11*, edited by A. M. Noll. Rowman & Littlefield Publishers, New York.

Coombs, Timothy W. (1999). *Ongoing Crisis Communication: Planning, Managing, and Responding*. Thousand Oaks, CA: Sage Publications.

Coombs, W. T. (2004). "Impact of Past Crises on Current Crisis Communication." *Journal of Business Communication 41*(3): 265–289.

Coopman, S. J., Hart, J. Hougland, J. G., & Billings, D. B. (1998). Speaking of God: The functions of church leader storytelling in Southern Appalachia in the 1950s [On-line]. *American Communication Journal, 1*(2). Available: www.americancomm.org.

Daly, J. (2004). "Personality and Interpersonal Communication." Pp. 133–180 in *Handbook of Interpersonal Communication*, edited by Mark L. Knapp and John A. Daly. Sage Publications, Thousand Oaks, CA.

Davis, K. M. and Ender, M.G. (1999). "The 1997 Red River Valley Flood: Impact on Marital Relationships." *Applied Behavioral Science Review* 7(2): 181–188.

Deluga, R. J. (1998). "Relationship of Transformational and Transactional Leadership with Employee Influencing Strategies." *Group and Organization Studies 13*: 456–467.

Dillard, J.P., Anderson, J.W. and Knoblock, L.K. (2002). "Interpersonal Influence." Pp. 423–474 in *Handbook of Interpersonal Communication*, edited by Knapp, M.L. and Daly, J.A. Sage Publications, Thousand Oaks, CA.

Drabek, T. E. (1985). "Managing the Emergency Response." *Public Administration Review 45*: 85–92.

Drabek, T. E. (1986). *Human System Response to Disaster: An Inventory of Sociological Findings*. Springer Verlag, New York.

Drabek, T. E. (1991). *Microcomputers in Emergency Management*. University of Colorado, Boulder, CO.

Drabek, T. E. and McEntire, D. A. (2002). "Emergent Phenomena and Multiorganizatinal Coordination in Disasters: Lessons from the Research Literature." *International Journal of Mass Emergencies and Disasters 20*(2): 197–224.

Dutton, W. H. and Nainoa, F. (2003). "The Social Dynamics of Wireless on September 11: Reconfiguring Access. Pp. 69–82 in *Crisis communications. Lessons from September 11*, edited by A. M. Noll. Rowman & Littlefield Publishers, New York.

Dynes, R. R. and Quarantelli, E.L. (1977). *Organizational Communications and Decision Making in Crises*. University of Delaware Disaster Research Center, Newark, DE.

Fairhurst, G. (2001). "Dualisms in Leadership Research." Pp. 379–439 in *The New Handbook of Organizational Communication. Advances in Theory, Research, and Methods*, edited by Fredric M. Jablin and Linda L. Putman. Sage Publications, Thousand Oaks, CA.

Fearn-Banks, K. (Ed.). (2002). *Crisis Communication. A Casebook Approach*. Lawrence Erlbaum Associates, London.

Feldman, S. P. (1990). "Stories as Cultural Creativity: On the Relation between Symbolism and Politics in Organizational Change." *Human Relations 43*(9): 809–828.

Finet, D. (2001). "Socio-Political Environments and Issues." Pp. 270–290 in *The New Handbook of Organizational Communication. Advances in Theory, Research, and Methods*, edited by Fredric M. Jablin and Linda L. Putman. Sage Publications, Thousand Oaks, CA.

Fischer, H. W., III. (1998). "The Role of the New Information Technologies in Emergency Mitigation, Planning, Response and Recovery." *Disaster Prevention and Management* 7(1): 28–37.

Fisher, Walter, R. (1987). *Human Communication as a Narration: Toward a Philosophy of Reason, Value, and Action*. Columbia, SC: University of Sourth Carolina Press.

Fulk, J. and Collins-Jarvis, L. (2004). "Wired Meetings: Technological Mediation of Organizational Gatherings." Pp. 624–663 in *The New Handbook of Organizational Communication. Advances in Theory, Research, and Methods*, edited by Fredric M. Jablin and Linda L. Putman. Sage Publications, Thousand Oaks, CA.

Granatt, Mike. (2004). "On Trust: Using Public Information and Warning Partnerships to Support the Community Response to an Emergency." *Journal of Communication Management, 8*: 354–365.

Gross, A. G. and Walzer, A. (1997). "The Challenger Disaster and the Revival of Rhetoric in Organizational Life. *Argumentation 11*: 75–93.

Hagen, C. A., Ender, M.G., Tiemann, K.A. and Hagen Jr., C.O. (1999). "Graffiti on the Great Plains: A Social Reaction to the Red River Valley Flood of (1997)." *Applied Behavioral Science Review* 7(2): 145–158.

Huxman, S. S. and Bruce, D.B. (1995). "Toward a Dynamic Generic Framework of Apologia: A Case Study of Dow Chemical, Vietnam, and the Napalm Controversy." *Communication Studies 46*(1–2): 57–88.

Jacobson, T. L. and Won Yang Jang. (2004). "Mediated War, Peace and Civil Society." Pp. in *Handbook of International and Intercultural Communication*, edited by William B. Gudykunst and Bella Mody. Sage Publications, Thousand Oaks, CA.

Jenkins, S. R. (1997). "Coping and Social Support Among Emergency Dispatchers: Hurricane Andrew." *Journal of Social Behavior & Personality 12*(1): 201–216.

Johnson, G. M. (1992). "Subordinate Perceptions of Superior's Communication Competence and Task Attraction Related to Superior's Use of Compliance-Gaining Tactics." *Western Journal of Communication 56*: 54–67.

Jorgensen-Earp, C. R. and Lanzilotti, L.A.. (1998). "Public Memory and Private Grief: The Construction of Shrines at the Sites of Public Tragedy." *Quarterly Journal of Speech 84*: 150–170.

Krug, G. J. (1993). "The Day the Earth Stood Still: Media Messages and Local Life in a Predicted Arkansas Earthquake." *Critical Studies in Mass Media 10*: 273–285.

Larson, G.. S. (2003). "A Worldview of Disaster: Organizational Sensemaking in a Wildland Firefighting Tragedy." *American Communication Journal 6*(2).

Liebenau, J. (2003). "Communication During the World Trade Center Disaster: Causes of Failure, Lessons, and Recommendations." Pp. 45–54 in *Crisis communications: Lessons from September 11*, edited by A. M. Noll. Rowman & Littlefield Publishers, New York.

Maines, D. R. (1993). "Narrative's Moment and Sociology's Phenomenon: Toward a Narrative Sociology." *The Sociological Quarterly 34*: 17–38.

Miller, G. R. and Burgoon, M.. (1978). "Persuasion Research: Review and Commentary." Pp. 29–47 in *Communication Yearbook 2* edited by B. Rubin. Transaction, New Brunswick NJ.

Monge, P. R. and Noshir S. Contractor. (2004). "Emergence of Communication Networks." Pp. 440–502 in *The New Handbook of Organizational Communication. Advances in Theory, Research, and Methods*, edited by Fredric M. Jablin and Linda L. Putman. Sage Publications, Thousand Oaks, CA.

Mumby, D. K. (2001). "Power and Politics." Pp. 585–623 in *The New Handbook of Organizational Communication. Advances in Theory, Research, and Methods*, edited by Jablin, F.M. and Putman, L.L.. Sage Publications, Thousand Oaks, CA.

Niles, M. (2004, November). Lessons learned from NASA: Deliberation processes in organization that can lead to disaster. Paper presented at the National Communication Association's annual meeting in Chicago, Illinois.

Noll, A. M. (Ed.). (2003). *Crisis Communications. Lessons from September 11*. Rowman & Littlefield Publishers, New York.

Perry, D. C., Taylor, M., and Doerfel, M.L. (2003). "Internet-Based Communication in Crisis Management." *Management Communication Quarterly 17*(2): 206–232.

Phillips, B. D. (1996). "Creating, Sustaining, and Losing Place: Homelessness in the Context of Disaster." *Humanity and Society 20*: 94–101.

Polletta, F. 1998. "'It was like a fever . . .' Narrative and Identity in Social Protest." *Social Problems 45*(2): 137–159.

Quarantelli, E.. L. (1982). Sheltering and Housing After Major Community Disasters: Case Studies and General Observations. University of Delaware Disaster Research Center, Newark, DE.

Quarantelli, E. L. (1997). "Ten Criteria for Evaluating the Management of Community Disasters." *Disasters 21*:39–56.

Quigley, D., D. Handy, R. Goble, V. Sanchez, and P. George. (2000). "Participatory Research Strategies in Nuclear Risk Management or Native Communities." *Journal of Health Communication 5*: 305–331.

Ravault, René-Jean. (2003). "Is there a Bid Laden in the Audience? Considering the Events of September 11 as a Possible Boomerang Effect of the Globalization of U.S. Mass Communication." Pp. 205–212 in *Crisis communications. Lessons from September 11*, edited by A. M. Noll. Rowman & Littlefield Publishers, New York.

Rogers, E. M., Dearing, J.W. , Rao, N., Campo, S., Meyer, G., Betts, G.J.F., and Casey, M.K. (1995). "Communication and Community in a City Under Siege: The AIDS Epidemic in San Francisco." *Communication Research 22*(6): 664–678.

Roloff, M. E. and Soule, K.P. (2002). "Interpersonal Conflict: A Review." Pp. 475–528 in *Handbook of Interpersonal Communication*, edited by Mark L. Knapp and John A. Daly. Sage Publications, Thousand Oaks, CA.

Rosenthal, U., M. T. Charles, and P. T. Hart, P. T. (1989). *Coping with Crises: The Management of Disasters, Riots and Terrorism*. Charles C Thomas Publisher, Springfield, IL.

Rowland, R. C. (1986). "The Relationship Between the Public and the Technical Spheres of Argument: A Case Study of the Challenger Seven Disaster." *Central States Speech Journal 37*(3): 136–146.

Scanlon, J. (1996-1997). "Human Behaviors in Disaster: The Relevance of Gender." *Australian Journal of Emergency Management 11*(4): 2–7.

Seeger, M. W., Sellnow, T.L. and Ulmer, R.R. (1998). "Communication, Organization, and Crisis." *Communication Yearbook 21*: 231–275.

Segrin, C.. 1993. "The Effects of Nonverbal Behavior on Outcomes of Compliance-Gaining Attempts." *Communication Studies 44*: 169–187.

Seibold, D. R. and Shea, B.C.. (2004). "Participation and Decision Making." Pp. 664–703 in *The New Handbook of Organizational Communication. Advances in Theory, Research, and Methods*, edited by Fredric M. Jablin and Linda L. Putman. Sage Publications, Thousand Oaks, CA.

Sellnow, T. L. (1993). "Scientific Argument in Organizational Crisis Communication: The Case of Exxon." *Argumentation and Advocacy 30*: 28–42.

Sellnow, T. L., Matthew W. Seeger, and Robert R. Ulmer. (2002). "Chaos Theory, Informational Needs, and Natural Disasters." *Journal of Applied Communication Research 30*(4): 269–292.

Sorensen, J. and Mileti, D. (1987). "Programs that Encourage the Adoption of Precautions Against Natural Hazards. Reviews and Evaluation." Pp. 208–230 in *Taking Care: Why People Take Precautions*, edited by N. Weinstein. Cambridge University Press, New York.

Sorensen, J. H. (2000). "Hazard Warning Systems: Review of 20 Years of Progress." *Natural Hazards 1*(2): 119–125.

Sutcliffe, K. M. (2004). "Organizational Environments and Organizational Information Processing." Pp. 197–230 in *The New Handbook of Organizational Communication. Advances in Theory, Research, and Methods*, edited by Fredric M. Jablin and Linda L. Putman. Sage Publications, Thousand Oaks, CA.

Tierney, K. J., Lindell, M.K. and Perry, R.W. (2001). *Facing the Unexpected. Disaster Preparedness and Response in the United States.* Joseph Henry Press, Washington, D.C.

Ting-Toomey, Stella, and Oetzel, John C. (2001). *Managing Intercultural Conflict Effectively.* Thousand Oaks, CA: Sage.

Ting-Toomey, S. and Otzel, J. (2004). "Cross-Cultural Face Concerns and Conflict Styles. Current Status and Future Directions." Pp. 143–164 in *Handbook of International and Intrcultural Communication*, edited by William B. Gudykunst and Bella Mody. Sage Publications, Thousand Oaks, CA.

Toulmin, L. M., Givans, C.J., and Steel, D.L. (1989). "The Impact of Intergovernmental Distance on Disaster Communications." *International Journal of Mass Emergencies and Disasters 7*(2): 116–132.

Tyler, L. (1992). "Ecological Disaster and Rhetorical Response Exxon's Communications in the Wake of the Valdez Spill." *Journal of Business and Technological Communication 6*(2): 143–171.

Walther, J. B., and Parks, M.R.. (2001). "Cues Filtered Out, Cues Filtered In: Computer-mediated Communication and Relationships." Pp. 529–563 in *Handbook of Interpersonal Communication*, edited by Mark L. Knapp and John A. Daly. Sage Publications, Thousand Oaks, CA.

Waugh, W. L., Jr. (2000). *Living with Hazards Dealing with Disasters.* M. E. Sharpe, London.

Weick, K. E. and Browning, L.D. (1986). "Argumentation and Narration in Organizational Communication." *Journal of Management 12*: 243–259.

Williams, D. E. and Treadaway, G. (1992). "Exxon and the Valdez Accident: A Failure in Crisis Communication." *Communication Studies 43*(1): 56–64.

Yelvington, K. A. (1997). "Coping in a Temporary Way: The Tent Cities." Pp. 92–115 in *Hurricane Andrew: Ethnicity, Gender and the Sociology of Disasters*, edited by W. G. Peacock, B. H. Morrow, and H. Gladwin. Rutledge, London.

Chapter 22

Business Crisis and Continuity Management

Gregory L. Shaw

ABSTRACT

This chapter is focused on the private sector organizations (businesses) that support the economy at the individual, family, community, local, state and national levels. However, even with this focus, the framework and principles of for profit business crisis and continuity management (BCCM) are applicable to all organizations, be they private, public or not-for-profit. Organizations exist to provide products and/or services to their customers and should strive to maintain and/restore this capability, even in the face of highly disruptive events. Regardless of the terminology chosen as the title for organizational continuity, crisis and continuity management or continuity of operations, continuity is a strategic responsibility and function for all organizations if they are to survive and prosper.

Central to the development and maintenance of a comprehensive organizational continuity program is an understanding of the myriad functions supporting continuity and their interdependencies. Recent efforts to develop a national standard as contained in the *NFPA 1600 Standard on Disaster/Emergency Management and Business Continuity Programs, 2004 Edition*, is a starting point, but falls short of the detail necessary to prescribe true standards.

As an alternate to the NFPA 1600 program description, a visual framework of BCCM, with definitions is presented and explained as the foundation of an enterprise wide program of BCCM. The framework was developed to be simple enough to be understandable at all levels of an organization, yet complete enough to support the case for functional integration and management to multiple stakeholders including boards of directors, executive level managers, stock owners, and customers. The framework supporting function of risk management and its subfunctions is explained to demonstrate the applicability and benefit of the business specific functions of business area analysis and business impact analysis to any organization.

INTRODUCTION

"Business" is not just the purview of the private sector. All organizations, be they private sector, public sector or not-for-profit provide products and/or services to their customers. Along with the delivery of products and/or services, all organizations also share the possibility of disruptive events that have impacts ranging from mere inconvenience and short-lived disruption of operations to the very failure of their ability to deliver their products and/or services which are the very nature of their business. Accordingly, organizational functions supporting business disruption prevention, preparedness, response and recovery such as risk management, contingency planning, crisis management, emergency response, and business resumption and recovery are established and resourced based upon the organization's perception of its relevant environments and the risks within those environments.

Individually, these functions can contribute to the protection of an organization and its business line. However, efficiency and effectiveness demand their integration and coordination into a comprehensive program of business crisis and continuity management. A logical starting point for accomplishing this integration is a visual framework and explanation that identifies the business crisis and continuity management supporting functions and their relationship to one another. Such a framework and its explanation are presented in this chapter. The framework, as presented, may appear quite different from the widely recognized Federal Emergency Management Agency model for Comprehensive Emergency Management which includes the phases of mitigation, preparedness, response and recovery, but the underlying philosophy and approach of both are actually quite similar and complementary.

THE TERM BUSINESS CRISIS AND CONTINUITY MANAGEMENT (BCCM)

Because of the many inconsistencies in terminology found in the contemporary literature of the business community the hybrid term business crisis and continuity management has been coined and introduced as a title for an organization wide strategic program and process. It is necessary to include a brief discussiòn of the creation and choice of this term since much of the current literature and business practices use the individual titles crisis management or business continuity management separately and often interchangeably as an umbrella term for the multiple functions and processes supporting the mitigation of and response to business disruption.

United States-based organizations such as Disaster Research Institute International (DRII, 2004), ASIS International (ASIS, 2004), and the Association of Contingency Planners (ACP, 2004) use the terms Business Continuity Management or Business Continuity Planning as their umbrella for multiple functions and processes including crisis management. The United Kingdom based Business Continuity Institute also employs the term Business Continuity Management as its overall program title. However, noted experts such as Ian Mitroff (Mitroff and Pauchant, 1992; Mitroff, 2001) and Stephen Fink (Fink, 1986) emphasize crisis management as the unifying structure and term for strategic business protection, response and recovery and include business continuity as one of many supporting functions.

Despite the difference in terminology, there is little debate in the business continuity and crisis management literature that crisis management, business continuity management, and their supporting functions need to be thoroughly integrated in support of overall business management. Business Continuity Management: Good Practices Guidelines explains the inconsistency in terminology by stating "Crisis Management and BCM [Business Continuity Management] are not seen as mutually exclusive albeit that they can of necessity stand alone based on the type of event. It is fully recognized that they are two elements in an overall business continuity process and frequently one is not found without the other" (Smith, 2002).

Thus, in an attempt to emphasize the interrelatedness and equal importance of crisis management and business continuity management, Business Crisis and Continuity Management has been chosen as the umbrella term and is defined as:

> Business Crisis and Continuity Management – "The business management practices that provide the focus and guidance for the decisions and actions necessary for a business to prevent, prepare for, respond to, resume, recover, restore and transition from a disruptive (crisis) event in a manner consistent with its strategic objectives" (Shaw and Harrald, 2004).

THE EVOLUTION OF BCCM

Business Crisis and Continuity Management, as a recognized business program, has evolved over the past twenty-plus years from a technology centric disaster recovery function dealing almost exclusively with data protection and recovery to a much wider holistic and enterprise wide supporting focus (Wheatman, Scott and Witty, 2001). Despite some strides to evolve BCCM into a profession including a widely accepted common body of knowledge and terminology, standards of performance, and certification process, progress has been slow and is hampered by the fact that BCCM, though generally recognized as a strategic function, remains a discretionary program for all but the most highly regulated business sectors such as the financial sector and healthcare sector. Even within these regulated sectors, standards of performance for all BCCM supporting functions may not be recognized and specified in sufficient detail to insure a truly comprehensive and integrated program.

As Ian Mitroff concludes from his extensive research in the area of business crisis management (his umbrella term for an integrated BCCM program), most businesses do not have an adequate crisis management program, supported by corporate culture, individual and organizational level expertise, infrastructure and plans and procedures to fully understand, prepare for, and manage the crises they may face (Mitroff, 1992). Mitroff has since updated his conclusions in the 2001 book, Managing Crises Before they Happen where he states that "The vast majority of organizations and institutions have not been designed to anticipate crises or to manage them effectively once they have occurred. Neither the mechanics nor the basic skills are in place for effective CM" (Mitroff, 2001). Mitroff's conclusions are further supported by the results of the 2001 Business Continuity Readiness Survey, jointly conducted by Gartner, Inc. Executive Programs and the Society for Information Management that found "Less than 25 percent of Global 2000 enterprises have invested in comprehensive business continuity planning" (Gartner, 2002).

This trend in BCCM acceptance is changing, however. The reality of business is that increasing and dynamic natural, technological and human induced threats, business complexity, government regulation, corporate governance requirements, and media and public scrutiny demand a comprehensive and integrated approach to business crisis and continuity management. Classic natural, technological and human induced events such as Hurricane Andrew (1992), the Northridge Earthquake (1994), the Exxon Valdez oil spill (1989), the Bhopal chemical release (1984), the World Trade Center attack of 1993, and the Tylenol poisoning case (1982) have provided lessons learned that emphasize each of these factors and the need for coordination and cooperation within and between organizations, and between all levels of government, the private and not-for-profit sectors.

These lessons have not been lost by many businesses that have reached the conclusion that integrated BCCM should be viewed as an investment rather than an additional cost that detracts from profits and have implemented their vision of comprehensive programs. The United States Business Roundtable, an association of business chief executive officers of leading corporations with the stated objective of improving public policy, explicitly recognizes the role of the Board of Directors and Management in the area of corporate governance in general, including specific business crisis and continuity management responsibilities. The Roundtable's white paper Principles of Corporate Governance charges the Board of Directors to periodically review management's plans for business resiliency and designate management level responsibility for business resiliency. Within the scope of business resiliency various functions are specifically mentioned and include business risk assessment and management, business continuity, physical and cyber security, and emergency communications (The Business Roundtable, 2002). However, lacking recognized standards and incentives, many businesses still consider BCCM as a burdensome cost that receives minimal and even no support.

The tragic events of September 11th, 2001 and the implications for businesses directly and indirectly impacted by the events have further reinforced the need for enterprise wide coordination of the multiple functions supporting business crisis and continuity management. Studies following the attacks of September 11th, 2001, such as the 9/11 Commission study and report have engaged the United States government, at all levels, in the process of recognizing the responsibilities of the private sector and encouraging the private sector to take adequate steps to protect people, property and business operations. Further steps, including mandated standards, may well follow beyond the current level of encouragement and voluntary compliance.

With roughly 80 percent of America's critical infrastructure managed by the private sector (The Conference Board, 2003), The National Strategy for the Physical Protection of Critical Infrastructures and Key Assets recognizes that the "private sector generally remains the first line of defense for its own facilities," and encourages private sector owners and operators to "reassess and adjust their planning, assurance and investment programs to better accommodate the increased risk presented by deliberate acts of violence" (The National Strategy, 2003). The most recent versions of the National Response Plan (January, 2005) and the National Incident Management System (March, 2004) include the private sector in all phases of crisis and emergency awareness, prevention, preparedness, response and recovery planning and operations. The National Response Plan explicitly charges the private sector to enhance overall readiness (NRP, 2004).

Supporting this goal of improved private sector readiness and intra and intersector coordination, the 9/11 Commission chartered the American National Standards Institute (ANSI) to develop a consensus on a national standard for preparedness for the private sector (9/11 Commission 2004). Based upon its collaboration with the National Fire Protection Association (NFPA) and the research of the 9/11 Commission, the "American National Standards Institute (ANSI) recommended to the 9-11 Commission that the National Fire Protection Association Standard, NFPA 1600 Standard on Disaster/Emergency Management and Business Continuity Programs, be recognized as the national preparedness standard (ISHN, 2004)." The 9-11 Commission report contains the following recommendation concerning private sector emergency preparedness and business continuity:

> We endorse the American National Standards Institute's recommended standard for private preparedness. We were encouraged by Secretary Tom Ridge's praise of the standard, and urge the Department of Homeland Security to promote its adoption. We also encourage the insurance and credit-rating industries to look closely at a company's compliance with the ANSI standard in assessing its insurability and creditworthiness. We believe that compliance with the standard should define the standard of care owed by a company to its employees and the public for legal purposes. Private-sector preparedness is not a luxury; it is a cost of doing business in the post-9/11 world. It is ignored at a tremendous potential cost in lives, money, and national security (9/11 Commission 2004).

Following from the 9/11 Commission Report, The Intelligence Reform and Terrorism Prevention Act of 2004, signed into law on December 18, 2004 specifically states in Section 7305 – Private Sector Preparedness, that:

> Preparedness in the private sector and public sector rescue, restart, and recovery of operations should include, as appropriate –
> (a) a plan for evacuation;
> (b) adequate communications capabilities; and
> (c) a plan for continuity of operations (IRTPA, 2004).

The Act goes on to state that the NFPA 1600 standard "establishes a common set of criteria and terminology," and charges the Department of Homeland Security to "work with the private, as well as government entities" (IRTPA, 2004). The Sense of Congress included in the Act falls short of mandating national standards for the private sector, but does encourage the adoption of voluntary standards such as those included in NFPA 1600.

The implications of the Act and the evolution of national standards on the private sector will certainly evolve over a period of time; however, there is already high level conjecture and discussions that compliance with NFPA 1600 will be established as an acceptable "legal standard of care" owed by businesses to their employees and the general public and will serve as a "safe harbor" to minimize potential legal liability. Compliance with NFPA 1600 may also find its way into insurance considerations including insurability, premium pricing, and deductible levels. Additionally, proof of adequate "preparedness" is increasingly finding its way into contractual agreements between the public and private sectors and between private sector businesses. Such requirements gained prominence in the preparations for Y2K, but lacked any real standard to demonstrate compliance. NFPA 1600 standards, though voluntary, appear to be the foundation of widely accepted national standards. Legal protection, insurance savings and contract requirements are certainly incentives for "preparedness" for all businesses and may be supplemented by additional measures such as tax savings and other forms of preferential treatment for business to business and business to government interactions.

NFPA 1600 STANDARD

The NFPA 1600 Standard on Disaster/Emergency Management and Business Continuity Programs (2004 edition) has gained national level attention and prominence as a result of the 9/11 Commission study and report, however, its development pre dates the events of September 11th, 2001. The original NFPA 1600 standards, published in 1995, focused on Recommended Practice for Disaster Management. The 2000 Edition, updated in the 2004 Edition, expanded the focus to a "total program approach for disaster/emergency management and business continuity programs (NFPA, 2004)." Lacking a visual framework of the functions comprising an integrated program of Disaster/Emergency Management and Business Continuity, NFPA 1600 specifies 15 program elements as displayed in Figure 22.1.

The intent of this chapter is not to be overly critical of NFPA 1600, but to recommend areas of improvement. NFPA 1600, the result of a consensus process representing multiple constituencies from all sectors, is a logical and necessary first step in the development of national standards written at a level of detail that can be used to define and measure compliance. As presented in the current edition (2004) of the document provides relatively broad descriptions of the program elements with minimal detail and is open to very liberal interpretation as to what actually comprises compliance at the program and program element level. A listing of the program elements is useful, but a graphical presentation of the elements, their hierarchy and interdependency could assist in the understanding and marketing of a comprehensive program that truly integrates the component parts. Additionally, NFPA 1600 defines a Business Continuity Program as:

> Business Continuity Program – An ongoing process supported by senior management and funded to ensure that the necessary steps are taken to identify the impact of potential losses, maintain viable recovery strategies and recovery plans, and ensure continuity of services through personnel training, plan testing, and maintenance. (NFPA 1600)

1. General
2. Law and Authorities
3. Hazard Identification, Risk Assessment and Impact Analysis
4. Hazard Mitigation
5. Resource Management
6. Mutual Aid
7. Planning
8. Direction, Control and Coordination
9. Communications and Warning
10. Operations and Procedures
11. Logistics and Facilities
12. Training
13. Exercises, Evaluations, and Corrective Actions
14. Crisis Communication and Public Information
15. Finance and Administration

Figure 22.1. NFPA 1600 2004 Edition Disaster/Emergency Management and Business Continuity Programs Elements.

This choice of a definition stresses preparedness, response and recovery with no mention of prevention and the linkage of the program to overall organizational goals. The definition of an overall Business Crisis and Continuity Management program presented earlier in this chapter provides this necessary emphasis and relegates reactive Business Continuity to its appropriate supporting function role.

A Framework for Integrated BCCM

Consistent with the philosophy of an integrated BCCM program is the need for a visual framework identifying the component functions and their relationship to one another. A visual framework should be simple enough to be understandable at all levels of an organization, yet complete enough to support the case for functional integration and management to multiple stakeholders including boards of directors, executive level managers, stock owners and customers. Such a framework, the synthesis of several existing frameworks as described in the paper The Core Competencies Required of Executive Level Business Crisis and Continuity Managers (Shaw and Harrald, 2004), is presented as Figure 22.2. This framework displays a hierarchy of the functions (from top to bottom) and the temporal nature of each (from left to right).

It must be emphasized that the BCCM framework, as presented, is in no way intended to prescribe a model organization chart for any business. It is merely the representation of multiple functions that require integration and coordination for the sake of program effectiveness and efficiency. Definitions for each of the functions are provided as a common point of understanding since there is significant disparity in the various glossaries of Business Crisis Management and Business Continuity Management found in sources such as NFPA 1600, The Business Continuity Institute, Disaster Recovery Institute International, and the Business Contingency Planning Group.

Enterprise Management – The systemic understanding and management of business operations within the context of the organization's culture, beliefs, mission, objectives, and organizational structure.

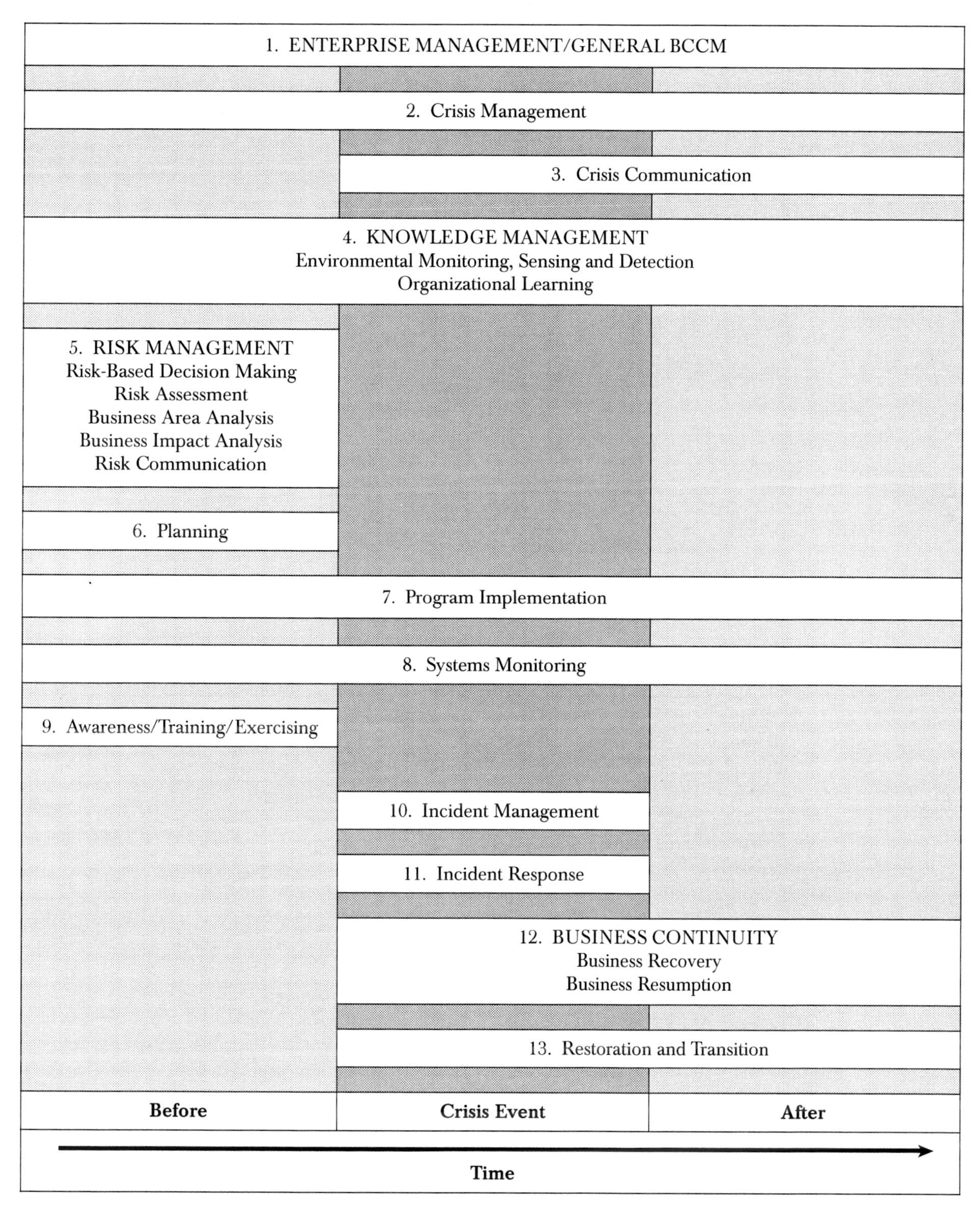

Figure 22.2. Business Crisis and Continuity Management Framework.

Crisis Management – The coordination of efforts to control a crisis event consistent with strategic goals of an organization. Although generally associated with response, recovery and resumption operations during and following a crisis event, crisis management responsibilities extend to pre-event mitigation, prevention and preparedness and post-event restoration and transition.
Crisis Communication – All means of communication, both internal and external to an organization, designed and delivered to support the Crisis Management function.
Knowledge Management – The acquisition, assurance, representation, transformation, transfer and utilization of information supporting Enterprise Management.
Risk Management – The synthesis of the risk assessment, business area analysis, business impact analysis, risk communication and risk-based decision-making functions to make strategic and tactical decisions on how business risks will be treated – whether ignored, reduced, transferred, or avoided.
Planning – Based upon the results of risk management and within the overall context of enterprise management, the development of plans, policies and procedures to address the physical and/or business consequences of residual risks which are above the level of acceptance to a business, its assets and its stakeholders. Plans may be stand alone or consolidated but must be integrated.
Program Implementation – The implementation and management of specific programs such as physical security, cyber security, environmental health, occupational health and safety, etc. that support the Business Crisis and Continuity Management (BCCM) program within the context of Enterprise Management.
Systems Monitoring – Measuring and evaluating program performance in the context of the enterprise as an overall system of interrelated parts.
Awareness/Training/Exercising – A tiered program to develop and maintain individual, team and organizational awareness and preparedness, ranging from individual and group familiarization and skill-based training through full organizational exercises.
Incident Management – The management of operations, logistics, planning, finance and administration, safety and information flow associated with the operational response to the consequences/impacts (if any) of a crisis event.
Incident Response – The tactical reaction to the physical consequences/impacts (if any) of a crisis event to protect personnel and property, assess the situation, stabilize the situation and conduct response operations that support the economic viability of a business.
Business Continuity – The business specific plans and actions that enable an organization to respond to a crisis event in a manner such that business functions, sub-functions and processes are recovered and resumed according to a predetermined plan, prioritized by their criticality to the economic viability of the business.
Restoration and Transition – Plans and actions to restore and transition a business to "new normal" operations following a crisis event.

Even though this framework of integrated BCCM and its accompanying definitions of supporting functions is similar to NFPA 1600 in falling short of the details for defining true standards, it is proposed as a more comprehensive basis for actual standards development. The actual examination of those details within the framework and all of the functions goes beyond the scope of this chapter. The next section will, however, briefly describe the function of Risk Management and its supporting sub-functions and its applicability to all organizations in all sectors.

RISK MANAGEMENT

Risk management is the foundation of a comprehensive BCCM program and drives the decisions impacting all of the other functions contained in the framework. Although portrayed as occurring in the time period before a specific crisis event, risk management is a continual and iterative process. It requires dialogue with multiple stakeholders, and monitoring and adjustment in light of changes to the environment and the economic, public relations, political and social impacts of BCCM-related decisions. All organizations in all sectors operate with constrained resources and have the responsibility to allocate available resources in a manner that best supports overall enterprise wide goals and objectives.

The protection of personnel, property and reputation and the ability to recover, resume and restore business operations according to a reasoned and defendable plan are inherent in these goals and objectives and follow from a risk-based decision-making process. A private sector model of risk management, including the supporting steps of business area analysis and business impact analysis, can provide an analytic approach to risk-based decision making that is also applicable and potentially beneficial to public and not-for-profit organizations.

Business area analysis, in its basic form is an understanding of the products and services provided by a business and how they are produced and delivered. Depending on the complexity of the business and the product and service, business area analysis can be as simple as merely observing and documenting how the business operates. At the other extreme, it can involve the decomposition of business functions to the process and even task level to understand interdependencies and points of potential failure. Removed from the context of a Business Crisis and Continuity Management program, business area analysis is still a necessary component of business operations in all sectors and supports general business efficiency and effectiveness. Regardless of the complexity of the business area analysis process, what is important is that decision makers fully understand their business and what is necessary (critical) to deliver its products and services.

Business impact analysis matches the results of risk assessment (the identification of hazards and a determination of their probability of occurrence and the consequences of their occurrence) to the business area analysis to determine the impacts of the hazards on business operations and to identify the potential interventions (controls) to protect business operations based upon their criticality. Taken together, the business area analysis and business impact analysis provide an analytic and economic basis for risk-based decision making and the allocation of resources supporting the overall risk management function. Beyond this analysis, other very legitimate considerations including political, social and environmental realities exist that impact the risk management process. They are, however, overlays to business area and business impact analysis and should not be the starting point for making risk management based decisions.

Risk management also includes the sub function of risk communication which is an essential component of risk management. Risk communication is a two way exchange of the risk related information, concerns, perceptions, and preferences within an organization and between an organization and its external environment that ties together overall enterprise management with the risk management function. The two seminal National Research Council documents, Improving Risk Communications (National Academy Press, Washington, DC, 1989) and Risk: Informing Decisions in a Democratic Society. The National Academy Press, Washington, DC (1996) provide a comprehensive description of risk communication and perception, their application and lessons learned, and the derivation of principles and guidelines applicable to organizations from all sectors.

CONCLUSION

This chapter describes Business Crisis and Continuity Management as a strategic program with supporting functions and sub-functions that must be integrated for the sake of overall effectiveness and efficiency. A functional framework and definitions are presented to visualize the structure and interdependencies of the components of such a program and are proposed as a logical basis for developing national standards for organizations from the private, public and not-for-profit sectors. All organizations, from all sectors, are in fact businesses to the extent that they provide products and/or services to their customers. Protection of the ability to provide these products and/or services is a strategic imperative that must be understood and supported at all levels of any organization.

REFERENCES

Association of Contingency Planners – International. (2004). Web Site. Oak Creek, WI. http://www.acp-international.com/.

ASIS Commission on Guidelines. (2003). Chief Security Officer (CSO) Guideline. Alexandria, Va. http://www.asisonline.org/guidelines/guidelineschief2003.pdf

ASIS Commission on Guidelines. (2004). Business Continuity Guideline: A Practical Approach for Emergency

Preparedness, Crisis Management, and Disaster Recovery. Draft Guideline. Alexandria, VA. (July 12). http://www.asisonline.org/guidelines/guidelinesbusinesscon.pdf

Barton, L. (1993). *Crisis in Organizations: Managing and Communicating in the Heat of Chaos.* Cincinnati, OH, South-Western Publishing Co.

Borge, D. (2004). *The Book of Risk.* New York: John Wiley and Sons, Inc.

The Business Round Table. (2002). *Principles of Corporate Governance.* A White Paper from the Business Roundtable.

Continuity Central. (2004). *What's Under the Business Continuity Umbrella?* (July 14). http://www.continuitycentral.com.

Cronin, K. P. (1993). Legal Necessity. Disaster Recovery World II [CD ROM]. *Disaster Recovery Journal.* St. Louis, MO, DRII.

Continuity Central. (2003). *Developing a Comprehensive Open-Source Business Continuity Model.* London, UK: Continuity Central. (June 27). http://www.continuitycentral.com/feature017.htm Last accessed (August 14, 2004).

Department of Homeland Security. (2004). *National Incident Management System* (NIMS). Washington, DC. (March 1).

Department of Homeland Security. (2005). *National Response Plan (NRP).* Washington, DC. (June 30).

Disaster Recovery Institute International. (2004). *Introduction and Professional Practices for Business Continuity Professionals.* DRI International. Falls Church, VA. http://www.drii.org.

Drabek, T. and Hoetmer, G. (Editors). (1991). Emergency Management Principles and Practice for Local Government. Washington, DC, ICMA.

Federal Emergency Management Agency. (1996). *Emergency Management Guide for Business and Industry.* Washington, DC, FEMA.

Fink, S. (1986). *Crisis Management: Planning for the Inevitable.* Authors Guild Backprint Edition, 1986.

Gartner 2002 press release. (2004). *Gartner Says That Less Than 25 percent of Global 2000 Enterprises Have Invested in Comprehensive Business Continuity Planning.* (October 8). http://www3.gartner.com/5_about/press_releases/2002_10/pr20021008a.jsp

Harrald, J. R. (1998). *A Strategic Framework for Corporate Crisis Management.* The International Emergency Management Conference 1998 (TIEMS '98) Proceedings. Washington, DC, TIEMS.

Hiles, A. (2002). *Business Continuity: Best Practices.* Rothstein Associates Inc. Brookfield, CT.

Hiles, A. (2002). *Enterprise Risk Assessment and Business Impact Analysis: Best Practices.* Brookfield, CT., Rothstein Associates Inc.

Industrial Safety and Hygiene News (ISHN) Online. (2004). NFPA 1600 to Become the National Preparedness Standard? (April 30). http://www.ishn.com/CDA/ArticleInformation/news/news_item/0,2169,123889,00.html

Laye, J. (2002). *Avoiding Disaster: How to Keep Your Business Going When Catastrophe Strikes.* J. Wiley and Sons, Inc. Hoboken, NJ.

Lerbinger, O. (1997). *The Crisis Manager – Facing Risk and Responsibility.* Mahwah, NJ., Lawrence Erlbaum Associates.

Mitroff, I. and Pauchant, T. C. (1992). *Transforming the Crisis-Prone Organization.* San Francisco, CA., Jossey-Bass, Inc.

Mitroff, I.. (2001). *Managing Crises Before They Happen: What Every Executive and Manager Needs to Know About Crisis Management.* New York, NY., Amaco.

9/11 Commission. (2004). *9/11 Commission Report.* Washington, DC, U. S. Government Printing Office.

NFPA. (2004). *NFPA 1600 Standard on Disaster/Emergency Management and Business Continuity Programs.* Edition. Quincy, MA. (2004).

National Research Council. (1989). *Improving Risk Communications.* Washington, DC, National Academy Press.

National Research Council. (1996). *Understanding Risk: Informing Decisions in a Democratic Society.* Washington, DC, National Academy Press.

Saraco, D. (1999). *White Paper – BC Management: A Marriage of Craft and Technology.* Irvine, CA., MLC & Associates, Inc.

Shaw, G. L. and Harrald, J.. R. (2004). Required Competencies for Executive Level Business Crisis and Continuity Managers. *Journal of Homeland Security and Emergency Management.* January.

Smith, D., J. Editor. (2002). *Business Continuity Management: Good Practices Guidelines.* London, England: The Business Continuity Institute. http://www.thebci.org

Standards of Australia Ltd. (2002). *A Handbook on Business Continuity Management: Preventing Chaos in a Crisis.* Sydney, Australia: Consensus Books.

Standards of Australia Ltd. (2003). Draft Business Continuity Handbook. Sydney, Australia.

United States Government. (2005). Intelligence Reform and Terrorism Prevention Act of 2004. Section 7305. Private Sector Preparedness. Washington, DC.

Wheatman, V., Scott, D., and Witty, R.. (2001). *Aftermath: Business Continuity Planning.* Gartner Top View. AV-14-5138. (September 21). http://www.gartner.com.

White House Administrative Office. (2003). *National Strategy for the Physical Protection of Critical Infrastructures and Key Assets.* Washington, DC. (February).

Chapter 23

Information Disasters and Disaster Information: Where Information Science Meets Emergency Management

Tisha Slagle Pipes

ABSTRACT

Information moves society – and with the flow of electronic information, the interfaces for information sharing are continuously becoming more diverse. Information Science offers systems and technologies to connect people worldwide while Emergency Management proffers methods to secure information that has become increasingly more vulnerable to destruction. This exacerbated vulnerability of information to disasters, combined with society's dependence on information, warrants the integration of the disciplines of Information Science and Emergency Management. Together, these disciplines can improve existing practices to prevent and mitigate information disasters. They can also ensure the usefulness of the information that flows among victims, responders, and members of emergency management organizations before, during, and after disasters.

INTRODUCTION

Information flows across space and time in unpredictable ways, creating new structures and forms as the situation requires (McDaniel, 1997).

Unprocessed information is intangible and nonconsumable, yet a plentiful resource that can be refined and used as a public or private good. *Information* is inherently more abundant than most resources because it is found in every person, place, and thing – it is the entirety of known data, facts and ideas. Information, in my opinion, is any meme, message, or meaning that influences, directly or indirectly, how persons understand their situations. It is the principle element of omniscience, and therefore the resource from which all knowledge is extracted. *Knowledge* includes units of systematic subjects, noted for their oneness, objectivity, respected social implications, usefulness, and resistance to obsolescence. Knowledge is mined and refined into the integrated disciplines the world calls *wisdom* – valued public goods like anthropology, information technology (IT), medical research, and universal religion (Cleveland, 1982).

As unprocessed public goods, information flows between and among people and groups in the form of verbal, non-verbal, or written interactions – whether memes, messages, or meanings – that serve as precursors to problem-solving and decision making. Interactions instigated directly or indirectly by a disaster could be deemed *disaster information.*

As processed public goods, information – whether a meme, message, or meaning – influences the lives of those who experience it. When life-sustaining or life-fulfilling information is absent, inaccessible, or useless because it is inaccurate or interrupted as the result of a hazard – natural, civil, or technological (Table 23.1), the persons affected may be said to be experiencing an *information disaster.* An information disaster hinders the access to or effective use of disaster information.

Table 23.1
CATEGORIES OF HAZARDS TO INFORMATION

Technological	*Natural*	*Civil*
chemical, electrical, nuclear	earthquake, flood, hurricane	cold war, cyber-terrorism, information warfare, terrorism, war

Information is a vital public good whether processed or unprocessed. How people encounter information, a phenomenon called *information-seeking behavior* or *information behavior* by information scientists, is the subject of extensive research (Case, 2002). The study of *disaster information behavior* –the actions or attitudes that affect encountering, needing, finding, choosing, or using disaster information – appears to be scant or absent in the literature.

This deficiency in the study of disaster information behavior may exist because studying information behavior involves field study – an option not always available to researchers in times of disaster. In addition, many researchers cannot afford the time and expense demanded by qualitative research, the preferred approach to effective information behavior studies. A further challenge for researchers is the inherent elusiveness of information itself. The form it assumes or the direction it will flow is not always apparent (Burlando, 1994). What is apparent, however, is that information, as the essence of all knowledge, and subsequently the essence all wisdom – is the basis for all disciplines of study, including information science and emergency management. Its pervasiveness alone demands interdisciplinary observation.

INFORMATION DISASTERS AND DISASTER INFORMATION

The Study of Information Disasters

Unprocessed information is impervious. It does not deplete with use or corrode with time. However, people can forget it or disregard it, and *representations* (Table 23.2) of it can be easily lost or destroyed. These intangible or tangible surrogates that hold and/or display information are quite vulnerable to disaster. Hazards – in the form of terrorism, vandalism, heating/air conditioning failure, user error, computer viruses, hackers, power failures, cyberterrorism, information warfare, cultural power struggles, or even careless or impulsive law-making/enforcement – all threaten the security and effectiveness of information. Because all organizations house information, it is imperative that all organizations implement disaster recovery plans that include recovery of information vital to the existence of the organization.

Studies that focus on the disruption and destruction of information have become more prevalent, especially in the management fields where chronicled information is vital to management operations. Useful human and/or artificially transmitted messages were recorded as early as 3000 B.C.E. when the Sumerians created and stored common cuneiform symbols by inscribing them into soft clay with a stylus. The Sumerians, as have societies since, used common symbols with technology to transfer information (Drucker, 1995). Information Science (IS) studies have shown that for information to be managed effectively, people must employ a premise from sociology – for example, culturally accepted standards and symbols – with technology – for example, stylus and clay or keyboard and computer. Otherwise, information cannot be physically or electronically organized, stored, processed, recorded, disseminated, preserved, or retrieved. Because of the urgency to preserve and retrieve informational records, organizations are incorporating information preservation into their business continuity plans (Shaw, 2005). A sub-discipline of IS, librarianship, has long implemented these disaster recovery plans, (DiMattia, 2001; Muir and Shenton, 2002; Ruyle and Schobernd, 1997; Tennant, 2001) to protect and preserve the physical and electronic representations of information in library holdings.

Table 23.2
COMMON REPRESENTATIONS OF INFORMATION

artifact	footprint	sign
code	hyperlink	signal
calculation	maps	summary
datum	model	synonym
diagram	photo/image	text
email	recording	thoughts
film	replica	title/name
fingerprint	secret	voice

The Study of Disaster Information Flow

Determining how information flows among organizations before, during, and after disasters can lead to new models of sound practice for Emergency Management (EM) practice to adopt. The continued omission of the study of information flow may allow the implementation of unsound practices and hastily enacted policies and decisions. IS methods from information flow research, including systems theory and small group interaction, may hold particular application for further study of information flow in EM.

The study of disaster information flow has been virtually ignored by IS researchers, despite its importance in EM and society. Research regarding *information flow* – the human and/or artificial information transactions that affect decisions – is of especial interest to EM where decisions affect the well-being of whole communities. EM decision-makers determine who is heard or not heard and what is done or not done regarding disaster planning and response – a vital public service that impacts communities socially, economically, and legally. People reach decisions through the processes of information flow during formal or informal meetings.[1] Information flow in meetings of EM organizations may or may not be conducive to optimal disaster management; and researchers have not provided conclusive evidence either way. It is imperative that EM researchers know if methods employed in decision-making – the result of the information flow – are increasing or decreasing the vulnerability of a community to disasters.

EM concentrates on the preparedness, response, recovery, and mitigation of disasters. McEntire (2004b) defines disasters as the "disruptive and/or deadly and destructive outcome or result of physical or human-induced triggering agents when they interact with and are exacerbated by vulnerabilities from diverse but overlapping environments." Teams within EM organizations may struggle for long periods – or be forced to decide quickly how best to approach disasters. During these times of decision-making, the members of a team participate, either consciously or unconsciously, in creating and modifying information flow. Productive information flow is vital to ensure that EM teams reach sensible decisions. Sensible decisions aid in the prevention and mitigation of disasters.

HISTORY OF "IS"

Information scientists historically seek solutions to problems regarding information in the broad disciplines of technology and sociology. The birth of this blend of technology and sociology in IS can be attributed to inspiration from "As We May Think," an article written by Vannevar Bush at the close of the second World War. Bush, a respected MIT scientist and director of the United States (U.S.) Wartime Office of Scientific Research and Development, believed that the scientists who had been busy devising methods to defeat U.S. enemies would now have time to devise methods to mitigate the chaos already evidenced by the explosion of information. He predicted scientific and social disaster if scientists did not address "the massive task of making more accessible a bewildering store of knowledge" (Bush, V., 1945a).

Bush had a suggestion – a technological knowledge management system in the form of a machine that would emulate human thought using "association of ideas." The *Memex* would link thoughts "in accordance with some intricate web of trails carried by the cells of the brain" – a concept remarkably similar to contemporary hypertext (1945a)! The postwar scientists were unsurprisingly fascinated with Bush's proposal and accepted the technological challenge.

Fortunately, Congress funded the scientists, with incentive from President Theodore Roosevelt who enlisted Bush to write a report to justify the financial support. Bush's report to Roosevelt, "Science the Endless Frontier" (1945b), provided the basis for the creation of the National Science Foundation (NSF) by means of the NSF Act of 1950. One of the Act's mandates was "to further the full dissemination of information of scientific value consistent with the national interest" (P.L. 81-507), a plan that eventually

1. Interestingly, there are 1.375 million non-profit organizations in the U.S. with 11 million meetings being held daily (Weitzel and Geist, 1998). In spite of this incredible number of meetings – and the importance of smooth information flow in and among EM organizations – public management organizations conduct few field studies in information flow.

led to the study of information flow that generates important decisions.

IS: Technology and Sociology

NSF scientists quickly developed two major IS directions – technologically-based information retrieval and sociologically-based human information behavior – and by the 1960s, a few researchers were defining the term IS. When the American Documentation Institute, founded in 1937, decided to change its name to the American Society for Information Science, definitions abounded. Borko (1968) wrote one of the most enduring definitions, one that roots IS firmly in *technology* by stating that it is "an interdisciplinary science that investigates the properties and behavior of information, the forces that govern the flow and use of information, and the techniques, both manual and mechanical, of processing information for optimal storage, retrieval and dissemination" (Borko, 1968).

Researchers gradually revised the more technologically-based definitions to reflect IS roots in *sociology*. The IS scope would be defined by Wersig and Nevelling (1975) who wrote that "transmitting knowledge to those who need it" is a "social responsibility." Belkin and Robertson (1976) would continue the technology-sociology theme by stating that the purpose of IS is to "facilitate communication of information between humans." Eleven years later, Vickery and Vickery (1987) emphasized the role of sociology in IS by identifying IS as "the study of the communication of information in society." Buckland and Liu (1998) would once again combine technology and sociology when they asserted that IS "is centered on the representation, storage, transmission, selection (filtering, retrieval), and the use of documents and messages, where documents and messages are created for use by humans." Bates (1999) clarified, however, by writing that IS is primarily, but not solely focused, "on *recorded* information and people's relationship to it." With all the progress in determining the definition of IS, however, the definition of *information* – the focus of IS – remained somewhat elusive.

What is Information

Information theorists Shannon and Weaver (1948) believed that "information is the reduction of uncertainty," and yet, ironically, finding a clear definition of information still seems to stump both researchers and readers of IS. Information has been defined within many disciplines by those who sometimes oversimplify or overcomplicate its meaning – nevertheless, IS researchers agree that information is fundamental to all disciplines for communication, and it must therefore be preserved, organized, and easily retrieved (Buckland, 1991; Ratzan, 2004). Information may be described as a representation of a message that is processed into something valuable so that it may be applied in a practical context. This description, however, suggests that the *value* of information has somehow been previously established. So, how, then, is the value of information determined?

The Value of Information

The value of information is best determined by what Repo (1983) calls *value-in-use* – "a benefit the user obtains from the use and the effect of the use." Value-in-use is subjective and specific to a user – so the value of information could be defined simply as *contingent upon its usefulness to an individual.* The value of information therefore is relative to the level of satisfaction directly or indirectly received from an information good, service, or resource.

Consider, for instance, contrasting views of those who receive a stack of 1820s newspapers from a ghost town. The *litterbug* casually tosses the papers outside – to the litterbug, the papers are *trash to be burned.* The *recycler* carefully collects the papers in a bag – to the recycler, the papers are *cash to be earned.* The *librarian* gladly accepts the papers from the recycle shop – to the librarian, the papers are *documents that must be sorted.* The *professor* delightedly inquires about the papers from the library – to the professor, the papers are *history to be reported.* The value of information is therefore determined by its user and its intended application.

The Sciences of Information

How, why, what, and where information is applied are questions investigated within the framework of several information studies – a truth that often identifies information *science* as information *sciences.* Whether it is appropriate to label the field of IS as singular or plural is another argument (Webber, 2003), however; IS is undeniably interdisciplinary (Machlup and Mansfield, 1983) with problems studied through four major interdisciplinary relations including: cognitive science, communications, computer science, and librarianship (Appendix A) (Saracevic, 1999).

IS as a Meta-Discipline

IS enables people to find information – a need based on psychological needs for survival and fulfillment. Finding sought-after information can change human perception by relieving anxiety, fulfilling a goal, realizing a need, or actualizing a concept. IS has dedicated years of research to training people how to find information and thereby enhance problem-solving and decision-making – helping to reduce uncertainty and change an individual's image of reality (Case, 2002).

Education, mass communications, and philosophy/theology also have distinctive relationships with information. Education is the teaching and learning of information; mass communications is the discovery and transmission of information; philosophy/theology is the search for true information. In fact, IS has application to all disciplines and is therefore more appropriately defined as a *meta-disciplinary* science (Bates, 1999). After all, IS is concerned with information, a resource (Cleveland, 1982) common to *all* disciplines and coincidentally, responsible for the creation of *bibliometrics*, the major quantitative method used to analyze *interdisciplinarity* among fields (Morillo and Gómez, 2003).

Bibliometrics uses *content analysis*, a method that includes comparing the frequency (F) of terms between disciplines, for example IS and EM. Content analysis, in this case, becomes a preliminary survey to determine whether researchers have initiated integration within disciplines (Ruben, 1992). A cursory examination of IS and EM journals identifies major terms commonly found in titles of articles in IS and EM from 2002 through July 2005. The IS journals are *African Journal of Library, Archives and Information Science, Journal of Human-Computer Studies, Journal of Librarianship and Information Science, Journal of the Society of Archivists, Library & Information Science Research,* and *International Journal of Human-Computer Studies.* The EM journals are *Disaster Prevention and Management, Disasters, International Journal of Emergency Management, Journal of Contingencies & Crisis Management,* and *Natural Hazards Review* (Table 23.3).

The infrequency of terms within journal titles indicates that studies of IS and EM have experienced little integration. IS and EM do however, share a substantial interest in the study of disasters in at least two distinct aspects: information disasters and disaster information flow. EM theories may be used to frame information disasters and vulnerabilities while IS theories may be used to study collaborative decisions by identifying patterns of information flow during the phases of a disaster.

EM THEORY, INFORMATION, AND DISASTERS

The purpose of EM is to minimize vulnerabilities to hazards that cause disasters. Vulnerabilities, according to McEntire, are "high levels of risk and susceptibility coupled with a low degree of resistance and resilience" that exacerbate potential hazards – "triggering agents" and cause disasters (2004b). Although it may be impossible to prevent or diminish hazards,

Table 23.3
FREQUENCY (F) OF IS/EM TERMS IN EM/IS JOURNAL TITLES 2002–2005

IS terms found in EM Journal Titles	F	EM terms found in IS Journal Titles	F
information (sharing, system, technology)	11	disaster	5
communication (or coordination or collaboration or interaction)	4	emergency or hazard(s)	2
Total	15	Total	7

it is possible to reduce the vulnerability to hazards that lead to technological, natural, or civil disasters (Table 23.3). For example, government officials may be able to reduce the vulnerability to terrorism through heightened "protection of borders and infrastructures or improvement in the prevention of weapons of mass destruction" (McEntire, 2004b).

Information and Technological Disasters

Disasters caused by technology are "the most difficult to predict" and are "largely unforeseeable" (Chapman, 2005). Tragically, many technological disasters result from seldom-inspected or outdated technology. A faulty computer surge protector is blamed for a fire that swept through Minnesota's Hastings County Library in 1993, destroying 80 percent of the library's hard-copy and electronic documents. The flames demolished the building and left the remaining documents damaged by soot, smoke, and water (Bolger, 2003). Since then, Hastings County has implemented disaster recovery plans to prevent and/or prepare for disasters.

Information and Natural Disasters

Reducing vulnerability to natural disasters may be more challenging. During the 1997 renovation of the Colorado State University Library (CSUL), library administrators stored approximately 462,000 volumes in the basement. During this temporary storage period, a flash flood enveloped the 77,000 square foot basement with 10 feet of water. Approximately 450,000 items – one-fourth of the library's holdings – were damaged or destroyed, including the library's vast newspaper collection. Although the flood was the world's fourth largest library disaster in the twenthieth century, there was at least one positive consequence. Camila Alire, CSUL librarian, and other CSUL staff wrote *Library Disaster Planning and Recovery Handbook*, considered by critics to be a valuable disaster practitioner's guide (Dugan, 2001; Williams, 2000).

Information and Civil Disasters

Unfortunately, civil disasters are prevalent throughout human history. At least three versions of the destruction[2] of the ancient Library of Alexandria may be found in Egyptian history; however, all versions attribute the great Library's demise to civil disaster (Chesser, 2005). Ptolomy II Soter, successor to Alexandria the Great, built the Library in 283 B.C. and vowed to use any means to amass all the books of the world (Heller-Roazen, 2002). The Library did acquire nearly one-half million books – an unequaled collection for its era – before its ruin in the first century (Erskine, 1995).

The threat of modern civil disasters has increased since the September 11, 2001 terrorist attacks (9/11) – a fact that makes tangible and intangible (electronic or human) information and information issues considerably more vulnerable to disaster (Comfort, 2005; Dory, 2003-04; The 9/11 Commission, 2004). Lawmakers have since enacted grave changes to the tenets of the U.S. Constitution (USA PATRIOT Act[3])

2. Some historical accounts attribute the Library's obliteration to Julius Caesar during the Roman civil war. While chasing the mutinous General Pompey, Caesar ordered his soldiers to burn the Egyptian fleet. Some accounts report that the fire spread into Alexandria and into the Library. Other accounts blame the destruction on a religious dispute around 391 A.D. Alexandrian Jews were attempting to burn down a Christian church when flames engulfed the nearby Library. The third rendition of the loss of the Library involves the Moslems who seized Alexandria in 640 A.D. Accusers blame the Moslem leader, Caliph Omar for burning all of the books upon hearing that the Library contained "all the knowledge of the world." Caliph Omar supposedly said about the Library's holdings that "they will either contradict the Koran, in which case they are heresy, or they will agree with it, so they are superfluous" (Chesser, 2005).

3. Attorney General John Ashcroft, who urged Congress during the aftermath of 9/11 to expand governmental powers in order to fight terrorism more aggressively, instigated the legislation leading to the PATRIOT Act. The PATRIOT Act quickly became law October 21, 2001 without the customary consultation and hearings of Congressional committees. The PATRIOT Act expanded the authority of law enforcement agencies to allow access to records previously protected by open records law. The PATRIOT Act, up for reenactment in 2005, made changes to many laws that can apply to library records and confidentiality. The amendments expanded the authority of law enforcement agencies. Agents can use wiretaps without making sure the target is actually using the phone to be tapped (Section 206). They can also access library circulation records, Internet use records, and registration information (electronic or printed) using gag orders and without demonstrating probable cause (Section 215). They can also monitor library computer use including Internet, email, IP addresses and Web page URLs (Section 216).

regarding access to information (Jaeger, Bertot and McClure, 2003). Both lawmakers and private citizens have made public information less accessible – so terrorists cannot access it – and private information more accessible – in case the information belongs to terrorists. The threat to public information and information privacy is an ongoing hazard, one that librarianship has struggled to mitigate (Appendix B).

Information in the Equation for Disasters

Librarians and library educators hold the preservation of information sacred and consider libraries – the primary keepers of recorded information – to be the last visage of a free public education for all citizens (Totten, 2005). Clearly, all disciplines benefit from the preservation and access to information, a sound reason for investigating information disasters. The study of vulnerability provides an excellent basis for future research in information disasters (as well as information flow regarding disasters). Information disasters may be portrayed using McEntire's equation for disasters: *hazard* (*triggering agent*) + *vulnerability* = *disaster* (2004a) (Appendix C).

IS THEORY, INFORMATION, AND DISASTERS

The flow of information within the management of disasters can be investigated using several methods found in the interdisciplinary domains of IS. Many theories have evolved including theories of individual and collective information behavior (information seeking and processing). Collective information behavior has been studied in the context of group research (information flow in both task and emergent groups). *Task groups* – individuals who accept a collective charge to form decisions and/or solve problems – dominate research of information flow in groups. *Emergent groups* – individuals that meet incidentally and collaborate – have appeared in recent IS research with an emphasis on conversational problem-solving (O'Connor, Copeland and Kearns, 2003). *Emergent behavior* – a more intense form of problem-solving – has been the subject of some EM studies within the context of disaster scenes (Drabek and McEntire, 2003).

Information Flow and Small Group Studies

Knowing and testing the varying properties of information flow in groups may be vital to the success of EM teams at the local, state, and national level, within all the disaster phases (preparedness, response, recovery, and mitigation) identified by Drabek (1986). Emergent groups that exist at disaster scenes warrant study, as do EM decision-making groups that contribute to the future health and survival of our governments, communities, and citizens. Within the last decade, the Communications discipline has introduced several ethnographic studies of group information flow, although the study of groups has traditionally been performed in laboratory settings. These contrived experiments cannot reveal the properties of real group information flow.

Although the small group[4] remains the oldest and most prevalent of the concepts in all social organization (Fisher, A., 1974), the disciplines that study information flow in small groups are diverse and disconnected. Research of information flow in groups has matured despite independent studies by scholars in psychology, sociology, management, communication, education, social work, political science, public policy, urban planning, and IS. The absence of convergence within the fields, however, has not prevented small group research from accumulating enough solid theory in the past 50 years to establish its own discipline of study (Poole, 2004).

The formal study of information flow in groups can be traced to 1898 when psychologist Norman Triplett tested the hypothesis that the presence of others in a group would facilitate the problem solving of an individual (Hare, 1962). By studying group behavior, educators and politicians believe people can collectively solve common problems in the communities (Gouran, 2003a).

4. A group is a collection of at least three people who interact with each other, display interdependence, establish roles within an open communcation system, have a sense of unity and identity, maintain norms, and share common motives or goals with the intention of making a decision (Brilhart, 1978; Ellis and Fisher, 1994; Hare, 1962; Harris and Sherblom, 2002).

Group Decision-Making

A major influence in 1910 on the study of group discussion and decision-making was the well-known book *How We Think* by the distinguished philosopher, John Dewey. Dewey's model is still the most widely-used model for directing the information flow toward problem solving and decision-making in IS studies of groups (Table 23.4).

GROUP PROPERTIES

Small group research became recognized in the late 1940s and early 1950s by an increasing number of references in the social science literature (Ellis and Fisher, 1994; Fisher, 1974; Gouran, 1999; Hare, 1962; Harris and Sherblom, 2002; Hartley, 1997). Although there are many more, 13 major properties, identified through theoretical studies and worthy of further study, may be depicted as an acronym, G.R.O.U.P. D.Y.N.A.M.I.C.S. (Appendix D) Understanding these properties of real groups in action provides descriptive and prescriptive methods that may enhance decision-making capabilities in EM organizations.

The G.R.O.U.P. D.Y.N.A.M.I.C.S. properties emerged because of studying the information flow in groups as a process. The group as a process led to groups being studied from the systems approach (Bales, 1999; Gouran, 2003b; Harris and Sherblom, 2002).

Information Flow and Systems Theory

Open systems of interaction, originally applied to biological systems, provides a compelling symbolic foundation for an IS study of EM decision-making. Many IS studies found basis for systems of information flow using Shannon and Weaver's 1948 communications/information systems model (Figure 23.1) to describe group communication systems.[5]

Lewin, Lippitt, and White (1939) introduced the systems approach to studying the group information process; however, scholars did not embrace the theory until Bales (1950) enhanced it. Bales compared groups – like those found in EM – to an open system that is, from inception to outcome, a cyclical process – dynamic, continuous, and evolving (Hare, 1962; Harris and Sherblom, 2002). EM organizations can be viewed as a subsystem within the larger social system – an open system. An open system is a set of interrelated *components* that operate together as a whole with three major elements: *input*, *process*, and *output*. Multiple subsystems of transactions, called *processes*, characterize a system of interaction, called *information flow*. These processes are between and among people, *components* who continually and simultaneously send *output* and receive *input*. The purpose of these transactions is to achieve a mutual goal, a successful *outcome* (Bales, 1950) (Table 23.5).

The Communications discipline investigates face-to-face and virtual systems of information flow and decisions. It has provided group information flow with a

Table 23.4
DEWEY'S REFLECTIVE THINKING MODEL

1. Determine what information is needed for understanding the issue at hand.
2. Access and gather the available information.
3. Gather the opinions of reliable sources in related fields.
4. Synthesize the information and opinions.
5. Consider the synthesis from all perspectives and frames of reference.
6. Finally, create some plausible temporary meaning that may be reconsidered and modified as more relevant information and opinions are learned. (Dewey, 1910)

5. Shannon and Weaver, 1948, Modified. Claude Shannon, a research scientist at Bell Telephone Company, attempted to maximize telephone line capacity with minimal distortion. He probably intended his mathematical theory of signal transmission for use with telephone technology only.

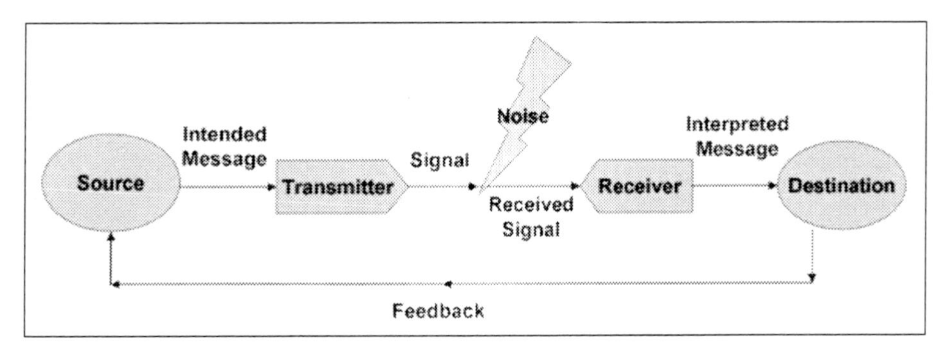

Figure 23.1

respectable position in formal research (Poole, 2004). Studies include groups as systems (Hare, 2003; Lester, Ready, Hostager and Bergmann, 2003; Mabry, 1999; McGrath, Arrow and Berdahl, 2000; Moon et al., 2003; Prekop, 2002), group interaction (information flow) (Chernyshenko, Miner, Baumann and Sniezek, 2003; Kelly and Loving, 2004; Wilkinson and Fung, 2002; Wittenbaum et al., 2004), virtual group interaction (Balthazard, Potter and Warren, 2004; Benbunan-Ficha, Hiltz and Turoff, 2003; Dasgupta, 2003), group decisions (Condon, Golden and Wasil, 2003; Ellis and Fisher, 1994; Poole, 1985; Slezak, 2000; Sunwolf and Seibold, 1999), and virtual group decisions (Alge, Wiethoff and Klein, 2003; Benbunan-Ficha, Hiltz and Turoff, 2003; Burkhalter, Gastil and Kelshaw, 2002).

IS studies of information behavior, flow, and processes as systems could contribute to the success of EM organizations, who interact by consulting and researching national organizations – for example, the Federal Emergency Management Agency (FEMA) – to gain information (input), and then make suggestions (output) to local EM members and EM researchers to solicit feedback. The feedback then becomes further input and shapes the subsequent suggestions made by all EM information contributors until a decision is reached. Systems theory is also the framework for a newer theory, the Bona Fide Group Perspective (BFGP) that can be used to demonstrate the relationship of EM as a whole to information flow. BFGP is one of four contemporary theories that describe information flow (Table 23.6).

The systems metaphor (Appendix E), however, remains the basis for textbook study of group information flow (Arrow, Poole, Henry, Wheelan and Moreland, 2004; Bales, 1999). The IS application of systems theory overlaps with communications, cognition, small group research, management, and therefore demonstrates the potential for integration into EM research.

CURRENT RESEARCH

Interestingly, the very nature of information prompts integration among ALL disciplines – especially

Table 23.5
INFORMATION FLOW SYSTEMS

System elements	*Example in EM*
process	preparation, response, recovery, mitigation
component	police, firefighters, emergency medical technicians, meteorologists, volunteer groups, building inspectors, politicians, local/state/federal officials, businesses, nearby jurisdictions
outcome	disaster prevention, disaster mitigation

Table 23.6
FOUR MAJOR CONTEMPORARY THEORIES OF INFORMATION FLOW IN GROUPS

Functional	Functional Theory is a normative approach to explaining group information flow. The focus of functional theory are inputs and processes, elements grounded in systems theory (Wittenbaum et al., 2004) and Dewey's reflective thingking model.
Structuration	Structuration Theory is a way to explain the pattern of relationship between the group system of information flow and the group structure, the rules and resources members use to maintain the group system (Poole, 1985).
Symbolic Convergence Theory (SCT)	Symbolic Convergence Theory (SCT) explains the information flow process of group members as they form group identity. The group assuming its own identity significantly improves descision-making (Bormann, 1982).
BFGP	BFGP is a theory created by Putnam and Stohl (1990) that embraces the relationship between groups and their larger social systems. BFGP proposes two major ideas: (1) groups have permeable and fluid boundaries with shifting borders, and (2) groups are embedded in and interdependent with their environment, their larger systems.

information management and technology – to produce knowledge using libraries, computers, email, and software for statistical analysis, database creation, and all information systems. The events of the September, 2001 terrorist attacks, however, have initiated current research in IT to improve information sharing among governmental organizations and enable efficient communications interoperability among emergency response organizations.

Decision Aids

Several decision aids (Appendix A) have been developed for the management of disasters including:

- CAMEO (Computer-Aided Management of Emergency Operations),
- ALOHA (Area Location of Hazardous Atmospheres),
- E-Team (created to manage every phase of a crisis),
- GIS (geographic information systems).

However, there is the need for new or the extension of existing theories that might enable information technologists to anticipate more fully the needs addressed by disaster management software.

Information Sharing

The Intelligence Reform and Terrorism Prevention Act of 2004 was intended to mobilize IT for counterterrorism information sharing (Dizard, 2004). The law created two influential positions: Director of National Intelligence and Director of the National Counterterrorism Center – both entrusted with the task of increasing information sharing (Office of the Press Secretary, 2004). Senator Susan Collins introduced the final version of the bill. She said the Commission found that "various agencies had pieces of the puzzle that [if assembled] might have allowed them to prevent the attacks. . . the bill will foster a new culture of information sharing in the intelligence community" (Dizard, 2004).

Information sharing is also being addressed by a few individual efforts including the March 2004 introduction of OSIS (Open Specification for Sensitive Information Sharing) by RAINS (Regional Alliances for Infrastructure and Network Security). RAINS, a not-for-profit public/private partnership that has promised to advance groundbreaking technology for homeland security, created OSIS for the safe sharing of sensitive information across state, local and national security systems (Appendix F). Unfortunately, the nation has not introduced a consistent strategy to address information sharing nationwide.

The U.S. National Commission on Libraries and Information Science (NCLIS), however, has proposed an unusual solution to President George W. Bush and Congress. Trust and Terror is an NCLIS proposal that envisions public libraries as an information center for crisis information dissemination and management. NCLIS claims that public libraries provide an appropriate forum for crisis information dissemination because the public considers libraries trustworthy sources that are already efficiently structured, aware of cultural diversity, many times employ multilingual staff, and accessible to local communities. Although libraries also can offer the information from anywhere in the world in real time in numerous formats (NCLIS, 2002), new law would be needed to authorize and equip libraries for access to secure information related to terrorism – the most prominent civil hazard currently threatening the security of information.

Communications Interoperability

Failed communications interoperability contributed to America's complacency during the 9/11 terrorist attacks. Consequently, thousands of civilians died alongside hundreds of first responders: emergency personnel including police, emergency medical technicians, and fire fighters, who were trained and willing to save lives.

Don Eddington, chief of the Center for IT Integration at the Defense Information Systems Agency admitted that "DOD (the Department of Defense) couldn't talk to state officials; state officials couldn't talk to city officials" (Onley, 2002). Unfortunately, first and second responder organizations had adopted many different information systems for their specific information sharing needs – and some that were ready to use were never implemented. Fire fighters, police, and other emergency personnel at the Pentagon and in New York City could not find common radio frequencies to communicate – cell phone networks flooded frequencies and further hindered information flow in the hours following the attacks (Riley, 2003).

The 9/11 Commission was enlisted to research and report the situations and events surrounding the attacks. The Commission found that civilians, fire fighters, police officers, emergency medical technicians, and emergency management professionals demonstrated "steady determination and resolve under horrifying, overwhelming conditions. . . . Their actions saved lives and inspired a nation. . . ." However, the Commission also found that the "Port Authority's response was hampered by the lack of standard operating procedures and radios capable of enabling multiple commands to respond to an incident in unified fashion." The Commission made the following recommendation:

> Make homeland security funding contingent on the adoption of an incident command system to strengthen teamwork in a crisis, including a regional approach. Allocate more radio spectrum and improve connectivity for public safety communications, and encourage widespread adoption of newly developed standards for private-sector emergency preparedness – since the private sector controls 85 percent of the nation's critical infrastructure (The 9/11 Commission, 2004).

Communications interoperability among officials from community first responders to high-level information security officers within the federal government is a major concern with the growing threats of terrorism and cyberterrorism. Recommendations for the development of shareable information systems have emerged from both public and private institutions. Creative solutions for integrating information technologies provide U.S. leaders with choices and challenges – for instance, what do they choose and how do they choose it? Congress passed the Homeland Security Act in November, 2002 specifically to address these and other questions about shareable information (Appendix G). Private and public sectors are busy introducing a mishmash of information-sharing products including software for:

- three-dimensional mapping of cities;
- disaster management simulations
- analysis of phone calls and other communications to help first responders make better decisions in emergencies;
- interpretation of garbled speech recordings;
- extraction of unstructured text;
- discovery of non-obvious relationships (background checks deluxe); and
- disparate systems queries (police, courthouse, homegrown databases, etc.) (Batzler, 2002; Mena, 2004).

Serious submissions are subject to SAFETY (Support Anti-Terrorism by Fostering Effective Technologies Act of 2002) guidelines (DHS Press Office, 2003). IS should collaborate with EM to further these goals.

RECOMMENDATIONS FOR FUTURE RESEARCH

IS researchers should consider studying disasters in the light of information and communication systems theory, collective information processing (Appendix D), knowledge management, decision-making and chaos theory. Chaos theory would provide a fitting foundation for the study of information flow in disasters. Disasters appear chaotic, yet chaos theory argues that there is order in chaos – the order is just not apparent because it is so complex.

Researchers might also consider more ethnographic studies of disaster information flow at the scene of real disasters. Field studies could greatly enhance the information behavior theory of emergent groups, information sharing among first, second and third responders, and governmental agencies at all levels.

Developing and testing information sharing network structures for disaster management would be instrumental in helping all involved in EM understand the routes of disaster information. EM could learn from whom, to whom, and how members can change those routes to enhance and expedite their important interactions. A sensitive information exchange technology that allows the display of information on heterogeneous networks across the world could eventually enable EM to send disaster information to all personal digital assistants or cell phones. Information sharing technology has the potential to make people safer, healthier, and more knowledgeable – chief rationales for expanding IT.

More technology, however, would bring more work to EM teams. Learning new systems and software and being better connected means more communication – all time-consuming activities to implement programs that may or may not be more efficient. "The vulnerability of complex networked systems, together with potential ways of using data resources to speed up recovery almost certainly will increasingly preoccupy emergency planning staff in some areas" (Stephenson and Anderson, 1997).

IS should therefore more aggressively address the possible negative impacts of the IT on disaster planning and research. Joy's speculations that humans may fall victim to their own technology (2000) is worthy of deeper investigation, as is Mesthene's opinion that human's "technical prowess always seems to run ahead of his ability to deal with and profit from it." EM researcher, Quarantelli (1997) raises provocative questions addressing the implementation of IT without preliminary, robust testing. He asserts that the information/communication revolution has at least ten inadvertent liabilities for disaster planning and management. These areas warrant intensive attention. How can we claim to mitigate disasters if the very methods we use exacerbate them? Quarantelli inspires many information-related suspicions.

1. If IT provides all persons possible with IT that connects them to disasters, will they be helpful or, as untrained professionals, become additional hazards preventing the trained professionals from doing their jobs?
2. Will the new IT provide too many choices for technology? or: too much information? or: lose information or be so dynamic that the information is outdated the second it is transferred?
3. Will the hackers and cyber-terrorists be as updated as the legitimate IT providers?
4. Will messages lose the richness only found in face-to-face communication?
5. Will the addition of Web-like platforms impede typically hierarchical information flow?
6. Will fad-like methods for dealing with disasters spread across the Internet before they can be tested?
7. Will safety and ergonomic guidelines be realized before possibly hazardous IT is implemented?
8. Will the increase of IT, and its computer representations, increase the likelihood of even more computer-system related disasters?

These questions emphasize the need for more research regarding new technology use with information disasters and disaster information. Studies should focus on disaster warning communication systems, disaster mitigation for information systems, and knowledge management for information disaster preparedness.

The disaster of 9/11 left many businesses devoid of information that was critical to their daily operations. Massive destruction obliterated electronic and hard copy client lists, sales records, billing information, and contracts. Neither sophisticated IT nor well-developed disaster recovery plans could prepare organizations for the permanent loss of knowledge – knowledge recorded on paper, electronically, and in the minds of the victims who lost their lives. IS/EM collaboration should thus focus on several aspects of disaster information flow including EM team problem-solving and decision-making; communications interoperability – both humanly and artificially produced at the local, state, national, and international levels; sensitive and/or vital information sharing among governmental entities, and the "problematic aspects of the information/communication revolution" as introduced by Quarantelli (1997). Also imperative is the convergence of disciplines in the research of the elusive concept of vulnerability (McEntire, 2004). Theoretical integration of IS and EM in these and other areas can only serve to improve all phases of disaster management from preparedness to response to recovery to mitigation.

CONCLUSION

Two distinct problems are evident in both IS and EM research – *information disasters* and the flow of *disaster information.* The impact of these problems demands *extensive* studies; however, the duality of the problems – *information* and *disasters* – demands *integrated* studies.

Vannevar Bush emphasized the need for integration among disciplines when he bemoaned the "growing mountain of research . . ." as studies became more diverse. He also felt that investigators were "staggered by the findings and conclusions of thousands of other workers . . ." with no time ". . . to grasp, much less to remember . . ." other researchers' contributions.

The events of 9/11 instigated a revival of Vannevar Bush's challenge to launch the "massive task of making more accessible a bewildering store of knowledge" (Bush, 1945a). Bush's 60-year-old recommendation is surprisingly similar to a recommendation by the 9/11 Commission in its 2004 report to the nation.

The U.S. government has access to a vast amount of information. But it has a weak system for processing and using what it has. The system of need to know should be replaced by a system of need to share" (The 9/11 Commission, 2004).

The need to share information among disciplines and governments cannot be met by IS or EM alone. Integrated research is vital to minimize the vulnerabilities to information disasters and consequently diminish disaster's inherent disruptions to life. Integrated research is also vital to maximize the effectiveness of disaster information flow among EM organizations and thereby facilitate the preservation of life. Possible repercussions from information disasters and ineffective disaster information flow necessitate the integration of IS and EM. The stability and survival of lives may depend on it.

APPENDIX A
SOME PROBLEMS STUDIED WITHIN INFORMATION SCIENCES

Subject Area and Problems	*Contributions to Research with Applications to EM*
Cognitive Science: expert systems; knowledge bases; hypertext; human-computer interaction **Artificial Intelligence:** software to emulate human intelligence **Semiotics:** Signs, both individually and grouped in sign systems, and includes the study of how meaning is transmitted and understood	• human factors (ergonomics); robot studies • vulnerability analysis • intelligence agency communications interoperability; cyber-terrorism; information warfare
Communications: Information flow (information sending, receiving), and information sharing **Cybernetics:** communication, feedback, and control mechanisms in living systems and machines **Telecommunications:** Distance electronic information exchange **Systemics:** Relationships of systems-transactions, processes, inputs, outputs, especially information/communication systems	• defense simulation; telemedicine • decision systems; online database systems and related telecommunications and networking technologies; specialized search functionalities; large machine-readable databases for the dissemination and graphic representation of disaster-related information; GIS (Geographical Information Systems); CAMEO (Computer-Aided Management of Emergency Operations) ALOHA (Area Location of Hazardous Atmospheres); E-Team (manages every phase of a crisis)
Computer Science: manipulation and storage of document records in electronic information storage, processing, and retrieval systems; information management; databases	• computer hardware and software to manipulate documents and document records for emergency, disaster, and crisis information storage and retrieval systems
Information Science: information behavior, information processing, information retrieval, information storage, information dissemination	• user information seeking, needs, preferences-relevance and utility assessment • development of standards for processing and communication of information; monitoring of the national information infrastructure (human, technological, materials and financial) to ensure maintenance of information systems and services related to the public interest • protocols (procedures) for information
Library Science: Acquisition, cataloging, classification, and preservation of information **Bibliometrics:** All quantitative aspects and models of communication, storage, dissemination and retrieval of scientific information **Citation analysis:** Citation frequency and patterns in scientific journals article citations; implications for how and where an author's work is subsequently cited **Co-citation analysis:** Literature coherence and changes over a period of time; maps oeuvres and their authors relationship to other oeuvres and authors **Content analysis:** Thesauri and frequency of terms, co-word/co-authorship/co-citation analysis for interdisciplinarity purposes **Cybermetrics:** Quantitative aspects of the construction and use of information resources, structures and technologies on the whole Internet drawing on bibliometric and informetric approaches **Informetrics:** Quantitative aspects of information in any form, not just records or bibliographies, and in any social group, not just scientists **Webometrics:** Quantitative aspects of the construction and use of information resources, structures and technologies on the Web draws on bibliometric and informetric approaches	• holdings of past research • study of all published literature and its usage in all disciplines to ensure scholarly productivity and communication in disaster research • indexing, citation indexing, keyword indexing, text analysis and natural language searching systems to aid researchers and practitioners in the finding of information vital to disaster situations • extensive development of these sub-disciplines and specialties to aid researchers interested in investigating interdisciplinarity • formal logic (Boolean operators AND, OR, and NOT) to improve database searching • formulation of national information policies related to issues of privacy, security, regulating dissemination, access, intellectual property, acceptable use • concept of "library" as an unbiased holder of all information regardless of content, political or moral implications

APPENDIX B
LIBRARIANSHIP AND INFORMATION

Librarianship typically protects and preserves recorded information, as well as advocates the rights and privacy of the information of American citizens. Public libraries and universities lobby for changes in law to benefit personal information rights of individuals. Librarians and educators – many who consider the violation of privacy of information rights to be disastrous – have been at the forefront of the fight for information privacy for a couple of centuries. Many support the First, Fourth, and Fifth Amendments from the 1791 Bill of Rights, quote Warren and Brandeis' renowned 1890 article, *The Right to Privacy* and uphold the Freedom of Information Act (1966), the Federal Privacy Act of 1974, and the Electronic Privacy Act of 1986 where Congress realized the need to protect private citizen records collected by the government.

THE BILL OF RIGHTS

The Bill of Rights was passed by the United States Congress on September 25, 1789. However more than a century would pass before the judicial branch recognized and implemented the right to privacy implications stated in the First, Fourth, and Fifth Amendments. The First amendment addresses the right to privacy by giving citizens the freedom of speech, interpreted by law to mean that citizens have the freedom to express their thoughts, views, and preferences without fear of retribution. The Fourth Amendment protects citizens' privacy by ensuring that their personal possessions are not searched or seized without warning and without "probable cause." The Fifth Amendment protects citizens' private property from seizure for public use.

THE RIGHT TO PRIVACY

Warren and Brandeis introduced the idea that people have the freedom and right, as American citizens, to expect that not only their tangible possessions, but also their intangible personal information – what they think, believe, say, and read – to be safe from public intrusion. If citizens disclose their thoughts in a private place, they have the expectation that their private thoughts, their private information remains in that private place.

FREEDOM OF INFORMATION ACT (1966)

The FOIA was enacted to allow persons to request access to federal agency records or information. In response to the FOIA, states adopted their own open records acts governing public access to state and local records. However, exceptions were made in most states, including Texas, to protect the confidentiality of library user records. In 1973, Texas enacted the Texas Open Records Act, later to be revised as the Texas Public Information Act of 1995 (The Act). The Act allowed the public access to all government entity records, except for records containing personal information about individuals. Library records – which includes database search records; circulation records; interlibrary loan records; other personally identifiable uses of library materials, facilities, programs or services; and information obtained in reference interviews – were exempt from disclosure except under certain circumstances. Before the USA PATRIOT Act, American citizens were free to walk into a library, pick out a magazine, sit down and read, and walk out again without anyone knowing who they are, where they live, or what they chose to read. These freedoms were understood unless a court issued a subpoena showing probable cause that the disclosure of their records was necessary to protect the public safety; the record was evidence of a crime; or the record was evidence against a particular person who committed a crime.

FEDERAL PRIVACY ACT OF 1974

The FOIA allowed access to government-held records. Some government-held records contained confidential information about individuals. Congress passed the Privacy Act in 1974 to ensure the protection of individual privacy from data collected by the government. The law allows individuals to view, copy, and correct their own records. It also prevents agencies from sharing data.

(cont'd.)

APPENDIX B

LIBRARIANSHIP AND INFORMATION *(Cont'd.)*

ELECTRONIC PRIVACY ACT OF 1986

The Electronic Communications Privacy Act of 1986 (ECPA) updated wiretapping laws for digital communications. It banned the capture of communications between network points – it protected electronic communications while they are en route. However, in 1996, the FBI urged Congress to pass the Communications Assistance for Law Enforcement Act of 1994 (Digital Telephony Act), a law that forces telecommunications carriers to design their systems so that law enforcement agencies can tap into them if necessary.

Librarianship and Online Information

Librarians also face decisions regarding library patrons' information privacy on the Internet. Threats to online information include the USA PATRIOT Act, cookies, COPPA, and CIPA.

COOKIES

Library user's activities can be tracked using tiny text files, called cookies. Because cookies reveal where users visit, when they visit, how long they stay, what links they click, what purchases they make, and any preferences they may have set during the session, most libraries find it unethical to retain permanently the information saved in cookies.

CHILDREN'S ONLINE PRIVACY PROTECTION ACT (COPPA) (2000)

The Children's Online Privacy Protection Act (COPPA), went into effect April 21, 2000 (Federal 1999). COPPA requires that commercial Web sites must have documented parental consent to collect "personally identifiable information (including an e-mail address) from children (COPPA 2002). However, this law does not mean that librarians must reveal to a parent what a child views on the Internet, or even what a child reads.

CHILDREN'S INTERNET PROTECTION ACT (CIPA) (2000)

The Children's Internet Protection Act (CIPA) is a law passed by the federal government in December, 2000 to speak to concerns regarding children's access to the Internet from schools and libraries. CIPA requires that institutions that receive federal E-Rate or Library Services and Technology Act (LSTA) funds) filter all of its Internet terminals to block access to sites defined as obscene, child pornography, or harmful to minors.

The American Library Association challenged the law in a Philadelphia district court in May 2002. The court ruled that CIPA violated first amendment rights of library users, however, the government appealed the decision. On June 23, 2003, the U.S. Supreme Court reversed the lower court decision and upheld CIPA ruling that CIPA does not violate the First Amendment.

APPENDIX C

MCENTIRE'S DISASTER EQUATION FOR INFORMATION

Example problems	*Vulnerability*	+	*Hazard (Triggering Agent)*	=	*Disaster*
Unsealed storage room	Uncovered, unwrapped physical archives	+	Terrorism/Vandalism Heating/air conditioning failure Flood, tornado, hurricane, earthquake, etc.	=	Damaged or destroyed books, artifacts, other historically significant objects
Novice user	Unprotected or unduplicated files	+	User error: unintentional pressing of delete key	=	Loss or change of important document(s),
No virus protection			Virus attack	=	Loss or damage of important documents
Vague file security policy			Unauthorized user access	=	information leakage, theft of sensitive or private information
No/inferior firewall			Hacking	=	Stolen identity, credit information theft
Faulty wiring			Power failure/surge	=	Loss or damage of important documents
Poor ventilation			Heating/air conditioning/failure	=	Loss or damage to: documents; software; hardware
Haphazard attitude toward warning systems	Failed/thwarted warning system	+	Cyberterrorism (using computer networks in the service of terrorism)	=	Destruction and death from terrorist acts
Adversaries to information holders	Systems holding sensitive information	+	Information warfare	=	Propaganda causing destruction: denial of vital information; reception of false information
Ambiguous Information/ knowledge policies/laws	Cultural differences	+	Cultural power struggles	=	Limitation of human and/or artificial knowledge (databases); inventory lists
Ambiguous/absent document/information sharing standards	Communication inoperability	+	Power failure/surge as a result natural hazard; terrorism; information security regulations	=	Missing, inaccurate withheld, or selective information
Information scarcity (i.e., Digital Divide)	No/limited access to information warning systems				
Sections 215 and 216 of the USA PATRIOT Act	Accessible sensitive private information	+	Careless or impulsive law enforcement	=	Breach of anonymity, autonomy, privacy rights

APPENDIX D

G.R.O.U.P. D.Y.N.A.M.I.C.S.: MAJOR PROPERTIES OF TASK GROUPS

Goals	Groups in EM organizations are impacted by the goals throughout every phase of the disaster management's decision process. A goal is the desired final status of the situation, the consensus opinion, recommendation, or mandate (Anderson, Riddle and Martin, 1999). The intensity of an individual's commitment to the group's goal is dependent upon the ability of the members to give and receive information that will help them attain the group's goal (Allen and Meyer, 1990).
Roles	"A role is a set of communicative behaviors performed by an individual and involves the behaviors performed by one member in light of the expectations that other members hold toward these behaviors" (Ellis and Fisher, 1994). As EM groups work out their own roles in cooperation with other team members, each member takes on a role that differentiates him or her from the other group members. A major influence on role development is the openness of the group to outside advice. Group roles form because of intergroup interaction, but more quickly identified through outside input. As open systems, outside influences not only impact the shaping of the roles, but also the shaping of the incoming and outgoing information.
Openness	Although EM organizations are open systems, they must also draw on their system's ability to consider the integrity of the input. Propp (1999) sets forth this filtered input and coordinated output of information in her Distillation Model of CIP (collective information processing). CIP is a distillation process that progresses from a substantial collective knowledge base to a distilled information base that is purged of irrelevant or unsound information. A decision is then brought forth based on the final collective information. Propp describes four developmental stages in the Model: 1) individual knowledge base – knowledge that each member brings to the group concerning the task; 2) group knowledge base – collective knowledge available to a group as a whole; 3) communicated information base – information exchanged and shaped through group discussion; 4) final collective information base – information accepted and utilized by a group to come to a decision (Propp, 1999).
Unity	This separate identity – a group within an EM organization for instance, assumes the qualities of a system. Groups, as systems, demonstrate unity in three ways: *wholeness* (Appendix E) , *groupness*, and *synergy* (Mabry, 1999) . *Groupness* was mentioned for the first time in 1967 by John Brilhart. Brilhart described groupness as a property only found in real groups – groups of individuals who perceive themselves as a group. According to Brilhart, groupness evolves slowly – and is developed as the group becomes cohesive (Ellis and Fisher, 1994). More research could show what events or situations accelerate or decelerate the development of unity. The other aspect of unity is found in *synergy* – from the Greek word sunergos, which means "working together." Synergy empowers groups to make better decisions. The combination of ideas generated by brainstorming is an example of synergy – the outcome is usually greater than a simple summation of individual ideas (Harris and Sherblom, 2002). Unity fosters compromise, cooperation, and consensus. Unified groups begin to adopt methods to facilitate their performance as a group: they adopt procedures to facilitate discussion, analyses, creativity, and agreement.
Procedures	Although formal procedures are time-consuming, there is sufficient evidence to suggest that formal discussion and problem-solving procedures improve group performance (Poole, M., 1991). Further in-depth study could determine what procedures to use for specific situations. One assumption is that the choices are largely dependent upon how culturally diverse the members are. The philosophical approach to group formation and problem solving varies between cultures – and task groups like those found in EM organizations, are becoming more and more culturally diverse with the increased participation of ethnic minorities.

(cont'd.)

APPENDIX D

G.R.O.U.P. D.Y.N.A.M.I.C.S.: MAJOR PROPERTIES OF TASK GROUPS *(Cont'd.)*

Diversity	The most significant difference among cultures is attributed to value differences between individualistic and collectivistic cultures. When working in task groups, people from individualistic Western cultures, like the United States, tend to concentrate primarily on the task dimension and secondarily on the social dimension. Conversely, people from collectivist cultures – East Asia, Latin America, and Africa – tend to concentrate primarily on the social dimension and secondarily on the task dimension (Jetten, Postmes and McAuliffe, 2002; Sosik and Jung, 2002). Individualistic and collectivistic values also affect group interactions and consequently, group outcomes. Individualistic members cultivate task roles early on whereas collectivistic members cultivate social roles first, however Oetzel (2001) did not find that groups composed of both individualistic and collectivistic members had communication problems. Surprisingly, even in individualistic cultures like the United States, EM team members can become unusually devoted to the social dimension, particularly groups with a highly powerful or persuasive leader who encourages yea-saying – blind support of the members to every view or suggestion of the leader.
Yea-saying	Yea-saying, also called *groupthink* was outlined by Janis (1982) who believed that certain conditions were indicative of a tendency to promote the urgency of a quick consensus. Janis conditions for groupthink to occur are: cohesive decision-makers; isolated/insulated group – no external influences; members with similar backgrounds and attitudes; provocative, stressful situations and outside pressures. Although the groupthink theory has been said to lack generalizability (Chen and Lawson, 1996; Park, 2000), Janis' theory has been used to entertainingly describe many U.S. presidential decision fiascoes including Roosevelt's complacency before Pearl Harbor, Truman's invasion of North Korea, Kennedy's Bay of Pigs failure, Johnson's escalation of the Vietnam War, and Nixon's Watergate ignominy. (If Janis updated his presidential study, he might include Carter's political asylum for the Shah of Iran, Reagan's Iran Contra affair, Ford's pardon of Nixon, George Bush's leadership in the Gulf War, Clinton's approval of the Branch Davidian raid, and George W. Bush's rush into Iraq.) Now that groupthink is a relatively known term, a current study of high-profile leaders could be very enlightening. Groupthink is highly observable in EM, especially when the decisions impact nations. Equally as powerful, yet nearly invisible, are social norms, the most influential form of group control.
Norms	Norms are "regular patterns of behavior or thinking that come to be accepted in a group as the usual way of doing things (Keyton, 1999)" Unlike rules, which are explicit guidelines of behavior, norms are implicit guidelines of behavior that emerge as the group evolves. These conventions, though unsaid, are powerful enough to shape group members' conduct, viewpoints, and interaction. Norms become apparent early in the group's formation. They are typically developed as members observe each other and become cohesive. What is acceptable and not acceptable is just understood as group members come to know each other. Norms and rules determine how, when, and why decisions are made; they also are instrumental in whether a leader emerges from the membership or if the group members maintain equal status. The understood rules dictate how an EM team evaluates and allocates authority – a concept that suggests leadership, as well as power.
Authority	Studies show that most groups need a leader to plan meetings and empower members to carry out tasks. Leaders are also ultimately responsible for the management of conflict. So called "leaderless" groups usually have an unofficial leader that gradually assumes the leadership role (Brown and Miller, 2000; Ellis and Fisher, 1994) This "leader emergence" is evident in many groups, especially groups that must make critical decisions. Leader emergence is found to be common in emergent groups at disaster scenes.

(cont'd.)

APPENDIX D

G.R.O.U.P. D.Y.N.A.M.I.C.S.: MAJOR PROPERTIES OF TASK GROUPS *(Cont'd.)*

Authority *(cont'd.)*	Some outstanding leaders are faced with contentious groups: groups composed of dominators, aggressors, conformists, or naysayers. Most leaders, however, find that their greatest challenge is managing members who consistently conform to majority or minority influence.
Majority/minority	The majority/minority concept discussed here does not relate to consensus – it instead refers to tendencies of task group members to be influenced to conform to a majority opinion or minority opinion. Little research has been done in this area, however, studies have shown that minority dissent is many times as powerful in swaying group decision as is majority consent (Hartley, 1997). Interactions among group members and members of majority and minority subgroups should be examined to determine the communication similarities and dissimilarities. Recent research shows that having a small majority consensus – 52% to 48% – sways the undecided members toward the majority view as much as having a large majority consensus – 82% to 18% (Martin, Gardikiotis and Hewstone, 2002). Interestingly, minority influences were posited to be stronger, regardless of the consensus size, if their views were highly distinctive from the majority views. A markedly atypical view is given even greater consideration by group members (Hartley, 1997).
Interdependence	Interdependence is apparent in the group system as members interact and respond to each other. The attributes of the members – personality, skills, attitudes – affect the experiences of all of the other members in the group. These attributes of all of the members also affect cohesiveness, relationships, and member satisfaction. Conversely, cohesiveness, relationships, and member satisfaction affect the behaviors and attitudes of all of group members. The output of the task dimension is productivity; the output of the social dimension is cohesiveness – which are also interdependent. The more productive the group is, the more cohesive it is. The more cohesive the group is, the more productive it is (Bonito, 2002; Ellis and Fisher, 1994; Harris and Sherblom, 2002; Keyton, 1999; Mabry, 1999; Meyers and Brashers, 1999).
Conflict	Conflict can be defined as recognition by all group members that there are differences, disagreements, contradictory or irreconcilable desires among group members (Sell, Lovaglia, Mannix, Samuelson and Wilson, 2004). Conflict can be exacerbated or diminished based on group behaviors during a heated discussion (Sillince, 2000). Conflict can be described as either affective or substantive. Affective conflict involves an emotional conflict or struggle that is usually based on selfish or personal issues. It may involve differences of opinions, interpretation of rules, or attitudes toward established norms. Substantive conflict involves intellectual opposition to the content of ideas or issues pertinent to the decision. It may involve bargaining, negotiations, or intellectual evaluation. The advantages to conflict outnumber the disadvantages.
Structure	Interdependence also exists in the flow of information between and among EM team members. As the members sharing of information becomes more organized, a structure becomes evident. This structure includes both intangible and tangible frameworks that organize group interaction. Two forms of group structure are networks and proxemics. Networks are links between and among members that develop into recurrent patterns for the exchange of information. Proxemics, sometimes called group ecology, is concerned with how group members arrange, use and are affected by physical space in their interaction with others.

APPENDIX E
GROUPS AS OPEN SYSTEMS

System concepts	*System*	*EM organization*
Wholeness	• Every component of a system affects and is affected by every other component.	• Every member affects and is affected by every other member.
	• The system, though it has many components, assumes an identity as a unit.	• The organization, though it has many members, assumes an identity as a unit.
	• A change in one component effects changes in all other components.	• A change in one member effects changes in all other members.
	• The whole is different from the sum of the parts.	• The outcome of collaboration is different from the combined outcomes of the same people working alone.
Openness	• An open system self-regulates: it receives processes new information, then discards what it does not need to survive.	• A group self-regulates: it receives and examines new information for relevance and reliability, then discards the irrelevant and unreliable.
	• It freely exchanges information with the environment.	• Group members interact outside the group, freely exchanging ideas.
Structure	• All systems have spatial relationships: components that are above, below, beside, behind, or facing.	• Group members have spatial relationships as explained by positions within a communication network.
Function	• Each component of a system has a function that complements the other components.	• Each member in a group has a role that complements other group roles.
Evolution	• Systems continuously transform through interactive processes.	• Groups continuously transform as they interact.
Interdependence	• System components depend on other components for proper functioning and replenishing.	• Group members depend on other group members to fulfill their goals.
Feedback	• A system sustains and adapts through feedback: responses to input and output.	• A group sustains and adapts through the cycle of interaction and feedback among its members.

(Bales, 1999; Fisher, A., 1974; Hare, A. Paul, 1962; Mabry, 1999)

APPENDIX F
OSIS/RAINS BACKGROUND

OSIS is based on RAINS established Connect & Protect program, a wide area network project that connects schools, government agencies and other organizations in Portland, Oregon (Fisher, D., 2004). Connect & Protect program – a cooperative effort between RAINS and the City of Portland's Bureau of Emergency Communications (BOEC) – allows the conveyance of real-time emergency information among more than 60 local public safety stakeholders including 911 centers, schools, hospitals, hotels, and banks. Alerts can be received through PCs, personal digital assistants or cellular phones (RAINS, 2004). BOEC was the first 9-1-1 Dispatch Center in the country to participate in this program. Oregon's 9-1-1 center simultaneously implemented Connect & Protect with RAINS-Net in Portland, in August 2003. RAINS designed the program to be a "scalable, affordable model" (below) and therefore usable anywhere – locally or nationally in cities and rural areas.

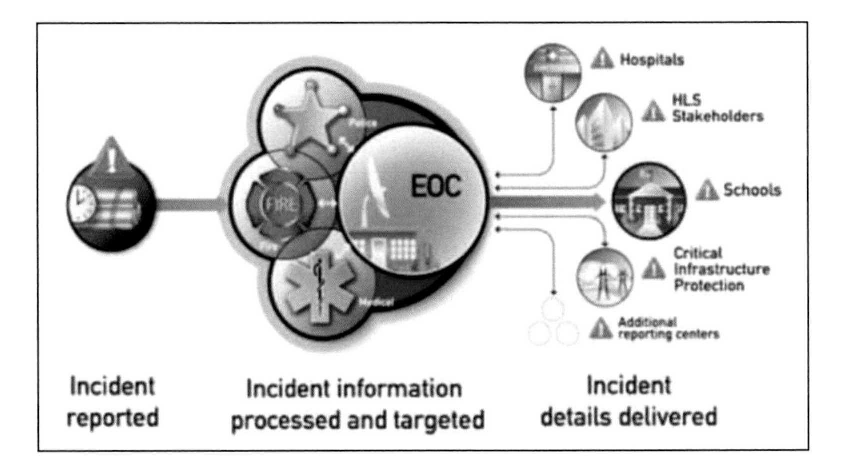

RAINS began as a result of 9/11, 2001. Several Oregon companies united as Oregon RAINS, a regional emergency response network. RAINS grew to include more than 60 companies and six research universities by the time it deployed RAINS-Net in 2003 (Jackson, 2004).

Since then it has recruited agencies in two states, Virginia and Washington – RAINS is optimistic that OSIS will attract at least 10 states in the near future. Already it is the first "statewide system in the country that will be able to send emergency alerts securely online" (Robinson, 2003).

RAINS hopes that OSIS can become the nation's prototype for sensitive information exchange (Robinson, 2004). Wyatt Starnes, co-founder of RAINS, confirmed that OSIS encourages the use of specialized Web Services and additional standards including XML, Common Alert Protocol, WS-Security, WS-Security Policy, WS-Trust, SAML – but it does not force users to follow specific form when implementing these services (What's next? 2004). RAINS believes its major advantage of OSIS is its non-proprietary approach – it can be used by both government and private organizations who need to share sensitive information without foregoing free market competition (ESRI, 2004; RAINS-Gauge, 2004). Currently, all government and first responder organizations using the RAINS-Net system are employing OSIS (What's next? 2004).

APPENDIX G

HOMELAND SECURITY ACT (2002)

SECTION 201: DIRECTORATE FOR INFORMATION ANALYSIS AND INFRASTRUCTURE PROTECTION

To review, analyze, and make recommendations for improvements in the policies and procedures governing the sharing of law enforcement information, intelligence information, intelligence-related information, and other information relating to homeland security within the Federal Government and between the Federal Government and State and local government agencies and authorities.

The major purpose of the Act was to establish the U.S. Department of Homeland Security (DHS) whose "primary mission is to protect our Homeland" (Bush, G. W., 2002). The DHS outlined seven priorities for completion in 2005. Priorities 1, 2, and 5 specifically address tactics to develop shareable information systems.

PRIORITY 1

The DHS intends to establish a Homeland Security Information Network (HSIN), a computer-based counterterrorism communications network designed to strengthen the two-way flow of threat information that will be available to all 50 states, 5 territories, Washington, D.C., and 50 other major urban areas. Its major function will be to prevent terrorist attacks, but it may also serve as a tool during crisis management. (DHS Press Office, 2004).

PRIORITY 2

The DHS intends to initiate measures that will significantly improve interoperability among fire fighters, police officers and other emergency personnel who need to be in contact and share equipment during catastrophes.

PRIORITY 5

The DHS intends to implement the National Incident Management System (NIMS). NIMS will offer a "consistent nationwide template to enable federal, state, local and tribal governments and private-sector organizations to work together effectively to prepare for, prevent, respond to, and recover from a terrorist attack or other major disaster" (United States Department of Homeland Security, 2005).

REFERENCES

Alge, B.J., Wiethoff, C. and Klein, H.J. (2003). When does the medium matter? Knowledge-building experiences and opportunities in decision-making teams. *Organizational Behavior and Human Decision Processes, 91*(1), 26–37.

Allen, N. and Meyer, J. (1990). Organizational socialization tactics: A longitudinal analysis of links to newcomer's commitment and role orientation. *Academy of Management Journal, 33*(4), 847–858.

Anderson, C., Riddle, B. and Martin, M. (1999). Socialization processes in groups. In L. Frey (Ed.), *The handbook of group communication theory & research* (pp. 139–163).

Arrow, H., Poole, M., Henry, K.B, Wheelan, S. and Moreland, R. (2004). Time, change, and development: The temporal perspective on groups. *Small Group Research, 35*(1), 73–105.

Bales, R. (1950). *Interaction process analysis: A method for the study of small groups.* Addison-Wesley, Cambridge, MA.

Bales, R. (1999). *Social interaction systems: Theory and measurement.* New Brunswick, NJ: Transaction Publishers.

Balthazard, P., Potter, R.E., and Warren, J. (2004). Expertise, extraversion and group interaction styles as performance indicators in virtual teams: How do perceptions of it's performance get formed? *ACM SIGMIS Database, 35*(1), 41–64.

Bates, M.J. (1999). The invisible substrate of information science. *Journal of the American Society for Information Science, 50*(12), 1043.

Batzler, L. (2002 November 3). Industry's emerging role in homeland defense. Government Computer News. Retrieved April 10, 2005 from http://www.gcn.com/vol1_no1/daily-updates/20485-1.html

Belkin, N.J. and Robertson, S.E. (1976). Information science and the phenomenon of information. *Journal of the American Society for Information Science* (pre-1986), *27*(4), 197–204.

Benbunan-Ficha, R., Hiltz, S.R. and Turoff, M. (2003). A comparative content analysis of face-to-face vs. Asynchronous group decision making. *Decision Support Systems, 34*(4), 457–469.

Bolger, L. (2003). Scared or prepared? Disaster planning makes the difference. *Information Outlook, 7*(7), 25–31.

Bonito, J. (2002). The analysis of participation in small groups: Methodological and conceptual issues related to interdependence. *Small Group Research, 33*(4), 412–438.

Borko, H. (1968). Information science: What is it? *American Documentation* (pre-1986), *19*(1), 3.

Bormann, E. (1982). The symbolic convergence theory of communication: Applications and implications for teachers and consultants. *Journal of Applied Communication Research, 10*(1), 50–61.

Brilhart, J. (1978). *Effective group discussion.* Dubuque, IA: Brown.

Brown, T. and Miller, C. (2000). Communication networks in task-performing groups: Effects of task complexity, time pressure, and interpersonal dominance. *Small Group Research, 31*(2), 131–157.

Buckland, M. (1991). Information as thing. *Journal of the American Society for Information Science* (1986–1998), *42*(5), 351.

Buckland, M. and Liu, Z. (1998). History of information science. In T. B. Hahn & M. K. Buckland (Eds.), *Historical studies in information science* (pp. 272–295). Information Today, Inc, Medford, NJ.

Burkhalter, S., Gastil, J. and Kelshaw, T. (2002). A conceptual definition and theoretical model of public deliberation in small face-to-face groups. *Communication Theory, 12*(4), 398–422.

Burlando, T. (1994). Chaos and risk management. *Risk Management, 4,* 54–61.

Bush, G.W. (2002). *Department of homeland security.* The White House, Washington, D.C.

Bush, V. (1945a). As we may think. *Atlantic Monthly, 176*(July), 101–108.

Bush, V. (1945b). Science the endless frontier. Retrieved from http://www.nsf.gov/about/history/vbush1945.htm

Case, D.O. (2002). *Looking for informaton: A survey of research on information seeking, needs, and behavior.* Academic Press, London.

Chapman, J. (2005). Predicting technological disasters: Mission impossible? *Disaster Prevention and Management: An International Journal 14,* no. 3.

Chen, Z. and Lawson, R. (1996). Groupthink: Deciding with the leader and the devil. *Psychological Record, 46*(4), 581–591.

Chernyshenko, O.S., Miner, A.G., Baumann, M.R.and Sniezek, J.A. (2003). The impact of information distribution, ownership, and discussion on group member judgment: The differential cue weighting model. *Organizational Behavior and Human Decision Processes, 91*(1), 12–25.

Chesser, P. (2005). eHistory.com. Retrieved July 31, 2005, from Ohio State University Web site: http://ehistory.osu.edu/world/articles/ArticleView.cfm?AID=9

Cleveland, H. (1982). Information as a resource. *The Futurist, 16*(December), 34–39.

Comfort, L.K. (2005). Risk, security, and disaster management. *Annual Review of Political Science, 8*(1), 335–356.

Condon, E., Golden, B. and Wasil, E. (2003). Visualizing group decisions in the analytic hierarchy process. *Computers & Operations Research, 30*(10), 1435–1445.

Dasgupta, S. (2003). The role of controlled and dynamic process environments in group decision making: An exploratory study. *Simulation & Gaming, 34*(54–69).

Dewey, J. (1910). *How we think*. Boston: Houghton Miflin Company.

DHS Press Office. (2003). Department of homeland security. Fact Sheet: Safety Act - Partnering With American Entrepreneurs in Developing New Technologies to Protect the Homeland. Retrieved from http://www.dhs.gov/dhspublic/interapp/press_release/press_release_0439.xml

DHS Press Office. (2004). Department of homeland security. Department of Homeland Security Implements Information Exchange System for G-8 Summit Events. Retrieved from http://www.dhs.gov/dhspublic/display?content=3649

DiMattia, S.S. (2001). Planning for continuity. *Library Journal, 126*(19), 32.

Dizard, W.P. (12/13/04). Final intelligence reform bill stresses data sharing. Government Computer News. Retrieved April 14, 2005 from http://www.gcn.com/23_34/news/28097-1.html

Dory, A.J. (2003-04). American civil society: The U.S. Public and homeland security. *The Washington Quarterly, 27*(1), 37–52.

Drabek, T.E. (1986). *Human system responses to disaster: An inventory of sociological findings*. Springer-Verlag, New York.

Drabek, T.E. and McEntire, D.A. (2003). Emergent phenomena and the sociology of disaster: Lessons, trends and opportunities from the research literature. *Disaster Prevention and Management: An International Journal 12*, no. 2.

Drucker, J. (1995). *The alphabetic labyrinth: The letters in history and imagination*. Thames and Hudson: London.

Dugan, R.E. (2001). Book reviews. *Journal of Academic Librarianship, 27*(1), 61.

Ellis, D.G. and B.A. Fisher. (1994). *Small group decision making: Communication and the group process* (4th ed.). McGraw-Hill, New York.

Erskine, A. (1995). Culture and power in ptolemaic Egypt: The museum and Library of Alexandria. *Greece & Rome, 42*(1), 38–48.

ESRI. (2004). Environmental systems research institute news release. RAINS announces open specification for sensitive information sharing across state/local/national homeland security systems. Retrieved from http://www.esri.com/common/rains/news.html

Fisher, A. (1974). *Small group decision making*. New York: McGraw-Hill.

Fisher, D. (2004, March 29). National security spec advances. eWeek. Retrieved November 4, 2004 from http://www.eweek.com/article2/0,1759,1555316,00.asp

Gouran, D. (1999). Communication in groups: The emergence and evolution of a field of study. In L. Frey (Ed.), *The handbook of group communication theory & research* (pp. 3–36). Sage, Thousands Oaks, CA.

Gouran, D. (2003a). Communication skills for group decision making. In J. O. Greene & B. R. Burleson (Eds.), *Handbook of communication and social interaction skills*. Mahwah, NJ: Lawrence Erlbaum Associates.

Gouran, D.S. (2003b). Communication skills for group decision making. In J. O. Greene & B. R. Burleson (Eds.), *Handbook of communication and social interaction skills*. Mahwah, NJ: Lawrence Erlbaum Associates.

Hare, A.P. (1962). *Handbook of small group research* (2nd ed.). Collier Macmillan, New York.

Hare, A.P. (2003). Roles, relationships, and groups in organizations: Some conclusions and recommendations. *Small Group Research, 34*(2), 123–154.

Harris, T. and J. Sherblom. (2002). *Small group and team communication* (2nd ed.). Allyn & Bacon, Boston.

Hartley, P. (1997). *Group communication*. London and New York: Routledge.

Heller-Roazen, D. (2002). *Tradition's destruction: On the Library of Alexandria*. October (100), 133.

Jackson, W. (2004, March 29). Coalition has specs for secure data-sharing. Washington Technology. Retrieved November 6, 2004 from http://www.wtonline.com/news/1_1/state/23116-1.html

Jaeger, P.T., Bertot, J.C. and McClure, C.R. (2003). The impact of the USA patriot act on collection and analysis of personal information under the foreign intelligence surveillance act. *Government Information Quarterly, 20*(3), 295.

Janis, I.L. (1982). *Groupthink: Psychological studies of policy decisions and fiascoes*, (2nd ed.). Houghton Mifflin, Boston.

Jetten, J., Postmes, T. and McAuliffe, B. (2002). We're all individuals: Group norms of individualism and collectivism, levels of identification and identity threat. *European Journal of Social Psychology, 32*, 189–207.

Joy, B. (2000). Why the future doesn't need us. *Wired, 8*(4).

Kelly, J.R. and Loving, T.J.. (2004). Time pressure and group performance: Exploring underlying processes in the attentional focus model. *Experimental Social Psychology, 40*(2), 185–198.

Keyton, J. (1999). Relational communication in groups. In L. Frey (Ed.), *The handbook of group communication theory & research* (pp. 192–224). Sage, Thousand Oaks, CA.

Lester, S.W., Ready, K.J., Hostager, T.J. and Bergmann, M. (2003). The human side of group support systems: Influences on satisfaction and effectiveness. *Journal of Managerial Issues, 15*(3), 317–337.

Lewin, K., Lippitt, R. andWhite, R. (1939). Patterns of aggressive behavior in experimentally created "social climates." *Journal of Social Psychology, 10*, 271–299.

Mabry, E. (1999). The systems metaphor in group communication. In L. Frey (Ed.), *The handbook of group communication theory & research* (pp. 71–91). Sage, Thousand Oaks, CA.

Machlup, F. and Mansfield, U. (1983). *The study of information: Interdisciplinary messages.* John Wiley & Sons, New York.

Martin, R., Gardikiotis, A. and Hewstone, M. (2002). Level of consensus and majority and minority influence. *European Journal of Social Psychology, 32*, 645–665.

McDaniel, R. (1997). Strategic leadership: A view from quantum and chaos theories. *Health Care Management Review, 22*(1), 21–37.

McEntire, D.A. (2004a). Development, disasters and vulnerability: A discussion of divergent theories and the need for their integration. *Disaster Prevention and Management: An International Journal 13*, no. 3.

McEntire, D.A. (2004b). The status of emergency management theory: Issues, barriers, and recommendations for improved scholarship. Paper presented at the FEMA Higher Education Conference, Emmitsburg, MD.

McGrath, J.E., Arrow, H, and Berdahl, J.L. (2000). The study of groups: Past, present, and future. *Personality and Social Psychology Review, 4*(1), 95–105.

Mena, J.S. (2004). Homeland security as catalyst. *Intelligent Enterprise* (Vol. 7, pp. 28–33): CMP Media LLC.

Meyers, R. and Brashers, D. (1999). Influence processes in group interaction. In L. Frey (Ed.), *The handbook of group communication theory & research.* Sage, Thousand Oaks, CA.

Moon, H., Conlonb, D.E., Humphrey, S.E., Quigley, N., Devers, C.E. and Nowakowski, J.M. (2003). Group decision process and incrementalism in organizational decision making. *Organizational Behavior and Human Decision Processes, 92*(1–2), 67–79.

Morillo, F. and Gómez, M.B.I. (2003). Interdisciplinarity in science: A tentative typology of disciplines and research areas. *Journal of the American Society for Information Science and Technology, 54*(13), 1237–1249.

Muir, A. and Shenton, S. (2002). If the worst happens: The use and effectiveness of disaster plans in libraries and archives. *Library Management, 23*(3), 115.

NCLIS. (2002). Trust and terror: A proposal of the national commission on libraries and information science (nclis) to expand the role of U.S. Libraries in crisis information dissemination and management. Retrieved July 29, 2005 from http://www.nclis.gov/info/trust/Trustand Terror.brochure.English.IFLA.pdf

O'Connor, B., Copeland, J. and Kearns, J. (2003). *Hunting and gathering on the information savanna: Conversations on modeling human search abilities.* Scarecrow Press, Inc, Lanham, MD.

Oetzel, J. (2001). Self-construals, communication processes, and group outcomes in homogenous and heterogeneous groups. *Small Group Research, 32*(1), 19–54.

Office of the Press Secretary. (2004). The White House. President Signs Intelligence Reform and Terrorism Prevention Act. Retrieved from http://www.whitehouse.gov/news/releases/2004/12/20041217-1.html

Onley, D.S. (2002 September 11). Better communications could have saved lives, homeland official says. Government Computer News. Retrieved April 3, 2002 from http://www.gcn.com/vol1_no1/daily-updates/19977-1.html

Park, W.W. (2000). A comprehensive empirical investigation of the relationships among variables of the groupthink model. *Journal of Organizational Behavior, 21*, 873–887.

Poole, M. (1985). Group decision-making as a structurational process. *Quarterly Journal of Speech, 71*(1), 74–102.

Poole, M. (1991). Procedures for managing meetings: Social and technological innovation. In R. A. Swanson & B. O. Knap (Eds.), *Innovative meeting management* (pp. 53–109). 3M Meeting Management Institute, Austin, TX.

Poole, M. (2004). Interdisciplinary perspectives on small groups. *Small Group Research, 35*(1), 3–16.

Prekop, P. (2002). A qualitative study of collaborative information seeking. *Journal of Documentation, 58*(5), 533–547.

Propp, K. (1999). Collective information processing. In L. Frey (Ed.), *The handbook of group communication theory & research* (pp. 225–250). Sage, Thousand Oaks, CA.

Putnam, L. and Stohl, C. (1990). Bona fide groups: A reconceptualization of groups in context. *Communication Studies, 41*, 248–265.

Quarantelli, E.L. (1997). Problematical aspects of the information\ communication revolution for disaster planning and research: Ten non-technical issues and questions. *Disaster Prevention and Management, 6*(2), 94.

RAINS-Gauge. (2004, April 8). Spotlight on rains open specification. The RAINS-Gauge. Retrieved November 5, 2004 from http://www.rainsnet.org/press/newsletter_vol2_special.asp

RAINS. (2004). Open specification. Open spec based on proven program. Regional Alliances for Infrastructure and Network Security (RAINS). Retrieved from http://www.rainsnet.org/programs/open_spec.asp

Ratzan, L. (2004). *Understanding information systems: What they do and why we need them.* Chicago: American Library Association.

Repo, A. (1983). The dual approach to the value of information: An appraisal of use and exchange values. *Information Processing and Management, 22*(5), 373–383.

Riley, B. (9/10/2003). Information sharing in homeland security and homeland defense: How the department of defense is helping. *Journal of Homeland Security.* Retrieved April 4, 2005 from http://www.homelandsecurity.org/journal/articles/displayArticle.asp?article=97

Robinson, B. (2003, April 2). Rains showcases secure info. FCW.com. Retrieved November 12, 2004 from http://www.fcw.com/geb/articles/2003/0331/web-oregon-04-02-03.asp

Robinson, B. (2004, October 7). Oregon pushes for dhs dollars. FCW.com. Retrieved November 12, 2004 from http://www.fcw.com/geb/articles/2003/1006/web-rains-10-07-03.asp

Ruben, B.D. (1992). The communication-information relationship in system-theoretic perspective. *Journal of the American Society for Information Science* (1986–1998), *43*(1), 15.

Ruyle, C.J. and E.M. Schobernd. (1997). Disaster recovery without the disaster. *Technical Services Quarterly, 14*(4), 13.

Saracevic, T. (1999). Information science. *Journal of the American Society for Information Science, 50*(12), 1051–1063.

Sell, J., M. Lovaglia, E. Mannix, C. Samuelson, and R. Wilson. (2004). Investigating conflict, power, and status within and among groups. *Small Group Research, 35*(1), 44–72.

Shannon, C. and W. Weaver. (1948). *The mathematical theory of communication.* University of Illinois Press, Chicago.

Shaw, G.L. (2005). Business crisis and continuity management. In D. A. McEntire (Ed.), *Disciplines and disasters.* Emmitsburg, MD: Emergency Management Institute.

Sillince, J.A. (2000). Rhetorical power, accountability and conflict in committees: An argumentation approach. *Journal of Management Studies, 37*(8), 1125–1156.

Slezak, S.L.K., Naveen. (2000). The effect of organizational form on information flow and decision quality: Informational cascades in group decision making. *Journal of Economics & Management Strategy, 9*(1), 115–156.

Sosik, J. and Jung, D. (2002). Work-group characteristics and performance in collectivistic and individualistic cultures. *The Journal of Social Psychology, 142*(1), 5–23.

Stephenson, R. and Anderson, P.S. (1997). Disasters and the information technology revolution. *Disasters, 21*(4), 305–334.

Sunwolf and Seibold, D. (1999). The impact of formal procedures on group processes, members, and task outcomes. In L. Frey (Ed.), *The handbook of group communication theory & research.* Thousand Oaks, CA: Sage.

Tennant, R. (2001). Coping with disasters. *Library Journal, 126*(19).

The 9/11 Commission. (2004). Final report of the national commission on terrorist attacks upon the United States. Washington, D.C.: US Government.

Totten, H.L. (2005). Libraries and education. In T. Pipes Lecture. Denton, TX.

United States Department of Homeland Security. (2005). Dhs organization. Priorities for Second Year. Retrieved from http://www.dhs.gov/dhspublic/display?theme=10&content=3240

Vickery, B.C. and Vickery, A. (1987). *Information science in theory and practice.* London: Butterworths.

Webber, S. (2003). Information science in 2003: A critique. *Journal of Information Science, 29*(4), 311–330.

Weitzel, A. and Geist, P. (1998). Parliamentary procedure in a community group: Communication and vigilant decision making. *Communication Monographs, 65*(9), 244–259.

Wersig, G. and Nevelling, U. (1975). The phenomena of interest to information science. *Information Scientist, 9,* 127–140.

What's next? (2004). New security specification targets government. *Processor, 26*(17), 23.

Wilkinson, I.A.G. and. Fung, I.Y.Y. (2002). Small-group composition and peer effects. *International Journal of Educational Research, 37,* 425–447.

Williams, W.W. (2000). Library disaster planning and recovery handbook (book review). *Library Journal, 125*(11), 123.

Wittenbaum, G., Hollingshead, A., Paulus, P., Hirokawa, R., Ancona, D., Peterson, R. et al. (2004). The functional perspective as a lens for understanding groups. *Small Group Research, 35*(1), 17–43.

Chapter 24

Making Sense of Consilience: Reviewing the Findings and Relationships among Disciplines, Disasters and Emergency Management

David A. McEntire and Sarah Smith

ABSTRACT

This concluding chapter reviews the findings pertinent to disasters and emergency management from the standpoint of each of the disciplines presented in this book. It reviews the status of knowledge in each particular field and uncovers opportunities to develop future research in those areas. The chapter also reiterates that each discipline is heavily dependent upon others for the purpose of theory generation and policy guidance. Finally, the chapter points out that the concept of vulnerability is important to each discipline interested in disasters and emergency management.

INTRODUCTION

As can be seen throughout this volume, the convergence of disciplines around disasters and emergency management is increasingly recognized among both scholars and professionals who are involved in this important field of study and area of activity. And yet, ironically, there is not a great deal of literature that addresses the contributions each academic field makes to disaster research and the implications this has for practitioners. With a few notable exceptions, work in one discipline has for too long remained aloof from that of another. For this reason the editor and contributing authors to this book found it imperative to assess multi- and interdisciplinary viewpoints about disasters, emergency management and related concepts.

With the above in mind, this concluding chapter reviews the findings pertinent to disasters and emergency management from the standpoint of each discipline presented in this book. It reviews the status of knowledge in each particular field and uncovers opportunities to develop future research in those areas. The chapter also reiterates that each discipline is heavily dependent upon others for the purpose of theory generation and policy guidance. Finally, the chapter points out that the concept of vulnerability is important to each discipline interested in disasters and emergency management.

REVIEW OF FINDINGS

The chapters in this book expand our knowledge of disasters and provide numerous recommendations for those who study or work in emergency management. Contributing scholars convey lucid histories of their respective disciplines and expound upon important concepts, issues, trends and dilemmas. The status of understanding has been exposed and gaps in research have been identified. The following section reviews some of the most pertinent findings of each discipline in chronological order of presentation.

In his chapter on the "Geographic Study of Disaster," Jim Kendra reveals that "geographers are concerned with the distribution of various kinds of social, biological, and geomorphological phenomena over space" (2007, p. 16). In the broadest sense, geographers are interested in studying the relationship of social, physical, and technological systems. It is therefore logical that these scholars were among the first scholars to study hazards, risks and disasters. While their focus has mainly been on natural hazards (and to a lesser extent technological hazards), the discipline has recently become more involved with the increased threat of terrorism (see Cutter, Richardson and Wilbanks, 2003). Although subject matter has changed throughout the years, a constant and increasingly important aspect of geography relates to Geographic Information Systems (GIS). Kendra notes "GIS provides information for decision makers, and theoretical value, helping to validate models of human environment interaction" (2005, p. 16). GIS is therefore regarded to be fundamental for effective spatial analysis.

While geographers have helped to generate important theoretical perspectives about disasters (including the human ecology school), Kendra suggests that geographers must refocus their efforts for the benefit of people. The implication of his assessment is that it is not enough to study the complex physical causes of earthquakes or landslides; geographers must ensure that their knowledge of hazards has bearing on disaster and emergency management policy. Kendra also notes that more needs to be learned about global warming. The main concern is that we are having difficulty knowing the extent of human impact on this phenomenon in comparison to naturally occurring fluctuations in temperature over time. Another area ripe for investigation deals with rising disaster losses. We do not have a clear understanding of the degree to which hazardousness may be increasing or shifting across locations. In addition, Kendra raises some interesting questions about what a hazard really is, and he encourages more research about the topics of ambiguity and surprise.

In his chapter on Meteorology and Emergency Management, Kent McGregor provides a basic, but vitally important review, of fundamental meteorological processes for those interested in emergency management. McGregor states that there is a very close relationship between his discipline and disasters. Meteorologists have the vital responsibility for predicting and alerting the public of impending natural hazards. Meteorology is also important during several types of disasters as wind direction and relative humidity have a significant impact on response and recovery activities as well as the safety and well-being of emergency workers and victims alike. This is especially the case for wildfire disasters.

In the future, meteorologists need to develop new ways to alert the public of adverse weather. This may include using the internet or cell phones to announce weather-related hazards. New models are needed to understand complex weather phenomena including the formation and behavior of tornadoes. McGregor also recommends that more studies be conducted about global weather patterns including El Niño/La Niña. He agrees with Kendra that additional attention on the causes and consequences of global warming is warranted.

Ana Maria Cruz' discussion of "Engineering Contributions to the Field of Emergency Management" underscores two important benefits for emergency management. These include the "setting of codes and standards, and the actual design and construction of infrastructure used to prevent damage and losses caused by hazards" (Cruz, 2007, p. 49). Therefore, the engineer's assistance to emergency management occurs primarily, but not solely, in the mitigation phase of disasters. That is say, engineers try to reduce the impact of a disaster by strengthening building code regulations or by developing levees and flood-walls to be used in areas of high risk. Of course, the later types of structural mitigation devices can be extremely problematic as we have recently witnessed in New Orleans.

Engineers' main thrust of study has been in regards to earthquake mitigation. However, engineers have investigated other types of disasters (e.g., tornadoes and hurricanes at the Texas Tech Wind Engineering and Research Institute). Engineering efforts have also been centered around impacts on buildings and lifeline systems (Heaney et al., 2000). Nevertheless, engineering activities have not fully taken into consideration any secondary or indirect impacts of hazards. This includes transportation disruption, loss of power, broken water and gas lines, hazardous materials releases and fires, and the fact that emergency response is hampered due to resulting isolation from surrounding communities. For this reason, Dr. Cruz' work on conjoint natural and technological hazards is extremely valuable.

In the future, Dr. Cruz recommends scholars begin to tackle the enigma of defining "acceptable risk." What is the proper balance between living in extremely hazardous areas and trying to limit loss? This is a question that must be addressed by emergency management scholars. Dr. Cruz also recognizes the need to improve the successful adoption and enforcement of building codes for further disaster reduction.

Of all the disciplines discussed in this book, sociology has devoted the most time to studying how humans respond to disasters. Thomas Drabek's review of the literature illustrates that sociologists have studied individuals and their social units, ranging from families to organizations and communities. Such work helps researchers understand the nature of disaster, the values of the community that have bearing on such events, the impact of mass emergencies on stability and change, how humans react to collective stress, and alternative role and structural arrangements of disaster organizations. Other major contributions of sociologists include their questioning of disaster mythology, their expositions on emergent groups and behavior, and their recommendations for disaster planning. This does not discount the role of sociology in generating novel methodological innovations however (Drabek, 2005, p. 13).

Regarding the future, Drabek reiterates the findings of a conference which had the purpose of celebrating the forthieth anniversary of the founding of the Disaster Research Center. He recommends that researchers learn more about the effects of globalization and development on disasters. He also encourages additional interdisciplinary work and acknowledges that "alternative theoretical perspectives for the future of emergency management should be elaborated on, encouraged, and compared" (Drabek, 2005, p. 22).

Scanlon's chapter, "Research about the Mass Media and Disaster," reveals that research about the media and disasters has been performed in two areas: by those interested in social science and others in mass communications. Although there is still insufficient information about the media's role in disasters, scholarship has uncovered several important lessons. Studies suggest that the media is heavily interested in reporting about disasters, and that they do warn the public and keep them informed as the disaster unfolds. In spite of their important role, the media also complicates responses at times by adding to convergence, perpetuating disaster myths, and treating victims with insensitivity.

Opportunities for improving media reporting concern their need to have disaster plans that will enable them to operate effectively under disaster situations with the increased demand placed on their resources. Scanlon's research also suggests that much more needs to be learned about the media's relation to modern terrorism. In light of 9/11, scholars involved in journalism need investigate the type of material they should report, recognizing that terrorists will also be available to receive that information, and that their portrayals may have an impact on terrorist activity (as witnessed by the recent protests and attacks over Danish cartoons of the Prophet Mohammed).

Much like sociologists, Gibbs and Montagnino notes in her chapter that psychologists have played an important role in understanding how humans react emotionally in the aftermath of a disaster. However, while sociologists focus more on groups as a level of analysis, psychologists give greater attention to individuals. A particular focus of this discipline is on the trauma resulting from disasters. In general, people are adaptive and can cope or deal with stress and loss. However, individuals might suffer from Post Traumatic Stress Disorder, especially when there has been an unpredictable, long-lasting event of mass violence, horror, or terror (Gibbs, 2005, pp. 10–11).

The chapter on psychology reviews the process of critical incident stress debriefing (CISD). Although research provides praise for the strategy, there are also studies that are critical of the treatment. This topic will therefore remain an important point of discussion among scholars in the future. Furthermore, there is also a lack of information about the benefit of professional psychologists versus paraprofessionals and "which kinds of interventions work best for which problems" (Gibbs and Montagnino 2007, 104). Some of the newest types of treatments, Eye Movement Desensitization and writing tasks, for instance, will require additional academic attention.

Anthropology is another discipline that is critical of CISD approaches. Doug Henry's chapter also indicates that the understanding of culture is extremely important for a comprehension of disasters, and he asserts that his discipline contributes much to the research about such occurrences in developing nations. Anthropological studies find that cultures are generally able to cope after disasters, although resettlement can be somewhat destabilizing. Research in this discipline also questions the appropriateness and effectiveness of international disaster relief operations, pointing out that dependence can be created through well-meaning recovery efforts.

What anthropologists need to learn more about is how cultural beliefs affect responses to disasters. This includes not only post-disaster activities, but the very definition of acceptable risk and how this influences mitigation and preparedness policies. Henry also encourages additional ethnographic research that is ethically sound.

The social work chapter reiterates many of the findings presented earlier by others in this volume. For instance, Zakour notes that scholars interested in social work define disasters primarily thorough notions of social disruption, excessive demands and collective stress (Zakour 2007 125). His chapter also questions the value of post-traumatic stress interventions. Zakour does acknowledge that the goal of social work is to prevent social, physical, and mental suffering, and to effectively serve disaster victims by coordinating volunteer agencies. In this sense, social work is closely aligned with non-profit activities in disasters and emergency management.

Zakour's chapter points out a number of areas that deserve additional investigation. Researchers need to learn more about what prompts volunteerism in the field of social work. Also, "most disaster research in the United States has studied middle-class populations, and it has not been clear to what extent research finding transfer to cross-cultural or international settings" (Zakour 2007 132). This is especially problematic in that many "effective methods of helping disaster victims through social services are not feasible in cross-cultural and international settings" (Zakour 2007 132). Furthermore, Zakour believes "more research is needed to assess the impact of acute and chronic environmental disasters on rural and small communities, which often contain high percentages of low-income residents" (2007, 137).

In the chapter, "Disaster Policy and Management in an Era of Homeland Security," Sylves discusses the importance of politics and how they relate to disasters. He notes that presidents since Eisenhower's time have declared disasters to free up funding for affected areas. One problem with disaster declarations is that there is not a concrete definition of what a disaster is. Consequently, decisions on what to do have fluctuated dramatically over time. Sylves also conveys the fact that federal declarations influence people's perceptions about the federal share of disaster losses. This could be one of many reasons why local and state governments do not do enough to mitigate against them in the first place.

Another major finding provided by the discipline of political science is that disasters have a dramatic effect on public policy. The attacks on 9/11 are an excellent example of these types of "focusing events" (Birkland, 1997). The nation's attention has shifted away from natural hazards toward the threat of terrorism. This has resulted in additional plans and national strategies (e.g., NRP and NIMS). However, Sylves notes that changes in policy may be problematic even when well-intentioned. Public policy tends to be reactive and may even be overreactive at times. The creation of the Department of Homeland Security has gutted FEMA, and this and an overreliance incident command for terrorist attacks may hurt our ability to deal with other types of disasters (i.e., Hurricane Katrina).

Sylves' work on presidential declarations raises a host of questions that will need to be addressed in future studies (2005, p. 8). Another major gap in political science research is that it is unknown if disaster policies are adequate until a disaster strikes. His chapter also advocates additional studies about the benefit of the Department of Homeland Security, NIMS, and incident command. Therefore, it is "unknown if natural hazards emergency planning has been helped or hurt by the federal emphasis on terrorism after 9/11" (Sylves, 2005, p. 39).

The chapter by Waugh shows that the public administration is directly involved in emergency management, but he notes that most public administrators and scholars in this discipline have traditionally had little or no experience and interest in emergency management. Nevertheless, "the discipline of public administration provides a foundation for emergency management educational programs, and the discipline is increasingly associated with emergency management research" (Waugh 2007, 163).

Public administration generates numerous lessons about the causes of rising disaster losses, blame placement after devastating events, and networking among key participants in emergency management. Scholars in this field may also help public officials use scarce budgets more efficiently and effectively. Studies in this field also improve knowledge about "decision-making, leadership, communication, interpersonal relations, group dynamics" (Waugh 2007, 166).

There are several questions that deserve further investigation by those studying public administration. Research reveals that we need to learn more about how disaster related policies impact societal values, government processes, and economic conditions. Waugh's work also suggests that we need to better understand the "management" aspect of emergency management. It is still unknown how much to invest in emergency management and homeland security to have the optimum outcome. The questions "what is effectiveness?" and "how much investment . . . is enough?" will need to be answered by those from the discipline of public administration.

The chapter on "International Relations and Disasters" mentions several links between this discipline and emergency management. McEntire shows that the profession of emergency management is in many ways an outgrowth of international affairs, and that international relations scholars have produced knowledge about international organizations involved in disaster relief. This discipline has also generated a great deal of knowledge about decision-making in crisis situations which may have some applicability to disasters. The need for a global approach to disasters, and the impact of epistemic communities, also illustrate the importance of this discipline for emergency management.

Scholars of international relations should help generate knowledge about the culture of Islamic extremists as well as the similarities and differences between national security during the Cold War and homeland security today. Because international relations as a discipline has gone through a period of self-reflection, it may provide some unique epistemological insights for the emerging discipline of emergency management. International relations may help us to better understand the nature of our subject matter, the alternative methods for studying it, and the impact of assumptions and values on our findings.

In the chapter "Comparative Politics and Disasters," McEntire and Mathis reveal that comparative politics is the study of political systems and processes around the world. Although comparativists have yet to fully engage the subject of disasters, the discipline may help to develop findings about alternative cultural views about emergency management, the impact of class relations on disasters, ways to increase political support for mitigation, how to improve intergovernmental relations, and the potential drawbacks of unchecked development.

However, the greatest potential contribution of comparative politics to disaster studies is in the area of methodology (McEntire and Mathis 2007, 180). By carefully examining a small number of cases, our knowledge about disasters and emergency management can be greatly enhanced. Additional comparative studies on emergency management institutions around the world are definitely needed. Comparison will help identify universal or semi-universal principles and other best practices to reduce disasters in the United States or abroad.

The discipline of management has the goal, as John Pine notes, of effectively establishing organizations and implementing decisions. In terms of disasters, the focus of this discipline on strategic planning and systems theory may lead to better management and improved coordination among the many units participating in emergency management. Pine uses former FEMA Director, James Lee Witt, as a perfect example of total quality management in action. He also suggests that the rational decision making model may not be appropriate for the uncertain and dynamic nature of disasters. Emergency managers must therefore be flexible in their approach to contingencies.

To improve the study and practice of emergency management, Pine stresses the need for education and organizational learning. He also recommends closer ties between "the Department of Homeland Security, the business community, as well as local and state operations" (Pine 2007 203). There is no doubt that we must work to enhance the management capabilities of future emergency managers.

Dreyer's chapter on gerontology illustrates that the elderly have unique needs in time of disaster. This discipline underscores how older adults react in prior disasters and indicates that unique challenges are faced because of their health in addition to financial and other conditions. Studies in this area reveal that evacuation, communication and the prevention of illnesses take on special features when considering the elderly in disasters. One of the major findings is that "nursing facilities often are overlooked . . . and generally are not incorporated into disaster-relief plans" (Dreyer [2005] quoting Saliba, Buchanan and Kington 2007, 210). This is a finding that has taken on extra meaning since the appalling loss of life among the elderly in Hurricanes Katrina and Rita.

The chapter on gerontology also exposes the fact that much needs to be known about the aged in disaster situations. For instance, what requires additional research is the emergency procedures for older adults living independently since they are more likely to have medical issues or physical limitations. Scholars need to learn more about the mindset of older people who depend on others during disasters. More studies are needed to understand the reaction of elderly to extreme temperature conditions, fires and nursing home evacuations, and the use of the aged as volunteers in disaster situations.

Richard Bissell reminds in his chapter on public health that the disasters that have killed the most people throughout history have been epidemics. He also notes that the threat of bio-terrorism makes the links between public health and emergency management vital today. Some of the important contributions of public health to emergency management center on the human-environment relationship, triage procedures, joint planning and disaster decision-making.

One thing that we need to learn more about is how to improve collaboration between public health and emergency management officials. For instance, communication must be addressed since the vocabulary used in the public health sector is sometimes different than that of emergency managers. There is also a need for more research on the "development and utilization of mechanisms for conducting rapid needs assessments in disease outbreaks, instead of relying on the much less illustrative damage assessment currently used by emergency management personnel" (Bissell 2007, 219). In addition, more studies need to be conducted to promote horizontal information sharing rather than just vertical communication. This would enable critical information to be accessed beyond the public health sector to those who need it. Successful intervention strategies need to be identified, and the processes of quarantine are worthy of further examination.

Louden's chapter about the criminal justice system reiterates the fact that the police have major roles during a disaster. They not only assist with traffic control and life safety issues, but they are likewise involved in criminal investigation and prosecution. Lowden's work quotes that criminal justice may also help us better understand riots, sieges, hostage situations and terrorist attacks.

Recent events have given ample reason to study the relation between criminal justice and emergency management. Louden notes that there was a great rift between police and fire departments prior to 9/11 and this resulted in many operational problems when the World Trade Center was attacked. The ongoing conflict between police and fire officials brings up the unanswered question as to who is or should be in charge after a disaster (although a more complete inquiry would possibly be: who is or should be in charge of what after a disaster, and how should the various organizations interact to achieve optimal results?). Hurricane Katrina has also opened up a new research agenda. Much more needs to be understood about police planning and operations, how humans behave in disaster situations, and what the police can do to better deal with violence and looting activity (when and if it occurs).

In the economics chapter, Terry Clower reveals that his discipline helps generate knowledge regarding damage assessment, disaster declarations, insurance provision, and impacts of disasters upon the economy. For instance, the economist's role is to "determine the affected area's losses in terms of employment income and indirect losses such as loss of business activity due to reduced activities at damaged firms or loss of income in secondary and tertiary employment" (Clower 2007, 237). Economists employ a variety of useful methods to determine the direct and indirect effects of disasters too (Clower 2007, 240). In spite of these contributions, economists are hardly ever called to respond to the disaster site to estimate the physical damage of the disaster.

Studies in this discipline have shown that areas afflicted with disasters are fairly resilient in regaining a GDP that was similar to economic performance prior to the event. However, Clower's research illustrates that terrorism has a significant effect on macroeconomic performance over the long term. Therefore, more studies will be needed on the impact of this type of hazard. In addition, researchers should attempt to better comprehend rising disaster losses, acceptable risk, and policies that create tax burdens for those who live in less-hazardous areas.

Nicholson's chapter illustrates that law has had a substantial impact on the direction of emergency management. His research argues that law determines how we define disasters, and that the legal field has certainly been influenced by recent policy pertaining to terrorism and homeland security. While there is an obvious and close relationship between law and emergency management, emergency managers and lawyers do not seem to mix until they meet in court. Nicholson suggests that emergency managers need to regard statutory liability as a serious issue. He also asserts that emergency managers must learn more about legal terminology so they can communicate more effectively with legal counsel. He also notes that additional research is needed on the impact of NFPA 1600 and EMAP.

In his chapter, John Labadie asserts that emergency management and environmental management are inherently related. The study of environmental management is an evolving field of academia that catapulted to the forefront with the publication of Rachel Carson's *Silent Spring* (1965). The major finding of such research is that industrialization is having a negative impact upon the environment. Therefore, there is a need to implement the "precautionary principle." This discipline also underscores the potential for environmental disasters that result from natural hazards (e.g., a flood that causes a hazardous materials release).

The chapter on the environment illustrates that there are many questions that deserve further investigation. For example, Labadie indirectly asks about the impact of homeland security policies on environmental protection. He also suggests the need to focus on long-term disasters instead of acute hazards. Finally, Labadie points out a close relation between environmental management and the phases of emergency management. However, one might indicate that the relationship between sustainability and disaster response deserves further investigation since the link is not always strong or self evident.

The chapter on communication studies by Richardson and Byers reveals that scholars in this area have mainly looked at rhetorical examination of political discourse or the impact of persuasion and decision-making. While these latter topics may have positive impact on emergency management, there have been insufficient studies on communication in disaster situations. The research that exists does note that communication is often a problem in disaster, that technology and the Internet may exacerbate such challenges, and that the ineffective use of the media may generate public relations nightmares.

Richardson and Byers believe that we need to understand more about how interorganizational communications and rational decision-making unfold in disasters. Also, researchers should study the role of communication in creating or being a source of disaster. They ask, "what do communications contribute to issues such as mob mentality, riots, and panicked evacuations" (2007, p. 279)? With the increase of globalization, we also need to learn more about different cultures – particularly terrorists and their way of thought and method for sharing information. Fortunately, "qualitative techniques, such as long-term observations, or ethnography, interviewing, focus groups, and document and artifact analysis provide valuable means of data collection and capture subjects' own words, perceptions, and experiences" (Richardson and Byers 2007, p. 280).

Shaw's chapter on business continuity management reminds that every organization has a responsibility for providing products and services, and to care for their employees and customers. For this reason, it is imperative that businesses also consider the impact of disasters and identify what they will do to prevent and better deal with them. Shaw illustrates that there has been some disagreement about what to call such functions in the private sector. However, his research illustrates that "the vast majority of organizations and institutions have not been designed to anticipate crises or to manage their consequences effectively once they have occurred" (Mitroff, 2001). This lack of planning appears to be changing, though, since disasters create hardships for the private sector or can easily put corporations out of business as has been witnessed by several types of events including 9/11.

Ironically, there is a dearth of knowledge about

the private sector. More needs to be learned about the impacts of a disaster on the economy. A great opportunity lies in the area of assessing the impact of numerous government documents and other standards on business continuity management (e.g., NRP, NFPA 1600), and how best to improve crisis communication, risk management, incident response, and business continuity, among other functions, in the corporate world. By identifying those factors, the private sector will be more equipped to prevent disasters and help individuals and communities deal with their adverse impacts.

Tisha Slagle Pipes' chapter on information management provides an excellent review of the development of this discipline in relation to technology and sociology. She also notes the close relation between information and disasters, suggesting that poor information management may lead to such disruptive events and that correct and complete information are always valued by emergency managers. The major findings of her discipline include the fact that there are many threats to information and information management, and that the effective flow of information may facilitate better decisions in emergency management.

There are a number of gaps that must be filled regarding information management and disasters. For instance, how does information flow among emergent volunteer groups? How can information sharing be improved among various levels of government and across different organizations and jurisdictions? Also, what are the possible implications of the restriction of information on disasters in light of the 9/11 terrorist attacks and the U.S. Patriot Act? Her findings, and those of the other scholars, suggest that we have made impressive strides in our knowledge about disasters and emergency management. However, there are numerous research opportunities that will need to be addressed in the future.

RELATIONS AMONG THE DISCIPLINES

Besides providing updated information about the findings in each field of study and remaining gaps of knowledge, the authors of this book have helped to develop a better understanding of the multi- and interdisciplinary nature of academia in general and disaster studies in particular. Their research reveals that all disciplines are multidisciplinary in nature, that emergency management also spans numerous disciplines, that each discipline relies on the findings of others to improve theoretical understanding and policy recommendations. The chapters likewise reveal many different disciplines provide similar findings about disasters.

All Disciplines are Multi- or Interdisciplinary

One of the apparent but perhaps underappreciated conclusions of this book is that many – if not all – disciplines are multi- or interdisciplinary in nature. This fact was noted in several of the chapters in this volume. Jim Kendra asserts, for instance, that "geography is quintessentially interdisciplinary" and "its concerns with the intersection of social, physical, and technological and political/legal systems means that it shares areas of interest, knowledge, and methods with many other fields of study" (2005 24).

Louden's comments in his chapter on criminal justice are similar to Kendra's. He states, "in our society criminal justice is perhaps the ultimate multidisciplinary discipline. At a minimum, aspects of the law, political science, public health, public management, psychology, and sociology influence the practical, tactical and legal activities of criminal justice system agencies on a daily basis" (Louden 2007, 224).

Bill Waugh likewise comments that "the boundaries of the public administration disaster literature are very broad and overlap considerably with other disciplines, including, for example, political science, business administration, criminal justice, psychology, history, geography, medicine, civil engineering, and sociology" (2007, 163). He further adds that "sorting out the contributions of public administration from those of other fields is difficult largely because public administration is an interdisciplinary field" (Waugh, 2005, p. 167).

There are a plethora of other examples of how distinct disciplines have influenced the direction and content of what have supposedly been single or isolated fields of study:

- Henry declares that anthropology is closely related to several disciplines including sociology, psychology, economics, geography, history, archaeology, and gender studies (2007, p. 111).

- Public health relies on scholarship from biology, chemistry, physics, psychology, epidemiology, biostatistics, sociology, anthropology, and psychology (Bissell 2007, 213).
- Zakour asserts that social work has historically drawn on research in psychology and sociology (2007, 138).
- "International relations has a close relation to political science, history, comparative politics and other disciplines in the social sciences" (McEntire 2007, 170).
- "Management also grew from many disciplines, especially from engineering (scientific management), psychology, sociology, and quantitative methods" (Pine 2007, 2002).
- Environmental management includes environmental science, environmental engineering, ecology, and related disciplines (Labadie 2005, 261).
- Information science has been defined as a "meta-disciplinary science." Quoting Borko (1968), Slagle Pipes says that this field is "an interdisciplinary science that investigates the properties and behavior of information, the forces that govern the flow and use of information, and the techniques, both manual and mechanical, of processing information for optimal storage, retrieval and dissemination" (2007, 296).

As can be readily noted, disciplines are in reality artificial constructs for the production of scientific knowledge. Disciplines typically combine approaches, knowledge and findings from several different fields of academia.

The Study of Disasters and Emergency Management is Interdisciplinary

Another apparent lesson to be drawn from the findings of this text is that disasters and emergency management require a multi- or interdisciplinary approach. Cruz asserts, for instance, that "disasters are complex events, which result from a combination of factors including urbanization, population growth and environmental degradation" (2007, 57). Along similar lines, Pine observes that "emergency management today is a complex function involving public safety and security, business affairs, public and information affairs, information systems administration, communication technologies, mapping sciences and hazard modeling, legal affairs, and coordination with numerous other organizations" (2007, 196). He declares that "emergency management draws from many disciplines and [this] suggests that emergency management is an interdisciplinary process" (Pine 2007, 202).

Numerous other scholars also make similar observations about the interdisciplinary study of disasters and emergency management. For instance, geography is closely related to sociology because of its focus on human ecology – the interaction of people with the natural environment. McGregor makes it known that meteorology's study of disaster is directly related to geography as well as hydrology, engineering, sociology, and journalism. This is because severe weather may impact the flooding of rivers and streams, damage residential and commercial structures, influence people's evacuation behavior, and lead the media to warn the population and report on the response to the disaster.

Sylves agrees with Waugh that political science is closely related to public administration because of the attention given to agenda setting. He also suggests that his discipline has affinity to law and sociology since policy is based on federal legislation and because disaster declarations may influence human behavior. The findings of political scientists share many similarities to those of economists. Clower's chapter is really about the political economy of emergency management, rather than about the economic consequences of disasters alone.

Other cases of overlap are equally evident in the study of disasters and emergency management. Dreyer's chapter on gerontology shows a direct relationship with psychology, public health, and economics. Elderly persons affected by disaster are likely to have emotional, physical and financial problems and struggles. Gibbs' chapter indicates that people's psychological reaction to disaster is directly related to the individual's age, race, religion, social and economic status, and gender. Therefore, the study of psychology is related to gerontology, sociology, and economics. The disciplines of communications, journalism and information sciences likewise appear to rely on each other to produce research regarding disasters.

Disciplines Rely on Others For Theoretical and Policy Improvement

Research in this book indicates that a single discipline cannot improve research and practical application alone. Instead, each discipline that investigates emergency management requires the assistance of other fields of study to fully understand disasters. There are, as Alexander notes, several approaches to disasters: geographical, sociological, anthropological, medical or epidemiological, and technical (Kendra 2007, 24). However, no single approach can fully help our understanding of disasters and emergency management.

For instance, Cruz argues that "the need for multi-hazard approaches to disaster management is increasingly called for" (Cruz 2007, 56). Her assertion is worthy of consideration for two reasons. First, there are, as was noted in the first chapter, numerous hazards that emergency managers have to be prepared to deal with. This includes winter storms, industrial fires and terrorist attacks among others. Second, any specific hazard may trigger additional hazards (e.g., an earthquake may trigger landslides, a hurricane may spawn a tornado, and a train derailment may lead to a hazardous materials spill). Thus, geography, meteorology, and engineering may not be able to provide all necessary information about each particular hazard or any combination thereof by themselves.

The focus on hazards should not overshadow other issues pertaining to disasters though. This brings up a second reason why it is imperative to approach emergency management from a multi- or interdisciplinary perspective. As previously noted, Jim Kendra observes in his chapter that geographers sometimes forget the humans they are supposed to be serving (2007, p. 25). In other words, scientific understanding of the physical characteristics of geological hazards is necessary, but insufficient if not acted upon. Geographers must consequently ensure that the knowledge they produce gets into the hands of policymakers and that effective decisions are made by politicians for the public good. This would suggest that geographers should align themselves with scholars of political science and public administration. Another example, again from geography, suggests that this discipline provides a great deal of information pertinent to mitigation activities (see Kendra's discussion of questions addressed by geographers, 2007, 19). However, geography may have less to say about managing disasters after the hazard event occurs. For instance, what should be done if people have chosen to settle in hazardous areas and an earthquake has just injured numerous individuals? This is a question that may not be totally pertinent to geography. This critique is not meant to demean geography as researchers in this discipline have no doubt been instrumental in disaster research and alerting professional emergency managers of impending hazards.

Cruz has made a similar assertion about the need for a broader approach in her chapter on engineering. In order to increase the usefulness of research in her discipline, engineers need to work closely with scholars in the fields of geology, geography, meteorology, environmental science, sociology, psychology, and public administration. She suggests that a unified effort on the part of several disciplines will go a long way to establish more effective building codes and enforce these standards in such as way as to reduce the destruction of life, property and the environment.

There are additional examples from this book of how and why disciplines augment, clarify, or enhance emergency management scholarship when the findings of others are taken into account:

- "Sociologists studying disasters frequently have integrated both theory and methodological tools reflective of other disciplines into their work" (Drabek 2007, 67).
- "Clearly, sociologists, engineers, psychologists, geographers, and others involved in emergency management and disaster policy research are recognizing the need to provide policy- and program-relevant information as well" (Waugh 2007, 167).
- "If we ever thought it was 'acceptable' that an emergency manager did not know much about the health sector and how it responds to threats and real events, that time abruptly and permanently disappeared with the recognition of bioterrorism as a serious hazard" (Bissell 2007, 213).
- "The sociology of disasters has been a major foundation on which social work research in disasters is based" (Zakour 2007, 135).
- "The relationship between law and emergency management may be characterized as one of mutual need" (Nicholson 2007, 254).

The Findings of Certain Disciplines are Similar to Those of Others

Another related point generated by the scholars of this book is that diverse disciplines have, on several occasions, come up with similar findings. Although there are obvious cases where research or practice diverges and lacks integration (e.g., see excellent chapters by Scanlon on the lack of integration among journalism and disasters; Bissell on the lack of collaboration among public health and emergency management officials; and Louden on the cultural gap between law enforcement and fire fighting), there are other cases where findings converge.

For instance, Kendra, McGregor and Labadie each note the need to improve our understanding about global warming. Several scholars, including Drabek, Gibbs, Henry, Zakour, and Dreyer, discuss how emergency managers must better care for special populations. Kendra, Richardson and Byers, and Slagle Pipes also suggest the importance of improving decision-making among public officials involved in disasters. The political impediments and difficulty of enforcing policy has been noted by Kendra and Cruz.

There are many other topics where there seems to be agreement for improved theory or practice:

- Drabek, McEntire, McEntire and Mathis, and Henry agree that a global focus on disasters is required.
- Complexity, chaos, ambiguity and surprise are subjects that Kendra and Pine believe merit additional attention.
- Kendra, Cruz, Drabek, McEntire and Mathis, and Clower each recommend further attention be given to development and the concept of sustainability.
- Drabek, Scanlon, Richarson and Byers reiterate the crucial role of the media in disasters.
- Kendra and McEntire imply that post-modern theory and epistemology deserve further consideration.
- The nature of terrorism and the impact of the Department of Homeland Security are recognized as important subjects according to Drabek, Scanlon, Waugh, McEntire, Sylves, Labadie and Nicholson.
- Concerns about the freedom of information have been expressed by Kendra, Waugh, and Slagle Pipes.
- Natech disasters were identified as important subjects by Kendra, Cruz and Labadie.
- Myths about human behavior have been identified by Drabek, Louden and others.
- Ways to improve compliance were offered by Cruz, Waugh, Richardson and Byers.
- The relation among information, communication and decision making was discussed by Bissell, Richardson and Byers, and Slagle Pipes.
- The need for more proactive and preventive activities was identified by Gibbs and Montagnino, Cruz, McEntire, and many other scholars.
- Economic issues in disasters were discussed by Shaw and Clower.
- Nicholson and Shaw recommend studies on the advantages and results of NFPA 1600.
- Use of methods or improvement in research techniques were discussed by Drabek, McEntire and Mathis, Clower, and Richardson and Byers.
- The benefit of PTSD was called into question by Zakour, Henry, and Gibbs and Montagnino.
- Cross cultural issues was seen as an area ripe for further research by McEntire, McEntire and Mathis, Zakour, Henry, and Richardson and Byers.
- Studies into slow onset and long running disasters were recommended by McGregor, Henry, Labadie.
- Kendra, Pine, Cruz and Slagle Pipes recognized the value of enhancing our understanding about the impact of technology on disasters.
- Risk and methods to determine acceptable risk were discussed by Henry, Clower and Shaw.

Thus, all disciplines are related to each other to varying degrees. Scholars need to integrate their research efforts and share their findings with other disciplines to better understand emergency management. There are also numerous areas where disaster scholarship seems to be converging.

MULTI- AND INTERDISCIPLINARY INTEREST IN VULNERABILITY

As can be seen in the foregoing discussion, research on disasters and emergency management agrees on many important topics, subjects and issues. One particular areas where there seems to be a great deal of interest is in the concept of vulnerability. Many disciplines appear to view vulnerability in terms of the characteristics that make individuals and groups prone to disasters or less able to deal with their consequences. Others see the concept in broader terms, acknowledging the wide range of variables that have a bearing on community vulnerability. Some of the contributing authors do not fully or directly discuss this concept, although others in their respective disciplines have done so or could engage the subject more fully.

Those That View Vulnerability in Terms of Socioeconomic Status

There are numerous disciplines that explore the vulnerability of people, and these studies are highly critical of social, political and economic structures. The chapter on geography cites Hewitt's well-known book, *Interpretations of Calamity* (1983). The major point of this book (which was developed from the social geographer's standpoint) is that "being exposed to hazard . . . could not entirely be ascribed to the consequence of bad decisions, but of choice that were constrained by the social and economic conditions" (Kendra 2007, 19).

Drabek's chapter on sociology reveals that scholars in this discipline have challenged the dominant view of disasters, thereby exposing their "root causes" (2005, p. 10). He continues, "rather than accept differential exposures and losses by the politically weak, be they female, aged, or ethnic minorities, those adopting this paradigm question the status quo. They ask, 'Why must the patterns of greed and financial corruption continue to perpetuate so-called disasters wherein those most vulnerable are disproportionately hurt?'" (Drabek 2007, p. 64).

Gibbs and Montagnino's chapter takes up many of the issues discussed by other scholars. They show that "vulnerability factors include, but are not limited to, socioeconomic status (SES), available resources, previous level of psychopathology, age, social/family factors, gender and ethnicity" (2007, 99). However, they also note that first responders and disaster workers may be adversely impacted by PTSD because of extreme physical and emotional toll of their work (2007, 99).

The discipline of anthropology, according to Henry, has produced similar findings regarding the concept of vulnerability. He states that cultural institutions determine who is most vulnerable in the social system (Henry 2007, 113). His research indicates that "ethnic minorities, disempowered castes or classes, religious groups, or occupations" may be more vulnerable than others (Henry 2007, 112).

Zakour's findings are almost identical to those of the others listed above. He says that "vulnerability at the individual level refers to social structural factors which increase individuals' probability of suffering long-term and serious social, psychological, and health problems after a disaster" (Zakour 2007, 126). He agrees that poverty and social isolation are two important factors that lead to vulnerability. He also notes that "older individuals, people of color, recent immigrants, and children" as well as the isolated that lack social capital are especially vulnerable (Zakour 2007, 126).

Dreyer's chapter on gerontology resembles the others that discuss social vulnerability. Nevertheless, she focuses specifically on age as a determinant of vulnerability and asserts that "physical condition . . . can affect an older person's ability to escape a disaster safely, thus making them more susceptible to the effects of a natural disaster" (Dreyer 2007, 207). She also mentions that vulnerability is complicated when age interacts with race, income levels and psychological response.

Those That View Vulnerability as a Product of Many Additional Factors

There are other disciplines that examine vulnerability more broadly – in terms of property, places or processes. Cruz's chapter on engineering suggests that several government programs seek to improve "techniques to reduce seismic vulnerability of facilities and systems (e.g., through the adoption of updated seismic building codes and better construction practices)" (2007, 51). She also notes that engineers "assess [the] vulnerability of lifelines to earthquakes" (Cruz 2007, 51).

McEntire's research on international relations and comparative politics illustrates a close relation to vulnerability. Regarding the former discipline, he concludes that terrorists' ability and willingness to use weapons of mass destruction would make the United States vulnerable to this type of activity as would an inability to defend itself or deal effectively with adverse consequences (McEntire 2007, 175). In his chapter with Mathis, he notes that some nations are more vulnerable than others. They also state that development activities may positively or negatively affect vulnerability (McEntire and Mathis 2007, 182). Furthermore, their research illustrates that education, funding, technology, culture and other variables influence a country's vulnerability.

The discussion of public health by Bissell shows relation to the concept of vulnerability. He views vulnerabilities as a weakness they may determine if someone becomes ill or dies from exposure to an infectious disease (Bissell 2007, 215). He also affirms that "vulnerability is a combination of external factors, the hazards to which we are exposed, and internal factors, such as the status of our immune response systems or, at a community level, the design of our structures of status of our public safety services" (Bissell 2007, 218).

Labadie's chapter on environmental science illustrates that people's choices – "farming practices, use and procurement of fuels, selection of building materials and sites, etc. – significantly affect their vulnerability to environmental disasters" (2007, 264). He also notes that structural devices (for flooding) may increase vulnerability because "hazards can surpass the margin of safety provided by technological solutions" (Labadie 2007, 265). Vulnerability is a product of this as well as environmental degradation such as deforestation, resource depletion, and so forth.

Law is also regarded as a determinant of vulnerability. Nicholson cites Handmer and Monson who define vulnerability as a "multi-faceted concept incorporating issues of livelihood, housing, security, and gender, among many others" (2007, 251). Nicholson observes that "the link between vulnerability and law exists when law sets out rights to adequate housing and livelihood" and he believes that "constraints of public and private law, social norms, custom, and international law are posited as having the potential to regulate vulnerability" (2007, 251-252). The attempt to limit legal liability through "litigation mitigation" is also a way to address vulnerability in Nicholson's view (2007, 255).

Slagle Pipes coverage of information science suggests that information is vulnerable to disasters. She quotes Stephenson and Anderson who assert that "the vulnerability of complex networked systems, together with potential ways of using data resources to speed up recovery almost certainly will increasingly preoccupy emergency planning staff in some areas" (Slagle Pipes 2007, 304). In addition, decision-making, because it is based on information flow, may increase or decrease a community's vulnerability to disaster.

Those That do Not Focus on Vulnerability Explicitly But Could

There are a few chapters in this book that do not discuss vulnerability directly. McGregor's discussion of meteorology illustrates that warning, evacuation and sheltering functions have a great impact upon the safety and well-being of citizens (2007). However, there are other discussions of meteorology that give greater attention to the positive impact of warnings on the protection of life and property (see Golden and Adams, 2005; Sorensen, 2000). Therefore, meteorologists agree that "improved warning reduces vulnerability" (Salter et al. 1993, 119).

Scanlon's chapter on journalism illustrates that the media may also have a bearing on vulnerability, although he does not address this point specifically. However, he does assert that mass communications are critical to public safety. In addition, the literature on terrorism indicates that the media may have an unrecognized impact on the potential for politically motivated acts of violence (Combs, 2000). Britton's discussion about disaster explicitly underscores how warning systems (and by association the media) have impact on disaster vulnerability (1986, 256).

Sylves does not mention vulnerability much in his piece (2007, 157). However, his work suggests that the policy decisions made after 9/11 have had a major impact upon emergency management in the United States. Specifically, he implies that the Department of Homeland Security may limit the possibility of terrorist attacks, but he raises disturbing questions about the impact this has or will have on natural disaster reduction. Others have more directly noted the influence of political decisions on disaster vulnerability (Wisner, et. al. 2004). Olson, Olson and Gawronski, in their study of mitigation policy after the Loma Prieta earthquake, draw similar conclusions:

> When all is said and done, when all the scientific and engineering studies have been completed, when all the technological options have been specified, when all the affected populations have been considered, and when the costs and benefits of the various policy options have been detailed to the extent possible, it is a community's political system that decides authoritatively who will get how much life safety and who will pay for it (Olson, Olson and Gawronski, 1998, p. 175).

Waugh's work parallel's that of Sylves, which is to be expected since there is such a close relation between public administration and political science. His research does not address vulnerability in an explicit manner either. Nevertheless, there can be little doubt that public administrators have a dramatic impact upon the vulnerability of a community. For instance, Olson, Olson and Gawronski's work underscores the fact that the ordinance created by city council in Oakland after the Loma Prieta earthquake had a direct bearing on building vulnerability (1998, p. 156). Disaster vulnerability is thus determined, in part, by the effectiveness and efficiency of policies and programs made in each jurisdiction.

Pine does not mention vulnerability in any depth, although he does define it as "susceptibility to hazard, disaster or risk" and equates it to a "measure of resilience" (2007, p. 202). Regardless, there can be no doubt that management activities – whether they be in the public, private or non-profit setting – influence the degree of vulnerability, particularly when there is uncertainty and ambiguity. His research suggests that systems theory may help to identify the plethora of factors that create disaster vulnerability.

In contrast to Pine, Louden's chapter on Criminal justice is more directly related to vulnerability. He believes that "correctional institutions present special problems and concerns when exposed to disasters" (Louden, 2007, p. 227). He admits that, in light of the "number of individuals incarcerated in jails and prisons throughout the country, the potential for disasters impacting on the correctional population is highly probable" (Louden, 2007, p. 227). His findings recommend additional planning measures for prison populations – a finding that seems closely related to those of the social vulnerability school.

Clower's chapter on economics only mentions the concept of vulnerability in passing (2007, p. 246). It is clear that poverty has an impact upon vulnerability, as has already been discussed above. For instance, it is well-known that limited income reduces choices for safe housing and the purchase of insurance. Economic development, and the fact that some people win and others lose in disasters (Scanlon, 1988), would logically have a bearing on vulnerability as well. Studies of insurance also indicate that "governments can help immensely by sponsoring research into patterns of event severity and frequency, and how to avoid or alleviate damage to particularly vulnerable structures or components" (Dlugolecki, 1993, p. 431).

Communication is not linked overtly to vulnerability in Richardson and Byers chapter. Nonetheless, there are logical – if not apparent – ties between communication studies and vulnerability. For instance, poor or inaccurate communication among intelligence agencies made us vulnerable to the 9/11 attacks. In addition, poor communication put emergency workers in jeopardy as in New York after the attacks on the World Trade Center. This point was emphasized as a vulnerability in the 9/11 Commission Report (2002, p. 280) and in numerous other after-action reports.

Shaw's coverage of business continuity does not identify a visible relation to vulnerability. However, if individuals and communities can be vulnerable to disasters it is follows that businesses can be vulnerable as well. Interestingly, Webb, Tierney, and Dahlhamer's research on businesses discusses a variety of factors that augment their vulnerability (2000, p. 86). Steps taken to prevent a disaster or prepare for their consequences, as well as the influence of NFPA 1600, would logically have a bearing on their vulnerability. Thus, the researchers in this book and many other scholars tend to view vulnerability as an important concept in the study of disasters and practice of emergency management.

CONCLUSION

Considering the number of disciplines represented in this book and the diversity of subjects addressed, it is difficult to make sense of Wilson's notion of "consilience" (see Chapter 1). Consequently, this text should be viewed as a continuation of our understanding of disasters and emergency management, rather than a definitive conclusion on these matters. Nevertheless, there are several inescapable lessons that can be drawn from the knowledgeable authors of this text.

First, and foremost among them, it is readily apparent that the study of disasters and emergency management involves multiple disciplinary perspectives. It is obvious that each discipline contributes to our understanding of these deadly, destructive and disruptive events as well as the emerging profession which has the responsibility of reducing and reacting to them. Nevertheless, there are many gaps within each discipline that must be addressed in future studies, and the authors of this text have identified the research agenda in each area that must be pursued.

A second major lesson is that disasters are complex phenomena that require a comprehensive approach by those working in emergency management. Disasters occur at the intersection of the physical and social environments, and are they a product of technological, engineering, political, economic, psychological, cultural, physiological and other variables. Therefore, it will be imperative to accept broader perspectives in the future. In Henry's view, "a holistic approach examines the complex interrelationships between humans, culture, and their environment, from the human actions that may cause of influence the severity of disaster, to the position of social vulnerability that defines disaster impact, to the range of socio-cultural adaptations and responses, including the impact of aid and the infusion of donor money" (2007, p. 111). Thus, the value and sole use of simple, linear thinking must be seriously questioned by those interested in the disaster and emergency management field if progress is to be made.

A third and related point is that disaster and emergency management scholars should aspire to "develop new theory or adapt old theory to produce manageable policy" (Sylves as cited by Drabek, 2007, p. 69). According to Cutter, Richardson and Wilbanks, the "greatest challenge . . . is to stretch our minds beyond familiar research questions and specializations so as to be innovative, even ingenious, in producing new understandings" (2003, p. 4). This implies that interdisciplinary approaches will be more vital for our comprehension of disasters and emergency management than individual or even multidisciplinary approaches of the past.

Finally, it should be noted that there are concepts that may help us to integrate our individual assessment of disaster studies and emergency management. This book makes it is clear that vulnerability is of paramount importance. While significant disagreement still exists among disciplines regarding future research priorities, the vast majority of disciplines discussed in this book show a direct relation to this concept. Even in cases where the relationship was less visible, the work of others helps to illustrate the salience of this concept for all academic disciplines. A focus on vulnerability may therefore capture more disciplines and variables than the hazards perspective, which has dominated the study of disasters since its inception. Consequently, students of emergency management may wish to converge on the concept of vulnerability as a way to effectively promote interdisciplinary scholarship. As an example, geographers discuss both hazards and vulnerability, but psychologists discuss vulnerability and give little – if any – attention to hazards. More importantly, giving priority to vulnerability reduction may be the only way to reduce the occurrence and severity of disasters since it puts responsibility directly on humans and not the physical environment alone.

In conclusion, and returning to the analogy in the first chapter, the vessel of disaster scholarship and harbor of reduced disasters truly do require "all hands on deck." Understanding vulnerability and working towards its diminution will be imperative if disasters are to be minimized in both frequency and severity. It is hoped that this book will move all disciplines and the practice of emergency management in this direction.

REFERENCES

Birkland, T.A. (1997). *After Disaster: Agenda Setting, Public Policy, and Focusing Events.* Washington, D.C.: Georgetown University Press.

Bissell, R.A. (2007). "Public Health and Medicine in Emergency Management." In McEntire, D.A. (ed.) *Disciplines, Disasters and Emergency Management: The Convergence of Concepts Issues and Trends From the Research Literature.*

Britton, N.R. (1986). "Developing an Understanding of Disaster." *Australian New Zealand Journal of Sociology* *22*(2): 254–271.

Clower, T.L. (2007). "Economic Applications in Disaster Research, Mitigation and Planning." In McEntire, D.A. (ed.) *Disciplines, Disasters and Emergency Management: The Convergence of Concepts Issues and Trends From the Research Literature.*

Combs, C.C. (2003). *Terrorism in the Twenty-First Century.* Prentice Hall: New Jersey.

Cruz, A.M. (2007). "Engineering Contributions to the Field of Emergency Management." In McEntire, D.A. (ed.) *Disciplines, Disasters and Emergency Management: The Convergence of Concepts Issues and Trends From the Research Literature.*

Cutter, S.L., Richardson, D.B., and Wilbanks, T.J. (2003). *The Geographical Dimensions of Terrorism.* New York: Routledge.

Dlugolecki, A.F. (1993). "The Role of Commercial Insurance in Alleviating Natural Disaster." In Merriman, P.A. and Browitt, C.W.A. (Eds.), *Natural Disasters: Protecting Vulnerable Communities.* London: Thomas Telford.

Drabek, T.E. (2007). "Sociology, Disasters and Emergency Management: History, Contributions, and Future Agenda." In McEntire, D.A. (ed.) *Disciplines, Disasters and Emergency Management: The Convergence of Concepts Issues and Trends From the Research Literature.*

Dreyer, K. (2007). "Gerontology and Emergency Management: Discovering Pertinent Themes and Functional Elements within the Two Disciplines." In McEntire, David A. (ed.) *Disciplines, Disasters and Emergency Management: The Convergence of Concepts Issues and Trends From the Research Literature.*

Gibbs, M. (2007). "Disasters: A Psychological Perspect." In McEntire, D.A. (ed.) *Disciplines, Disasters and Emergency Management: The Convergence of Concepts Issues and Trends From the Research Literature.*

Golden, J.H. and Adams, C.R. (2000). "The Tornado Problem: Forecast, Warning, and Response." *Natural Hazards Review 1*(2): 107–118.

Heaney, James P., Jon Peterka, Members, ASCE, and Leonard T. Wright. (2000). " Research Needs for Engineering Aspects of Natural Disasters." *Journal of Infrastructure Systems, 6*(1):4–14.

Henry, D. (2007). "Anthropological Contributions to the Study of Disaster." In McEntire, D.A. (ed.) *Disciplines, Disasters and Emergency Management: The Convergence of Concepts Issues and Trends From the Research Literature.*

Kendra, J.M. (2007). "Geography's Contributions to Understanding Hazards and Disasters." In McEntire, D.A. (ed.) *Disciplines, Disasters and Emergency Management: The Convergence of Concepts Issues and Trends From the Research Literature.*

Labadie, J.R. (2007). "Environmental Management and Disasters: Contributions of the Discipline to the Profession and Practice of Emergency Management." In McEntire, David A. (ed.) *Disciplines, Disasters and Emergency Management: The Convergence of Concepts Issues and Trends From the Research Literature.*

Louden, R.J. (2007). "Who's in Charge Here? Some Observations on the Relationship Between Disasters and the American Criminal Justice System." In McEntire, D.A. (ed.) *Disciplines, Disasters and Emergency Management: The Convergence of Concepts Issues and Trends From the Research Literature.*

McEntire, D.A. (2007). "International Relations and Disasters: Illustrating the Relevance of the Discipline to the Study and Practice of Emergency Management." In McEntire, D.A. (ed.) *Disciplines, Disasters and Emergency Management: The Convergence of Concepts Issues and Trends From the Research Literature.*

McEntire, D.A. and Mathis, S. (2007). "Comparative Politics and Disasters: Assessing Substantive and Methodological Contributions." In McEntire, D.A. (ed.) *Disciplines, Disasters and Emergency Management: The Convergence of Concepts Issues and Trends From the Research Literature.*

McGregor, K.M. (2007). "Meteorology and Emergency Management." In McEntire, D.A. (ed.) *Disciplines, Disasters and Emergency Management: The Convergence of Concepts Issues and Trends From the Research Literature.*

Mitroff, Ian I. (2001). *Managing Crises Before They Happen: What Every Executive Manager Needs to Know About Crisis Management.* Amaco: New York.

National Commission on Terrorist Attacks Upon the United States. (2002). *9/11 Commission Report.* New York: W.W. Norton & Company.

Nicholson, W.C. (2007). "Emergency Management and Law." In McEntire, D.A. (ed.) *Disciplines, Disasters and Emergency Management: The Convergence of Concepts Issues and Trends From the Research Literature.*

Olson, R.S., Olson, R.A. and Gawronski, V.T. (1998). "Night and Day: Mitigation Policymaking in Oakland, California Before and After the Loma Prieta Earthquake. *International Journal of Mass Emergencies and Disasters 16*(2): 145–179.

Pine, J.C. (2007). "The Contributions of Management Theory and Practice to Emergency Management." In McEntire, D.A. (ed.) *Disciplines, Disasters and Emergency Management: The Convergence of Concepts Issues and Trends From the Research Literature.*

Richardson, B.K. and Byers, L. (2007). "Communication Studies and Emergency Management: Common Ground, Contributions, and Future Research Opportunities for Two Emerging Disciplines." In McEntire, D.A. (ed.) *Disciplines, Disasters and Emergency Management: The Convergence of Concepts Issues and Trends From the Research Literature.*

Salter, J., Bally, J., Elliott, J., Packham, D. (1993). "Vulnerability and Warnings." In P.A. Merriman and C.W.A. Browitt (eds.), *Natural Disasters: Protecting Vulnerable Communities.* Thomas Telford: London.

Scanlon, Joseph. (2007). "Research about the Mass Media and Disaster: Never (Hardly Ever) the Twain Shall Meet." In McEntire, D.A. (ed.) *Disciplines, Disasters and Emergency Management: The Convergence of Concepts Issues and Trends From the Research Literature.*

Scanlon, J. (1988). "Winners and Losers: Some Thoughts about the Political Economy of Disasters." *International Journal of Mass Emergencies and Disasters 6*(1): 47–63.

Shaw, G.L. (2007). "Business Crisis and Continuity Management." In McEntire, David A. (ed.) *Disciplines, Disasters and Emergency Management.*

Slagle Pipes, T. (2007). "Information Disasters and Disaster Information: Where Information Science Meets Emergency Management." In McEntire, D.A. (ed.) *Disciplines, Disasters and Emergency Management: The Convergence of Concepts Issues and Trends From the Research Literature.*

Sorensen, J.H. (2000). "Hazard Warning Systems: Review of 20 Years of Progress." *Natural Hazards Review 1*(2): 119–125.

Sylves, R.T. (2007). "U.S. Disaster Policy and Management in an Era of Homeland Security." In McEntire, D.A. (ed.) *Disciplines, Disasters and Emergency Management: The Convergence of Concepts Issues and Trends From the Research Literature.*

Waugh, W.L. Jr. (2007). "Public Administration, Emergency Management, and Disaster Policy." In McEntire, D.A. (ed.) *Disciplines, Disasters and Emergency Management: The Convergence of Concepts Issues and Trends From the Research Literature.*

Webb, G.R., Tierney, K.J. and Dahlhamer, J.M. (2000). "Business and Disasters: Empirical Patterns and Unanswered Questions." *Natural Hazards Review 1*(2): 83–90.

Wisner, B., Blakie, P., Cannon, T. and Davis, I. (2004). *At Risk: Natural Hazards, People's Vulnerability and Disasters.* Routledge: New York.

Zakour, M.J. (2007). "Social Work and Disasters." In McEntire, D.A. (ed.) *Disciplines, Disasters and Emergency Management: The Convergence of Concepts Issues and Trends From the Research Literature.*

Index

D

E